An Introduction to
and Probability

Springer
*New York
Berlin
Heidelberg
Barcelona
Budapest
Hong Kong
London
Milan
Paris
Singapore
Tokyo*

J.C. Taylor

An Introduction to Measure and Probability

With 12 Illustrations

Springer

J.C. Taylor
Department of Mathematics and Statistics
McGill University
805 Sherbrooke Street West
Montreal, Quebec H3A 2K6
Canada

Library of Congress Cataloging-in-Publication Data
Taylor, J.C. (John Christopher), 1936–
  An introduction to measure and probability/J.C. Taylor.
    p.   cm.
  Includes bibliographical references and index.
  ISBN 0-387-94830-9 (pbk.:alk. paper)
  1. Measure theory. 2. Probabilities. I. Title.
QA325.T39 1996
519.2—dc20   96-25447

Printed on acid-free paper.

© 1997 Springer-Verlag New York, Inc.
All rights reserved. This work may not be translated or copied in whole or in part without the written permission of the publisher (Springer-Verlag New York, Inc., 175 Fifth Avenue, New York, NY 10010, USA), except for brief excerpts in connection with reviews or scholarly analysis. Use in connection with any form of information storage and retrieval, electronic adaptation, computer software, or by similar or dissimilar methodology now known or hereafter developed is forbidden.
The use of general descriptive names, trade names, trademarks, etc., in this publication, even if the former are not especially identified, is not to be taken as a sign that such names, as understood by the Trade Marks and Merchandise Marks Act, may accordingly be used freely by anyone.

Production managed by Hal Henglein; manufacturing supervised by Jeffrey Taub.
Camera-ready copy prepared from the author's $\mathcal{A}_{\mathcal{M}}\mathcal{S}$-TEX files.
Printed and bound by Maple-Vail Book Manufacturing Group, York, PA.
Printed in the United States of America.

9 8 7 6 5 4 3 2 (Corrected second printing, 1998)

ISBN 0-387-94830-9 Springer-Verlag New York Berlin Heidelberg SPIN 10671968

I DEDICATE THIS BOOK TO ALL MY TEACHERS
IN GRATITUDE AND THANKFULNESS

Teach me Your way, O Lord
and I will walk in Your truth.
Psalm 86, verse 11

PREFACE

The original version of this book was an attempt to present the basics of measure theory, as it relates to probability theory, in a form that might be accessible to students without the usual background in analysis. This set of notes was intended to be a "primer" for measure theory and probability — a cross between a workbook and a text — rather than a standard "manual". They were intended to give the reader a "hands-on" understanding of the basic techniques of measure theory together with an illustration of their use in probability as illustrated by an introduction to the laws of large numbers and martingale theory. The sections of the first five chapters not marked by an asterisk are the current version of the original set of notes and constitute the essential core of the book.

Since the main goal of the core of this book is to be a "fast-track", self-contained approach to martingale theory for the reader without a solid technical background, many details are put into exercises, which the reader is advised to do if a basic mastery of the subject is the goal.

The present book retains the basic purpose and philosophy of the original set of notes but has been expanded to make it of more use to statisticians by adding a sixth chapter on weak convergence, which is an introduction to this very large circle of ideas as well as to the central limit theorem.

In addition, having in mind its possible use by students of analysis, several sections in this book branch off from the central core to examine the Radon–Nikodym theorem, functions of bounded variation, Fourier series (very briefly), and differentiation theory and maximal functions. Measure theory for σ-finite measures reduces in many respects to the study of finite measures and hence (after normalization) to the study of probabilities. From this point of view, this book can serve as an introduction to measure theory.

Each chapter, except for Chapter V, concludes with a section of additional exercises in which various ancillary topics are explored. They may be omitted on first reading and are not used in the core of this text. Much of the material in the exercises may be found in the standard texts on measure theory or probability that are cited. The author is of the opinion that it is more useful to fill in details for many of these items than to read completed arguments but is well aware that it is more time-consuming! Note that the additional exercises and other sections marked by asterisks may be omitted without prejudice to understanding the core of these notes.

There are six chapters: the first two chapters constitute the basic introduction; and the remaining four explain some uses of measure theory and integration in probability. They have been written to be "self-contained", and as a result, some background information from analysis is scattered throughout. While this may no doubt appear to be excessive to the "well-prepared" reader, it is my hope that with this approach the technical details of the subject will be made accessible to the "well-motivated" reader who may not have all the appropriate background. In other words, I hope to empower readers who wish to learn the subject but who have not come to this point by a standard route. As in many things, a user-friendly attitude seems to me to be very desirable. Whether these notes exhibit that is obviously up to you, the reader, to judge.

As stated before, the main purpose of these notes is to explain and then make use of the main tools and techniques of the theory of Lebesgue integration as they apply to probability. To begin, Chapter I shows how to construct a probability measure on the σ-algebra $\mathcal{B}(\mathbb{R})$ of Borel subsets of the real line from a distribution function F. Then Chapter II defines integration with respect to a probability measure and proves the main theorems of integration theory (e.g., the theorem of dominated convergence). Also, it defines Lebesgue measure on the real line and explains the relation between the Lebesgue integral and the Riemann integral. These two chapters (excepting for §2.7 and §2.8 on the Radon–Nikodym theorem and functions of bounded variation, respectively) constitute the essential part of the central core of the book, and the reader is advised to be familiar with them before proceeding further.

Chapter III is concerned with independence and product measures. It deals mainly with finite products, although the existence of countable products of probability measures is proved following ideas of Kakutani (and independently Jessen) as an illustration of the use of measure-theoretic techniques. This is the first technically difficult result in these notes, and the reader can skip the details (as indicated) without any loss of continuity. The extension of these ideas to prove the existence of Markov chains associated with transition kernels on the real line is made in §3.5.

Chapter IV discusses the main convergence concepts for random variables (i.e., measurable functions) and proves Khintchine's weak law and Kolmogorov's strong law of large numbers for sums of independent, identically distributed (i.i.d.) integrable random variables. The chapter closes with a discussion of uniform integrability and truncation.

The last chapter of the original version, Chapter V, covers conditional expectation and martingales. While sufficient statistics are discussed in §5.3 as an application of conditional expectation, the bulk of the chapter is an introduction to martingales. Doob's optional stopping theorem (finite case) and the martingale convergence theorem (for discrete martingales) are proved. The chapter concludes with Doob's proof of the Radon–Nikodym theorem on countably generated σ-algebras and the backward martingale proof of Kolmogorov's strong law.

The final chapter, Chapter VI, examines weak convergence and the central limit theorem. It is independent of Chapter V and can be read after having worked through §4.1 and §4.3 of Chapter IV together with the beginning of §4.2. Tightness is emphasized, and an attempt is made to enhance the mathematical focus of the standard discussions on weak convergence. The discussion of uniqueness and the continuity theorem for the characteristic function follows Feller's and Lamperti's approach of using the Gaussian or normal densities. The central limit theorem is first discussed in the i.i.d. case and concludes with the case of a sequence or triangular array satisfying the Lindebergh condition.

The reader may find the references to standard texts and sources on probability distracting. In that case, please ignore them. They are there to encourage one to look at and explore other treatments as well as to acknowledge what I have learned from them. In addition, if you are currently using one of them as a text, it may be helpful to have some connections pointed out. References are mainly to books listed in the Bibliography, and the occasional reference to an article is done with a footnote. The book by Loève [L2] has many references in it and will be helpful if the reader wants to explore the classical literature. For information and references on martingales, Doob's book [D1] has lots of historical notes in an appendix and an extensive bibliography. Since these notes emphasize the connection between analysis and probability, the reader wishing to go further into the subject should consult the recent book [S2] by Stroock, which has a decidedly analytic flavour. Finally, for a more abstract version of many of the topics covered in these notes and for detailed historical notes, the book by Dudley [D2] is recommended.

For several years since 1986, versions of these notes have been used at McGill University (successfully, from my perspective) to teach a one-semester course in probability to a mixed audience of statisticians and electrical engineers. Depending on the preparation of the class, most of the original version of this book was covered in a one-semester course. These notes have also been used to teach an introductory course on measure theory to honours students (in this case, the course was restricted to the first four chapters with the sections on infinite product measures and Markov chains omitted as well as §4.5 and §4.6).

I would like to thank all the students who have (unwittingly) participated in the process that has led to the creation of this set of notes; my wife, Brenda MacGibbon of UQAM, for her enthusiastic encouragement, my colleagues Minggao Gu and Jim Ramsay of McGill University and Ahmid Bose of Carleton University for their support and encouragement, as well as Regina Liu of Rutgers and D. Phong of Columbia University. I thank Jal Choksi of McGill University and Miklós Csörgő of Carleton University for useful conversations on various points. In addition, I wish to thank all my colleagues and friends who have helped me in many, many ways over the years. I also wish to thank Mr. Masoud Asgharian, Mr. Jun Cai, and Ms. Gulhan Alpargu for their diligent search for misprints and errors, and Mr. Erick Lamontagne for help in final editing. Finally, I also wish to thank my copy editor, Kristen Cassereau, and my editor at Springer, John Kimmel, for their patience and professional help.

J.C. Taylor

CONTENTS

Preface	vii
List of Symbols	xiii
Chapter I. Probability Spaces	**1**
1. Introduction to $\mathbb{R}$	1
2. What is a probability space? Motivation	5
3. Definition of a probability space	9
4. Construction of a probability from a distribution function	16
5. Additional exercises*	26
Chapter II. Integration	**29**
1. Integration on a probability space	29
2. Lebesgue measure on $\mathbb{R}$ and Lebesgue integration	44
3. The Riemann integral and the Lebesgue integral	49
4. Probability density functions	55
5. Infinite series again	58
6. Differentiation under the integral sign	59
7. Signed measures and the Radon–Nikodym theorem*	60
8. Signed measures on $\mathbb{R}$ and functions of bounded variation*	71
9. Additional exercises*	78
Chapter III. Independence and Product Measures	**86**
1. Random vectors and Borel sets in $\mathbb{R}^n$	86
2. Independence	89
3. Product measures	95
4. Infinite products	110
5. Some remarks on Markov chains*	119
6. Additional exercises*	131
Chapter IV. Convergence of Random Variables and Measurable Functions	**137**
1. Norms for random variables and measurable functions	137
2. Continuous functions and L^p*	149
3. Pointwise convergence and convergence in measure or probability	167
4. Kolmogorov's inequality and the strong law of large numbers	176
5. Uniform integrability and truncation*	180

6. Differentiation: the Hardy–Littlewood maximal
   function* 186
7. Additional exercises* 199

**Chapter V. Conditional Expectation and
an Introduction to Martingales** **210**
1. Conditional expectation and Hilbert space 210
2. Conditional expectation 217
3. Sufficient statistics* 226
4. Martingales 229
5. An introduction to martingale convergence 238
6. The three-series theorem and the Doob decomposition 241
7. The martingale convergence theorem 245

**Chapter VI. An Introduction to
Weak Convergence** **250**
1. Motivation: empirical distributions 250
2. Weak convergence of probabilities:
   equivalent formulations 251
3. Weak convergence of random variables 255
4. Empirical distributions again: the Glivenko–Cantelli
   theorem 260
5. The characteristic function 262
6. Uniqueness and inversion of the characteristic function 266
7. The central limit theorem 273
8. Additional exercises* 281
9. Appendix* 284

Bibliography **291**

Index **293**

LIST OF SYMBOLS

$\complement A$, 6
$\mathfrak{A}$, 6
$\overline{A}$, 15
A^c, 6
$a \wedge b$, 5
$a \vee b$, 5
$\mathfrak{B}(\mathbb{R})$, 14
$\mathfrak{B}(\mathbb{R}^2)$, 87
$\mathfrak{B}(\mathbb{R}^n)$, 87
$\mathfrak{B}(E)$, 131
$\overline{\mathfrak{B}(\mathbb{R})}^\mu$, 82
$b(n, p)$, 141
$\mathcal{C}_b(E)$, 153
$\mathcal{C}_c(\mathbb{R})$, 150
$\mathcal{C}_c(O)$, 150
$\mathcal{C}(E)$, 151
$\mathcal{C}_b(\mathbb{R})$, 153
$\mathcal{C}_c(\mathbb{R}^n)$, 152
δ_0, 26
$d(F_1, F_2)$, 257
D_h, 258
$DF^+(x)$, 194
$DF^-(x)$, 194
$DF_+(x)$, 194
$DF_-(x)$, 194
$\operatorname{dist}(x, C)$, 150
$E \Delta B$, 78
ε_0, 26
ε_a, 26
$\{E_n \text{ i.o.}\}$, 148
E^+, 64

xiii

E_α^+, 66
E^-, 64
E_α^-, 66
$E[s]$, 30
$E[X]$, 38
$E[X|Y=y]$, 220
$E[X|\mathfrak{G}]$, 212
$(\mathfrak{F}_t)_{t\in I}$, 229
$\check{f}$, 158
$\mathfrak{F}$, 9
$\mathfrak{F}_1 \otimes \mathfrak{F}_2$, 97
$\mathfrak{F}_1 \times \mathfrak{F}_2$, 97
$\otimes_{n=1}^\infty \mathfrak{F}_n$, 113
ϕ_c, 181
$\times_{n=1}^\infty \mathfrak{F}_n$, 113
$f \star_{2\pi} g$, 160
$f \cdot X$, 232
$f \star g$, 157
$f_1 \star f_2$, 109
$f_1 \star \mathbf{P}_2$, 108
$F_n \xrightarrow{w} F$, 251
$F_n \Rightarrow F$, 251
$F_n(x,\omega)$, 250
$\mathfrak{F}_T$, 234
$H_n(x,t)$, 80
$5I$, 193
$\int s(\omega)\mathbf{P}(d\omega)$, 30
$\int s d\mathbf{P}$, 30
$\int X d\mathbf{P}$, 38
$\int_0^{+\infty} f(x)dx$, 54
$\int_a^b \varphi(x)dx$, 50
K_n, 160
$L^1(\Omega, \mathfrak{F}, \mathbf{P})$, 41
ℓ_1, 58
ℓ_2, 59
ℓ, 217

ℓ_p, 143
ℓ_∞, 143
$\liminf_n a_n, \liminf\limits_{n} a_n$, 33
$\limsup_n a_n, \limsup\limits_{n} a_n$, 33
L, 137
$L(\pi, \varphi)$, 49
L^1, 41
$L^1(\mathbb{N}, \mathfrak{P}(\mathbb{N}), dn)$, 58
L^2, 139
L^∞, 144
L^p, 141
$L^p(\Omega, \mathfrak{F}, \mu)$, 146
$L^p(\Omega, \mathfrak{F}, \mathbf{P})$, 146
$\mathbf{M}_2(x_0, dx)$, 123
$\mathbf{M}_2(x_0, dx_1, dx_2)$, 123
$\mathbf{M}_n(x_0, dx_1, dx_2, \ldots, dx_n)$, 128
$\mathfrak{M}$, 25
$\mu_1 \times \mu_2 \times \cdots \times \mu_n$, 104
M^*, 239
$\langle \, , \, \rangle, (\, , \,)$, 59
$\mathbb{N}$, 1
$\mathbf{N}$, 120
$\mathbf{N}(x_0, dx_1)$, 123
ν^+, 65
ν^-, 65
$\nu << \mu$, 66
$\| \, \|$, 58
$n(x)$, 79
$n_t(x)$, 79
$(\Omega, \mathfrak{F}, \mathbf{P})$, 9
$(\Omega_1 \times \Omega_2, \mathfrak{F}_1 \times \mathfrak{F}_2, \mathbf{P}_1 \times \mathbf{P}_2)$, 99
$(\Omega_1 \times \Omega_2, \mathfrak{F}_1 \times \mathfrak{F}_2, \mu_1 \times \mu_2)$, 105
$(\Omega_1 \times \Omega_2 \times \cdots \times \Omega_n, \mathfrak{F}_1 \times \mathfrak{F}_2 \times \cdots \times \mathfrak{F}_n, \mathbf{P}_1 \times \mathbf{P}_2 \times \cdots \times \mathbf{P}_n)$, 102
$(\times_{k=1}^\infty \Omega_k, \, \times_{k=1}^\infty \mathfrak{F}_k, \, \times_{n=1}^\infty \mathbf{P}_k)$, 118
$\times_{n=1}^\infty \Omega_n$, 113
$\mathbf{P}_1 \times \mathbf{P}_2$, 99

$\mathbf{P}$, 9
$\mathbf{P}(d\omega|B)$, 223
$\mathbf{P}^*$, 21
$\mathbf{P}[X \leq x|Y = y]$, 221
$\mathbf{P}_1 \otimes \mathbf{P}_2$, 99
$\mathbf{P}_1 \times \mathbf{P}_2 \times \cdots \times \mathbf{P}_n$, 102
$\mathbf{P}_1 \star \mathbf{P}_2$, 108
ψ^*, 187
$\times_{n=1}^{\infty} \mathbf{P}_n,$, 118
P_n, 175
$\mathbb{Q}$, 1
$\mathbf{Q}_2(dx_0, dx_1, dx_2)$, 124
$\mathbf{Q}_n(dx_0, dx_1, \ldots, dx_n)$, 125
$\mathbb{R}$, 1
$\mathfrak{R}$, 45
$\sigma(\mathfrak{C})$, 14
$\sigma(X)$, 140
$\sigma^2(X)$, 140
σ, 140
$\sigma(\{X_\iota \mid \iota \in F\})$, 110
$\sigma(\{X_\iota \mid \iota \in I\})$, 110
$\sigma(\{X_i \mid 1 \leq i \leq n\})$, 93
$\sigma(\{X_i \mid i \in F\})$, 93
σ^2, 140
S_n, 170
$\mathbb{T}$, 159
$U([a,b])$, 246
$U(\pi, \varphi)$, 49
$U^X([a,b])$, 246
$\overline{V}(E)$, 62
$\underline{V}(E)$, 65
$V(\Phi)$, 72
$V(\Phi, [a,b])$, 71
$V(\Phi, \pi)$, 71
$V^+(\Phi, \pi)$, 74
$V^-(\Phi, \pi)$, 74

$X_n \xrightarrow{d} X$, 255
$\| X \|_1$, 137
$\| X \|_\infty$, 144
$\| X \|_2$, 139
$\| X \|_p$, 141
$X \wedge Y$, 233
X^*, 237
X^+, 35
X^-, 35
x^+, 5
x^-, 5
$X \in \mathfrak{F}$, 32
X_c, 180
$X_n \xrightarrow{\mu\text{-}a.e.} X$, 167
$X_n \xrightarrow{a.e.} X$, 167
$X_n \xrightarrow{L^p} 0$, 145
$X_n \xrightarrow{m} X$, 168
$X_n \xrightarrow{pr} X$, 168
$X_n \xrightarrow{w} X$, 255
$X_n \Rightarrow X$, 255
$X_n \xrightarrow{\mathbf{P}\text{-}a.s.} X$, 167
$X_n \xrightarrow{a.s.} X$, 167
X_T, 235
$|X|$, 41
$|x|$, 5
$\mathbb{Z}$, 1

CHAPTER I

PROBABILITY SPACES

1. Introduction to $\mathbb{R}$

An introduction to analysis usually begins with a study of properties of $\mathbb{R}$, the set of **real numbers**. It will be assumed that you know something about them. More specifically, it is assumed that you realize that

(i) *they form a field (which means that you can add, multiply, subtract and divide in the usual way), and*
(ii) *they are (totally) ordered (i.e., for any two numbers $a, b \in \mathbb{R}$, either $a < b, a = b,$ or $a > b$) and $a \geq b$ if and only if $a - b \geq 0$. Note that $a < b$ and $0 < c$ imply $ac < bc$; $a \leq b$ imply $a + c \leq b + c$ for any $c \in \mathbb{R}$; $0 < 1$ and $-1 < 0$.*

Inside $\mathbb{R}$ lies the set $\mathbb{Z}$ of **integers** $\{\ldots, -2, -1, 0, 1, 2, 3, \ldots\}$ and the set $\mathbb{N}$ of **natural numbers** $\{1, 2, \ldots\}$. Also, there is a smallest field inside $\mathbb{R}$ that contains $\mathbb{Z}$, namely $\mathbb{Q}$, the set of **rational numbers** $\{p/q : p, q \in \mathbb{Z} \text{ and } q \neq 0\}$. A real number that is not rational is said to be **irrational**.

It is often useful to view $\mathbb{R}$ as the points on a straight line (see Fig. 1.1).

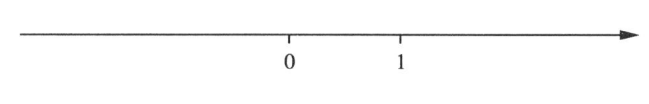

Fig. 1.1

Then $a < b$ is indicated by placing a strictly to the left of b, if the line is oriented by putting 0 to the left of 1. The field $\mathbb{Q}$ is also totally ordered, but it differs from $\mathbb{R}$ in that there are "holes" in $\mathbb{Q}$. To explain this, it is useful to make a definition.

Definition 1.1.1. *Let $A \subset \mathbb{R}$. A number b with the property that $b \geq a$ for all $a \in A$ is said to be an **upper bound** of A. If $b \leq a$ for all $a \in A$, b is said to be a **lower bound** of A.*

Example. Let $A = \{a \in \mathbb{R} \mid a < 1\}$. Then any number $b \geq 1$ is an upper bound of A. Here there is a smallest or **least upper bound** (l.u.b.), namely 1.

Exercise 1.1.2. Let $A = \{a \in \mathbb{Q} \mid a^2 < 2\}$. Show that if there is a least upper bound of A, it cannot belong to $\mathbb{Q}$. [*Hints:* show that, for small $p > 0$, $a^2 < 2$ implies that $(a+p)^2 < 2$ and $a^2 > 2$ implies that $(a-p)^2 > 2$; conclude that if b is a l.u.b. of A, then $b^2 = 2$; show that there is no rational number b with $b^2 = 2$.] One can view $\sqrt{2} \in \mathbb{R}$ as a number that "fills a hole in $\mathbb{Q}$". "Holes" exist because it is often possible to split the set $\mathbb{Q}$ into two disjoint subsets A, B so that (i) $a \in A$ and $b \in B$ implies that $a \leq b$; (ii) there is no l.u.b. of A in $\mathbb{Q}$ and no largest or **greatest lower bound** (g.l.b.) of B in $\mathbb{Q}$. A decomposition of $\mathbb{Q}$ satisfying (i) is called a **Dedekind cut** of $\mathbb{Q}$.

1.1.3. Axiom of the least upper bound. *Every subset A of $\mathbb{R}$ that has an upper bound has a least upper bound.*

This axiom is an extremely important property of $\mathbb{R}$. It will be taken for granted that the real numbers exist and have this property. (Starting with suitable axioms for set theory, one can show that (i) there exist fields with the properties of $\mathbb{R}$ and (ii) any two such fields are isomorphic (i.e., are the "same"); see Halmos [H2].)

Exercise 1.1.4. (1) Show that $\mathbb{N}$ has no upper bound. [*Hint:* show that if b is an upper bound of $\mathbb{N}$, then $b-1$ is also an upper bound of $\mathbb{N}$.]

(2) Let $\epsilon > 0$ be a positive real number. Show that there is a natural number $n \in \mathbb{N}$ with $0 < 1/n < \epsilon$. [*Hint:* use (1) and the fact that multiplication by a positive number preserves an inequality.]

This last exercise shows that the ordered field $\mathbb{R}$ has the **Archimedean property**, i.e., for any number $\epsilon > 0$ there is a natural number n with $0 < 1/n < \epsilon$.

If E and F are two sets, it will be taken for granted that the concept of a function $f : E \mapsto F$ is understood. When $E = \mathbb{N}$, a function $f : \mathbb{N} \to F$ is also called a **sequence of elements from F**. The function f in this case is often denoted by $(f(n))_{n \geq 1}$ or $(f_n)_{n \geq 1}$, where $f(n) = f_n$ is the value of f at n. When the domain of n is understood, $(f_n)_{n \geq 1}$ is often shortened to (f_n). Given a sequence $(f_n)_{n \geq 1}$, a **subsequence** of $(f_n)_{n \geq 1}$ is a sequence of the form $(f_{n_k})_{k \geq 1}$, where $n : \mathbb{N} \to \mathbb{N}$ is a strictly increasing function $k \to n_k$, e.g., $n_k = 2k$ for all $k \geq 1$. If the original sequence is written as $(f(n))_{n \geq 1}$, a subsequence may be indicated as $(f(n_k))_{k \geq 1}$ or $(f(n(k)))_{k \geq 1}$. The important thing is that in a subsequence one selects elements from the original sequence by using a strictly increasing function n.

Definition 1.1.5. *A sequence $(b_n)_{n \geq 1}$ of real numbers **converges to B** as $n \to +\infty$ if for any positive number $\epsilon > 0$,*

$$B - \epsilon < b_n < B + \epsilon, \quad \text{for all sufficiently large } n.$$

That is, $|b_n - B| \leq \epsilon$ if $n \geq n(\epsilon)$, where n is an integer depending on ϵ and the sequence (see Exercise 1.1.16 for the definition of $|a|$). This is denoted by writing $B = \lim_{n \to +\infty} b_n$.

*A sequence $(b_n)_{n \geq 1}$ of real numbers **converges to $+\infty$** if for any $N \in \mathbb{N}$, $b_n \geq N$, $n \geq n(N)$.*

Exercise 1.1.6. Let $(b_n)_{n\geq 1}$ be a non-decreasing sequence of real numbers, i.e., $b_n \leq b_{n+1}$ for all $n \geq 1$. Show that

(1) the sequence converges,
(2) the limit is finite if and only if the sequence has an upper bound,
(3) when the limit is finite, it equals the l.u.b. of $\{b_n \mid n \geq 1\}$.

Let $(b'_n)_{n\geq 1}$ be another non-decreasing sequence with $b_n \leq b'_n$ for all $n \geq 1$. Show that

(4) $\lim_n b_n \leq \lim b'_n$.

Definition 1.1.7. *Let $(a_n)_{n\geq 0}$ be a sequence of real numbers. Let $s_n = a_0 + a_1 + a_2 + \cdots + a_n$. The sequence $(s_n)_{n\geq 0}$ is called the **infinite series** $\sum_{n=0}^{\infty} a_n$, and the series is said to converge if $(s_n)_{n\geq 0}$ converges.*

Exercise 1.1.8. Let $(a_n)_{n\geq 0}$ be a sequence of positive real numbers. Show that

(1) the series $\sum_{n=0}^{\infty} a_n$ converges if and only if $\{s_n \mid n \geq 0\}$ has an upper bound,
(2) if $\sum_{n=0}^{\infty} a_n$ converges, then it converges to l.u.b. $\{s_n \mid n \geq 0\}$.

Finally, show that

(3) if $\sum_{n=0}^{\infty} a_n$ converges to a finite limit, then $\lim_{N\to\infty} \sum_{n=N}^{\infty} a_n = 0$.

Exercise 1.1.9. Suppose one has a random variable X whose values are non-negative integers. Let $(a_n)_{n\geq 0}$ be a sequence of positive real numbers. When can the probability that X is n equal ca_n where c is a fixed constant? What happens if $a_n = \frac{1}{n}$? (if $a_n = \frac{1}{n^2}$ or $a_n = \frac{1}{n \log n}$ or $\frac{n}{(n^3+2)}$?) This exercise begs the question: what is a random variable? For the time being, think of it as a procedure that assigns probabilities to certain outcomes. The definition is given in Chapter II, see Definition 2.1.6.

Exercise 1.1.10. A random variable X has a Poisson distribution with mean 1, if the probability that X is n is $e^{-1}/n!$ (see Feller [F1] for the Poisson distribution).

Proposition 1.1.11. (Exchange of order of limits). *Let $(b_{m,n})_{m,n\geq 1}$ be a double sequence of real numbers, i.e., a function $b : \mathbb{N} \times \mathbb{N} \to \mathbb{R}$. Assume that*

(1) $m_1 \leq m_2 \Rightarrow b_{m_1,n} \leq b_{m_2,n}$ *for all* $n \geq 1$,
(2) $n_1 \leq n_2 \Rightarrow b_{m,n_1} \leq b_{m,n_2}$ *for all* $m \geq 1$.

Then

$$\lim_{n\to+\infty}(\lim_{m\to+\infty} b_{m,n}) = \lim_{m\to+\infty}(\lim_{n\to+\infty} b_{m,n}) = \lim_{n\to+\infty} b_{n,n},$$

where an increasing sequence has limit $+\infty$ if it is unbounded.

Proof. By symmetry it suffices to verify the second equality. Now $b_{m,m} \leq b_{n,n}$ if $m \leq n$ and so $B = \lim_{n\to\infty} b_{n,n}$ exists, as does $B_m \stackrel{\text{def}}{=} \lim_{n\to\infty} b_{m,n}$.

Also, by Exercise 1.1.6, $B_m \le B$ as $b_{m,n} \le b_{n,n}$ when $n \ge m$. It follows from (1) and Exercise 1.1.6 that $B_{m_1} \le B_{m_2}$. Hence, $\lim_{m \to \infty} B_m = B'$ exists and is less than or equal to B (see Exercise 1.1.6 again).

If B is finite, let $\epsilon > 0$ and $n = n(\epsilon)$ be such that $B - \epsilon \le b_{n,n} \le B$ if $n \ge n(\epsilon)$. Let $m = n(\epsilon)$. Then $B_m = \lim_{n \to \infty} b_{m,n} \ge b_{n(\epsilon),n(\epsilon)} \ge B - \epsilon$. Hence, $B' = B$ as $B' \ge B_m$ for all m.

If $B = +\infty$ and $N \le b_{n,n}$ for $n \ge n(N)$, then $B_m = \lim_{n \to \infty} b_{m,n} \ge N$ if $m = n(N)$ and so $B' = +\infty$. □

Corollary 1.1.12. $\sum_{i=1}^{\infty}(\sum_{j=1}^{\infty} a_{ij}) = \sum_{j=1}^{\infty}(\sum_{i=1}^{\infty} a_{ij})$ if $a_{ij} \ge 0, 1 \le i,j$.

Proof. Let $b_{m,n} = \sum_{i=1}^{m}\sum_{j=1}^{n} a_{ij}$ and verify that the conditions of Proposition 1.1.11 are satisfied. □

Exercise 1.1.13. Decide whether Proposition 1.1.11 is valid when only one of (1) and (2) is assumed. In Corollary 1.1.12, what happens if $a_{ii} = 1, a_{i,i+1} = -1$ for all $i \ge 1$ and all other $a_{ij} = 0$?

Exercise 1.1.14. (See Feller [F1], p. 267.) Let $p_n \ge 0$ for all $n \ge 0$ and assume $\sum_{n=0}^{\infty} p_n = 1$. Let $m_r = \sum_{n=0}^{\infty} n^r p_n$. Show that $\sum_{r=0}^{\infty} \frac{m_r}{r!} t^r = \sum_{n=0}^{\infty} p_n e^{nt}$ for $t \ge 0$.

This brief discussion of properties of $\mathbb{R}$ concludes with a discussion of intervals.

Definition 1.1.15. A set $I \subset \mathbb{R}$ is said to be an **interval** if $x \le y \le z$ and $x, z \in I$ implies $y \in I$.

If an interval I has an upper bound, then $I \subset (-\infty, b]$, where $b = $ l.u.b. I and $(-\infty, b] = \{x \in \mathbb{R} \mid x \le b\}$. If it also has a lower bound, then $I \subset [a,b] = \{x \mid a \le x \le b\}$ if $a = $ g.l.b. I. A **bounded interval** I — one having both upper and lower bounds — is said to be

(1) a **closed interval** when $I = [a,b]$,

(2) an **open interval** when $I = (a,b) \stackrel{\text{def}}{=} \{x \mid a < x < b\}$ (often denoted by $]a,b[$),

(3) a **half-open interval** when $I = (a,b]$ or $[a,b)$, where $(a,b] \stackrel{\text{def}}{=} \{a \mid a < x \le b\}$ and similarly $[a,b) \stackrel{\text{def}}{=} \{a \mid a \le x < b\}$.

One also denotes $(a,b]$ by $]a,b]$ and $[a,b)$ by $[a,b[$.

An **unbounded interval** I is one of the following:

$$(-\infty, b) = \{x \mid x < b\};$$
$$(-\infty, b] = \{x \mid x \le b\};$$
$$(a, +\infty) = \{x \mid a < x\};$$
$$[a, +\infty) = \{x \mid a \le x\}; \text{ or}$$
$$(-\infty, +\infty) = \mathbb{R}.$$

Exercise 1.1.16. If $x \in \mathbb{R}$, define $|x| = x$ if $x \geq 0$ and $= (-1)x$ if $x < 0$. Let a, b be any two real numbers. Show that
  (1) $|a + b| \leq |a| + |b|$ (**the triangle inequality**),
  (2) conclude that $||a| - |b|| \leq |a - b|$ by two applications of the triangle inequality.

If a, b are two numbers, let $a \vee b$ denote their maximum, also denoted by $\max\{a, b\}$, and $a \wedge b$ denote their minimum, also denoted by $\min\{a, b\}$. Define x^+ to be $\max\{x, 0\} = x \vee 0$ and x^- to be $\max\{-x, 0\} = (-x) \vee 0$. Show that
  (3) $x^- = -(x \wedge 0)$,
  (4) $x = x^+ - x^-$,
  (5) $|x| = x^+ + x^-$,
  (6) $a \vee b = \frac{1}{2}\{a + b + |a - b|\}$, and
  (7) $a \wedge b = \frac{1}{2}\{a + b - |a - b|\}$.

Exercise 1.1.17. Verify the following statements for $-\infty < a < b < +\infty$:
  (1) $(a, b) = \cup_{n=N}^{\infty}(a, b - \frac{1}{n}]$, where $b - a > \frac{1}{N}$;
  (2) $[a, b) = \cap_{n=N}^{\infty}(a - \frac{1}{n}, b)$;
  (3) $[a, b] = \cap_{n=1}^{\infty}[a, b + \frac{1}{n})$.
Show that if $x_0 \in (a, b)$, then $(x_0 - \delta, x_0 + \delta) \subset (a, b)$ for some $\delta > 0$ (this implies that (a, b) is an open set; see Exercise 1.3.10).

Exercise 1.1.18. Let $0 < \alpha < 1$. If $p > 0$, show that $\alpha^{\frac{1}{p}} = e^{\frac{1}{p}\log \alpha}$ is an increasing function of p and that $\lim_{p \to \infty} \alpha^{\frac{1}{p}} = 1$.

Exercise 1.1.19. Let $\mathbb{R}$ be the union of two disjoint intervals I_1 and I_2. Show that
  (1) one of the two intervals is to the left of the other (either $x_1 \in I_1$ and $x_2 \in I_2$ always implies $x_1 < x_2$ or vice-versa), and
  (2) if neither of these two intervals is the void set, then $\sup I_1 = \inf I_2$ if I_1 is to the left of I_2.

Let (I_n) be an increasing sequence of intervals. Show that
  (3) if $I = \cup_n I_n$, then I is an interval.

Assume that each I_n above is unbounded and bounded below, i.e., there is $a_n \in \mathbb{R}$ with $(a_n, +\infty) \subset I_n \subset [a_n, +\infty)$. Assume that I has the same property. Show that
  (4) if $(a, +\infty) \subset I \subset [a, +\infty)$, then $\lim a_n = a$.

For further information on the real line and general background information in analysis, consult Marsden [M1] or Rudin [R4].

2. What is a Probability Space? Motivation

A probability space can be viewed as something that models an "experiment" whose outcomes are "random" (whatever that means). There are often "simple" or "elementary outcomes" in the model (as points in an underlying set Ω and weights assigned to these outcomes that indicate the likelihood or probability of the outcome occurring (see Feller [F1]). The general outcome or "event" is often a collection of "elementary outcomes". For example, consider the following.

Example 1.2.1. The "experiment" consists of rolling a fair six-sided die two times. The "elementary outcomes" could be taken to be ordered pairs $\omega = (m, n)$, where m and n are integers from 1 to 6. The set Ω of elementary outcomes may be taken to be the set of all such ordered pairs (it is usual to denote this set as the Cartesian product $\{1, 2, \ldots, 6\} \times \{1, 2, \ldots, 6\}$). The set of all events may be taken as the collection of all subsets of Ω, denoted by $\mathfrak{P}(\Omega)$. If each elementary outcome ω is assigned weight $\frac{1}{36}$, then one may define the probability $\mathbf{P}(A)$ of an event A by $\mathbf{P}(A) = \sum_{\omega \in A} \mathbf{P}(\{\omega\}) = \frac{|A|}{|\Omega|} = \frac{|A|}{36}$, where $|A|$ denotes the number of elements in A. If the die is not fair — the probability of getting either a $1, 2, 3$, or a 4 is $\frac{1}{8}$ and for example that of getting either a 5 or a 6 is $\frac{1}{4}$ — then the basic probabilities or weights of the elementary outcomes will need to be altered to correspond to the new situation.

For elementary situations as in Example 1.2.1, it suffices to consider a so-called **finitely additive probability space**. This is a triple $(\Omega, \mathfrak{A}, \mathbf{P})$, where Ω is a set (corresponding intuitively to the set of "elementary outcomes", $\mathfrak{A}$ is a collection of subsets of Ω (the "events") with certain "algebraic" properties that make it into a **Boolean algebra** of subsets of Ω, and for each event $A \in \mathfrak{A}$ there is a number $\mathbf{P}(A)$ assigned that lies between 0 and 1 (the **probability** of the occurrence of A). More explicitly, to say that $\mathfrak{A}$ is a Boolean algebra means that the collection $\mathfrak{A}$ of subsets satisfies the following conditions:

($\mathfrak{A}_1$) $\Omega \in \mathfrak{A}$;
($\mathfrak{A}_2$) $A_1, A_2 \in \mathfrak{A}$ implies that $A_1 \cup A_2 \in \mathfrak{A}$; and
($\mathfrak{A}_3$) $A \in \mathfrak{A}$ implies that $A^c \in \mathfrak{A}$, where $A^c \stackrel{\text{def}}{=} \{\omega \in \Omega \mid \omega \notin A\} \stackrel{\text{def}}{=} \complement A$.

The statement that $\mathbf{P}$ is a probability means that it is a function defined on $\mathfrak{A}$ with the following properties:

(P_1) $\mathbf{P}(\Omega) = 1$;
(P_2) $0 \leq \mathbf{P}(A) \leq 1$; and
(FAP_3) $A_1 \cap A_2 = \emptyset \Rightarrow \mathbf{P}(A_1 \cup A_2) = \mathbf{P}(A_1) + \mathbf{P}(A_2)$.

It is not hard to see that Example 1.2.1 is a finitely additive probability space.

Some simple consequences of the properties of a Boolean algebra $\mathfrak{A}$ and a finitely additive probability defined on it are given in the next exercise.

Exercise 1.2.2. Show that
   (1) $A_1, A_2 \in \mathfrak{A}$ implies that $A_1 \cap A_2 \in \mathfrak{A}$ and $A_1 \cap A_2^c \in \mathfrak{A}$ (one often denotes $A_1 \cap A_2^c$ by $A_1 \backslash A_2$ or $A_1 \cap \complement A_2$),

2. WHAT IS A PROBABILITY SPACE? MOTIVATION

(2) $\mathbf{P}(\emptyset) = 0$,
(3) $A_1 \subset A_2$ implies that $\mathbf{P}(A_1) \leq \mathbf{P}(A_2)$,
(4) $\mathbf{P}(A_1 \cup A_2) \leq \mathbf{P}(A_1) + \mathbf{P}(A_2)$,
(5) $\mathbf{P}(A_1 \cup A_2) + \mathbf{P}(A_1 \cap A_2) = \mathbf{P}(A_1) + \mathbf{P}(A_2)$,
(6) $\mathbf{P}(\cup_{k=1}^n A_k) = \sum_{k=1}^n \mathbf{P}(A_k \setminus \cup_{i=1}^{k-1} A_i) \leq \sum_{k=1}^n \mathbf{P}(A_k)$.

Remark 1.2.3. In Exercise 1.2.2 (6), a union of sets $\cup_{k=1}^n A_k$ is converted to a disjoint union $\cup_{k=1}^n A'_k$, where $A'_k = A_k \setminus \cup_{i=1}^{k-1} A_i$. This is a standard device or trick that is often used, especially for countable unions $A = \cup_{k=1}^\infty A_k$.

Here is another example of an "experiment" with "random" outcomes.

Example 1.2.4. What probability space is it natural to use to discuss the probability of choosing a number at random from [0,1]? What is the probability of choosing a number from $(\frac{1}{3}, \frac{3}{4}]$? Clearly, one should take Ω to be [0,1] and $\mathfrak{A}$ to be the collection of finite unions of intervals contained in [0,1] (so that $\mathfrak{A}$ is a Boolean algebra containing intervals and their unions). Show that $\mathfrak{A}$ is a Boolean algebra. How do you define $\mathbf{P}$ on $\mathfrak{A}$?

Example 1.2.5. Continuing with the same probability space as in Example 1.2.4, suppose that one wants to discuss the probability of selecting at random a number x with the following property: it does not lie in $(\frac{1}{3}, \frac{2}{3})$ — i.e., the middle third of [0,1] — nor does it lie in the middle third of either $[0, \frac{1}{3}]$ or $[\frac{2}{3}, 1]$ — and so on, infinitely often. This describes a subset C of [0,1], an event.

Look at the complementary event C^c: it is the disjoint union of middle third intervals; $C^c = (\frac{1}{3}, \frac{2}{3}) \cup [(\frac{1}{9}, \frac{2}{9}) \cup (\frac{7}{9}, \frac{8}{9})] \cup [(\frac{1}{3^3}, \frac{2}{3^3}) \cup (\frac{7}{3^3}, \frac{8}{3^3}) \cup (\frac{19}{3^3}, \frac{20}{3^3}) \cup (\frac{25}{3^3}, \frac{26}{3^3})] \cup \ldots$.

Let q be the probability of C^c. Then one sees that

(1) $q \geq \frac{1}{3}$,
(2) $q \geq \frac{1}{3} + \frac{1}{9} + \frac{1}{9} = \frac{1}{3} + \frac{2}{9}$,
(3) $q \geq \frac{1}{3} + \frac{2}{9} + \frac{4}{27}$,
$\vdots$
(n) $q \geq \frac{1}{3} + \frac{2}{9} + \frac{4}{27} + \cdots + \frac{2^{n-1}}{3^n}$.

To verify this, one makes use of the principle of mathematical induction, stated below.

The principle of mathematical induction. *Let $P(n)$ be a proposition or statement for each $n \in \mathbb{N}$. If*

(1) *$P(1)$ is true and*
(2) *$P(n+1)$ is true provided $P(n)$ is true,*

then $P(n)$ is true for all n. (This principle amounts to saying that if $A \subset \mathbb{N}$ is such that (1) $1 \in A$ and (2) $n \in A$ implies $n+1 \in A$, then $A = \mathbb{N}$).

Now $\frac{1}{3}+\frac{2}{9}+\frac{4}{27}+\cdots+\frac{2^{n-1}}{3^n}$ is the nth partial sum of the series $\sum_{k=0}^{\infty} \frac{2^k}{3^{k+1}}$. This is a geometric series with $a_0 = \frac{1}{3}$ and $r = \frac{2}{3}$, and so the sum is $\frac{1}{3}(1-\frac{2}{3})^{-1} = 1$. Consequently, one expects the probability of choosing such an x from C to be zero!

Remark. The subset C of $[0,1]$ described in Example 1.2.5 is called the **Cantor set** or **Cantor discontinuum**. It contains no interval with distinct end points. Why? What would happen if instead of extracting middle thirds one removed the middle quarter at each stage?

Note that neither the Cantor set C nor its complement is in the Boolean algebra $\mathfrak{A}$ defined in Example 1.2.4, as the set C^c is the union of an infinite number of open intervals each of which is in the Boolean algebra $\mathfrak{A}$: it can be shown that C^c cannot be expressed as a finite union of sets from $\mathfrak{A}$ — this has to do with the fact that each of the intervals involved is a connected component of C^c, i.e., for any point x in one of these intervals I, the largest open interval that contains it coincides with I (see Exercise 1.3.11).

Example 1.2.6. (Coin tossing) Suppose one tosses a fair coin until a head occurs. What is the probability of this event? To begin with, what could one take as Ω, the set of "elementary outcomes"? Take Ω to be $\mathbb{N}$, where each $n \geq 1$ corresponds to a finite string of length n that commences with $n-1$ heads and concludes with a tail. Here probabilities may be assigned to each integer $n \geq 1$, namely $\frac{1}{2^n}$ for the string of length n. Since $\sum_{n=1}^{\infty} \frac{1}{2^n} = \frac{1}{2}\left(\frac{1}{1-\frac{1}{2}}\right) = 1$ this gives a probability space with $\mathfrak{A}$ taken to be all the subsets of Ω and $\mathbf{P}(A)$ defined to be $\sum_{n \in A} \frac{1}{2^n}$. Therefore, the probability of first obtaining a head on an even number toss is $\sum_{k=1}^{\infty} \frac{1}{2^{2k}} = \frac{1}{4}\left(\frac{1}{1-\frac{1}{4}}\right) = \frac{1}{3}$. In this example, you can verify that if $A = \cup_{n=1}^{\infty} A_n$ and $A_n \cap A_m = \emptyset$ when $n \neq m$, then $\mathbf{P}(A) = \sum_{n=1}^{\infty} \mathbf{P}(A_n)$. This is clearly a desirable property of a probability, but how can it be obtained in the context of the previous exercise, where $\Omega = [0,1]$?

Example 1.2.7. Suppose X is a random variable with unit normal distribution, i.e., the probability that $a < X \leq b$ is $\frac{1}{\sqrt{2\pi}} \int_a^b e^{-(\frac{x^2}{2})} dx = \mathbf{P}((a,b])$. What is the probability p that X takes values in the Cantor set? Imitating Example 1.2.5, one may compute the probability q of the complementary set. Then

(1) $q \geq \mathbf{P}((\frac{1}{3}, \frac{2}{3}))$,
(2) $q \geq \mathbf{P}((\frac{1}{3}, \frac{2}{3})) + \mathbf{P}((\frac{1}{9}, \frac{2}{9})) + \mathbf{P}((\frac{7}{9}, \frac{8}{9}))$, and
(n) $q \geq \mathbf{P}((\frac{1}{3}, \frac{2}{3})) + \mathbf{P}((\frac{1}{9}, \frac{2}{9})) + \mathbf{P}((\frac{7}{9}, \frac{8}{9})) + \cdots + \mathbf{P}((\frac{3^n-2}{3^n}, \frac{3^n-1}{3^n}))$.

Therefore, one expects the probability p to be

$1 - \lim_{n \to \infty} \{ \mathbf{P}((\frac{1}{3}, \frac{2}{3})) + \mathbf{P}((\frac{1}{9}, \frac{2}{9})) + \mathbf{P}((\frac{7}{9}, \frac{8}{9})) + \mathbf{P}((\frac{1}{3^3}, \frac{2}{3^3})) + \mathbf{P}((\frac{7}{3^3}, \frac{8}{3^3}))$
$+ \mathbf{P}((\frac{19}{3^3}, \frac{20}{3^3})) + \mathbf{P}((\frac{25}{3^3}, \frac{26}{3^3})) + \cdots + \mathbf{P}((\frac{3^n-2}{3^n}, \frac{3^n-1}{3^n})) \}.$

At this point, it is not so clear what the answer should be here. In fact it is zero! (See Proposition 2.4.1.)

Exercise 1.2.8. Write a more explicit formula for the probability that the above random variable X takes values in the set C_n, which results after the procedure for constructing the Cantor set has been applied n times. This amounts to getting a handle on this set by writing an explicit description of the intervals involved in the removal process.

[Hints: after completion of the nth stage of the procedure for constructing the Cantor set, one is left with 2^n intervals each of length $\frac{1}{3^n}$. They are all translates of the interval $[0, \frac{1}{3^n}]$. To describe C_n, it suffices to determine the left hand endpoints of these intervals. One may do this by observing that, for each integer n, if $k < 3^n$, then k has a unique expression as $k = \sum_{i=0}^{n-1} a_i 3^i$ with the $a_i \in \{0, 1, 2\}$. One may use this to write a triadic "decimal" expansion for the left hand endpoints of the intervals remaining at the nth stage. Show that they will be of the form $0.b_1 b_2 \ldots b_n$ with $b_i \in \{0, 2\}$, where $0.b_1 b_2 \ldots b_n = \frac{1}{3^n} \sum_{i=1}^{n} b_i 3^{n-i}$. Show that one may obtain them from the 2^{n-1} left hand endpoints occurring at the $(n-1)$st stage by first putting a zero in the "first position", i.e., by shifting the "decimal" over to the right by one place and inserting a zero and then doing the same thing but this time inserting a two.]

What Examples 1.2.5, 1.2.6, and 1.2.7 hint at is the following: while one often starts to construct a probability using some "obvious" definition for certain simple sets (those in $\mathfrak{A}$), it is soon useful and necessary to try to extend the probability to more complicated sets that are made up from those of $\mathfrak{A}$ by infinite operations. In addition the probability should not only be finitely additive (i.e., satisfy $(FAP3)$), but also countably additive, as its computation will often involve infinite series.

3. Definition of a Probability Space

Definition 1.3.1. $(\Omega, \mathfrak{F}, \mathbf{P})$ is said to be a **probability space** if Ω is a set, $\mathfrak{F}$ is a σ-**algebra** of subsets of Ω, and $\mathbf{P}$ is a (countably additive) **probability** on $\mathfrak{F}$. To say that $\mathfrak{F}$ is a σ-algebra means that the collection $\mathfrak{F}$ of subsets of Ω satisfies

($\mathfrak{F}_1$) $\Omega \in \mathfrak{F}$;
($\mathfrak{F}_2$) $A \in \mathfrak{F}$ implies $A^c \in \mathfrak{F}$; and
($\mathfrak{F}_3$) $(A_n)_{n \geq 1} \subset \mathfrak{F}$ implies $\cup_{n=1}^{\infty} A_n \in \mathfrak{F}$.

Furthermore, $\mathbf{P}$ is a function defined on $\mathfrak{F}$ that satisfies

(P_1) $\mathbf{P}(\Omega) = 1$;
(P_2) $0 \leq \mathbf{P}(A) \leq 1$ if $A \in \mathfrak{F}$; and
(P_3) $(A_n)_{n>1} \subset \mathfrak{F}$, and $A_n \cap A_m = \emptyset$ if $n \neq m$ implies
 $\mathbf{P}(\cup_{n=1}^{\infty} A_n) = \sum_{n=1}^{\infty} \mathbf{P}(A_n)$ (**countable additivity**).

Remarks. (1) A probability differs from a finitely additive probability by the important property of being **countably additive** (P_3) (also called **σ-additive**).

(2) A probability space is also a finitely additive probability space since countable additivity (P_3) implies finite additivity (FAP_3).

(3) A σ-algebra is also called a **σ-field**.

In Example 1.2.4, the basic way of computing probability was to assign length to intervals contained in $[0,1]$. Corresponding to this is a natural function F, which determines these probabilities: define $F(x)$ to be the length of $[0,x]$ if $0 \leq x \leq 1$. This function can be extended to all of $\mathbb{R}$ by setting $F(x) = 0$ when $x \leq 0$ and $F(x) = 1$, if $1 \leq x$. This extended function can be used to assign a probability to any interval $(a,b]$: namely, $\mathbf{P}((a,b]) = F(b) - F(a)$. This probability is the length of $(a,b] \cap [0,1]$. As a result, the "experiment" of Example 1.2.4 can also be modeled by using all the intervals of $\mathbb{R}$ as basic events, with the proviso that any interval disjoint from $[0,1]$ has probability zero. The function F is an example of a **distribution function** (see Definition 1.3.5).

Whenever one has a probability space $(\Omega, \mathfrak{F}, \mathbf{P})$ with $\Omega = \mathbb{R}$ and the σ-algebra $\mathfrak{F}$ contains every interval $(-\infty, x]$, then the probability determines a natural function in the same way: let $F(x) \stackrel{\text{def}}{=} \mathbf{P}((-\infty, x])$.

This function has the properties $(DF_1), (DF_2)$, and (DF_3) stated in the following proposition.

Proposition 1.3.2. *The function $F(x) = \mathbf{P}((-\infty, x])$ associated with a probability $\mathbf{P}$ has the following properties:*

(DF_1) $x \leq y$ implies $F(x) \leq F(y)$ (i.e., F is a non-decreasing function);

(DF_2) (i) $\lim_{x \to +\infty} F(x) = 1$ (i.e., for any positive number ϵ, $F(x) \geq 1 - \epsilon$ if x is large enough and positive); and

(ii) $\lim_{x \to -\infty} F(x) = 0$ (i.e., for any positive number ϵ, $F(x) \leq \epsilon$ if x is negative and $|x|$ is sufficiently large);

(DF_3) for any x, if (x_n) is a sequence that decreases to x, then the values $F(x_n)$ decrease to $F(x)$ (i.e., $F(x) = \lim_{n \to \infty} F(x_n)$).

Proof. As an exercise prove (DF_1) and (DF_2) using properties $(P_1), (P_2)$, and (P_3). [For the first part of (DF_2) use Exercise 1.1.8 (3) and the fact that $\mathbb{R} = \cup_{n=-\infty}^{+\infty} (n, n+1]$ to show that $1 = F(n) + \sum_{k=n}^{\infty} \mathbf{P}((k, k+1])$.]

The property (DF_3) says that for any x, the values of $F(x_n)$ as x_n approaches x from above (i.e., from the right) approach $F(x)$. The technical formulation of this statement is that $\lim_{n \to \infty} F(x_n) = F(x)$ if $x \leq x_{n+1} \leq x_n$ and $\lim_n x_n = x$. Since F is non-decreasing, it suffices to show that for any x, $F(x) = \lim_{n \to \infty} F(x + \frac{1}{n})$.

Let $A_n = (x, x + \frac{1}{n}]$. Then $A_{n+1} \subset A_n$ and $\cap_{n=1}^{\infty} A_n = \emptyset$. Now $A_1 = (x, x+1] = \cup_{k=1}^{\infty} (x + \frac{1}{(k+1)}, x + \frac{1}{k}]$ and $A_n = \cup_{k=n}^{\infty} (x + \frac{1}{k+1}, x + \frac{1}{k}]$. Hence, $\mathbf{P}(A_1) = \sum_{k=1}^{\infty} \mathbf{P}((x + \frac{1}{k+1}, x + \frac{1}{k}]) \leq 1$, and so $\sum_{k=n}^{\infty} \mathbf{P}((x + \frac{1}{k+1}, x + \frac{1}{k}]) =$

$\mathbf{P}(A_n) \to 0$ as $n \to +\infty$ since it is the tail end of a convergent series. Now $F(x + \frac{1}{n}) = F(x) + \mathbf{P}(A_n)$. Hence, $\lim_{n \to \infty} F(x + \frac{1}{n}) = F(x)$. □

Exercise 1.3.3. Let $(A_n)_{n \geq 1}$ be a sequence in $\mathfrak{F}$, where $(\Omega, \mathfrak{F}, \mathbf{P})$ is a probability space. Show that $\mathbf{P}(A_n) \to 0$ if (1) $A_n \supset A_{n+1}$ for all $n \geq 1$ and (2) $\cap_{n=1}^{\infty} A_n = \emptyset$.

The property (DF_3) of a distribution function is a reflection of the countable additivity of $\mathbf{P}$. It can be reformulated by saying that $F(x)$ is right continuous at every point x.

Definition 1.3.4. *A function $F : \mathbb{R} \to \mathbb{R}$ has a **right limit** λ at a point $a \in \mathbb{R}$ if, for any $\epsilon > 0$, there exists a $\delta > 0$, where $\delta = \delta(a, \epsilon)$, such that*

$$|F(x) - \lambda| < \epsilon \text{ whenever } a < x < a + \delta.$$

*The right limit λ is denoted by $F(a+)$. The function F is **right continuous** at a if $F(a) = F(a+)$. Similarly, one defines the **left limit** $F(a-)$ to be λ if, for any $\epsilon > 0$, there exists a $\delta > 0$, where $\delta = \delta(x_0, \epsilon)$, such that*

$$|F(x) - \lambda| < \epsilon \text{ whenever } a - \delta < x < a$$

*and F is **left continuous** at a if $F(a) = F(a-)$.*

Remark. When F is the distribution function of a probability, then $F(a-) = \mathbf{P}((-\infty, a))$ (see Exercise 1.4.14 (3)).

Definition 1.3.5. *A function $F : \mathbb{R} \to [0, 1]$ is said to be a **distribution function** if*

(1) *it is non-decreasing and right continuous; and*
(2) $\lim_{x \to -\infty} F(x) = 0$ *and* $\lim_{x \to +\infty} F(x) = 1$.

A distribution function F determines the probability of certain sets, namely the probability of $(a, b]$, where $\mathbf{P}((a, b]) \stackrel{\text{def}}{=} F(b) - F(a)$. This notion of probability can then be extended to the Boolean algebra $\mathfrak{A}$ generated by the intervals $(a, b]$, where $A \in \mathfrak{A}$ if and only if A is a finite union of intervals of the form $(a, b]$, with $-\infty \leq a \leq b \leq +\infty$, and by convention one takes $(-\infty, +\infty] = \mathbb{R}$. Note that $\mathfrak{A}$ is the smallest Boolean algebra containing all the intervals $(-\infty, x])$, $x \in \mathbb{R}$. A priori, there is no reason why the probability $\mathbf{P}$ on $\mathfrak{A}$ determined in this way by a distribution function F should come from a probability on a σ-algebra $\mathfrak{F}$ containing the Boolean algebra $\mathfrak{A}$. This raises the following issue.

Basic Problem 1.3.6. *Given a distribution function F on $\mathbb{R}$, are there a σ-algebra $\mathfrak{F} \supset \mathfrak{A}$ and a probability $\mathbf{P}$ on $\mathfrak{F}$ such that F is the distribution function of $\mathbf{P}$? Note that for such a probability $\mathbf{P}$, it follows that $\mathbf{P}((a, b]) = F(b) - F(a)$ whenever $a \leq b$?*

Remark. As will be seen later in Theorem 2.2.2, this basic problem has an important generalization: the distribution function F is replaced by any right continuous, non-decreasing function G on $\mathbb{R}$. Then the question becomes: is there a σ-algebra $\mathfrak{F} \supset \mathfrak{A}$, and is there a function μ on $\mathfrak{F}$ that behaves like a probability except that its value on $\Omega = \mathbb{R}$ is not forced to be 1? For example, if $G(x) = x$ for all x, then such a set function μ would compute the length of sets.

Returning to distribution functions, the rest of this chapter is devoted to the solution of Basic Problem 1.3.6: to show that every distribution function comes from a probability. As a first step, one has the following.

Exercise 1.3.7. Let $\mathfrak{A}$ be the collection of finite unions of intervals of the form $(a, b]$, where $-\infty \leq a \leq b \leq +\infty$, and by convention one takes $(-\infty, +\infty] = \mathbb{R}$. Verify that

(1) $\mathfrak{A}$ is a Boolean algebra, (i.e. it satisfies $(\mathfrak{A}_1), (\mathfrak{A}_2), (\mathfrak{A}_3)$ of §1.2), and

(2) there is one and only one finitely additive probability $\mathbf{P}$ on $\mathfrak{A}$ such that $\mathbf{P}((a, b]) = F(b) - F(a)$ for all $a \leq b$.

[*Hints:* (1) Show that any $A \in \mathfrak{A}$ can be written as a finite pairwise disjoint union of intervals of the form $(a, b]$; note that the union of two intervals of this type is an interval of this type if they are not disjoint and observe that $(a, b] \cap (c, d]$ is $\emptyset$ or $(a \vee c, b \wedge d]$. (2) Convince oneself of (2) by showing that if $A = \cup_{i=1}^n (a_i, b_i]$ with the intervals pairwise disjoint, then $\sum_{i=1}^n \{F(b_i) - F(a_i)\}$ is not dependent on the particular way A is written as a disjoint union. For example if $A = (0, 3] = (0, 1] \cup (1, 3] = (0, 2] \cup (2, 3]$, then $\mathbf{P}((0, 3]) = F(3) - F(0) = \{F(1) - F(0)\} + \{F(3) - F(1)\} = \mathbf{P}((0, 1]) + \mathbf{P}((1, 3])$. *Suggestion:* given two disjoint unions, make up a "finer" disjoint union by using the second one to "cut up" all the intervals of the first. Then observe that, if $(a, b] = \cup_{\ell=1}^L (c_\ell, d_\ell]$ with the intervals $(c_\ell, d_\ell]$ pairwise disjoint, after relabeling if necessary, one has $a = c_1 < d_1 = c_2 < \cdots < d_{L-1} = c_L < d_L = b$.]

Convention 1.3.8. *Until further notice, unless otherwise stated or the context makes it evident (see Definition 1.4.2), $\mathfrak{A}$ will denote the above Boolean algebra of finite unions of half-open intervals $(a, b] \subset \mathbb{R}$.*

Example 1.3.9. Here are three well-known distribution functions:

(1) for the **uniform distribution on [0,1]**, (see Fig. 1.2)

$$F(x) = \begin{cases} 0 & \text{if } x \leq 0, \\ x & \text{if } 0 \leq x \leq 1, \\ 1 & \text{if } 1 \leq x. \end{cases}$$

3. DEFINITION OF A PROBABILITY SPACE   13

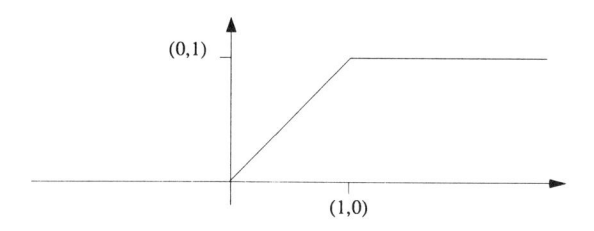

```
       (0,1)
          ┃  ┌─────────
          ┃ /
          ┃/
──────────┼──────────────▶
          ┃    (1,0)
```

Fig. 1.2

(2) for the **unit normal distribution** (see Fig. 1.3)

$$F(x) = \frac{1}{\sqrt{2\pi}} \int_{-\infty}^{x} e^{-x^2/2} dx.$$

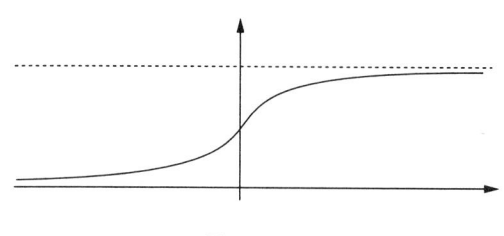

Fig. 1.3

(3) for the **Poisson distribution** with mean one (see Fig. 1.4)

$$F(x) = \begin{cases} 0, & x < 0, \\ \frac{1}{e} \sum_{0 \leq n \leq x} \frac{1}{n!}, & 0 \leq x. \end{cases}$$

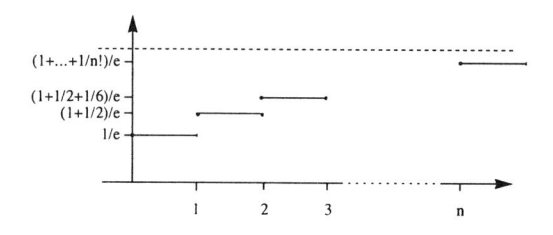

Fig. 1.4

Exercise 1.3.10. (a) Let Ω be a set and let $\mathfrak{C}$ be a collection of subsets of Ω. Show that there is a smallest σ-algebra of subsets of Ω that contains $\mathfrak{C}$. It is called the **σ-algebra generated by** $\mathfrak{C}$ and will be denoted by $\sigma(\mathfrak{C})$.

(b) Let $\Omega = \mathbb{R}$. Show that the smallest σ-algebra containing each of the following collections $\mathfrak{C}$ of subsets of $\mathbb{R}$ is the same:

(1) $\mathfrak{C} = \{(a,b] \mid a \leq b\}$;
(2) $\mathfrak{C} = \{(a,b) \mid a \leq b\}$;
(3) $\mathfrak{C} = \{[a,b) \mid a \leq b\}$;
(4) $\mathfrak{C} = \{[a,b] \mid a \leq b\}$;
(5) $\mathfrak{C}$ is the collection of **open** subsets G of $\mathbb{R}$; and
(6) $\mathfrak{C}$ is the collection of **closed** subsets F of $\mathbb{R}$.

The σ-algebra that results is called the **σ-algebra of Borel subsets of** $\mathbb{R}$ and is denoted by $\mathfrak{B}(\mathbb{R})$.

[Hints for (1) to (4) in (b): make use of Exercise 1.1.17.]

[Hints for (2),(5), and (6) in (b): for the two other collections of sets, two definitions are needed: a subset G of $\mathbb{R}$ is said to be **open** if $x_0 \in G$ implies that, for some $\epsilon > 0$, $(x_0 - \epsilon, x_0 + \epsilon) \subset G$; a subset F is said to be **closed** if F^c is open. As a result, the σ-algebras generated by collections (5) and (6) coincide. To show that collections (2) and (5) generate the same σ-algebras, it is necessary to know the following fact: every open set can be written as a countable union of open intervals (a fact that is part of the next exercise).]

Exercise 1.3.11. (a) Let $\mathfrak{O}$ be the collection of open subsets $\mathbb{R}$. Show that

(1) $\mathbb{R} \in \mathfrak{O}$,
(2) $G_1, G_2 \in \mathfrak{O} \Rightarrow G_1 \cap G_2 \in \mathfrak{O}$,
(3) the union of any collection of open sets is open.

(b) Show that if G is open and $x_0 \in G$, then there is a largest open interval $(a,b) \subset G$ that contains x_0.

[Hint: the union of a collection of intervals that contain a fixed point is itself an interval: recall Definition 1.1.15.]

(c) Let G be an open set and let $x_1, x_2 \in G$. Show that the largest open interval $I_1 \subset G$ that contains x_1 either equals I_2, the largest open interval $\subset G$ that contains x_2, or is disjoint from I_2.

(d) Let G be an open set. Show that G is a disjoint union of at most a countable number of open intervals. [Hint: suppose $G \subset (-1,1)$. Show that if G is expressed as a disjoint union of open intervals using (c), at most a finite number can have length $\geq 1/n$, where $n \geq 1$ is any fixed positive integer.]

Comment. It is standard mathematical terminology to call a collection $\mathfrak{O}$ of sets a **topology** if it satisfies (1), (2), and (3) in the above exercise. The complements of open sets are defined to be the closed sets, and it follows

3. DEFINITION OF A PROBABILITY SPACE

from (3) that for any set A, the intersection of all the closed sets containing it is a closed set. It is called the **closure** of A and is usually denoted by $\overline{A}$.

Digression on countable sets. It is about time to say what the word "countable" means. A set E is **countable** if all its elements can be labeled by natural numbers in a 1:1 way, i.e., if there is a function $c : \mathbb{N} \to E$ such that (i) $E = \{c(n) \mid n \in \mathbb{N}\}$, (ii) $c(n_1) = c(n_2)$ implies $n_1 = n_2$. A set is **at most countable** if it is either finite (i.e., it can be "counted" using $\{1, \ldots, n\}$ for some n) or countable (i.e., it can be "counted" using $\mathbb{N}$).

Given two sets A and B, **the Cartesian product** $A \times B \stackrel{\text{def}}{=} \{(a, b) \mid a \in A, b \in B\}$.

Proposition 1.3.12. *If A and B are countable, then $A \times B$, is countable.*

Proof. To begin with, $A \times B$ is clearly not finite. To "count" $A \times B$ is really the same as "counting" $\mathbb{N} \times \mathbb{N}$. This can be viewed as a set in the plane. Figure 1.5 explains how to "count" or "enumerate" $\mathbb{N} \times \mathbb{N}$:

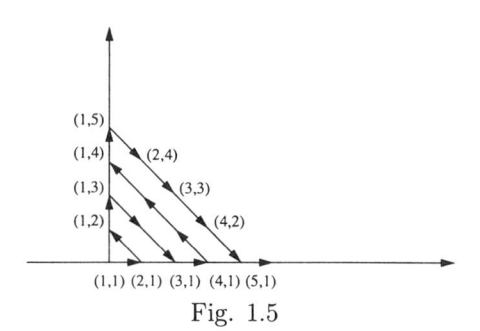

Fig. 1.5

Proposition 1.3.13. $\mathbb{Q}$ *is countable.*

Proof sketch. It suffices to show that $\{q \in \mathbb{Q} \mid q > 0\}$ is countable. Why? Look at the diagram of $\mathbb{N} \times \mathbb{N}$, and at each "site" (n, m) attach the rational n/m. It should then be clear how to count $\{q \in \mathbb{Q} \mid q > 0\}$! □

Proposition 1.3.14. (**Cantor's diagonal argument**) $(0, 1]$ *and hence* $\mathbb{R}$ *is not countable.*

Proof. If $0 < a \leq 1$, a has a decimal expansion as a=$0.a_1 a_2 \cdots a_n \cdots = \sum_{i=1}^{\infty} a_i/10^i$. If one eliminates decimals that terminate in an unbroken string of zeros, this decimal expression is unique (explain!).

Assume that the numbers in $(0, 1]$ can be enumerated, and write them in a sequence using only decimals that fail to terminate in an unbroken string of zeros:

$$a_1 = 0.a_{11}a_{12}a_{13}\cdots a_{1n}\cdots$$
$$a_2 = 0.a_{21}a_{22}a_{23}\cdots a_{2n}\cdots$$
$$\vdots$$
$$a_n = 0.a_{n1}a_{n2}a_{n3}\cdots a_{nn}\cdots.$$

Let $a = 0.a'_{11}a'_{22}a'_{33}\cdots a'_{nn}\cdots$, where $0 \neq .a'_{nn} \neq a_{nn}$ for each $n \neq 0$. In other words, use the diagonal entries of the above infinite table to make a number in $(0,1]$. Then a is not in the list since its decimal expansion fails to agree with that of any of the expansions for $a_1, a_2, \ldots$. This is a contradiction.

Since $(0,1]$ is not countable, $\mathbb{R}$ is not countable. □

This argument applies to any non-void interval since, for example, the function $f(x) = \frac{x-a}{b-1}$ maps $[a,b]$ in a 1:1 way onto $[0,1]$. As a result, any non-void open set is not countable, as it contains some interval $[a,b]$ with $a \neq b$. In particular, one has the following observation.

Exercise 1.3.15. Let C be any countable subset of $\mathbb{R}$. Then the closure $\overline{\mathbb{R}\backslash C}$ of $\mathbb{R}\backslash C$ equals $\mathbb{R}$. Show this by observing that $C \supset \complement\overline{\mathbb{R}\backslash C}$, which is an open set. This property of the complement $\mathbb{R}\backslash C$ of C is referred to by saying that C is **dense**. Show that a set D is dense in $\mathbb{R}$ if and only if every non-void open set contains a point of D.

Finally, one can verify the following fact.

Exercise 1.3.16. Let $E_1, \ldots, E_n$ be a finite collection of disjoint countable sets. Then $E = \cup_{i=1}^n E_i$ is countable. Let $E_1, \ldots, E_n, \ldots$ be a countable collection of countable sets. Then $E = \cup_{i=1}^\infty E_i$ is countable.

4. Construction of a Probability from a Distribution Function

Now consider the basic problem of constructing a probability from a distribution function F. Exercise 1.3.10 (b) shows that if one is to get a probability space $(\mathbb{R}, \mathfrak{F}, \mathbf{P})$ from F with $\mathfrak{F} \supset \mathfrak{A}$, then $\mathfrak{F}$ will have to contain the σ-algebra $\mathfrak{B}(\mathbb{R})$ of all Borel subsets of $\mathbb{R}$.

Furthermore, if $A \in \mathfrak{A}$ and $A = \cup_{n=1}^\infty A_n$, where the A_n are in $\mathfrak{A}$ and pairwise disjoint — for example, $(0,1] = \cup_{n=1}^\infty (\frac{1}{n+1}, \frac{1}{n}]$ — then, it will be necessary (as shown later) that

$$\mathbf{P}(A) = \sum_{n=1}^\infty \mathbf{P}(A_n)$$

if there is to be a probability $\mathbf{P}$ on a σ-algebra $\mathfrak{F} \supset \mathfrak{A}$. Recall that $\mathbf{P}$ is to be σ-additive.

Exercise 1.4.1. Show that $F(1) - F(0) = \sum_{n=1}^{\infty} \{F(\frac{1}{n}) - F(\frac{1}{n+1})\}$. [*Hint*: use Exercise 1.1.8 and the right continuity of F.]

Definition 1.4.2. *A finitely additive probability* $\mathbf{P}$ *on a Boolean algebra* $\mathfrak{A}$ *is said to be* σ-**additive** *if* $A = \cup_{n=1}^{\infty} A_n$ *implies*

$$\mathbf{P}(A) = \sum_{n=1}^{\infty} \mathbf{P}(A_n),$$

when A *in* $\mathfrak{A}$, *the sets* A_n *are all in* $\mathfrak{A}$, *and are pairwise disjoint.*

A finitely additive probability $\mathbf{P}$ on $\mathfrak{A}$ need not be σ-additive, as the following example shows.

Example. Define $\mathbf{P}$ on $\mathfrak{A}$ by setting $\mathbf{P}(A) = 0$ if A has an upper bound and $\mathbf{P}(A) = 1$ if A has no upper bound (remember that $\mathfrak{A}$ is a special Boolean Algebra — see Convention (1.3.8)). Show that $(\mathbb{R}, \mathfrak{A}, \mathbf{P})$ satisfies conditions $(\mathfrak{A}_1), (\mathfrak{A}_2), (\mathfrak{A}_3)(P_1), (P_2)$, and (FAP_3) in §1.2. Show that it is not σ-additive. To see this, try to calculate $\mathbf{P}(\mathbb{R})$ as $\sum_{n=-\infty}^{\infty} \mathbf{P}((n, n+1])$. Notice that this "probability" does not come from a distribution function. What is missing?

Returning to the problem of extending $\mathbf{P}$ from $\mathfrak{A}$ to $\mathfrak{B}(\mathbb{R})$, it will now be shown that if $\mathbf{P}$ on $\mathfrak{A}$ is determined by a distribution function, then it is σ-additive.

Exercise 1.4.3. Show that $\mathbf{P}$ on $\mathfrak{A}$ is σ-additive if and only if $(a, b] = \cup_{k=1}^{\infty} (c_k, d_k]$, with the $(c_k, d_k]$ pairwise disjoint, implies

$$\mathbf{P}((a,b]) = F(b) - F(a) = \sum_{k=1}^{\infty} \{F(d_k) - F(c_k)\}.$$

Theorem 1.4.4. *Let* F *be a distribution function on* $\mathbb{R}$. *Let* $\mathbf{P}$ *be the unique finitely additive probability on* $\mathfrak{A}$ *such that* $\mathbf{P}((a, b]) = F(b) - F(a)$ *whenever* $a \leq b$. *Then* $\mathbf{P}$ *is* σ-*additive on* $\mathfrak{A}$.

Remark. This theorem may appear obvious to you in view of Exercise 1.4.3. In a way it should. However, its actual justification depends on the Axiom 1.1.3.

Proof. By Exercise 1.4.3, it suffices to prove that if $(a, b] = \cup_{k=1}^{\infty} (c_k, d_k]$ with the intervals $(c_k, d_k]$ pairwise disjoint, then

(*) $$F(b) - F(a) = \sum_{k=1}^{\infty} \{F(d_k) - F(c_k)\}.$$

Now it is obvious that $F(b) - F(a) \geq \sum_{k=1}^{n} \{F(d_k) - F(c_k)\}$ since $(a, b] \supset \cup_{k=1}^{n} (c_k, d_k]$. Hence, by Exercise 1.1.8, $F(b) - F(a) \geq \sum_{k=1}^{\infty} \{F(d_k) - $

$F(c_k)\}$. Therefore, it is enough to verify the opposite inequality, and it is here that the Axiom 1.1.3 and the right continuity of the distribution function F come into play.

First, one shows that it suffices to verify (*) when $-\infty < a < b < +\infty$. Suppose, for example, that $-\infty = a < b < +\infty$. Then, for large n, one has $-n < b$. If $(-\infty, b] = \cup_{k=1}^{\infty}(c_k, d_k]$ with the intervals $(c_k, d_k]$ pairwise disjoint, then $(-n, b] = \cup_{k=1}^{\infty}(c_k \vee (-n), d_k]$. If (*) holds when the endpoints are both finite, then

$$F(b) - F(-n) = \sum_{k=1}^{\infty} F(d_k) - F(c_k \vee (-n)) \leq \sum_{k=1}^{\infty} F(d_k) - F(c_k).$$

Since $\mathbf{P}((-\infty, b]) = F(b)$ and $F(-n)$ converges to zero as n tends to $+\infty$, it follows that $F(b) \leq \sum_{k=1}^{\infty} F(d_k) - F(c_k)$, and hence $F(b) = \sum_{k=1}^{\infty} F(d_k) - F(c_k)$ (the opposite inequality is valid, as observed above). Similar arguments apply when $-\infty < a < b = +\infty$ and when $-\infty = a < b = +\infty$. This shows that the theorem holds provided it holds for intervals with finite endpoints.

Assume $-\infty < a < b < +\infty$, and choose any positive number $\epsilon > 0$. For each k, there is a positive number $e_k > 0$ such that

$$F(d_k) \leq F(d_k + e_k) \leq F(d_k) + \frac{\epsilon}{2^k}$$

because the distribution function F is increasing and right continuous. In other words,

$$\mathbf{P}((c_k, d_k]) \leq \mathbf{P}((c_k, d_k + e_k]) \leq \mathbf{P}((c_k, d_k]) + \frac{\epsilon}{2^k}.$$

Therefore,

$$\sum_{k=1}^{\infty} \{F(d_k + e_k) - F(c_k)\} = \sum_{k=1}^{\infty} \mathbf{P}((c_k, d_k + e_k]) \leq \sum_{k=1}^{\infty} \{F(d_k) - F(c_k)\} + \epsilon.$$

Note that $(a, b] = \cup_{k=1}^{\infty}(c_k, d_k] \subset \cup_{k=1}^{\infty}(c_k, d_k, +e_k)$, which is an open set. Using the right continuity once again (at $x_0 = a$), there is a positive number $e < b - a$ such that $F(a + e) \leq F(a) + \epsilon$.

In other words,

$$\mathbf{P}((a, b]) - \epsilon \leq \mathbf{P}((a + e, b]).$$

Since $[a + e, b] \subset (a, b] \subset \cup_{k=1}^{\infty}(c_k, d_k + e_k)$, the closed interval $[a + e, b]$ is contained in a countable union of open intervals. A very famous theorem (the **Heine–Borel theorem**) asserts that, as a result, the union of some finite number of the open intervals $(c_k, d_k + e_k)$ contains $[a + e, b]$: the proof will be given below and it uses the Axiom 1.1.3. Assuming the validity of

the Heine–Borel theorem, suppose that one has $[a+e, b] \subset \cup_{k=1}^{n}(c_k, d_k+e_k)$. Then
$$(a+e, b] \subset \cup_{k=1}^{n}(c_k, d_k + e_k] = A' \in \mathfrak{A},$$
and so
$$\mathbf{P}((a+e, b]) = F(b) - F(a+e) \leq \mathbf{P}(A')$$
$$\leq \sum_{k=1}^{n} \mathbf{P}((c_k, d_k + e_k]) = \sum_{k=1}^{n} \{F(d_k + e_k) - (c_k)\}$$
(recall that $\mathbf{P}(A') \leq \sum_{k=1}^{n} \mathbf{P}((c_k, d_k + e_k])$ by Exercise 1.2.2 (6). The conclusion is then that
$$F(b) - F(a) - \epsilon = \mathbf{P}((a, b]) - \epsilon \leq \mathbf{P}((a+e, b])$$
$$\leq \sum_{k=1}^{n} \{F(d_k + e_k) - F(c_k)\}$$
$$\leq \sum_{k=1}^{\infty} \{F(d_k + e_k) - F(c_k)\} \leq \sum_{k=1}^{\infty} \{F(d_k) - F(c_k)\} + \epsilon.$$

Since ϵ is any positive number, the desired inequality is proved. □

For completeness, the key theorem that was used in the proof of this result will now be proved.

Theorem 1.4.5 (Heine–Borel). *Let $[a, b]$ be a closed, bounded interval in $\mathbb{R}$. Assume that there is a collection of open intervals (a_ι, b_ι) whose union contains $[a, b]$. Then the union of some finite number of the given collection of open intervals also contains $[a, b]$.*

Proof. The idea of the proof is to see how large an interval $[a, x]$ with $a \leq x \leq b$ can actually be covered by a finite number of the open intervals (i.e., $[a, x] \subset$ some finite union of these intervals). One knows that for some i, say i_1, $a \in (a_{i_1}, b_{i_1})$. So, if $x = \min\{b, \frac{1}{2}(a+b_{i_1})\}$, then $[a, x] \subset (a_{i_1}, b_{i_1})$. Let H equal the set of x in the interval $[a, b]$ such that $[a, x]$ is contained in a finite number of the intervals. This is a set with an upper bound b. Note that if $x \in H$ and $a \leq y \leq x$, then $y \in H$. Also, if d is an upper bound of H, then $d < e \leq b$ implies $e \notin H$. By Axiom 1.1.3, H has an l.u.b. Call it c.

Exercise 1.4.6. (1) Show that $[a, c]$ is contained in a finite union of the open intervals. [*Hint:* c is in some interval.] (2) Show that if $[a, x]$ is contained in a finite union of the open intervals and $x < b$, then x is not an upper bound of H.

This exercise implies $c = b$, and so the theorem is proved. □

Remark. An equivalent form of the Heine–Borel theorem is the following result.

Theorem 1.4.7. *Let $(O_i)_{i \in J}$ be a family of open sets that covers the closed, bounded interval $[a, b] \subset \mathbb{R}$ (i.e., $\cup_{i \in J} O_i \supset [a, b]$). Then a finite number of the sets O_i covers $[a, b]$ (i.e., for some finite set $F \subset J$, $\cup_{i \in F} O_i \supset [a, b]$).*

Exercise 1.4.8. Show that Theorem 1.4.7 and Theorem 1.4.5 are equivalent.

The Heine–Borel theorem is so basic that the class of sets for which it is true is given a name.

Definition 1.4.9. *A set $K \subset \mathbb{R}$ is said to be **compact** if, whenever $K \subset \cup_{i \in J} O_i$, each O_i open, there is a finite set $F \subset J$ with $K \subset \cup_{i \in F} O_i$.*

The Heine–Borel theorem states that every closed and bounded interval is compact. Given this theorem, it is not hard to show that a set $K \subset \mathbb{R}$ is compact if and only if K is closed and bounded (see Exercise 1.5.6). For example, the Cantor set is compact!

Returning again to the extension problem for $\mathbf{P}$, it is clear that if $A = \cup_{n=1}^{\infty} A_n$, $A_n \in \mathfrak{A}$ for all n and A not necessarily in $\mathfrak{A}$, then $A \in \mathfrak{F}$. Also, by Remark 1.2.3, A can be written as a disjoint union of sets in $\mathfrak{A}$: one replaces each A_n by $A_n \setminus \cup_{i=1}^{n-1} A_i$. Consequently, if there is an extension, $\mathbf{P}(A) = \sum_{n=1}^{\infty} \mathbf{P}(A_n \setminus \cup_{i=1}^{n-1} A_i) \leq \sum_{n=1}^{\infty} \mathbf{P}(A_n)$. Without assuming the extension to be possible, one may define $\mathbf{P}^*(A)$ to be the greatest lower bound of $\{\sum_{n=1}^{\infty} \mathbf{P}(A_n) | A = \cup_{n=1}^{\infty} A_n, A_n \in \mathfrak{A} \text{ for all } n\}$. Then $\mathbf{P}^*(A)$ is an estimate for the value $\mathbf{P}(A)$ of a possible extension when $A \in \mathfrak{A}_\sigma$, the collection of sets that are countable unions of sets from $\mathfrak{A}$. Since $\mathbb{R} = \cup_{n=-\infty}^{+\infty} (n, n+1]$, every $E \subset \mathbb{R}$ is a subset of some set $A \in \mathfrak{A}_\sigma$. Hence if $E \subset A$, then one expects to have $\mathbf{P}(E) \leq \mathbf{P}^*(A)$. This motivates the definition of the following set function $\mathbf{P}^*$.

If $E \subset \mathbb{R}$, define $\mathbf{P}^*(E)$ to be the greatest lower bound of $\{\mathbf{P}^*(A) | E \subset A \in \mathfrak{A}_\sigma\}$, i.e.,

$$\mathbf{P}^*(E) \stackrel{\text{def}}{=} \inf_{\substack{A \in \mathfrak{A}_\sigma \\ A \supset E}} \mathbf{P}^*(A) = \inf\left\{\sum_{n=1}^{\infty} \mathbf{P}(A_n) | E \subset \cup_{n=1}^{\infty} A_n, A_n \in \mathfrak{A} \text{ for all } n\right\}.$$

Remark. The terms **infimum** (abbreviated to "inf") and **supremum** (abbreviated to "sup") are merely other words for "g.l.b." and "l.u.b.", respectively.

In general, the set function $\mathbf{P}^*$ does not behave like a probability because (P_3) need not hold! However, it has certain important properties which make it into what is called an **outer measure**. They are stated in the next definition.

4. CONSTRUCTION OF A PROBABILITY

Definition 1.4.10. *An* **outer measure** *on the subsets of* $\mathbb{R}$ *is a set function* $\mathbf{P}^*$ *such that*
  (1) $0 \leq \mathbf{P}^*(E)$ *for all* $E \subset \mathbb{R}$,
  (2) $E_1 \subset E_2$ *implies* $\mathbf{P}^*(E_1) \leq \mathbf{P}^*(E_2)$, *and*
  (3) $E = \cup_{n=1}^\infty E_n$ *implies* $\mathbf{P}^*(E) \leq \sum_{n=1}^\infty \mathbf{P}^*(E_n)$ (*i.e., it is* **countably subadditive**).

Proposition 1.4.11. *The set function* $\mathbf{P}^*$ *defined above is an outer measure with* $\mathbf{P}^*(E) \leq 1$ *for all* $E \in \mathbb{R}$. *Furthermore, since* $\mathbf{P}$ *is* σ-*additive on* $\mathfrak{A}$, $\mathbf{P}(A) = \mathbf{P}^*(A)$, *for all* $A \in \mathfrak{A}$.

Proof. Properties (1) and (2) of Definition 1.4.10 are obvious. Let $\epsilon > 0$ and, for each n, let $E_n \subset \cup_{k=1}^\infty A_{n,k}$ be such that $\sum_{k=1}^\infty \mathbf{P}(A_{n,k}) \leq \mathbf{P}^*(E_n) + \frac{\epsilon}{2^n}$ where the sets $A_{n,k} \in \mathfrak{A}$. Then $E = \cup_{n=1}^\infty E_n \subset \cup_{n=1}^\infty \cup_{k=1}^\infty A_{n,k}$, which is a countable union of sets from $\mathfrak{A}$. Hence,

$$\mathbf{P}^*(E) \leq \sum_{n=1}^\infty \sum_{k=1}^\infty \mathbf{P}(A_{n,k}) \leq \sum_{n=1}^\infty \left\{ \mathbf{P}^*(E_n) + \frac{\epsilon}{2^n} \right\} = \sum_{n=1}^\infty \mathbf{P}^*(E_n) + \epsilon.$$

If $A \in \mathfrak{A}$ is contained in $\cup_{n=1}^\infty A_n$, $A_n \in \mathfrak{A}$, $n \geq 1$, then $A = \cup_{n=1}^\infty A'_n$, where $A'_n = A \cap \left[A_n \setminus \cup_{k=1}^{n-1} A_k \right]$. Then, by the σ-additivity of $\mathbf{P}$ on $\mathfrak{A}$ (Theorem 1.4.4), $\mathbf{P}(A) = \sum_{n=1}^\infty \mathbf{P}(A'_n) \leq \sum_{n=1}^\infty \mathbf{P}(A_n)$ as $A'_n \subset A_n$, for all $n \geq 1$. This shows that $\mathbf{P}(A) \leq \mathbf{P}^*(A)$ and hence $\mathbf{P}(A) = \mathbf{P}^*(A)$ if $A \in \mathfrak{A}$. □

Remark. The fact that $\mathbf{P}$ and $\mathbf{P}^*$ agree on $\mathfrak{A}$ is crucial in what follows. This is why it is so important that $\mathbf{P}$ be σ-additive on $\mathfrak{A}$.

While $\mathbf{P}^*$ is defined for all subsets of $\mathbb{R}$, it is not necessarily a probability. This raises the problem as to whether there is a natural class of sets on which it is a probability. The following way of solving this problem is due to a well-known Greek mathematician, C. Carathéodory. He observed that (i) the sets in any σ-algebra $\mathfrak{G}$ containing $\mathfrak{A}$ have a special property (see (C) below) provided the outer measure $\mathbf{P}^*$ restricted to $\mathfrak{G}$ is a probability, and (ii) the collection of all sets with this property is in fact a σ-algebra, and the restriction of $\mathbf{P}^*$ to this σ-algebra is a probability.

First, notice that because of property (3) of Definition 1.4.10, for any two sets E and Q, one has $\mathbf{P}^*(E) \leq \mathbf{P}^*(E \cap Q) + \mathbf{P}^*(E \setminus Q)$. However, if $Q \in \mathfrak{A}$, then in fact

(C) $$\mathbf{P}^*(E) = \mathbf{P}^*(E \cap Q) + \mathbf{P}^*(E \cap Q^c),$$

for any set E because $\mathbf{P}$ and $\mathbf{P}^*$ agree on $\mathfrak{A}$.

To prove (C) for sets $Q \in \mathfrak{A}$, let $\epsilon > 0$ and let $A_n \in \mathfrak{A}$ for all $n \geq 1$ be such that $E \subset \cup_{n=1}^\infty A_n$ and $\sum_{n=1}^\infty \mathbf{P}(A_n) \leq \mathbf{P}^*(E) + \epsilon$. These sets A_n may be assumed to be pairwise disjoint by Remark 1.2.3. Further, $A_n \cap Q^c \in \mathfrak{A}$ for all $n \geq 1$, since $Q \in \mathfrak{A}$.

Now

$$\mathbf{P}^*(E \cap Q) + \mathbf{P}^*(E \cap Q^c) \leq \mathbf{P}^*\left(\cup_{n=1}^\infty (A_n \cap Q)\right) + \mathbf{P}^*\left(\cup_{n=1}^\infty (A_n \cap Q^c)\right)$$
$$\leq \sum_{n=1}^\infty \{\mathbf{P}(A_n \cap Q)\} + \sum_{n=1}^\infty \{\mathbf{P}(A_n \cap Q^c)\}$$
$$= \sum_{n=1}^\infty \{\mathbf{P}(A_n \cap Q) + \mathbf{P}(A_n \cap Q^c)\} = \sum_{n=1}^\infty \mathbf{P}(A_n) \leq \mathbf{P}^*(E) + \epsilon.$$

Therefore, $\mathbf{P}^*(E \cap Q) + \mathbf{P}^*(E \cap Q^c) \leq \mathbf{P}^*(E)$, proving (C).

Now suppose that $\mathfrak{G} \supset \mathfrak{A}$ is a σ-algebra and that $\mathbf{P}^*$ restricted to $\mathfrak{G}$ is a probability, say $\mathbf{R}$. Then, since a σ-algebra is a Boolean algebra and $\mathbf{R}$ is σ-additive on $\mathfrak{G}$, one could construct a new outer measure $\mathbf{R}^*$ from $\mathbf{R}$. As stated later in Exercise 1.5.4, in fact $\mathbf{R}^* = \mathbf{P}^*$. Therefore, it follows from what has just been proved for $\mathfrak{A}$ that condition (C) holds for all the sets in $\mathfrak{G}$.

This suggests that one should look at the collection $\mathfrak{F}$ of all sets for which condition (C) holds, i.e.,

$$\mathfrak{F} = \{Q \mid \mathbf{P}^*(E) = \mathbf{P}^*(E \cap Q) + \mathbf{P}^*(E \cap Q^c) \text{ for any } E \subset \mathbb{R}\}.$$

It will now be shown that

(i) $\mathfrak{F}$ is a σ-algebra (containing $\mathfrak{A}$), and
(ii) $\mathbf{P}^*$ restricted to $\mathfrak{F}$ is a probability.

Hence, $(\mathbb{R}, \mathfrak{F}, \mathbf{P}^*)$ is a probability space and $\mathfrak{F} \supset \mathfrak{A}$ (also, in view of Exercise 1.3.10 (b), $\mathfrak{F} \supset \mathfrak{B}(\mathbb{R})$ — the algebra of Borel sets).

It remains to verify (i) and (ii).

To verify (i), first note that $\Omega = \mathbb{R} \in \mathfrak{F}$ and that $A \in \mathfrak{F}$ implies $A^c \in \mathfrak{F}$. If $A_1, A_2 \in \mathfrak{F}$, then $A_1 \cup A_2 \in \mathfrak{F}$. To see this, let $E \subset \mathbb{R}$. Then, by (C), one has

$$\mathbf{P}^*(E) = \mathbf{P}^*(E \cap A_1) + \mathbf{P}^*(E \cap A_1^c).$$

Since, by (C),

$$\mathbf{P}^*(E \cap A_1) = \mathbf{P}^*(E \cap A_1 \cap A_2) + \mathbf{P}^*(E \cap A_1 \cap A_2^c),$$

and, again by (C),

$$\mathbf{P}^*(E \cap A_1^c) = \mathbf{P}^*(E \cap A_1^c \cap A_2) + \mathbf{P}^*(E \cap A_1^c \cap A_2^c),$$

this implies that

(1) $$\mathbf{P}^*(E) = \mathbf{P}^*(E \cap A_1 \cap A_2) + \mathbf{P}^*(E \cap A_1 \cap A_2^c) \\ + \mathbf{P}^*(E \cap A_1^c \cap A_2) + \mathbf{P}^*(E \cap A_1^c \cap A_2^c).$$

4. CONSTRUCTION OF A PROBABILITY

Since
$$E \cap (A_1 \cup A_2) = (E \cap A_1 \cap A_2) \cup (E \cap A_1 \cap A_2^c) \cup (E \cap A_1^c \cap A_2)$$

it follows from (1) and Definition 1.4.10 (3) that

$$\mathbf{P}^*(E) \geq \mathbf{P}^*(E \cap (A_1 \cup A_2)) + \mathbf{P}^*(E \cap A_1^c \cap A_2^c).$$

Hence, $A_1, A_2 \in \mathfrak{F}$ implies that $A_1 \cup A_2 \in \mathfrak{F}$.

By now it should be clear that $\mathfrak{F}$ is a Boolean algebra. Therefore, $\mathfrak{F}$ is a σ-algebra providing $\cup_{n=1}^{\infty} A_n \in \mathfrak{F}$ whenever the $A_n \in \mathfrak{F}$ are pairwise disjoint.

To verify this, one has to show that for any set $E \subset \Omega$,

$$\mathbf{P}^*(E) = \mathbf{P}^*(E \cap (\cup_{n=1}^{\infty} A_n)) + \mathbf{P}^*(E \cap (\cup_{n=1}^{\infty} A_n)^c),$$

when the $A_n \in \mathfrak{F}$ are pairwise disjoint. To do this, it will suffice to show that for all $n \geq 1$,

(2) $$\mathbf{P}^*(E) \geq \sum_{i=1}^{n} \mathbf{P}^*(E \cap A_i) + \mathbf{P}^*(E \cap (\cup_{n=1}^{\infty} A_n)^c).$$

The reason is that this inequality and Definition 1.4.10 (3) imply that

$$\mathbf{P}^*(E) \geq \sum_{n=1}^{\infty} \mathbf{P}^*(E \cap A_n) + \mathbf{P}^*(E \cap (\cup_{n=1}^{\infty} A_n)^c)$$
$$\geq \mathbf{P}^*(E \cap (\cup_{n=1}^{\infty} A_n)) + \mathbf{P}^*(E \cap (\cup_{n=1}^{\infty} A_n)^c).$$

In view of the validity of the opposite inequality, again by Definition 1.4.10 (3), one then has

$$\mathbf{P}^*(E) = \mathbf{P}^*(E \cap (\cup_{n=1}^{\infty} A_n)) + \mathbf{P}^*(E \cap (\cup_{n=1}^{\infty} A_n)^c), \text{ i.e., } \cup_{n=1}^{\infty} A_n \in \mathfrak{F}.$$

Now (2) holds if

(3) $$\mathbf{P}^*(E) \geq \sum_{i=1}^{n} \mathbf{P}^*(E \cap A_i) + \mathbf{P}^*(E \cap (\cup_{i=1}^{n} A_i)^c).$$

Hence, to verify (2), it suffices to prove this last inequality (i.e., inequality (3)) as $(\cup_{i=1}^{n} A_i)^c \supset (\cup_{i=1}^{\infty} A_i)^c$. Note that (3) is equivalent by Definition 1.4.10 (3) to the identity

(4) $$\mathbf{P}^*(E) = \sum_{i=1}^{n} \mathbf{P}^*(E \cap A_i) + \mathbf{P}^*(E \cap (\cup_{i=1}^{n} A_i)^c).$$

To prove (4), let $A_1, \ldots, A_n$ be pairwise disjoint and in $\mathfrak{F}$. Then, by applying the defining property of $\mathfrak{F}$ first to A_1 using E, and then to A_2 and using $E \cap A_1^c$ in place of E, it follows that

$$\mathbf{P}^*(E) = \mathbf{P}^*(E \cap A_1) + \mathbf{P}^*(E \cap A_1^c)$$
$$= \mathbf{P}^*(E \cap A_1) + \mathbf{P}^*(E \cap A_1^c \cap A_2) + \mathbf{P}^*(E \cap A_1^c \cap A_2^c)$$
$$= \mathbf{P}^*(E \cap A_1) + \mathbf{P}^*(E \cap A_2) + \mathbf{P}^*(E \cap A_1^c \cap A_2^c),$$

since $A_1 \cap A_2 = \emptyset$.

Hence, (4) holds for $n = 2$. Note that this follows immediately from formula (1) if $A_1 \cap A_2 = \emptyset$. Assume that (4) is true for $n - 1$ pairwise disjoint sets A_i, i.e., for any $E \subset \mathbb{R}$,

$$\mathbf{P}^*(E) = \sum_{i=1}^{n-1} \mathbf{P}^*(E \cap A_i) + \mathbf{P}^*(E \cap (\cup_{i=1}^{n-1} A_i)^c).$$

Apply the defining property of $\mathfrak{F}$ to A_n and use the set $E \cap (\cup_{i=1}^{n-1} A_i)^c$. It then follows that

$$\mathbf{P}^*(E \cap (\cup_{i=1}^{n-1} A_i)^c) = \mathbf{P}^*(E \cap (\cup_{i=1}^{n-1} A_i)^c \cap A_n) + \mathbf{P}^*(E \cap (\cup_{i=1}^{n-1} A_i)^c \cap A_n^c)$$
$$= \mathbf{P}^*(E \cap A_n) + \mathbf{P}^*(E \cap (\cup_{i=1}^{n} A_i)^c).$$

Therefore,

$$\mathbf{P}^*(E) = \sum_{i=1}^{n} \mathbf{P}^*(E \cap A_i) + \mathbf{P}^*(E \cap (\cup_{i=1}^{n} A_i)^c).$$

This completes the proof of (i). The proof of (ii) is given below.

Definition 1.4.12. *The σ-algebra $\mathfrak{F}$ of sets Q that satisfy*

(C) $$\mathbf{P}^*(E) = \mathbf{P}^*(E \cap Q) + \mathbf{P}^*(E \cap Q^c)$$

for all $E \subset \mathbb{R}$ is called the σ-algebra of $\mathbf{P}^*$-measurable sets.

The following theorem is the goal to which all these arguments have been leading.

Theorem 1.4.13. *Let F be a distribution function on $\mathbb{R}$.*
 (1) *Then there is a unique probability $\mathbf{P}$ on $\mathfrak{B}(\mathbb{R})$, the σ-algebra of Borel subsets of $\mathbb{R}$, such that $\mathbf{P}((a, b]) = F(b) - F(a)$ whenever $a \leq b$.*
 (2) *The σ-algebra $\mathfrak{F}$ of $\mathbf{P}^*$-measurable subsets contains $\mathfrak{B}(\mathbb{R})$, and $\mathbf{P}^*$ restricted to $\mathfrak{F}$ is a probability such that $\mathbf{P}^*((a, b]) = F(b) - F(a)$ whenever $a \leq b$.*

Furthermore, the σ-algebra $\mathfrak{F}$ of $\mathbf{P}^*$-measurable sets is the largest σ-algebra $\mathfrak{G}$ containing $\mathfrak{A}$ with the property that the restriction of $\mathbf{P}^*$ to $\mathfrak{G}$ is a probability.

Proof. First consider (2). It has been shown that $\mathfrak{F}$ is a σ-algebra. To prove that $\mathbf{P}^*$ restricted to $\mathfrak{F}$ is a probability, it will suffice to verify (P_3) of Definition 1.3.1 since $\mathbf{P}^*$ obviously satisfies (P_1) and (P_2). Let $(A_n)_{n \geq 1} \subset \mathfrak{F}$ be pairwise disjoint. Since $\mathbf{P}^*$ is an outer measure, it is countably subadditive, i.e., (3) of Definition 1.4.10 holds. Hence

$$\mathbf{P}^*(\cup_{n=1}^\infty A_n) \leq \sum_{n=1}^\infty \mathbf{P}^*(A_n).$$

To prove the reverse inequality, note that $\mathbf{P}^*(\cup_{n=1}^\infty A_n) \geq \mathbf{P}^*(\cup_{n=1}^N A_n) = \sum_{n=1}^N \mathbf{P}^*(A_n)$ in view of (4) (take $E = \cup_{n=1}^N A_n$). This completes the proof of (2) and of (ii) above.

To prove the first statement, let $\mathbf{P}_1$ and $\mathbf{P}_2$ be two probabilities on $\mathfrak{B}(\mathbb{R})$ such that $\mathbf{P}_1((a,b]) = \mathbf{P}_2((a,b]) = F(b) - F(a)$ whenever $a \leq b$. Let $\mathfrak{M} = \{A \in \mathfrak{B}(\mathbb{R}) \mid \mathbf{P}_1(A) = \mathbf{P}_2(A)\}$. Then, by Exercise 1.3.7, $\mathbf{P}_1$ and $\mathbf{P}_2$ agree on $\mathfrak{A}$.

Exercise 1.4.14. Let $(\Omega, \mathfrak{F}, \mathbf{P})$ be a probability space and let $(A_n)_{n \geq 1} \subset \mathfrak{F}$. Show that

(1) if $A_n \subset A_{n+1}$ for all n, then $\lim_{n \to \infty} \mathbf{P}(A_n) = \mathbf{P}(\cup_{n=1}^\infty A_n)$,
(2) if $A_n \supset A_{n+1}$ for all n, then $\lim_{n \to \infty} \mathbf{P}(A_n) = \mathbf{P}(\cap_{n=1}^\infty A_n)$.
[*Hint*: recall Exercise 1.3.3.]

If F is the distribution function of a probability $\mathbf{P}$ on $\mathfrak{B}(\mathbb{R})$, show that

(3) $F(x-) = \mathbf{P}((-\infty, x))$ for all $x \in \mathbb{R}$.

Exercise 1.4.15. Show that the set $\mathfrak{M}$ defined above has the following two properties (where $(A_n)_{n \leq 1} \subset \mathfrak{M}$):

(1) $A_n \subset A_{n+1}$ for all n implies $\cup_{n=1}^\infty A_n \in \mathfrak{M}$; and
(2) $A_n \supset A_{n+1}$ for all n implies $\cap_{n=1}^\infty A_n \in \mathfrak{M}$,

i.e., $\mathfrak{M}$ is a so-called **monotone class**.

Exercise 1.4.16. (**Monotone class theorem for sets**) Let $\mathfrak{M}$ be a monotone class that contains a Boolean algebra $\mathfrak{A}$. Show that $\mathfrak{M}$ contains the smallest σ-algebra $\mathfrak{F}$ containing $\mathfrak{A}$. [*Hints*: let $\mathfrak{M}_0$ be the smallest monotone class containing $\mathfrak{A}$ (why does it exist?). Let $A \in \mathfrak{A}$. Show that $\{B \in \mathfrak{M}_0 \mid B \cup A \in \mathfrak{M}_0\}$ is a monotone class. Conclude that $B \in \mathfrak{M}_0, A \in \mathfrak{A}$ imply $B \cup A \in \mathfrak{M}_0$. Now fix $B_1 \in \mathfrak{M}_0$ and look at $\{B \in \mathfrak{M}_0 \mid B_1 \cup B \in \mathfrak{M}_0\}$. Conclude as before that $B_1, B \in \mathfrak{M}_0$ implies $B_1 \cup B \in \mathfrak{M}_0$. Make a similar argument to show that $\{B^c \mid B \in \mathfrak{M}_0\}$ is $\mathfrak{M}_0$. Conclude that $\mathfrak{M}_0 \supset \mathfrak{F}$.]

The monotone class theorem has a function version, which is stated as Theorem 3.6.14. It gives conditions that ensure that a given collection $\mathcal{H}$ of bounded functions on Ω contains all the bounded random variables in the σ-algebra determined by a subset $\mathcal{C}$ of $\mathcal{H}$ that is closed under multiplication.

From these three exercises and Exercise 1.3.10 (b), one concludes that $\mathfrak{M} = \mathfrak{B}(\mathbb{R})$, as it contains $\mathfrak{A}$, and so (1) is established.

The last statement of the theorem is a consequence of Exercise 1.5.4 and the observation made following Proposition 1.4.11 that condition (C) holds for all the sets in $\mathfrak{G}$ if $\mathfrak{G}$ is a σ-algebra containing $\mathfrak{A}$ such that $\mathbf{P}^*$ restricted to $\mathfrak{G}$ is a probability. □

Remark 1.4.17. It is easily verified that the collection $\mathfrak{M}$ defined above has the following properties: (i) $\Omega \in \mathfrak{M}$ and (ii) if $A, B \in \mathfrak{M}$ with $A \subset B$, then $B \backslash A \in \mathfrak{M}$. Dynkin proved (see Proposition 3.2.6) that the smallest such system $\mathfrak{L}$ containing a collection $\mathfrak{C}$, of sets closed under finite intersections is the smallest σ-field containing $\mathfrak{C}$. This is another version of the monotone class theorem. It is proved in Exercise 3.2.5, part C, that any collection of sets $\mathfrak{L} \supset \mathfrak{C}$ satisfying (i) and (ii) also contains the smallest Boolean algebra $\mathfrak{A}$ containing $\mathfrak{C}$ if the collection $\mathfrak{C}$ is closed under finite intersections. Taking $\mathfrak{C}$ as $\{(a, b] \mid -\infty \leq a < b \leq +\infty\}$ Proposition 3.2.6 also proves the uniqueness of the probability $\mathbf{P}$ on $\mathfrak{B}(\mathbb{R})$ for which $\mathbf{P}((a, b]) = F(b) - F(a)$.

5. Additional exercises*

Exercise 1.5.1. Let $(\Omega, \mathfrak{F}, \mathbf{P})$ be a probability space. Show that
(1) $\mathbf{P}(\cup_{n=1}^{\infty} A_n) \leq \sum_{n=1}^{\infty} \mathbf{P}(A_n)$ (see Exercise 1.2.2 (6)); and
(2) $\mathbf{P}(\cup_{n=1}^{\infty} A_n) = 0$ if $\mathbf{P}(A_n) = 0$ for all $n \geq 1$.

Exercise 1.5.2. The so-called **Heaviside function** is the function H, where
$$H(x) = \begin{cases} 0, & x < 0, \\ 1, & x \geq 0. \end{cases}$$
Calculate for this distribution function the corresponding outer measure $\mathbf{P}^*$, and determine the σ-algebra $\mathfrak{F}$ of $\mathbf{P}^*$-measurable subsets of $\mathbb{R}$. [*Hint:* guess the σ-algebra $\mathfrak{F}$ and $\mathbf{P}^*$. Then see if they "work".]
The resulting measure, denoted by ε_0 or δ_0 is called the **Dirac measure at the origin** or **unit point mass at the origin**. Replace $H(x)$ by $H(x - a) \stackrel{\text{def}}{=} H_a(x)$, and let ε_a denote the resulting measure. Give the formula for $\epsilon_a(A), A \in \mathfrak{F}$, where $\mathfrak{F}$ is the corresponding σ-algebra of measurable sets. The measure ε_a is the **Dirac measure** or **unit point mass at** a.

Exercise 1.5.3. For the distribution function of the Poisson distribution (see Example 1.3.9 (3)), calculate the corresponding outer measure $\mathbf{P}^*$ and determine the σ-algebra $\mathfrak{F}$ of $\mathbf{P}^*$-measurable subsets of $\mathbb{R}$.

5. ADDITIONAL EXERCISES

Exercise 1.5.4. Let $\mathfrak{G} \supset \mathfrak{A}$ be a σ-algebra, and assume that $\mathbf{P}^*$ restricted to $\mathfrak{G}$ is a probability $\mathbf{R}$. Define the outer measure $\mathbf{R}^*$ by setting $\mathbf{R}^*(E) = \inf\{\sum_{n=1}^{\infty} \mathbf{R}(B_n) \mid E \subset \cup_{n=1}^{\infty} B_n, B_n \in \mathfrak{G} \text{ for all } n\}$. Show that
  (1) $\mathbf{R}^*(E) = \inf\{\mathbf{R}(B) \mid E \subset B, B \in \mathfrak{G}\}$,
  (2) $\mathbf{R}^*(E) \leq \mathbf{P}^*(E)$ for all $E \subset \mathbb{R}$,
  (3) $\mathbf{R}^*(E) = \mathbf{P}^*(E)$ for all $E \subset \mathbb{R}$. [Hint: $\mathbf{R}(B) = \mathbf{P}^*(B)$.]

Exercise 1.5.5. Show that Q is $\mathbf{P}^*$-measurable if and only if $\mathbf{P}^*(Q) + \mathbf{P}^*(Q^c) = 1$. This exercise may be done by going through the following steps.
  (1) Let Q, E be subsets of $\mathbb{R}$, and let (A_m), (B_m), and (C_p) be pairwise disjoint collections of sets from $\mathfrak{A}$ such that (i) $E \subset \cup_p C_p = C$ and (ii) $Q \subset \cup_m A_m = A$, $Q^c \subset \cup_n B_n = B$. Show that

$$\mathbf{P}^*(E \cap Q) + \mathbf{P}^*(E \cap Q^c) \leq \sum_{m,p} \mathbf{P}(A_m \cap C_p) + \sum_{n,p} \mathbf{P}(B_n \cap C_p)$$
$$\leq \sum_p \mathbf{P}(C_p) + \sum_{m,n,p} \mathbf{P}(A_m \cap B_n \cap C_p)$$
$$\leq \sum_p \mathbf{P}(C_p) + \sum_{m,n} \mathbf{P}(A_m \cap B_n).$$

  [Hint: for the second inequality use Exercise 1.2.2 (5).]
  (2) If (A_m) and (B_n) are two collections of pairwise disjoint sets from $\mathfrak{A}$ such that (i) $\mathbb{R} = A \cup B$, where $A = \cup_{m=1}^{\infty} A_m$ and $B = \cup_{n=1}^{\infty} B_n$, and (ii) $\sum_m \mathbf{P}(A_m) + \sum_n \mathbf{P}(B_n) \leq 1 + \varepsilon$, show that $\sum_{m,n} \mathbf{P}(A_m \cap B_n) \leq \varepsilon$.
  [Hint: make use of Exercise 1.2.2 (5) to compute $\mathbf{P}(A_m) + \mathbf{P}(B_n)$ and use the fact that $\mathbb{R} = \cup_{m,n} A_m \cup B_n$.]
  (3) If $\mathbf{P}^*(Q) + \mathbf{P}^*(Q^c) = 1$ and $E \subset \mathbb{R}$, show that for any $\varepsilon > 0$ one has $\mathbf{P}^*(E \cap Q) + \mathbf{P}^*(E \cap Q^c) \leq \mathbf{P}^*(E) + \varepsilon$.

Remarks. Result (2) says that $\mathbf{P}(A \cap B) \leq \varepsilon$, i.e., the overlap of A and B is small (relative to $\mathbf{P}$) if (i) and (ii) hold. It will help to realize that all the sets in $\mathfrak{A}_\sigma$ are Borel and that, for example, $\sum_m \mathbf{P}(A_m) = \mathbf{P}(A)$, where $\mathbf{P}$ is the probability on $\mathfrak{B}(\mathbb{R})$. It is also of some interest to realize that all the above computations can be done without knowing that $\mathbf{P}$ on $\mathfrak{A}$ has an extension as a probability to $\mathfrak{B}(\mathbb{R})$. As a result, the $\mathbf{P}^*$-measurable sets Q may be defined by the equation $\mathbf{P}^*(Q) + \mathbf{P}^*(Q^c) = 1$ and Theorem 1.4.13 proved using this definition (see Neveu [N1]).

Exercise 1.5.6. (The Bolzano–Weierstrass property)
This exercise characterizes the compact subsets of $\mathbb{R}$.

Part A. Let A be a subset of $\mathbb{R}$. Show that A is compact if and only if it is closed and bounded. [Hints: if A is closed and bounded, then for some

$N > 0$ one has $A \subset [-N, N]$; if $A \subset \cup_i O_i$, then $[-N, N] \subset \cup_i O_i \cup A^c$; now use the Heine–Borel Theorem 1.4.5. For the converse, observe $A \subset \cup_n(-n, n)$; and if $x_0 \in A^c$ and no open interval $(x_0 - \frac{1}{k}, x_0 + \frac{1}{k}) \subset A^c$, look at the open sets $[x_0 - \frac{1}{k}, x_0 + \frac{1}{k}]^c$.]

Part B. Let A be a compact subset of $\mathbb{R}$, and let $(x_n)_{n \geq 1}$ be a sequence of points of A. Let $S = \{x_n \mid n \geq 1\}$. If the sequence has no convergent subsequence, show that

(1) S is infinite [*Hint* what happens if it is finite?],
(2) S is closed [*Hint*: if $a \notin S$, show that some open interval centered at a is disjoint from S; otherwise what happens?],
(3) for each point s of S, there is an open interval centered at s containing no other point of S. [*Hint*: start off with the first point.]

Show that (3) contradicts the assumption that A is compact [use (2)]. Conclude that $(x_n)_{n \geq 1}$ has a convergent subsequence.

Part C. Let A be subset of $\mathbb{R}$, and assume that every sequence $(x_n)_{n \geq 1}$ of points in A has a subsequence that converges to a point of A. Show that

(1) A is closed [*Hint*: consider the hint for Part B (2)] ,
(2) A is bounded. [*Hint*: if a sequence converges it is bounded.]

This exercise emphasizes the importance of the **Bolzano–Weierstrass property** (see Royden [R3]) for a set: every sequence (in the set) has a convergent subsequence (convergent to a point in the set). It is equivalent to the property of compactness.

CHAPTER II

INTEGRATION

1. Integration on a Probability Space

In elementary probability, if Ω is finite, say $\Omega = \{1, 2, \ldots, n\}$, $\mathfrak{F}$ is the collection of all subsets of Ω, and if $\mathbf{P}$ gives weight $a_i \geq 0$ to $\{i\}$ with $\sum_{i=1}^{n} a_i = 1$, then the (mathematical) expectation $E[X]$ of a random variable X on Ω (i.e., a function $X : \Omega \to \mathbb{R}$) is defined to be $\sum_{i=1}^{n} X(i) a_i = \sum_{i=1}^{n} X(i) \mathbf{P}(\{i\})$. When $a_i = 1/n$ for each i, this expectation is the usual average value of X. Heuristically, this number is what we expect as the average of a large number of "observations" of X — see the weak law of large numbers in Chapter IV. Also, if X is non–negative, then $E[X]$ can be conceived as the "area" under the graph of X: over each i one may imagine a rectangle of width $\mathbf{P}(\{a_i\})$ and height $X(i)$; then $\sum_{i=1}^{n} X(i) \mathbf{P}(\{i\})$ is the sum of the "areas" of the rectangles that make up the set under the graph.

Now let $(\Omega, \mathfrak{F}, B)$ be an arbitrary probability space and let X be a function on Ω. For what functions X can an expectation $E[X]$ or average be defined? In other words, what functions X can be integrated or summed against $\mathbf{P}$?

For example, let $\Omega = \mathbb{R}$, $\mathfrak{F} = \mathfrak{B}(\mathbb{R})$, and $\mathbf{P}$ be the uniform distribution on $[0,1]$ (see Example 1.3.9 (1)). Recall that the distribution function F is given by

$$F(x) = \begin{cases} 0 & x < 0, \\ x & 0 \leq x < 1, \\ 1 & 1 \leq x. \end{cases}$$

Let $X(x) = \sin x$ if $x \in C$, the Cantor set (see the remark following Example 1.2.5), and $X(x) = 0$, if $x \notin C$. Can one integrate X using $\mathbf{P}$? The answer is "yes", and it turns out that the value of the integral is zero since $\mathbf{P}(C) = 0$ (see the calculation in Example 1.2.5).

The basic idea in integration is that if a function X takes a constant value a on a set A in $\mathfrak{F}$ and is zero elsewhere, then $\int X d\mathbf{P}$ should be $a\mathbf{P}(A)$, (i.e., when $a > 0$, it should be the "area" of the picture in $\Omega \times \mathbb{R}$ shown in Fig. 2.1, where A is indicated by a heavy line).

Fig. 2.1

Just as a distribution function F immediately determines a probability **P** on simple sets, namely the intervals $(a, b]$, so does a probability (more generally, a measure) determine an obvious notion of integral for certain simple functions. In both instances, the problem arises as to how to extend to more general situations.

Define the **characteristic or indicator function of a set** A to be the function that equals 1 on A and 0 on A^c. In these notes, it will be denoted by 1_A. (It is also often denoted by χ_A.)

Definition 2.1.1. *A function $s : \Omega \to \mathbb{R}$ is said to be a simple function or to be $\mathfrak{F}$-simple if it can be written as $s = \sum_{n=1}^{N} a_n 1_{A_n}$ with $a_n \geq 0$, and the sets $A_n \in \mathfrak{F}$ pairwise disjoint.*

Let $\mathfrak{F}_s^+$ denote the class of simple functions.

Proposition 2.1.2. *Let s be a simple function, and assume*

$$s = \sum_{n=1}^{N} a_n 1_{A_n} = \sum_{m=1}^{M} b_m 1_{B_m},$$

where the collections of sets $A_n, 1 \leq n \leq N$, and $B_m, 1 \leq m \leq M$, are both pairwise disjoint. Then

$$\sum_{n=1}^{N} a_n \mathbf{P}(A_n) = \sum_{m=1}^{M} b_m \mathbf{P}(B_m).$$

Proof. Note that s has a "natural" representation as $\sum_{n=1}^{N} a_n 1_{A_n}$, A_n pairwise disjoint and the a_n all distinct. To obtain it, consider the sets $\{\omega \mid s(\omega) = a\}$ as a varies over $\mathbb{R}$. If s is expressed as $\sum_{m=1}^{M} b_m 1_{B_m}$, then

$$\sum_{m=1}^{M} b_m \mathbf{P}(B_m) = \sum_{n=1}^{N} a_n \{ \sum_{m, b_m = a_n} \mathbf{P}(B_m) \}$$

$$= \sum_{n=1}^{N} a_n \mathbf{P}(A_n) \text{ as } \cup_m \{B_m \mid b_m = a_n\} = A_n. \quad \square$$

Remark. This result corresponds to Exercise 1.3.7 (2), for the probability **P** defined by a distribution function.

Definition 2.1.3. *Let $s \in \mathfrak{F}_s^+$. Define the **integral** or **expectation** of s with respect to **P** to be $\sum_{n=1}^{N} a_n \mathbf{P}(A_n)$ if $s = \sum_{n=1}^{N} a_n 1_{A_n}$, the sets $A_n \in \mathfrak{F}$ and pairwise disjoint. This number will be variously denoted by $E[s]$, $\int s d\mathbf{P}$, or $\int s(\omega) \mathbf{P}(d\omega)$.*

1. INTEGRATION ON A PROBABILITY SPACE

Remark. Proposition 2.1.2 shows that this definition makes sense, i.e., the value of $E[s]$ does not depend on the particular representation of s.

Proposition 2.1.4. (**Properties of simple functions and their integrals**) Let s, s_1, s_2, etc., denote simple functions, i.e., elements of $\mathfrak{F}_s^+$. The collection $\mathfrak{F}_s^+$ has the following properties:

(S_1) for all $\lambda \geq 0$, $\lambda s \in \mathfrak{F}_s^+$ and $\int (\lambda s) d\mathbf{P} = \lambda (\int s d\mathbf{P})$;
(S_2) $s_1 + s_2 \in \mathfrak{F}_s^+$ and $\int (s_1 + s_2) d\mathbf{P} = \int s_1 d\mathbf{P} + \int s_2 d\mathbf{P}$;
(S_3) $s_1 \leq s_2$ implies $\int s_1 d\mathbf{P} \leq \int s_2 d\mathbf{P}$, where $s_1 \leq s_2$ means that for all ω, $s_1(\omega) \leq s_2(\omega)$;
(S_4) $s_1, s_2 \in \mathfrak{F}_s^+$ implies $s_1 \wedge s_2$ and $s_1 \vee s_2 \in \mathfrak{F}_s^+$, where
$(s_1 \wedge s_2)(\omega) = \min\{s_1(\omega), s_2(\omega)\}$ and
$(s_1 \vee s_2)(\omega) = \max\{s_1(\omega), s_2(\omega)\}$; and
(S_5) if (s_n) is a sequence of $\mathfrak{F}$-simple functions with $s_n \leq s_{n+1} \leq s$ for all n, where s is $\mathfrak{F}$-simple and such that $s = \lim_n s_n$ (i.e., $\lim_n s_n(\omega) = s(\omega)$ for all ω), then $\lim_n \int s_n d\mathbf{P} = \int s d\mathbf{P}$.

Proof. (S_1) is obvious. It suffices to verify (S_2) for $s_1 = a 1_A$ and $s_2 = \sum_{n=1}^N a_n 1_{A_n}$. Then $s_1 + s_2 = a 1_B + \sum_{n=1}^N \{a_n 1_{A_n \setminus A} + (a_n + a) 1_{A_n \cap A}\}$, where $B = A \setminus \cup_{n=1}^N A_n$.

Exercise. Use this expression for $s_1 + s_2$ when $s_1 = a 1_A$ to verify that in this case
$$\int (s_1 + s_2) d\mathbf{P} = \int s_1 d\mathbf{P} + \int s_2 d\mathbf{P}.$$
Explain why it is enough to prove (S_2) in this case.

To prove (S_3), let $s = s_2 - s_1$. Then $s \in \mathfrak{F}_s^+$, as can be seen by writing a formula for s if $s_2 = \sum_{n=1}^N a_n 1_{A_n} \geq s_1 = \sum_{k=1}^K b_k 1_{B_k}$.

Exercise. Determine the desired formula. First, show that $\cup_{n=1}^N A_n \supset \cup_{k=1}^K B_k$ and then use the B_k to "cut up" each A_n and use this observation to define s.

Hence,
$$\int s_2 d\mathbf{P} = \int s_1 d\mathbf{P} + \int s d\mathbf{P} \geq \int s_1 d\mathbf{P}.$$

To show (S_4), first note that if $s = \sum_{n=1}^N a_n 1_{A_n}$ with the sets A_n pairwise disjoint, then $s = \vee_{n=1}^N a_n 1_{A_n}$, (i.e., $s(\omega) = \max\{a_n 1_{A_n}(\omega) \mid n = 1, \ldots, N\})$.

Exercise. Use this observation to verify that if $s_1, s_2 \in \mathfrak{F}_s^+$, then $s_1 \vee s_2 \in \mathfrak{F}_s^+$ (first let $s_1 = a 1_A$ and $s_2 = \sum_{n=1}^N a_n 1_{A_n}$).

Exercise. Let a and b be any two real numbers. Show that $a + b = a \vee b + a \wedge b$, where $a \vee b = \max\{a, b\}$ and $a \wedge b = \min\{a, b\}$.

(S_4) follows from the above two exercises and from the first part of (S_2).

The most difficult part of the proposition — and also the most important — is the last part. To prove (S_5), first assume $s = 1_A$. Choose $m \geq 1$, and let $A_{n,m} = \{\omega | s_n(\omega) \geq 1 - 1/m\}$. Since $s_n \uparrow 1_A$ (i.e., the functions s_n increase to 1_A), the sets $A_{n,m}$ increase with n: $A_{n,m} \subset A_{n+1,m}$ and $\bigcup_{n=1}^{\infty} A_{n,m} = A$.

Exercise. Verify this statement.

Returning to the proof, note that $(1 - 1/m)1_{A_{n,m}} \leq s_n \leq 1_A$. Hence, $(1 - 1/m)\mathbf{P}(A_{n,m}) \leq \int s_n d\mathbf{P} \leq \mathbf{P}(A)$. From the next exercise and Exercise 1.4.14, it follows that $\int s_n d\mathbf{P} \to \mathbf{P}(A) = \int s d\mathbf{P}$.

This proves (S_5) in case $s = a1_A$.

Exercise 2.1.5. (Properties of limits: see Definition 1.1.5)
Let (a_n) have limit A and (b_n) have limit B. Show that

(1) $(a_n + b_n)$ has limit $A + B$ (i.e., $\lim_n(a_n + b_n) = A + B$), and
(2) $(a_n b_n)$ has limit AB (i.e., $\lim_n a_n b_n = AB$).

Suppose that $A = B$ and that (x_n) is such that $a_n \leq x_n \leq b_n$. Show that

(3) (x_n) has limit A.

Now consider $s = \sum_{k=1}^{K} a_k 1_{A_k}$, where the A_k are pairwise disjoint. Since $0 \leq s_n \leq s$, it follows that $s_n = \sum_{k=1}^{K} s_n 1_{A_k}$, where $(s_n 1_{A_k})(\omega) = s_n(\omega)1_{A_k}(\omega)$. Now $\int s_n d\mathbf{P} = \sum_{k=1}^{K} \int s_n 1_{A_k} d\mathbf{P}$ and, by what has been proved, $\lim_{n\to\infty} \int s_n 1_{A_k} d\mathbf{P} = a_k \mathbf{P}(A_k)$. Therefore, $\lim_{n\to\infty} \int s_n d\mathbf{P}$ exists and equals $\sum_{k=1}^{K} a_k \mathbf{P}(A_k)$, which by definition is $\int s d\mathbf{P}$. □

The next definition determines the class of functions on $(\Omega, \mathfrak{F}, \mathbf{P})$ that can be nicely approximated by simple functions (see Proposition 2.1.11 (RV_7)). One defines the integral of such a function as the limit of the integrals of the approximating functions.

Definition 2.1.6. *A function* $X : \Omega \to \mathbb{R} \cup \{\pm\infty\}$ *such that, for all* $\lambda \in \mathbb{R}$, $\{\omega \mid X(\omega) < \lambda\} \in \mathfrak{F}$ *is called a* **random variable** *on the probability space* $(\Omega, \mathfrak{F}, \mathbf{P})$ *or a* **measurable function** *on* $(\Omega, \mathfrak{F}, \mathbf{P})$ *($\mathfrak{F}$-measurable if the σ-algebra $\mathfrak{F}$ needs to be indicated). Recall that one assumes* $-\infty < x < +\infty$ *for all* $x \in \mathbb{R}$.

Remarks. (1) The above concept can be extended to functions on a **measurable space**, i.e., a pair $(\Omega, \mathfrak{F})$, where Ω is a set and $\mathfrak{F}$ is a σ-algebra. Define $X : \Omega \to \mathbb{R} \cup \pm\infty$ to be a **measurable function** on the measurable space $(\Omega, \mathfrak{F})$ or an **$\mathfrak{F}$-measurable function**, if, for all $\lambda \in \mathbb{R}$, $\{\omega \mid X(\omega) < \lambda\} \in \mathfrak{F}$. Note that no probability is needed for the

1. INTEGRATION ON A PROBABILITY SPACE

definition. However, the use of the term "random variable" signifies that the underlying measurable space is equipped with a probability.

(2) The fact that X is a random variable w.r.t. $\mathfrak{F}$ or an $\mathfrak{F}$-measurable function will be denoted by writing "$X \in \mathfrak{F}$" or "X in $\mathfrak{F}$".

Exercise 2.1.7. Show that X is a random variable if and only if X has any one of the following properties:
(1) for all $\lambda \in \mathbb{R}$, $\{\omega \mid X(\omega) \leq \lambda\} \in \mathfrak{F}$;
(2) for all $\lambda \in \mathbb{R}$, $\{\omega \mid X(\omega) > \lambda\} \in \mathfrak{F}$;
(3) for all $\lambda \in \mathbb{R}$, $\{\omega \mid X(\omega) \geq \lambda\} \in \mathfrak{F}$.

Show that every $\mathfrak{F}$-simple function is a random variable.

Definition 2.1.8. Let $(a_n) \subset \mathbb{R} \cup \{\pm\infty\}$. The **limit supremum** or **limsup** of (a_n) is defined to be $\inf_n \{\sup_{m \geq n} a_n\}$. The **limit infimum** or **liminf** of (a_n) is defined to be $\sup_n \{\inf_{m \geq n} a_n\}$. These numbers "measure" the "oscillation at $+\infty$" of the sequence (a_n) and are denoted by $\limsup_n a_n$ (or $\limsup_n a_n$) and $\liminf_n a_n$ (or $\liminf_n a_n$), respectively.

Exercise 2.1.9. Show that for any sequence $(a_n) \subset \mathbb{R} \cup \{\pm\infty\}$,
(1) $\limsup_n a_n$ and $\liminf_n a_n$ exist, and
(2) $\liminf_n a_n \leq \limsup_n a_n$. [Hint: find some monotone sequences.]

Calculate $\limsup_n a_n$ and $\liminf_n a_n$ for the following sequences.
(3) $a_n = 1 - 1/n$ if n is even and $= 1/n$ if n is odd,
(4) $a_n = \sin(n\pi/2)$,
(5) $a_n = \sin n\pi$ if n is not divisible by and $a_n - n$ otherwise.

Exercise 2.1.10. Show that $\liminf_n a_n = \limsup_n a_n$ and is finite if and only if the sequence (a_n) converges in $\mathbb{R}$ to this common value.

Proposition 2.1.11. (Properties of random variables)
Let $X, Y, X_1, X_2, \ldots$, etc, be random variables on $(\Omega, \mathfrak{F}, \mathbf{P})$. Then

(RV_1) $\alpha \in \mathbb{R}$, $X \in \mathfrak{F}$ implies $\alpha X \in \mathfrak{F}$;

(RV_2) $X_1 + X_2 \in \mathfrak{F}$ if $X_i \in \mathfrak{F}, i = 1$ or 2, where $(X_1 + X_2)(\omega) \stackrel{\text{def}}{=} X_1(\omega) + X_2(\omega)$, providing the sum makes sense (i.e., where one observes the conventions, $\lambda + (\pm\infty) = \pm\infty$ if $\lambda \in \mathbb{R}$ and $(\pm\infty) + (\pm\infty) = \pm\infty$ while $(+\infty) + (-\infty)$ is not defined);

(RV_3) $X_1 \vee X_2$ and $X_1 \wedge X_2 \in \mathfrak{F}$, where
$(X_1 \vee X_2)(\omega) \stackrel{\text{def}}{=} \max\{X_1(\omega), X_2(\omega)\}$ for all $\omega \in \Omega$, and
$(X_1 \wedge X_2)(\omega) \stackrel{\text{def}}{=} \min\{X_1(\omega), X_2(\omega)\}$ for all $\omega \in \Omega$;

(RV_4) $(X_n) \subset \mathfrak{F}$ implies $\inf_n X_n$ and $\sup_n X_n \in \mathfrak{F}$, where
$(\inf_n X_n)(\omega) \stackrel{\text{def}}{=} \inf\{X_n(\omega) \mid n \geq 1\}$ for all $\omega \in \Omega$, and
$(\sup_n X_n)(\omega) \stackrel{\text{def}}{=} \sup\{X_n(\omega) \mid n \geq 1\}$ for all $\omega \in \Omega$;

(RV_5) $\liminf_n X_n$ and $\limsup_n X_n \in \mathfrak{F}$ if $(X_n)_{n \geq 1} \subset \mathfrak{F}$, where
$(\liminf_n X_n)(\omega) \stackrel{\text{def}}{=} \liminf_n X_n(\omega)$ for all $\omega \in \Omega$, and

$(\limsup_n X_n)(\omega) \stackrel{\text{def}}{=} \limsup_n X_n(\omega)$ for all $\omega \in \Omega$;

(RV$_6$) if $X = \lim_n X_n$ and $(X_n)_{n\geq 1} \subset \mathfrak{F}$, then $X \in \mathfrak{F}$, where
$(\lim_n X_n)(\omega) \stackrel{\text{def}}{=} \lim_n X_n(\omega)$ for all $\omega \in \Omega$;

(RV$_7$) if $X \in \mathfrak{F}$, and X non-negative, then there is a sequence $(s_n)_{n\geq 1}$ of simple functions with $s_n \leq s_{n+1}$ for all n and $X = \lim_n s_n$;

(RV$_8$) if $X, Y \in \mathfrak{F}$ and are both non-negative or both real-valued, then $XY \in \mathfrak{F}$.

Proof. (RV$_1$) is obvious in view of Exercise 2.1.7. (RV$_2$): Let $\lambda \in \mathbb{R}$. Let $A = \bigcup_{r \in \mathbb{Q}} \{\omega \mid X_1(\omega) < r\} \cap \{\omega \mid X_2(\omega) < \lambda - r\}$. Since $\mathbb{Q}$, the set of rational numbers, is countable, the set $A \in \mathfrak{F}$. Also, if $\omega \in A$, then $X_1(\omega) + X_2(\omega) < \lambda$ and so $A \subset \{X_1 + X_2 < \lambda\}$ (note that $\{X_1 + X_2 < \lambda\}$ denotes $\{\omega \mid X_1(\omega) + X_2(\omega) < \lambda\}$). To simplify matters, assume that X_1 and X_2 are real-valued and that $(X_1 + X_2)(\omega) = X_1(\omega) + X_2(\omega) < \lambda$. Let $\alpha_1 = X_1(\omega)$ and $\alpha_2 = X_2(\omega)$. The point $\omega \in A$ if there is a rational number r with $\alpha_1 < r$ and $\alpha_2 < \lambda - r$. Since $\alpha_1 + \alpha_2 < \lambda$ implies $\alpha_1 < \lambda - \alpha_2$, one wants to find a rational number r with $\alpha_1 < r < \lambda - \alpha_2$. This is possible in view of the following fundamental property of $\mathbb{R}$.

Exercise 2.1.12. Let $a < b$ be two real numbers. Then there is a rational number r with $a < r < b$. Show that, as a result, in every non-void open subset of $\mathbb{R}$, there is a rational number. (Recall that a subset A of $\mathbb{R}$ is said to be dense if it has this property (see Exercise 1.3.15): in other words, $\mathbb{Q}$ is dense in $\mathbb{R}$.) [*Hints*: choose an $n \geq 1$. The distance between consecutive rational numbers of the form $k/n, k \in \mathbb{Z}$, is $1/n$. Choose n so that $1/n < b - a$ (this is possible in view of Exercise 1.1.4). Now look at the numbers $k/n \leq a$ and those of this form $\geq b$. If there is no rational number between a and b, what happens?] For a related exercise, see Exercise 2.9.15.

The upshot of this is that $A = \{X_1 + X_2 < \lambda\}$ and so $X_1 + X_2 \in \mathfrak{F}$ if both are real-valued.

Exercise. Prove (RV$_2$) when the X_i are not necessarily finite. [*Hint:* let $\Lambda = \{X_1 > -\infty, X_2 > -\infty\}$. Show that $\Lambda \cap \{X_1 + X_2 < \lambda\} \in \mathfrak{F}$.]

Exercise. Prove (RV$_3$). [*Hint:* express $\{\omega \mid \max\{X_1(\omega), X_2(\omega)\} < \lambda\}$ in terms of $\{X_1 < \lambda\}$ and $\{X_2 < \lambda\}$.]

Exercise. Prove (RV$_4$). [*Hint:* express $\{\inf_n X_n < \lambda\}$ in terms of the sets $\{X_n < \lambda\}, n \geq 1$.]

Exercise. Prove (RV$_5$). [*Hint:* use (RV$_4$).]

Exercise. Prove (RV$_6$). [*Hint:* use Exercise 2.1.10.]

The verification of (RV$_7$) is a bit more exciting. To understand the argument, Fig. 2.2 is useful:

1. INTEGRATION ON A PROBABILITY SPACE

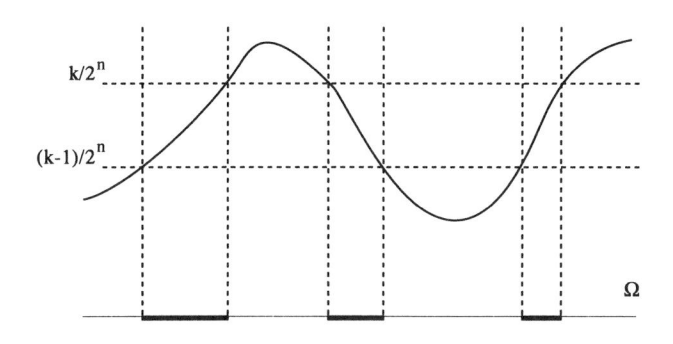

Fig. 2.2

The set where $\frac{(k-1)}{2^n} \leq X < \frac{k}{2^n}$ is indicated by heavy lines.

For each $n \geq 1$, divide $\mathbb{R}$ into intervals using the dyadic rationals $\frac{k}{2^n}$, $k \in \mathbb{Z}$. Define

$$s_n(\omega) = \begin{cases} \frac{(k-1)}{2^n} & \text{if } \frac{(k-1)}{2^n} \leq X(\omega) < \frac{k}{2^n} \text{ and } 1 \leq k \leq 2^{2n}, \\ 2^n & \text{if } 2^n \leq X(\omega), \text{ i.e.,} \end{cases}$$

$$s_n = \sum_{k=1}^{2^n} \frac{(k-1)}{2^n} 1_{\{\frac{(k-1)}{2^n} \leq X < \frac{k}{2^n}\}} + 2^n 1_{\{2^n \leq X\}}.$$

Then each s_n is simple and $s_n \leq X$ for all n. Since the dyadic rational $\frac{(2k-1)}{2^{n+1}}$ divides the interval $[\frac{(k-1)}{2^n}, \frac{k}{2^n}]$ in half, $\frac{(k-1)}{2^n} \leq X(\omega) < \frac{k}{2^n}$ implies $s_{n+1}(\omega) \geq \frac{(k-1)}{2^n} = s_n(\omega)$, and so $s_n \leq s_{n+1} \leq X$ for all n.

Let $s = \lim_{n \to \infty} s_n$. Then $s \leq X$. If $X(\omega) < +\infty$, then $s(\omega) = X(\omega)$ since $|s_n(\omega) - X(\omega)| < \frac{1}{2^n}$ for large enough n since $X(\omega) < 2^n$ for large n. If $X(\omega) = +\infty$, then $s_n(\omega) = 2^n$ for all n and so $s(\omega) = +\infty$. Therefore, $s = X$.

Exercise. Prove (RV_8). [*Hint*: first verify it for products of $\mathfrak{F}$-simple functions; then observe that each real-valued X can be written as $X = X^+ - X^-$, where $X^+ \stackrel{\text{def}}{=} X \vee 0$ and $X^- \stackrel{\text{def}}{=} -(X \wedge 0) = (-X) \vee 0$ (see Exercise 1.1.16).] □

When the probability space $(\Omega, \mathfrak{F}, \mathbf{P})$ has a specific description, it is often useful and important to know that certain functions are random variables. For example, if $\Omega = \mathbb{R}$, and $\mathfrak{F} = \mathfrak{B}(\mathbb{R})$, then every continuous function is measurable.

Definition 2.1.13. A function $f : \mathbb{R} \to \mathbb{R}$ is **continuous at** $x_0 \in \mathbb{R}$ if for any $\epsilon > 0$ there is a $\delta = \delta(x_0, \epsilon) > 0$ such that

$$|f(x) - f(x_0)| < \epsilon \quad \text{when } |x - x_0| < \delta.$$

A function $f : \mathbb{R} \to \mathbb{R}$ is **continuous** if it is continuous at every point of $\mathbb{R}$.

A function $f : E \subset \mathbb{R} \to \mathbb{R}$, (i.e., defined on a subset E of $\mathbb{R}$), is **continuous at** $x_0 \in E$ if for any $\epsilon > 0$ there is a $\delta = \delta(x_0, \epsilon)$ such that

$$|f(x) - f(x_0)| < \epsilon \quad \text{when } |x - x_0| < \delta \text{ and } x \in E.$$

It is **continuous on** E if it is continuous at every point of E.

Exercise 2.1.14. Let $f : \mathbb{R} \to \mathbb{R}$. (1) Show that f is continuous at x_0 if and only if for any open set O containing $y_0 = f(x_0)$, the set $f^{-1}O = \{x \mid f(x) \in O\}$ contains an open interval about x_0 (such a set is called a **neighbourhood of x_0**).

(2) Show that $f : \mathbb{R} \to \mathbb{R}$ is continuous if and only if for any open set $O \subset \mathbb{R}$, $f^{-1}O$ is also open.

Remember that a set O is open if and only if $a \in O$ implies O contains an open interval about a (i.e., if and only if O is a neighbourhood of each of its points).

Given the above results, one has the following result.

Proposition 2.1.15. Let $(\mathbb{R}, \mathfrak{F}, \mathbf{P})$ be a probability space, and assume $\mathfrak{F} \supset \mathfrak{B}(\mathbb{R})$. Then every continuous function $X : \mathbb{R} \to \mathbb{R}$ is a random variable, i.e., is measurable.

Proof. $\{\omega \mid X(\omega) < \lambda\}$ is open since $(-\infty, \lambda)$ is open. Every open set is a Borel set (see Exercise 1.3.10 (b)). □

Definition 2.1.16. A function $\varphi : \mathbb{R} \to \mathbb{R}$ is called a **Borel function** if it is Borel-measurable, i.e., if it is a measurable function on $(\mathbb{R}, \mathfrak{B}(\mathbb{R}))$.

Proposition 2.1.17. (**Composition of measurable functions**) Let X be a finite random variable or measurable function on $(\Omega, \mathfrak{F}, \mathbf{P})$, and let φ be a Borel function. Then the function $\varphi \circ X$ defined by $(\varphi \circ X)(\omega) = \varphi(X(\omega))$ for all $\omega \in \Omega$ is again a random variable.

In particular, if $B \in \mathfrak{B}(\mathbb{R})$, $X^{-1}(B) \stackrel{\text{def}}{=} \{\omega \mid X(\omega) \in B\} \in \mathfrak{F}$, as it is $\{\omega \mid (1_B \circ X)(\omega) > 0\}$. Hence, X is a random variable if and only if $X^{-1}(B) \in \mathfrak{F}$ for all $B \in \mathfrak{B}(\mathbb{R})$.

Proof. $\{\omega \mid (\varphi \circ X)(\omega) < \lambda\} = \{\omega \mid X(\omega) \in \{x \mid \varphi(x) < \lambda\}\}$. Now $\{x \mid \varphi(x) < \lambda\} = \varphi^{-1}((-\infty, \lambda)) \in \mathfrak{B}(\mathbb{R})\}$, so the result will be proved if $\{\omega \mid X(\omega) \in B\} \in \mathfrak{F}$ whenever B is a Borel set (i.e., if the proposition is proved for $\varphi = 1_B$, $B \in \mathfrak{B}(\mathbb{R})$).

Exercise 2.1.18. If $A \subset \mathbb{R}$, define $X^{-1}(A) = \{\omega \mid X(\omega) \in A\}$. Show that
(1) $X^{-1}(A^c) = \left(X^{-1}(A)\right)^c$,
(2) $X^{-1}(A_1 \cap A_2) = X^{-1}(A_1) \cap X^{-1}(A_2)$,
(3) $X^{-1}(\cup_{n=1}^\infty A_n) = \cup_{n=1}^\infty X^{-1}(A_n)$.

Let $\mathfrak{G} = \{A \subset \mathbb{R} \mid X^{-1}(A) \in \mathfrak{F}\}$. Show that
(4) $\mathfrak{G}$ is a σ-algebra.

Given this exercise, the fact that X is a random variable implies that $\mathfrak{G} \supset \{(-\infty, \lambda] \mid \lambda \in \mathbb{R}\}$. Hence, in view of Exercise 1.3.10 (b), $\mathfrak{G} \supset \mathfrak{B}(\mathbb{R})$. Clearly, if $\mathfrak{G} \supset \mathfrak{B}(\mathbb{R})$, it follows that X is a random variable since the intervals $(-\infty, \lambda]$ are Borel sets. □

As an almost immediate consequence, one has the following important result.

Proposition 2.1.19. *Let X be a finite random variable on a probability space $(\Omega, \mathfrak{F}, \mathbf{P})$. Then there is a unique probability $\mathbf{Q}$ on $\mathfrak{B}(\mathbb{R})$ such that its distribution function F satisfies $F(x) = \mathbf{P}[X \leq x]$ for all $x \in \mathbb{R}$.*

Proof. Let $B \in \mathfrak{B}(\mathbb{R})$. Define $\mathbf{Q}(B)$ to be $\mathbf{P}(X^{-1}(B))$. Since X^{-1} preserves the σ-algebra operations by Exercise 2.1.18, it follows automatically that $\mathbf{Q}$ is a probability.

Clearly, $\mathbf{Q}((-\infty, x]) \stackrel{\text{def}}{=} F(x)$ is $\mathbf{P}[X \leq x]$, and the uniqueness of $\mathbf{Q}$ follows from Theorem 1.4.13. □

Remark. One way to prove this is to start from the distribution function of X, which is $F(x) = \mathbf{P}[X \leq x]$, and then appeal to Theorem 1.4.13 for the existence of $\mathbf{Q}$. The point of the above argument is that since one has a probability available, namely $\mathbf{P}$, one obtains $\mathbf{Q}$ by a straightforward set-theoretic formal procedure. In analysis, one refers to $\mathbf{Q}$ as the **image** of $\mathbf{P}$ under the measurable map X. This procedure, which produces an image measure, is very general and is not at all restricted to the situation of the above proposition. It will appear again in Proposition 3.1.10 when random vectors are considered.

Definition 2.1.20. *The probability $\mathbf{Q}$ that occurs in Proposition 2.1.19 is called the **distribution** (or probability **law**) of X.*

Remarks 2.1.21. (1) Every probability $\mathbf{Q}$ on $\mathfrak{B}(\mathbb{R})$ is the distribution of some random variable; namely $X(x) = x$ on $(\mathbb{R}, \mathfrak{B}(\mathbb{R}), \mathbf{Q})$.

(2) Statisticians are usually interested only in the distribution of a random variable X; its domain of definition is often of little interest to them. The construction of random variables with specified properties is a mathematical problem that for statistical purposes may often be taken for granted.

Returning to the main theme, what can be said about the integral of a random variable X, its so-called expectation $E[X]$? To define this, one uses property (RV_8) of Proposition 2.1.11 of a positive random variable. The next result allows $E[X]$ to be defined by approximation.

Proposition 2.1.22. *Let X be a non-negative random variable, and let $(s_n)_{n\geq 1}$ and $(s'_m)_{m\geq 1}$ be two increasing sequences of simple functions with $\lim_n s_n = X = \lim_m s'_m$. Then*

$$\lim_n \int s_n d\mathbf{P} = \lim_m \int s'_m d\mathbf{P}.$$

Proof. Let $t_{n,m} = s_n \wedge s'_m$. By (S_4) and (S_5) of Proposition 2.1.4, $t_{n,m} \in \mathfrak{F}_s^+$ and $\lim_{m\to\infty} \int t_{n,m} d\mathbf{P} = \int s_n d\mathbf{P}$. Hence

$$\lim_{n\to\infty} \int s_n d\mathbf{P} = \lim_{n\to\infty}\left[\lim_{m\to\infty} \int t_{n,m} d\mathbf{P}\right]$$
$$= \lim_{m\to\infty}\left[\lim_{n\to\infty} \int t_{n,m} d\mathbf{P}\right] \quad \text{(by Proposition 1.1.11)}$$
$$= \lim_{m\to\infty} \int s'_m d\mathbf{P}. \quad \square$$

Definition 2.1.23. *Let X be a non-negative random variable on a probability space $(\Omega, \mathfrak{F}, \mathbf{P})$, and let $(s_n)_{n\geq 1}$ be an increasing sequence of simple random variables with $X = \lim_n s_n$. Define $\int X d\mathbf{P}$ to be $\lim_n \int s_n d\mathbf{P}$. It is called the **integral of X with respect to $\mathbf{P}$** or the **expectation of X** and is also denoted by $E[X]$.*

Remarks. (1) This definition makes sense because of Proposition 2.1.22.
(2) The integral $\int X d\mathbf{P}$ can also be defined as the supremum of $\{\int s d\mathbf{P} \mid s \text{ simple}, 0 \leq s \leq X\}$ (see Exercise 2.9.8).

The expectation of X may be $+\infty$ for $X \geq 0$. If $E[X] < +\infty$, then X is said to be **integrable**, and one writes $X \in L^1(\Omega, \mathfrak{F}, \mathbf{P})$ to indicate this.

Example 2.1.24.
(1) Let $\Omega = \mathbb{N} \cup \{0\}$, and let $\mathfrak{F}$ be the collection $\mathfrak{P}(\Omega)$ of all subsets of Ω. A probability $\mathbf{P}$ is defined on $\mathfrak{F}$ by the numbers $p_n = \mathbf{P}(\{n\})$ provided that $\sum_{n=0}^{\infty} p_n = 1$. A real-valued random variable X on $(\Omega, \mathfrak{F}, \mathbf{P})$ may be viewed as a sequence $(a_n)_{n\geq 0}$, i.e., $a_n = X(n)$. When X is non-negative, the simple functions s_n defined by $s_n(i) = X(i) = a_i$, $0 \leq i \leq n$, and $s_n(i) = 0$ for $i > n$ have the property that $s_n \uparrow X$. Further, $\int s_n d\mathbf{P} = \sum_{i=0}^{n} a_i p_i$, and so $E[X] = \sum_{i=0}^{\infty} a_i p_i$. Consequently, on this probability space integration amounts to "summing" infinite series.

(2) Let $\mathbf{P}$ be the **Poisson distribution with mean** $\lambda > 0$, i.e., $p_n = e^{-\lambda}(\frac{\lambda^n}{n!})$ if $n \geq 0$. Show that (i) if $X(n) = n$, for all $n \geq 0$, then $E[X] = \lambda$, and (ii) if $X(n) = n!$, for all $n \geq 0$, then $E[X] < +\infty$ if and only if $0 < \lambda < 1$.

(3) In Feller's classic book [F1], the definition of mathematical expectation for a non-negative X is the extension of part (1) of this example obtained by replacing Ω with the countable set $\{x_n \mid n \geq 1\}$ of values of X. This means that the mathematical expectation is defined by means of the distribution of X; in other words, $E[X] = \sum_{i=0}^{\infty} X(i) p_i = \sum_{n=1}^{\infty} x_n \mathbf{P}(\{i \mid X(i) = x_n\})$ (see Lemma 3.5.3).

Proposition 2.1.25. (Properties of the integral for non-negative X)

(E_1) $\lambda \geq 0$ implies $E[\lambda X] = \lambda E[X]$.
(E_2) $E[X_1 + X_2] = E[X_1] + E[X_2]$ if X_1 and X_2 are non-negative random variables.
(E_3) $0 \leq X_1 \leq X_2$ implies $E[X_1] \leq E[X_2]$.
(E_4) **(Principle of monotone convergence)** If (X_n) is a sequence of non-negative random variables X_n such that $0 \leq X_n \leq X_{n+1} \leq X$ for all n and $\lim_n X_n = X$, then $E[X] = \lim_n E[X_n]$.
(E_5) **(Fatou's lemma)** $E[\liminf_n X_n] \leq \liminf_n E[X_n]$ for any sequence (X_n) of non-negative random variables X_n.

Proof. (E_1) is obvious since $s_n \uparrow X$ implies $\lambda s_n \uparrow \lambda X$, where as before $s_n \uparrow X$ means that $s_n \leq s_{n+1} \leq X$ for all n and $\lim_n s_n(\omega) = X(\omega)$ for all $\omega \in \Omega$. (E_2) is also obvious since $s_n \uparrow X_1$, $t_n \uparrow X_2$ implies $s_n + t_n \uparrow X_1 + X_2$ and hence $E[X_1 + X_2] = \lim_{n \to \infty} E[s_n + t_n] = \lim_{n \to \infty} (E[s_n] + E[t_n]) = \lim_{n \to \infty} E[s_n] + \lim_{n \to \infty} E[t_n] = E[X_1] + E[X_2]$ (even if one of the limits is $+\infty$).

To prove (E_3), let

$$X(\omega) = \begin{cases} X_2(\omega) - X_1(\omega) & \text{if } X_1(\omega) < +\infty, \\ 0 & \text{if } X_1(\omega) = +\infty. \end{cases}$$

Then X is a random variable and $X_2 = X + X_1$ (remember that X_1, X_2 are non-negative and can take infinite values). Then, by (E_2), $E[X_2] = E[X] + E[X_1] \geq E[X_1]$.

The proof of (E_4) is less trivial. First, observe that X is a random variable as $X = \lim_n X_n$. Note that $E[X_n] \leq E[X_{n+1}] \leq E[X]$ for all n and so $\lim_{n \to \infty} E[X_n] \leq E[X]$.

Let $(s_{n,k})_{k \geq 1}$ be an increasing sequence of simple functions with $\lim_k s_{n,k} = X_n$ for each $n \geq 1$.

Consider the following infinite array:

$$
\begin{array}{ccccc}
s_{1,1} & s_{1,2} & s_{1,3} & \cdots \; s_{1,n} \cdots & \to \quad X_1 \\
s_{2,1} & s_{2,2} & s_{2,3} & \cdots \; s_{2,n} \cdots & \to \quad X_2 \\
\vdots & \vdots & \vdots & \vdots & \vdots \\
s_{m,1} & s_{m,2} & s_{m,3} & \cdots \; s_{m,n} \cdots & \to \quad X_m \\
s_{(m+1),1} & s_{(m+1),2} & s_{(m+1),3} & \cdots \; s_{(m+1),n} & \to \quad X_{m+1} \\
\vdots & \vdots & \vdots & \vdots & \vdots
\end{array}
$$

If one replaces the entry at the (m, n) position with the maximum of all the simple functions $s_{k,\ell}$ for $1 \leq k \leq m, 1 \leq \ell \leq n$ and calls this simple function $t_{m,n}$, then it is clear that

(1) $\qquad s_{m,n} \leq t_{m,n} \leq X_m, \qquad$ for all m, for all n;

(2) $\qquad \left. \begin{array}{c} t_{m,n} \leq t_{m,(n+1)} \\ t_{m,n} \leq t_{(m+1),n} \end{array} \right\} \qquad$ for all m, for all n.

Hence, $\lim_{n \to \infty} t_{m,n} = X_m$ for all m, and the sequence $(t_{m,m})$ is increasing. Let $Y = \lim t_{m,m}$. Then $Y = X$ since by Proposition 1.1.11, $\lim_m t_{m,m}(\omega) = \lim_m \lim_n t_{m,n}(\omega) = \lim_m X_m(\omega) = X(\omega)$.

Consequently, $E[X] \geq \lim_m E[X_m] = \lim_m \lim_n E[t_{m,n}]$, which by another application of Proposition 1.1.11 equals $\lim_m E[t_{m,m}]$. Since by Proposition 2.1.22, $\lim_m E[t_{m,m}] = E[X]$, this proves (E_4).

Fatou's lemma (E_5) is obtained by applying the principle of monotone convergence (E_4). Let $U_n = \inf_{m \geq n} X_m$. Then $\liminf_n X_n = \lim_n U_n \stackrel{\text{def}}{=} U$ and $U_n \leq U_{n+1}$ for all n. Hence $E[U] = \lim_n E[U_n]$ by (E_4). Now $U_n \leq X_m$ if $m > n$ and so $E[U_n] \leq E[X_m]$ if $m > n$. Therefore, $E[U_n] \leq \inf_{m \geq n} E[X_m] \leq \liminf_n E[X_n]$, and so $E[U] \leq \liminf_n E[X_n]$. $\square$

Remarks.

(1) Fatou's lemma is an extremely useful estimating tool.

(2) Exercise 2.1.24 (1) is an illustration of (E_4), the principle of monotone convergence.

Exercise 2.1.26. Show that if X is a non-negative integrable random variable, then $\mathbf{P}[X = +\infty] = 0$. [Hint: let $A = \{\omega \mid X(\omega) = +\infty\}$. Then $s_n \geq 2^n 1_A$, where s_n is defined in Proposition 2.1.11 (RV_8).]

Exercise 2.1.27. Let Y and Z be two random variables on $(\Omega, \mathfrak{F}, \mathbf{P})$, and let $A \in \mathfrak{F}$. Show that $Y 1_A + Z 1_{A^c}$ is a random variable, where by convention $0 \cdot (\pm \infty) = 0$. Note that the sum is always defined since for any ω at most one of $Y(\omega) 1_A(\omega)$ and $Z(\omega) 1_{A^c}(\omega)$ is infinite.

Exercise 2.1.28. A random variable N such that $\mathbf{P}[N \neq 0] = 0$ will be called a **null variable** or **null function**. Show that
  (1) $\mathbf{P}(A) = 0$ if and only if 1_A is a null function,
  (2) if N is a non-negative null variable, then $E[N] = 0$,
  (3) every non-negative null variable N is integrable with $E[N] = 0$, and $|N|$ is a null variable if N is a null variable,
  (4) if $E[|X|] = 0$, then X is a null variable or function, where $|X|(\omega) = |X(\omega)|$ for all $\omega \in \Omega$, (note that $|x| = x^+ + x^-$ by Exercise 1.1.16).

Exercise 2.1.29. Let Y be a non-negative integrable random variable, and let $A \in \mathfrak{F}$ be such that $\mathbf{P}(A) = 1$. Show that
  (1) $E[Y] = E[Y 1_A] = E[U]$, where $U = Y 1_A + Z 1_{A^c}$ and Z is any non-negative random variable, and
  (2) $\int_B Y d\mathbf{P} = \int_B Y 1_A d\mathbf{P} = \int_B U d\mathbf{P}$ for any $B \in \mathfrak{F}$.

Exercise 2.1.29 shows that modifying (in a measurable way) a non-negative random variable Y on a set of probability zero does not change its expectation or its integral over any set B in the σ-algebra $\mathfrak{F}$. In view of Exercise 2.1.28, any non-negative integrable random variable X may be modified to be (say) zero on $\{X = +\infty\}$ without changing its expectation or integral over any set in $\mathfrak{F}$. So when dealing with questions involving integration of non-negative integrable random variables, one may as well assume that they are real-valued, i.e., are everywhere finite. It is therefore convenient to make the following convention.

Convention 2.1.30. *If X is a non-negative integrable random variable, it will be assumed that X is real-valued.*

To integrate arbitrary finite random variables X, it suffices to express them in terms of non-negative finite random variables.

Definition 2.1.31. *A finite random variable X is said to be **integrable** if $X = X_1 - X_2$, where X_i, $i = 1, 2$, are finite, non-negative integrable random variables. If X is an integrable random variable with $X = X_1 - X_2$ and X_i non-negative and integrable, then $E[X] \stackrel{\text{def}}{=} E[X_1] - E[X_2]$ (i.e., $\int X d\mathbf{P} \stackrel{\text{def}}{=} \int X_1 d\mathbf{P} - \int X_2 d\mathbf{P}$). Let $L^1 = L^1(\Omega, \mathfrak{F}, \mathbf{P})$ denote the set of finite integrable random variables.*

Remarks 2.1.32.
  (1) The definition of $E[X]$ makes sense if it is independent of the expression of X as a difference of non-negative integrable random variables. Let $X = X_1 - X_2 = Y_1 - Y_2$, where the X_i and the Y_i are non-negative integrable random variables. Then $X_1 + Y_2 = Y_1 + X_2$ and by (E_2), $E[X_1] + E[Y_2] = E[Y_1] + E[X_2]$. Hence, $E[X_1] - E[X_2] = E[Y_1] - E[Y_2]$ since these are real (and hence) finite numbers.
  (2) For any random variable X, if $X \vee 0 = X^+$ and $X \wedge 0 = -X^-$, then X will also be said to be **integrable** if X^+ and X^- are integrable whether

they are finite or not, and one sets $E[X] \stackrel{\text{def}}{=} E[X^+] - E[X^-]$. Let $Y(\omega) = X(\omega)$ if $|X(\omega)| < \infty$ and $= 0$ otherwise. Then, Y is integrable in the sense of Definition 2.1.31, and $E[Y] = E[Y^+] - E[Y^-] = E[X^+] - E[X^-]$. This extended notion of integrability is equivalent to the following: there is a set $A \in \mathfrak{F}$ with $\mathbf{P}(A) = 1$ and two non-negative random variables X_1 and X_2, both finite on A and with $E[X_i] < \infty$, $i = 1, 2$, such that $X(\omega) = X_1(\omega) - X_2(\omega)$ for all $\omega \in A$. As a result, a random variable X is integrable in the above extended sense if and only if after being modified to equal zero on a set of probability zero, the resulting random variable is integrable in the sense of Definition 2.1.31. In order to comply with standard usage, one writes $X \in L^1$ if X is integrable in the above sense. The above remark explains the connection between this usage and the requirement of Definition 2.1.31 that the random variables in $L^1(\Omega, \mathfrak{F}, \mathbf{P})$ be finite.

(3) In the context of Example 2.1.24 (1), where $\Omega = \mathbb{N} \cup \{0\}$, $\mathfrak{F}$ is the collection of all subsets of Ω, and $\mathbf{P}(\{n\}) = p_n$ for all n, a random variable X is integrable if and only if the series $\sum_{n=0}^{\infty} X(n) p_n$ is absolutely convergent. For such an X, $E[X] = \sum_{n=0}^{\infty} X(n) p_n$. In Feller [F1], X is said to have **finite expectation** if it is integrable.

Proposition 2.1.33. *Let $(\Omega, \mathfrak{F}, \mathbf{P})$ be a probability space. Then*
  (1) *$L^1(\Omega, \mathfrak{F}, \mathbf{P})$ is a real vector space and the function $X \to E[X]$ is linear (i.e., it is a linear functional);*
  (2) *$X \in L^1$ if and only if $|X| \stackrel{\text{def}}{=} X^+ + X^- \in L^1$ and $|E[X]| \leq E[|X|]$;*
  (3) *$|X| \leq Y$, $Y \in L^1$ implies $X \in L^1$.*

Proof. (1) Let $X = X_1 - X_2$ and $Y = Y_1 - Y_2$. Then $X + Y = (X_1 + Y_1) - (X_2 + Y_2) \in L^1$ if $X, Y \in L^1$. Let $X = X_1 - X_2 \in L^1$ and $\lambda \in \mathbb{R}$. Then $\lambda X = \lambda X_1 - \lambda X_2 \in L^1$ if $\lambda \geq 0$ and $\lambda X = (-\lambda) X_2 - (-\lambda) X_1 \in L^1$ if $\lambda < 0$. It is clear that $E[\lambda X + \mu Y] = \lambda E[X] + \mu E[Y]$ if $X, Y \in L^1$ and $\lambda, \mu \in \mathbb{R}$.

(2) If $X = X_1 - X_2 \in L^1$, $X_i \geq 0$, then $X^+ \leq X_1$ and $X^- \leq X_2$. Hence $X^{\pm} \in L^1$ and so $|X| = X^+ + X^- \in L^1$. Conversely, if $|X| \in L^1$, $X^{\pm} \in L^1$ and so $X = X^+ - X^- \in L^1$. It follows from the triangle inequality (Exercise 1.1.16) that $|E[X]| = |E[X^+] - E[X^-]| \leq E[X^+] + E[X^-] = E[|X|]$. Therefore, $|E[X]| \leq E[|X|]$ if $X \in L^1$.

(3) $|X| = X^+ + X^- \leq Y \in L^1$ implies $E[X^+], E[X^-] < \infty$. □

Everything is now in place (barring the three exercises that follow) for a proof of Lebesgue's famous theorem of dominated convergence.

Theorem 2.1.34. *(Theorem of dominated convergence, first version) Let (X_n) be a sequence of random variables dominated in modulus by an integrable random variable Y (i.e., for all n, $|X_n| \leq Y \in L^1$).*
  If $X = \lim_n X_n$, then
  (1) *$X \in L^1$ and*
  (2) *$E[X] = \lim_{n \to \infty} E[X_n]$.*

Proof. (1) $|X_n| \to |X|$ by Exercise 2.1.35 and so $|X| \leq Y \in L^1$. Hence by Proposition 2.1.33 (3), $X \in L^1$.

(2) The proof uses Fatou's lemma. Note that $X_n + Y \geq 0$ for all n. Hence by Fatou's lemma (i.e., (E_5) of Proposition 2.1.25), $E[\liminf_n (Y + X_n)] \leq \liminf_n E[Y + X_n] = \liminf_n (E[Y] + E[X_n]) = E[Y] + \liminf_n E[X_n]$ in view of Exercise 2.1.36.

Applying this exercise again, one has that $\liminf_n (Y + X_n) = Y + \liminf_n X_n = Y + X$. Hence, $E[Y] + E[X] \leq E[Y] + \liminf_n E[X_n]$ and so $E[X] \leq \liminf_n E[X_n]$.

Since $\lim_n(-X_n) = -X$ and $|-X_n| = |X_n| \leq Y$, it follows from what has been proved that $-E[X] = E[-X] \leq \liminf_n E[-X_n]$. By Exercise 2.1.37, $\liminf_n E[-X_n]$ equals $-\limsup_n E[X_n]$ as $E[-X_n] = -E[X_n]$. Hence, $-E[X] \leq -\limsup_n E[X_n]$ and so $\limsup_n E[X_n] \leq E[X] \leq \liminf_n E[X_n]$. By Exercise 2.1.10, this implies that $E[X] = \lim_n E[X_n]$. □

Exercise 2.1.35. If (a_n) is a sequence of real numbers that converges to A, then $(|a_n|)$ converges to $|A|$. [*Hint*: use Exercise 1.1.16.]

Exercise 2.1.36. Let (a_n) be any sequence of real numbers. Then if a is a real number, $\lim_n \inf(a_n + a) = (\lim_n \inf a_n) + a$. [*Hint*: remember that $\lim_n \inf a_n$ is the lower bound for the oscillation at ∞ of the sequence (a_n).]

Exercise 2.1.37. Let (a_n) be any sequence of real numbers. Then $\lim_n \inf(-a_n) = -(\lim_n \sup a_n)$.

If N is a null variable, $E[|N|] = 0$. So if $X \in L^1$ in the sense of Remark 2.1.32 (2), then $X \pm N \in L^1$ and $E[X \pm N] = E[X]$. Consequently, to decide whether a random variable X is integrable in the sense of Remark 2.1.32 (2), it suffices to know X modulo a null variable N, i.e., $X \pm N$. This leads to the final version of Lebesgue's theorem.

Theorem 2.1.38. (Theorem of dominated convergence) *Let (X_n) be a sequence of random variables dominated in modulus almost surely (usually written as "a.s.") by an integrable random variable Y (i.e., for all n, $\mathbf{P}(\{\omega \mid |X_n(\omega)| \leq Y(\omega)\}) = 1$).*

If $X = \lim_n X_n$ a.s. (i.e., $\mathbf{P}(\{\omega \mid X(\omega) = \lim_n X_n(\omega)\}) = 1$), then

(1) $X \in L^1$, *and*
(2) $E[X] = \lim_{n \to \infty} E[X_n]$.

Proof. Let $F = \{\omega \mid X(\omega) = \lim_{n \to \infty} X_n(\omega)\}$, and let $F_n = \{\omega \mid |X_n(\omega)| \leq Y(\omega)\}$. Then $\Omega_1 = F \cap [\cap_{n=1}^\infty F_n] \in \mathfrak{F}$ and $\mathbf{P}(\Omega_1) = 1$ (see Exercise 1.5.1 (2)). If U is a random variable on $(\Omega, \mathfrak{F}, \mathbf{P})$, let $\tilde{U}$ be defined by setting $\tilde{U}(\omega) = U(\omega)$ if $\omega \in \Omega_1$ and 0 otherwise. Then $U = \tilde{U} + N$, N a null variable, and $\tilde{U}$ is in L^1 if U is in L^1 in the sense of Remark 2.1.32 (2). The initial version of the theorem, Theorem 2.1.34, applies to $(\tilde{X}_n)_{n \geq 1}$ and

$\tilde{X}$. Since $X = \tilde{X} + N_1$, N_1 a null variable, $X \in L^1$ and $E[X] = E[\tilde{X}] = \lim_n E[\tilde{X}_n] = \lim_n E[X_n]$. □

Remark. In analysis, the phrase "almost surely" is replaced by "almost everywhere" and consequently "a.s." by "a.e."

Remark 2.1.39. The hypotheses of the theorem of dominated convergence not only imply that $E[X_n] \to E[X]$ as $n \to \infty$: they also imply that $E[|X_n - X|] \to 0$ as $n \to \infty$. It suffices to observe that $|X_n - X| \to 0$ a.e. and that $|X_n - X| \leq 2Y \in L^1$. Conversely, since $|E[X_n] - E[X]| = |E[X_n - X]| \leq E[|X_n - X|]$, it follows that $E[X_n] \to E[X]$ as $n \to \infty$ if $E[|X_n - X|] \to 0$ as $n \to \infty$. In other words, there is an equivalent form of the theorem of dominated convergence, which is stated below.

Theorem 2.1.40. (**Theorem of dominated convergence: equivalent form**) *Let (X_n) be a sequence of random variables dominated in modulus almost everywhere (usually written as a.e.) by an integrable random variable Y (i.e., for all n, $\mathbf{P}(\{\omega \mid |X_n(\omega)| \leq Y(\omega)\}) = 1$).*
If $X = \lim_n X_n$ a.e. (i.e., $\mathbf{P}(\{\omega \mid X(\omega) = \lim_n X_n(\omega)\}) = 1$), then

(1) *$X \in L^1$, and*
(2) *$\lim_{n \to \infty} E[|X_n - X|] = 0$.*

Remarks. (1) The convergence of $E[|X_n - X|] \to 0$ is what is later referred to as **convergence in L^1**, i.e., X_n converges to X in L^1 if $E[|X_n - X|] \to 0$ (see Definition 4.1.23).

(2) This theorem gives a very important sufficient condition in order that one may pass to the limit when integrating to get $E[\lim_n X_n] = \lim_n E[X_n]$. It is by no means necessary, since for non-decreasing sequences of non-negative random variables, the result is always valid by monotone convergence. Later on, in Chapter IV, a necessary and sufficient condition is given for this result to be true (see Remark 4.5.10).

2. LEBESGUE MEASURE ON $\mathbb{R}$ AND LEBESGUE INTEGRATION

(A) σ-finite measures.

The preceding discussion of probability and integration makes very little use of the fact that $\mathbf{P}$ is a probability. What is extremely important is that $\mathbf{P}$ is a function defined on a σ-algebra $\mathfrak{F}$ of subsets of a set Ω that has the following two properties:

(M_1) $0 \leq \mathbf{P}(A) \leq +\infty$ for all $A \in \mathfrak{F}$ and $\mathbf{P}(\emptyset) = 0$;
(M_2) if $A = \cup_{n=1}^{\infty} A_n$ with the sets $A_n \in \mathfrak{F}$ and pairwise disjoint, then
$\mathbf{P}(A) = \sum_{n=1}^{\infty} \mathbf{P}(A_n)$.

Definition 2.2.1. *A **measure**, also referred to as a **non-negative measure**, on a σ-algebra $\mathfrak{F}$ is a function μ satisfying (M_1) and (M_2). A measure μ on a σ-algebra $\mathfrak{F}$ of subsets of Ω is said to be **finite** if $\mu(\Omega) < \infty$ and is said to be **σ-finite** if $\Omega = \cup_{n=1}^{\infty} A_n$ with $\mu(A_n) < +\infty$ for all n. A* measure space $(\Omega, \mathfrak{F}, \mu)$ is defined to be a triple consisting of a set Ω, a σ-algebra $\mathfrak{F}$ of subsets of Ω, and a (non-negative) measure μ on $\mathfrak{F}$. It is said to be a **finite measure space** (respectively, a **σ-finite measure space**) if μ is finite (respectively, σ-finite).

It was shown in Chapter I (Theorem 1.4.13) that there is a one-to-one correspondence between probabilities on $\mathfrak{B}(\mathbb{R})$ and distribution functions F on $\mathbb{R}$. In the case of σ-finite measures μ on $\mathfrak{B}(\mathbb{R})$, there is an analogous result provided one assumes that $\mu((a,b])$ is finite whenever $-\infty < a < b < +\infty$. One replaces a distribution function F on $\mathbb{R}$ by a finite-valued, non-constant, non-decreasing, right continuous function G. Each such function determines a unique σ-finite measure on $\mathfrak{B}(\mathbb{R})$ as stated in the following result, and the measure determines G uniquely if one sets $G(0) = 0$. This corresponds to the solution for non-decreasing functions of the analogue of Basic Problem 1.3.6 for distribution functions (see the remark following its statement).

Theorem 2.2.2. *To each non-constant, non-decreasing, right continuous $G : \mathbb{R} \to \mathbb{R}$, there is a unique σ-finite measure μ on $\mathfrak{B}(\mathbb{R})$ with $\mu((a,b]) = G(b) - G(a)$ if $a < b$ are any two real numbers. Further, if $G(0) = 0$, then μ determines G:*

$$G(x) = \begin{cases} \mu((0,x]), & x > 0, \\ -\mu((x,0]), & x < 0. \end{cases}$$

One way to prove this theorem consists of looking a little carefully at the arguments used to prove Theorem 1.4.13 and "handwaving" a bit.

To begin, it is usual to replace $\mathfrak{A}$ by $\mathfrak{R}$, the collection of finite unions of finite intervals $(a,b]$. It is a **Boolean ring**: $A, B \in \mathfrak{R}$ implies that $A \cap B^c, A \cup B$, and $A \cap B$ all belong to $\mathfrak{R}$; so that A^c need not be in $\mathfrak{R}$ if $A \in \mathfrak{R}$, but its relative complement $B \cap A^c$ is in $\mathfrak{R}$ for all $B \in \mathfrak{R}$.

Exercise 1.3.7 has to be modified. There is a unique finite set function μ on $\mathfrak{R}$ that is finitely additive (i.e., (FAP_3) holds) such that $\mu((a,b]) = G(b) - G(a)$ for all $a < b$, and with $\mu(A) \geq 0$ for all $A \in \mathfrak{R}$.

Remark. One may also carry the whole discussion through using the Boolean algebra $\mathfrak{A}$. The price one pays is that the set function on $\mathfrak{A}$ may take $+\infty$ as a value; e.g., in the case of $G(x) = x$, which is used to define Lebesgue measure.

As every subset E of $\mathbb{R}$ is a subset of a countable union of sets from $\mathfrak{R}$, one may define the outer measure μ^* as

$$\mu^*(E) = \inf\{\sum_{n=1}^{\infty} \mu(A_n) \mid E \subset \cup_{n=1}^{\infty} A_n,\ A_n \in \mathfrak{R} \text{ for all } n \geq 1\}.$$

Since, as before, μ is σ-additive on $\mathfrak{R}, \mu(R) = \mu^*(R)$ for $R \in \mathfrak{R}$. Furthermore, Carathéodory's definition of μ^*-measurable sets is the same as that for $\mathbf{P}^*$-measurable sets and once again the class $\mathfrak{F}$ of μ^*-measurable sets is a σ-algebra, and the restriction of μ^* to $\mathfrak{F}$ is a measure on $\mathfrak{F}$.

Exercise 1.4.14 (1) remains true, but (2) is false unless $\mu(A_n) < +\infty$ for some n. To verify the uniqueness statement in Theorem 1.4.13, one uses the fact that $\mu((-n,n]) = G(n) - G(-n) < +\infty$. Then, if μ_1 and μ_2 are two measures on $\mathfrak{B}(\mathbb{R})$ that agree on intervals $(a,b]$, the set $\mathfrak{M}_n = \{A \cap (-n,n],\ A \in \mathfrak{B}(\mathbb{R}) \mid \mu_1(A \cap (-n,n]) = \mu_2(A_2 \cap (-n,n])\}$ is a monotone class of subsets of $(-n,n]$ that contains the Boolean algebra $\mathfrak{A}_n$ of subsets $R \cap (-n,n]$ of $(-n,n]$, where $R \in \mathfrak{R}$. Hence by Exercise 1.4.16, $\mathfrak{M}_n$ contains the smallest σ-algebra $\mathfrak{F}_n$ of subsets of $(-n,n]$ containing $\mathfrak{A}_n$.

Exercise 2.2.3. Show that $\mathfrak{F}_n$ equals $\{A \cap (-n,n] \mid A \in \mathfrak{B}(\mathbb{R})\}$. [Hint: show that $\{A \in \mathfrak{B}(\mathbb{R}) \mid A \cap (-n,n] \in \mathfrak{F}_n\}$ is a monotone class containing $\mathfrak{A}$.]

Consequently, $\mu_1(A) = \mu_2(A)$ since it follows from Exercise 1.4.14 (1), that for $i = 1, 2$, $\mu_i(A) = \lim_{n \to \infty} \mu_i(A \cap (-n,n])$, for all Borel sets A. □

(B) Lebesgue measure.

In particular, if $G(x) = x$ for all x, the resulting measure is called **Lebesgue measure**. It is commonly denoted by dx if x is the variable of integration and will also be sometimes denoted by λ. This measure computes the length of a set. If E is a Lebesgue measurable set, its Lebesgue measure is often denoted by $|\mathrm{E}|$ (i.e., $|\mathrm{E}| = \lambda^*(E)$, where λ^* is the outer measure corresponding to the measure λ on $\mathfrak{R}$ such that $\lambda((a,b]) = b - a$, if $a < b$).

Another way to prove Theorem 2.2.2 is outlined in the following exercise for the case of Lebesgue measure, i.e., when $G(x) = x$ for all $x \in \mathbb{R}$.

Exercise 2.2.4. For each $n \geq 1$, define the distribution function F_n by the formula

$$F_n(x) = \begin{cases} 0, & x \leq -n, \\ \frac{1}{2n}(x+n), & -n < x \leq n, \\ 1, & n < x. \end{cases}$$

This distribution function corresponds to the probability of choosing a number at random from $[-n,n]$, the uniform distribution on $[-n,n]$. It is also constructed by adding n and scaling by $\frac{1}{2n}$ the function G_n defined by

$$G_n(x) = \begin{cases} -n, & x \leq -n, \\ x, & -n < x \leq n, \\ n, & n < x. \end{cases}$$

Let $\mathbf{P}_n$ be the probability on $\mathfrak{B}(\mathbb{R})$ determined by the distribution function F_n.

Show that $\lambda_n \stackrel{\text{def}}{=} 2n\mathbf{P}_n$ is the unique positive measure μ on $\mathfrak{B}(\mathbb{R})$ such that
(1) $\mu((a,b]) = b - a$, if $-n < a < b \leq n$, and
(2) $\mu(-\infty, -n]) = \mu((n, \infty)) = 0$.

Show that

(3) if A is a Borel set and $A \subset (-n, n]$, then $\lambda_{n+1}(A) = \lambda_n(A)$.

Define μ on $\mathfrak{B}(\mathbb{R})$ by setting $\mu(A) = \sum_{n=1}^{\infty} \lambda_n(A_n)$, where $A_1 = A \cap (-1, 1]$, $A_{n+1} = (A \cap (-n-1, -n]) \cup (A \cap (n, n+1])$.

Show that

(4) μ is the unique positive measure ν on $\mathfrak{B}(\mathbb{R})$ such that $\nu((a,b]) = b - a$ if $-\infty < a < b < \infty$; in other words, that μ agrees with λ on $\mathfrak{B}(\mathbb{R})$.

With the approach of Exercise 2.2.4, there remains the problem of determining the σ-field $\mathfrak{F}$ of Lebesgue measurable sets. Let λ^* be the outer measure on $\mathfrak{P}(\mathbb{R})$ produced via the Carathéodory procedure applied to the function $G(x) = x$. Then, $\lambda^*_{|\mathfrak{B}(\mathbb{R})} = \mu$. Let $\mu^*(E) = \inf\{\mu(B) \mid E \subset B, B \in \mathfrak{B}(\mathbb{R})\}$. One may show, as in Exercise 1.5.4, that $\lambda^* = \mu^*$. Hence, the σ-algebra $\mathfrak{F}$ of Lebesgue measurable sets coincides with the σ-algebra of μ^*-measurable sets. Alternatively, one may use **completion** to extend the measure μ (which is Lebesgue measure on $\mathfrak{B}(\mathbb{R})$) to the Lebesgue measurable sets (see Exercises 2.9.11 and 3.3.13).

Remark. The basic idea of Exercise 2.2.4 is that given a positive measure μ on a σ-algebra $\mathfrak{F}$, to each set $A \in \mathfrak{F}$ with $0 < \mu(A) < \infty$ one can associate a probability $\mathbf{P}_A$ on $\mathfrak{F}$ by defining $\mathbf{P}_A(E) = \frac{1}{\mu(A)}\mu(E \cap A)$. In other words, by "normalizing" the measure μ on the subsets of A, one obtains a probability. When μ itself is a probability, the probability $\mathbf{P}_A$ is called **the conditional probability of E given A**. $\mathbf{P}_A(E)$ is usually denoted by $\mathbf{P}[E \mid A]$, and one has the familiar formula $\mathbf{P}[E \mid A] = \mathbf{P}[E \cap A]/\mathbf{P}[A]$.

Exercise 2.2.5. Carry through the procedure outlined in the previous exercise for any non-constant, non-decreasing, right continuous, finite function G on $\mathbb{R}$. Begin by choosing n large enough that $G(n) - G(-n) > 0$.

Exercise 2.2.6. If E is a Borel set, let $|E|$ denote its Lebesgue measure. Define $\mu(E) = |E \cap [0,1]|$. Show that μ is a probability and that its distribution function F is the first one of Example 1.3.9. Hence, μ is **the** probability on $\mathfrak{B}(\mathbb{R})$ corresponding to F.

(C) Integration with respect to a σ-finite measure μ.

Let $(\Omega, \mathfrak{F}, \mu)$ be a σ-finite measure space (Definition 2.2.1). Instead of being called random variables, the functions X that satisfy the condition in Definition 2.1.6 are called measurable functions and, as before, every X non-negative is the limit of an increasing sequence $(s_n)_{n \geq 1}$ of non-negative simple functions s_n (defined as before). For a simple function s, $\int s \, d\mu$

is defined as before. As in the case of a probability measure, $\lim_n \int s_n d\mu$ depends only on X, and hence one defines $\int X d\mu$ as this limit. The discussion of integration on a probability space applies without change to a σ-finite measure space. One writes $\int X d\mu$, $\int X(x)\mu(dx)$ or $\int X(x)d\mu(x)$ for the integral with respect to μ of $X \in L^1 = L^1(\Omega, \mathfrak{F}, \mu)$. The principle of monotone convergence, Fatou's lemma, and Lebesgue's theorem of dominated convergence all hold in the context of a σ-finite measure space.

In particular, if $\Omega = \mathbb{R}$ and $d\mu = dx$, the integral $\int X dx$ is called the **Lebesgue integral** of X. One often denotes $L^1(\mathbb{R}, \mathfrak{B}(\mathbb{R}), dx)$ by $L^1(dx)$ or $L^1(\mathbb{R})$.

If $E \subset \mathbb{R}$ is a Lebesgue measurable set, let $\mathfrak{B}(E) = \{A \cap E \mid A \in \mathfrak{B}(\mathbb{R})\}$. The set E determines a σ-finite measure space: $\Omega = E, \mathfrak{F} = \mathfrak{B}(E)$, and the measure $\mu(A \cap E) = |A \cap E|$ — this measure is called **Lebesgue measure on** E. Note that since the usual metric on $\mathbb{R}$ induces a metric on E, the σ-algebra $\mathfrak{B}(E)$ is also the same as the smallest σ-algebra of subsets of E that contains all the open subsets of the **metric space** E (Definition 4.1.25). This is because a subset O' of E is open in E if and only if $O' = O \cap E, O$ an open subset of $\mathbb{R}$ (see Exercise 3.6.3).

Exercise 2.2.7. Let $\mathfrak{F}$ denote the σ-algebra of Lebesgue measurable sets (i.e., the λ^*-measurable sets where λ^* is the outer measure corresponding to $G(x) = x$). Then one knows that $\mathfrak{F} \supset \mathfrak{B}(\mathbb{R})$. Show the following:

(1) if $E \in \mathfrak{F}$, then there is a Borel set $A \supset E$ with $|A| = |E|$ (where these values may be $+\infty$);

(2) if $E \in \mathfrak{F}$ and $|E| < \infty$, then there is a Borel set $A \supset E$ with $|A \backslash E| = 0$;

(3) if $E \in \mathfrak{F}$ and $|E| < \infty$, show that there are Borel sets B and A with $B \subset E \subset A$ with $|A \backslash E| = |E \backslash B| = 0$; [Hint: let $F_n = F \cap (-n, n], F \in \mathfrak{F}$, and verify (3) for E_n and E_n^c. Note that (3) says $1_B \leq 1_E \leq 1_A$ and $\int 1_B(x)dx = \int 1_A(x)dx$, cf. the situation in Proposition 2.3.4.]

(4) show that $E \in \mathfrak{F}$ if and only if there are Borel sets B and A with $B \subset E \subset A$ and $|A \backslash B| = 0$. [Hints: (only if) use (3); (if) use (2.3.6) or use the definition of a λ^*-measurable set.]

(5) let X be $\mathfrak{F}$-measurable and non-negative. Show that there are Borel measurable functions U and V such that (a) $0 \leq U \leq X \leq V$ and (b) $|\{U < V\}| = 0$.

Exercise 2.2.8. Let μ be a σ-finite measure on $\mathfrak{B}(\mathbb{R})$. If $b \in \mathbb{R}$, define $\mu_b(A) = \mu(A - b)$ for $A \in \mathfrak{B}(\mathbb{R})$, where $A - b = \{u - b \mid u \in A\}$. Show that

(1) μ_b is a σ-finite measure on $\mathfrak{B}(\mathbb{R})$,

(2) if λ denotes Lebesgue measure, then $\lambda_b((c, d]) = \lambda((c, d])$ for any $c < d$, and

conclude that

(3) $\lambda = \lambda_b$ for any $b \in \mathbb{R}$ (this property of Lebesgue measure is called

translation invariance and Lebesgue measure is said to be **translation invariant**).

Show that

(4) Lebesgue measure is the only translation invariant measure μ on $\mathfrak{B}(\mathbb{R})$ such that $\mu((0,1]) = 1$. [*Hint*: show that $\mu((0,r]) = r$ for any rational number $r > 0$.]

Remark 2.2.9. (A set that is not Lebesgue measurable) Not every subset of $\mathbb{R}$ is Lebesgue measurable. The standard example of a non-(Lebesgue)-measurable set E makes use of a countable dense subgroup G of $\mathbb{R}$; for example $\mathbb{Q}$, or the structurally simpler one that is discussed in Exercise 2.9.15. The sets $x + G, x \in \mathbb{R}$ are called the **cosets** of G and either are disjoint or else coincide: $(x_1 + G) \cap (x_2 + G) \neq \emptyset$ if and only if $x_1 - x_2 \in G$, in which case $x_1 + G = x_2 + G$. The set E is given by choosing exactly one point out of each of the cosets of G, an apparently obvious possibility but which technically speaking involves the use of the so-called axiom of choice (see [H2]). It follows that if $y_1, y_2, \in E$, then $y_1 - y_2 \notin G$. One proves that E is not Lebesgue measurable by contradiction: (i) if E is Lebesgue measurable, then $|E| \neq 0$ since $\mathbb{R} = \cup_{g \in G}(g + E)$ implies that $|\mathbb{R}| = \sum_{g \in G} |g + E|$ (recall G is countable), which equals zero if $0 = |E|$ since for any Lebesgue measurable set E, $|E| = |g + E|$; (ii) for any Lebesgue measurable set E with $|E| \neq 0$, the set $D = \{y_1 - y_2 \mid y_i \in E\}$ contains a non-void open set (see Exercise 4.7.18) and so, by the denseness of the countable group $D \cap G \neq \emptyset$, this contradicts the basic property of E, namely that, if $y_1, y_2 \in E$, then $y_1 - y_2 \notin G$.

Example 2.2.10. $X \in L^1 = L^1(\mathbb{R}, \mathfrak{B}(\mathbb{R}), dx)$ if $X(x) = e^{-(\frac{x^2}{2})}, x \in \mathbb{R}$. To see this, one can define a new function $Y(x) = e^{-(\frac{k^2}{2})}$ if $k \leq |x| < k+1$. Then, since $0 \leq X \leq Y$, it suffices to show $Y \in L^1$ by Proposition 2.1.33 (3) (as applied to a σ-finite measure space). Let $s_n = Y 1_{(-n,n)}$. Then $s_n \leq s_{n+1}$ and $\lim_n s_n = Y$. Also, the s_n are simple functions and $\int s_n(x) dx = 2 \sum_{k=1}^{n} e^{-(\frac{k^2}{2})}$. Since the series $\sum_{k=1}^{\infty} e^{-(\frac{k^2}{2})}$ converges, $\lim_{n \to \infty} \int s_n(x) dx = \int Y(x) dx < \infty$, and so $Y \in L^1$.

Note that this discussion does not compute $\int e^{-(\frac{x^2}{2})} dx$. It merely says that it is finite. To actually compute the integral, it is necessary to relate the Lebesgue integral to the Riemann integral.

3. THE RIEMANN INTEGRAL AND THE LEBESGUE INTEGRAL

Let $\varphi(x)$ be a bounded function on $[a,b]$. A **partition** π of $[a,b]$ is a finite sequence of points $a = t_0 < t_1 < \ldots < t_n = b$. A partition π' refines π if π' is given by adding more points to those that determine π.

Let $m_i = \inf\{\varphi(x) \mid t_i \leq x \leq t_{i+1}\}$ and $M_i = \sup\{\varphi(x) \mid t_i \leq x \leq t_{i+1}\}$. The **upper sum** $U(\pi, \varphi)$ is defined to be $\sum_{i=0}^{n-1} M_i(t_{i+1} - t_i)$ and the **lower** sum $L(\pi, \varphi)$ is defined to be $\sum_{i=0}^{n-1} m_i(t_{i+1} - t_i)$.

Definition 2.3.1. *A bounded function φ on $[a, b]$ is **Riemann integrable** if for any $\epsilon > 0$ there is a partition π with $U(\pi, \varphi) - L(\pi, \varphi) < \epsilon$.*

Proposition 2.3.2. *Assume φ is Riemann integrable on $[a, b]$. Then $\inf_\pi U(\pi, \varphi) = \sup_\pi L(\pi, \varphi)$.*

Proof. If π' refines π, then $L(\pi, \varphi) \leq L(\pi', \varphi) \leq U(\pi', \varphi) \leq U(\pi, \varphi)$ (one sees this by adding one point to π, then another, and so on). Given any two partitions π_1, π_2, there is a least partition π that refines π_1 and π_2. Hence, $L(\pi_1, \varphi) \leq L(\pi, \varphi) \leq U(\pi_2, \varphi)$. Consequently, $\inf_\pi U(\pi, \varphi) \geq \sup_\pi L(\pi, \varphi)$. The fact that φ is Riemann integrable implies that the difference between these numbers is less than ϵ, for any $\epsilon > 0$. Therefore, these numbers agree. □

Definition 2.3.3. *Let φ be a bounded function on $[a, b]$ that is Riemann integrable. The **Riemann integral of** φ over $[a, b]$ is defined to be $\inf_\pi U(\pi, \varphi) = \sup_\pi L(\pi, \varphi)$. It is denoted by $\int_a^b \varphi(x) dx$.*

Assume that $|\varphi(x)| \leq M$ for all $x \in [a, b]$. Let π be a partition of $[a, b]$ defined by $n + 1$ points t_i, and define

$$\ell_\pi(x) = \begin{cases} 0 & \text{if } x \leq a, \\ -M & \text{if } x = t_i, 0 \leq i \leq n+1, \\ m_i & \text{if } t_i < x < t_{i+1}, 0 \leq i \leq n, \\ 0 & \text{if } b < x, \end{cases}$$

and

$$u_\pi(x) = \begin{cases} 0 & \text{if } x \leq a, \\ M & \text{if } x = t_i, 0 \leq i \leq n+1, \\ M_i & \text{if } t_i < x < t_{i+1}, 0 \leq i \leq n, \\ 0 & \text{if } b < x. \end{cases}$$

These functions are Borel functions, $\ell_\pi \leq u_\pi$ and $|\ell_\pi(x)|, |u_\pi(x)| \leq M 1_{[a,b]} \in L^1$. These functions are therefore in $L^1(dx)$, and one has $L(\pi, \varphi) = \int \ell_\pi(x) dx$, $U(\pi, \varphi) = \int u_\pi(x) dx$. They are like simple functions, except that they are not necessarily positive.

Proposition 2.3.4. *Assume φ is bounded and Riemann integrable on $[a, b]$. Then there exists a sequence (π_n) of partitions with π_{n+1} a refinement of π_n, for all $n \geq 1$ such that*

$$\int_a^b \varphi(x) dx = \lim_{n \to \infty} U(\pi_n, \varphi) = \lim_{n \to \infty} L(\pi_n, \varphi).$$

Hence, there are Borel functions $f \leq \varphi 1_{[a,b]} \leq g$ that are in L^1 with

$$\int f(x) dx = \int g(x) dx.$$

3. THE RIEMANN INTEGRAL AND THE LEBESGUE INTEGRAL

Consequently, $\varphi 1_{[a,b]}$ is a Lebesgue measurable function that is integrable and (the Lebesgue integral) $\int \varphi(x) 1_{[a,b]}(x) dx = \int_a^b \varphi(x) dx$ (the Riemann integral).

Proof. For each n, there is a partition π'_n with $U(\pi'_n, \varphi) - L(\pi'_n, \varphi) < 1/n$. Let π_n be the smallest partition that refines $\pi'_1, \pi'_2, \ldots,$ and π'_n. Then,

$$L(\pi'_n, \varphi) \leq L(\pi_n, \varphi) \leq U(\pi_n, \varphi) \leq U(\pi'_n, \varphi)$$

and so

$$U(\pi_n, \varphi) - L(\pi_n, \varphi) < 1/n.$$

Hence,

$$\int_a^b \varphi(x) dx = \lim_{n \to \infty} U(\pi_n, \varphi) = \lim_{n \to \infty} L(\pi_n, \varphi).$$

Since π_{n+1} refines π_n, $\ell_{\pi_n} \leq \ell_{\pi_{n+1}} \leq u_{\pi_{n+1}} \leq u_{\pi_n}$. Let $f = \lim_{n \to \infty} \ell_{\pi_n}$ and $g = \lim_{n \to \infty} u_{\pi_n}$. These are Borel functions, with $f \leq \varphi 1_{[a,b]} \leq g$. Since $|\ell_{\pi_n}|, |u_{\pi_n}| \leq M 1_{[a,b]} \in L^1$, by dominated convergence (Theorem 2.1.38 applied to dx) $f, g \in L^1$ and $\int f(x) dx = \lim_{n \to \infty} \int \ell_{\pi_n}(x) dx = \lim_{n \to \infty} L(\pi_n, \varphi) = \int_a^b \varphi(x) dx = \lim_{n \to \infty} U(\pi_n, \varphi) = \lim_{n \to \infty} \int u_{\pi_n}(x) dx = \int g(x) dx$.

At this point, the full strength of Theorem 1.4.13 comes into play. So far no explicit use has been made of the σ-algebra $\mathfrak{F}$ of λ^*-measurable sets (the **Lebesgue measurable sets** if λ^* is the outer measure determined by $G(x) = x$) as opposed to the σ-algebra $\mathfrak{B}(\mathbb{R})$ of Borel subsets. Also, in the case of a probability, no explicit use has been made of the **P***-measurable sets that are not Borel measurable, where **P*** is the outer measure defined by a distribution function.

Look at $E = \{x \mid g(x) > f(x)\}$. This is a Borel set since g and f are Borel functions. Its Lebesgue measure is zero: let $E_n = \{x \mid g(x) \geq f(x) + 1/n\}$; as $g \geq f$ and $\int g(x) dx = \int f(x) dx$, $\int \{g(x) - f(x)\} dx = 0$; since $\int \{g(x) - f(x)\} dx \geq (1/n) \int 1_{E_n}(x) dx = (1/n) |E_n|$, it follows that $|E_n| = 0$ and hence, $|E| = 0$. Since $g \geq \varphi 1_{[a,b]} \geq f$, the function $\varphi 1_{[a,b]} = h$ is a function that agrees with a Borel function (say f or g) except possibly on the set E of Lebesgue measure zero.

Once one shows that $\varphi 1_{[a,b]}$ is measurable with respect to the σ-algebra $\mathfrak{F}$ of Lebesgue measurable sets, then by integrating on $(\mathbb{R}, \mathfrak{F}, dx)$, rather than on $(\mathbb{R}, \mathfrak{B}(\mathbb{R}), dx)$, the integral $\int \varphi(x) 1_{[a,b]}(x) dx$ is defined (in the sense of Lebesgue). Since $f \leq \varphi 1_{[a,b]} \leq g$, where $f, g \in L^1$ and $\int f(x) dx = \int g(x) dx$, the fact that $\int f(x) dx \leq \int \varphi(x) 1_{[a,b]}(x) dx \leq \int g(x) dx$ implies that $\int \varphi(x) 1_{[a,b]}(x) dx = \int_a^b \varphi(x) dx$.

To complete the proof of Proposition 2.3.4, it suffices to verify the Lebesgue measurability of $\varphi 1_{[a,b]}$.

Lemma 2.3.5. Let φ be a function on $\mathbb{R}$ such that there exists (1) a Borel set E with $|E| = 0$ and (2) a Borel function f with $\varphi = f$ on $\mathbb{R}\backslash E$. Then φ is Lebesgue measurable, i.e., is measurable with respect to the σ-algebra $\mathfrak{F}$ of Lebesgue measurable sets.

Proof. $\{\varphi < \lambda\} = \{\varphi < \lambda\} \cap (\mathbb{R}\backslash E) \cup \{\varphi < \lambda\} \cap E = \{f < \lambda\} \cap (\mathbb{R}\backslash E) \cap \{\varphi < \lambda\} \cap E$. Since f is Borel, $\{f < \lambda\} \in \mathfrak{B}(\mathbb{R})$ and so $\{f < \lambda\} \cap (\mathbb{R}\backslash E) \in \mathfrak{B}(\mathbb{R})$. Also, if $\{\varphi < \lambda\} \cap E = E_1$, it is a subset of a Borel set E of Lebesgue measure zero.

Sublemma 2.3.6. Every subset E_1 of a set E of Lebesgue measure zero is Lebesgue measurable and of measure zero.

Proof. Let $F \subset \mathbb{R}$ be any set. Since $\lambda^*(F \cap E_1) \leq |E| = 0$, $\lambda^*(F) \geq \lambda^*(F \cap E_1^c) = \lambda^*(F \cap E_1) + \lambda^*(F \cap E_1^c)$. As the opposite inequality is always true by Definition 1.4.10 (3), the result is proved. $\square$

The sublemma shows that $\{\varphi < \lambda\} \cap E \in \mathfrak{F}$, the σ-algebra of Lebesgue measurable sets. Hence, $\{\varphi < \lambda\} \in \mathfrak{F}$ for all $\lambda \in \mathbb{R}$, i.e., φ is Lebesgue measurable. $\square$

Remark. Sublemma 2.3.6 is stated for Lebesgue measure, but it applies to any probability associated with a distribution function and equally well to any positive σ-finite measure.

Exercise 2.3.7. (See Lusin's theorem (Exercise 4.7.9).) Let E be a Lebesgue measurable set and $f : E \to \mathbb{R}$ a function such that, for any $\epsilon > 0$, there is a closed set A with $A \subset E$ such that (i) the restriction of f to A is continuous (on A) (see Definition 2.1.13) and (ii) $|E\backslash A| < \epsilon$.

For each $n \geq 1$, let A_n be a closed set with $A_n \subset E$ such that (i) the restriction of f to A_n is continuous and (ii) $|E\backslash A_n| < \frac{1}{n}$. Show that

(1) $A_n \cap \{x \in E \mid f(x) \geq \lambda\}$ is closed for each λ,
(2) if $A \stackrel{\text{def}}{=} \cup_{n=1}^\infty A_n$, then $A \cap \{x \in E \mid f(x) \geq \lambda\}$ is a Borel set for each λ, and
(3) $|E\backslash A| = 0$.

Conclude that f is Lebesgue measurable, i.e., $\{x \in E \mid f(x) \geq \lambda\}$ is a Lebesgue measurable set for any λ. Note that Lusin's theorem (Exercise 4.7.9) states that the converse is true.

Remark. From now on, unless explicitly stated or the context makes it clear, if φ is a Lebesgue measurable function defined on an interval $[a, b]$ (or any of the other three bounded intervals with these endpoints), then $\int_a^b \varphi(x)dx$ denotes the Lebesgue integral $\int \varphi(x) 1_{[a,b]}(x) dx$ of the Lebesgue measurable function $\varphi 1_{[a,b]}$ on $\mathbb{R}$, which agrees with φ on $[a, b]$ and is zero on its complement.

In a calculus course, the Riemann integral is usually discussed only for a

3. THE RIEMANN INTEGRAL AND THE LEBESGUE INTEGRAL   53

function $\varphi(x)$ that is continuous on $[a,b]$. For such functions, the Riemann integral exists. A proof is outlined below.

2.3.8. Some properties of real-valued functions continuous on $[a,b]$.
(1) Let φ_1, φ_2 both be continuous on $[a,b]$. Then $\varphi_1 \pm \varphi_2$ and $\varphi_1 \varphi_2$ are continuous on $[a,b]$. If $\varphi(x) > 0$ for all x, $a \le x \le b$, and is continuous, then $1/\varphi$ is continuous on $[a,b]$.

(These facts are to be found in [M1] or [R4], for example. One can also try to prove them.)

(2) φ is bounded on $[a,b]$.

For each x, $a \le x \le b$, if one takes $\epsilon = 1$, by definition of continuity at x there is a $\delta = \delta(x) > 0$ such that $|\varphi(x) - \varphi(y)| < 1$ if $|x - y| < \delta(x)$ and $y \in [a,b]$; the open intervals $(x - \delta(x), x + \delta(x))$ cover $[a,b]$; by the Heine Borel theorem (Theorem 1.4.5), there is a finite number of points $x_1, \ldots, x_m$ such that $[a,b] \subset \cup_{i=1}^{m} (x_i - \delta(x_i), x_i + \delta(x_i))$. Hence, if $y \in [a,b]$, $|\varphi(y) - \varphi(x_i)| < 1$ for some $1 \le i \le m$. Consequently, $|\varphi(y)| \le 1 + \max_{1 \le i \le m} |\varphi(x_i)| = M$, i.e., φ is bounded. Alternatively, the continuous image of a compact set is compact and so bounded.

(3) φ is uniformly continuous on $[a,b]$: i.e., given $\epsilon > 0$, there exists $\delta > 0$ such that $|\varphi(x) - \varphi(y)| < \epsilon$ whenever $|x - y| < \delta$ and $\{x, y\} \subset [a,b]$. Note that the δ does not depend on x or y.

Proof. Choose $\epsilon > 0$ and for each x, let $\delta(x) > 0$ be such that $|\varphi(x) - \varphi(y)| < \epsilon/2$ if $|x - y| < \delta(x)$. The open intervals $(x - \delta(x)/2, x + \delta(x)/2)$ cover $[a,b]$. Let $[a,b] \subset \cup_{i=1}^{m}(x_i - \delta(x_i)/2, x_i + \delta(x_i)/2)$ $i = 1, \ldots, m$, where such a set of points x_i exists by virtue of the Heine-Borel theorem. Let $\delta = \min_{1 \le i \le m} \delta(x_i)/2$.

Let $|s - t| < \delta$ and $s, t \in [a,b]$. Then s is in some $[x_i - \delta(x_i)/2, x_i + \delta(x_i)/2]$ and by the choice of δ, $t \in (x_i - \delta(x_i), x_i + \delta(x_i))$.

Hence,

$$|\varphi(s) - \varphi(t)| \le |\varphi(s) - \varphi(x_i)| + |\varphi(x_i) - \varphi(t)| < \epsilon/2 + \epsilon/2 = \epsilon.$$

(4) If φ is continuous on $[a,b]$, there are points $x_m, x_M \in [a,b]$ such that $m = \varphi(x_m) \le \varphi(x) \le \varphi(x_M) = M$, $a \le x \le b$. In other words, φ attains its maximum and its minimum on $[a,b]$.

(1) Let φ_1, φ_2 both be continuous on $[a,b]$. Then $\varphi_1 \pm \varphi_2$ and $\varphi_1 \varphi_2$ are continuous on $[a,b]$. If $\varphi(x) > 0$ for all x, $a \le x \le b$ and is continuous, then $1/\varphi$ is continuous on $[a,b]$.

Proof. By (2) above, φ is bounded above and below. Let $M = \sup\{\varphi(x) \mid a \le x \le b\}$. The function $f(x) = M - \varphi(x)$ is continuous on $[a,b]$ by (1), and if it never vanishes, $1/f(x)$ is again continuous on $[a,b]$. It is bounded — see (2)— by a constant N. So $1/\{M - \varphi(x)\} \le N$ if $a \le x \le b$. Hence, $1/N \le M - \varphi(x)$, (i.e., $\varphi(x) \le M - 1/N$ if $a \le x \le b$). This contradicts the fact that $M = \sup\{\varphi(x) \mid a < x < b\}$. Therefore, $M - \varphi(x)$ vanishes somewhere. Applying this to $-\varphi$, $-m + \varphi(x)$ also vanishes somewhere.

Proposition 2.3.9. *Let φ be a continuous real-valued function on $[a,b]$. Then φ is Riemann integrable.*

Proof. By (2) above, φ is bounded. Let $\epsilon > 0$, and choose n so that $|s - t| < (b - a)/n$ implies $|\varphi(s) - \varphi(t)| < \epsilon$ (possible by (3)). Let π be given by dividing $[a, b]$ into intervals of length $(b - a)/n$. Then in view of (4), applied to each closed subinterval $[t_i, t_{i+1}]$, and the choice of n, $M_i - m_i < \epsilon$ for all i. Hence, $U(\pi, \varphi) - L(\pi, \varphi) < (b - a)\epsilon$. Since ϵ is arbitrary, φ is Riemann integrable.

Improper integrals and Lebesgue integrable functions.

The function $X(x) = e^{-(\frac{x^2}{2})}$ is continuous, even representable by a power series that converges uniformly. As a result, the Riemann integral $\int_0^R e^{-(\frac{x^2}{2})} dx$ exists.

Definition 2.3.10. $\int_0^{+\infty} f(x)dx = \lim_{R\to+\infty} \int_0^R f(x)dx$ *is called the* **improper Riemann integral of f over** $[0, +\infty)$, *if this limit exists (to simplify questions of existence of the integral over $[0, R]$, f is assumed continuous). If the improper integral $\int_0^{+\infty} |X(x)|dx < +\infty$, the improper integral $\int_0^\infty X(x)dx$ exists and is said to be* **absolutely convergent**. *If the improper integral $\int_0^\infty X(x)dx$ exists but $\int_0^\infty |X(x)|dx = +\infty$, the improper integral is said to be* **conditionally convergent**.

From calculus (see Marsden [M1], p. 322) one learns that $\int_0^{+\infty} e^{-(\frac{x^2}{2})}dx = \frac{\sqrt{2\pi}}{2}$ and so $\int_{-\infty}^\infty e^{-(\frac{x^2}{2})}dx = \sqrt{2\pi}$.

Proposition 2.3.11. *Let $X(x)$ be a non-negative continuous function on $[0, +\infty)$. If $\int_0^{+\infty} X(x)dx$ exists, then $X1_{[0,+\infty)} \in L^1$ and $\int X1_{[0,+\infty)}(x)dx = \int_0^{+\infty} X(x)dx$. When X is continuous but not necessarily non-negative, the improper integral $\int_0^{+\infty} X(x)dx$ can be conditionally convergent.*

Proof. Assume X is non-negative. Since $X1_{[0,+\infty)} = \lim_{N\to\infty} X1_{[0,N]}$ with $X1_{[0,N]} \leq X1_{[0,(N+1)]}$, and $\int_0^{+\infty} X(x)dx = \lim_{N\to+\infty} \int_0^N X(x)dx$, it follows from the principle of monotone convergence (Proposition 2.1.25 (E_4)) that

$$\int X1_{[0,+\infty)}(x)dx = \lim_{N\to\infty} \int X1_{[0,N]}(x)dx$$
$$= \lim_{N\to\infty} \int_0^N X(x)dx = \int_0^{+\infty} X(x)dx.$$

Let
$$X(x) = \begin{cases} \frac{\sin x}{x} & \text{if } x > 0, \\ 1 & \text{if } x = 0. \end{cases}$$

Then one can show that $|X(x)|1_{[0,+\infty)}(x) \notin L^1$, i.e., $X1_{[0,+\infty)} \notin L^1$ but that $\int^{+\infty} \frac{\sin x}{x} dx$ exists by using an argument similar to the argument used to show that an alternating series converges. □

Remark 2.3.12. When X is non-negative, then $\int_0^{+\infty} X(x)dx = \int X1_{[0,+\infty)}(x)dx$ even if this limit (the improper integral) is infinite.

Proposition 2.3.11 shows how to calculate the integral $\int e^{-(\frac{x^2}{2})}dx$. of the L^1-function $X(x) = e^{-(\frac{x^2}{2})}$. The value of this Lebesgue integral is $\sqrt{2\pi}$.

4. PROBABILITY DENSITY FUNCTIONS

Consider the distribution function $F(x) = \frac{1}{\sqrt{2\pi}} \int_{-\infty}^x e^{-(\frac{u^2}{2})} du$ of Example 1.3.9 (2). According to Theorem 1.4.13, there is a unique probability **P** on $\mathfrak{B}(\mathbb{R})$ with F as its distribution function.

Proposition 2.4.1. *For any Borel set E,*

$$\mathbf{P}(E) = \frac{1}{\sqrt{2\pi}} \int_{-\infty}^{+\infty} e^{-\frac{u^2}{2}} 1_E(u) du \stackrel{\text{def}}{=} \frac{1}{\sqrt{2\pi}} \int_E e^{-\frac{u^2}{2}} du.$$

In particular, if C is the Cantor set, then $\mathbf{P}(C) = 0$. (See Example 1.2.7.)

Proof. Define $\mu(E) = \frac{1}{\sqrt{2\pi}} \int_E e^{-(\frac{u^2}{2})} du$ for any Borel set E. Then $F(x) = \mu((-\infty, x])$, and the first statement follows from Theorem 1.4.13 once it is shown that μ is a probability.

Clearly, $0 \le \mu(E) \le 1$ and $\mu(\mathbb{R}) = 1$ since the Lebesgue integral over $\mathbb{R}$ equals the improper Riemann integral $\frac{1}{\sqrt{2\pi}} \int_{-\infty}^{+\infty} e^{-(\frac{u^2}{2})} du$, which equals 1.

Let $A = \cup_{n=1}^\infty A_n$ with the sets A_n pairwise disjoint and Borel. Let $B_n = \cup_{r=1}^n A_i$ and $Y_n(u) = \frac{1}{\sqrt{2\pi}} e^{-(\frac{u^2}{2})} 1_{B_n}(u) = \sum_{i=1}^n 1_{A_i}(u) \cdot \frac{1}{\sqrt{2\pi}} e^{-(\frac{u^2}{2})}$. Then $\lim_{n\to\infty} Y_n(u) = \frac{1}{\sqrt{2\pi}} e^{-(\frac{u^2}{2})} 1_A(u)$. Call this function Y.

Now $Y_n \le Y_{n+1}$. Thus, by the principle of monotone convergence, (E_4) of Proposition 2.1.25 applied to the Lebesgue integral, $\frac{1}{\sqrt{2\pi}} \int_A e^{-(\frac{u^2}{2})} du = \int Y(u) du = \lim_{n\to\infty} \int Y_n(u) du = \lim_{n\to\infty} \sum_{i=1}^n \frac{1}{\sqrt{2\pi}} \int_{A_i} e^{-(\frac{u^2}{2})} du$. In other words,

$$\mu(A) = \lim_{n\to\infty} \sum_{i=1}^n \mu(A_i) = \sum_{i=1}^\infty \mu(A_i).$$

To compute (C) in Example 1.2.7, notice that $|C| = 0$. Let s be a simple function. Then $\int_C s(x) dx = 0$ as s is a null function by Exercise 2.1.28. Hence, $\int_C X(x) dx = 0$ for any non-negative, measurable function X. □

This proposition illustrates a general fact. Namely, the integral of a non-negative integrable random variable X over a set $A \in \mathfrak{F}$, determines a set function $\nu(A)$ that is a measure (see Definition 2.2.1). The next exercise explains the details of the argument.

Exercise 2.4.2. Use the preceding arguments to show the following, where $(\Omega, \mathfrak{F}, \mu)$ is a σ-finite measure space.
(1) Let X be non-negative and μ-integrable, i.e., $X \in L^1(\Omega, \mathfrak{F}, \mu)$. Let $\nu(A) = \int_A X d\mu \stackrel{\text{def}}{=} \int 1_A(w) X(w) \mu(dw)$.
(2) Show that ν is a σ-finite measure on $\mathfrak{F}$ with $\nu(\Omega) = \int X d\mu$.
(3) Let X be a non-negative, measurable function, and let $E \in \mathfrak{F}$ be a set with $\mu(E) = 0$: show that $\int_E X d\mu = 0$.
(4) Show that $\int f d\nu = \int f X d\mu$ for any non-negative measurable function f.
(5) Show that $f \in L^1(\Omega, \mathfrak{F}, \nu)$ if and only if $fX \in L^1(\Omega, \mathfrak{F}, \mu)$ and $\int f d\nu = \int f X d\mu$.

Examples 2.4.3.
(1) Let $\lambda > 0$ and set $\mathbf{P}(E) = \frac{1}{\lambda} \int_{E \cap [0,+\infty)} e^{-\lambda x} dx$. Then $\mathbf{P}$ is a probability. This follows once it is observed that the function $X \in L^1$, with $\int X(x) dx = 1$, where

$$X(x) = \begin{cases} \frac{1}{\lambda} e^{-\lambda x} & \text{if } x \geq 0, \\ 0 & \text{if } x < 0. \end{cases}$$

A random variable with this distribution or law is said to be **exponentially distributed (with parameter λ)**.

(2) The function $X \in L^1$, where

$$X(x) = \begin{cases} \alpha^\nu x^{\nu-1} e^{-\alpha x} & \text{if } x \geq 0, \\ 0 & \text{if } x < 0, \end{cases}$$

and $\alpha, \nu > 0$.

The function $\gamma(t) = \int_0^{+\infty} u^{t-1} e^{-u} du$ is called the Gamma function. $\gamma(n+1) = n!$, $n = 0, 1, \ldots$. The probability $\mathbf{P} = \mathbf{P}_{\alpha,\nu}$ given by $\mathbf{P}(E) = \frac{1}{\gamma(\nu)} \int_E X(x) dx$ is called a **Gamma distribution**.

(3) Other familiar distributions from statistics are given by integrating other L^1-functions (e.g. Beta distributions, Student's t, etc., see Feller [F2] for additional examples).

These examples (should) raise the following question. When can a probability $\mathbf{P}$ on $\mathfrak{B}(\mathbb{R})$ be written as $\mathbf{P}(E) = \int_E X(x) dx$, $X \in L^1$?

In view of Exercise 2.4.2 (3), for this to happen it is necessary that $\mathbf{P}(E) = 0$ if $|E| = 0$. It is an important result that this condition is not only necessary but sufficient. In other words, a probability $\mathbf{P}$ on $\mathfrak{B}(\mathbb{R})$ is determined by a non-negative L^1-function of integral one — a so-called **probability density function** — if and only if $\mathbf{P}(E) = 0$ when $|E| = 0$. This theorem is a special case of the Radon–Nikodym theorem (Theorem

2.7.19) which will be proved later in §7 using the Hahn decomposition for signed measures and again at the end of these notes using martingale theory (see Corollary 5.7.4 and Exercise 5.7.6). Such a probability is said to be **absolutely continuous** with respect to Lebesgue measure λ. The (probability) density is called the **Radon–Nikodym derivative of P with respect to** λ.

The derivative of the distribution function F of a probability **P** exists almost everywhere (see Wheeden and Zygmund [W1], p. 111, Theorem 7.21). In particular, the derivative of the distribution function of an absolutely continuous probability **P** exists almost everywhere, as is shown in Chapter IV. In this case, the derivative X (or f) of the distribution function is the probability density function for **P**.

In general, the derivative f of a distribution function F is a non-negative integrable function. The probability **P** can then be written as the sum of the absolutely continuous measure ξ given by f and another so-called singular measure η, where a measure μ is said to be **singular** if there is a Borel set E with $|E| = 0$ and $\mu(E^c) = 0$. This decomposition of **P** (Theorem 2.7.22) is known as Lebesgue's decomposition theorem, and the measures ξ and η are unique. By normalizing these measures (i.e., multiplying them by scalars), it follows that every probability that is neither absolutely continuous nor singular is a unique convex combination of an absolutely continuous probability and a singular one (Exercise 2.7.25).

Another decomposition of a probability results from the fact that a distribution function has at most a countable number of jumps (Exercise 2.9.13), where a jump occurs at x if F is not continuous at x (i.e., if $F(x) \neq F(x-)$, where $F(x-)$ is defined in Definition 1.3.4). It is explained in Exercise 2.9.14 that every distribution function that is neither continuous nor discrete is a unique convex combination of a continuous distribution function and a so-called **discrete** one: a distribution function F is said to be discrete if the corresponding probability **P** is concentrated on a countable set D (i.e., $\mathbf{P}(D) = 1$); note that one may assume that $\mathbf{P}(\{x\}) > 0$ for all $x \in D$. Since a countable set has Lebesgue measure zero, every discrete probability is singular. Further, Exercise 2.4.4 explains how the corresponding distribution function increases only by jumps.

Continuous distribution functions do not necessarily determine absolutely continuous probabilities. In fact, a continuous distribution function F can have a derivative equal to 0 almost everywhere: a classic example, the Cantor–Lebesgue function (see Wheeden and Zygmund [W1], p. 35), is defined by using the complement of the Cantor set (see Exercise 2.9.16). The resulting probability is therefore singular with respect to Lebesgue measure. In the case of the Cantor–Lebesgue function, it lives on the Cantor set C (i.e., the Cantor set has probability equal to one).

If a continuous probability is not absolutely continuous, Lebesgue's decomposition theorem (Theorem 2.7.22) applied to it produces a singular continuous probability. As a result, every probability that is neither continuous nor discrete, neither absolutely continuous nor singular, is a unique convex combination of an absolutely continuous probability, a singular continuous probability, and a discrete probability (Exercise 2.9.14). Alternatively, one may obtain this result by decomposing the singular part of a probability using (Exercise 2.9.14) into a continuous part and a discrete part.

Hence, every probability is a convex combination of an absolutely continuous one, a continuous singular one, and a discrete one. This decomposition is in fact unique in the sense that the resulting measures are unique (see Exercise 2.9.14 (5)).

Exercise 2.4.4. Let F be the distribution function of a finite random variable X defined on a probability space $(\Omega, \mathfrak{F}, \mathbf{P})$. Let E be a Borel subset, and assume that $X(\Omega) \subset E$ (i.e., all the values of X lie in E). Let $\mathbf{Q}$ be the distribution of X, i.e., the unique probability on $(\mathbb{R}, \mathfrak{B}(\mathbb{R}))$ determined by F. Show that

(1) $\mathbf{Q}(E) = 1$, and
(2) $F(x) = \mathbf{P}[X \in E \cap (\infty, x]] = \mathbf{Q}(E \cap (\infty, x])$ for all $x \in \mathbb{R}$.

Assume that E is a countable set and that $\mathbf{P}[X = x] = \mathbf{Q}(\{x\}) > 0$ for each $x \in E$. Show that

(3) $F(x) = F(x-)$ if and only if $x \notin E$;
(4) conclude that F increases only by jumps and that at a point x of E the jump size is $\mathbf{Q}(\{x\})$.

5. INFINITE SERIES AGAIN

The analogue for $\mathbb{N}$ (or any countable set) of Lebesgue measure is often called counting measure. If $E \subset \mathbb{N}$, let $|E|$ denote the number of points in E. Then, $|\cdot|$ is a measure on $\mathfrak{P}(\mathbb{N})$ — the collection of all subsets of $\mathbb{N}$. The corresponding integral could be written as $\int X \, dn$.

The real-valued functions a on $\mathbb{N}$ are just the real-valued sequences $(a_n)_{n \geq 1}$.

Exercise 2.5.1. Let $a = (a_n)_{n \geq 1}$ be ≥ 0. Show that $\sum_{n=1}^{\infty} a_n$ is the integral of a with respect to counting measure dn on $\mathbb{N}$.

Exercise 2.5.2. Let ℓ_1 denote the space $L^1(\mathbb{N}, \mathfrak{P}(\mathbb{N}), dn)$. Show that $a \in \ell_1$ if and only if $\sum_{n=1}^{\infty} a_n$ is absolutely convergent.

Definition 2.5.3. Let V be a real vector space. A **norm** on V is a function $\| \ \|: V \to \mathbb{R}^+$ such that

(1) for all $u, v \in V, \| u + v \| \leq \| u \| + \| v \|$;
(2) for all $u \in V$ and $\lambda \in \mathbb{R}, \| \lambda u \| = |\lambda| \| u \|$; and
(3) $\| u \| = 0$ implies $u = 0$.

Exercise 2.5.4. Show that ℓ_1 is a real vector space and that the function $a \to \sum_{n=1}^{\infty} |a_n| \stackrel{\text{def}}{=} \| a \|_1$ is a norm on ℓ_1.

Exercise 2.5.5. Let ℓ_2 denote the set of sequences $a = (a_n)_{n \geq 1}$ such that $\sum_{n=1}^{\infty} a_n^2 < +\infty$ (the set of **square summable sequences**). Show that
(1) ℓ_2 is a real vector space, and
(2) $a, b \in \ell_2$ implies $ab \in \ell_1$, where $(ab)_n = a_n b_n$, $n \geq 1$.

Definition 2.5.6. *Let V be a real vector space. An **inner product** on V is a function $\langle\, ,\, \rangle : V \times V \to \mathbb{R}$ such that*
(1) *for all $u, v \in V$, $\langle u, v \rangle = \langle v, u \rangle$,*
(2) *for all $u_1, u_2, v \in V$, $\langle u_1 + u_2, v \rangle = \langle u_1, v \rangle + \langle u_2, v \rangle$,*
(3) *for all $u, v \in \ell_2, \lambda \in \mathbb{R}$, $\langle \lambda u, v \rangle = \lambda \langle u, v \rangle$,*
(4) *for all $u \neq 0 \in \ell_2$, $\langle u, u \rangle > 0$.*

One often denotes $\langle u, v \rangle$ by (u, v). In addition, if the vector space is a complex vector space, then an inner product is complex-valued, and the above conditions are satisfied, with the important difference that (1) now states that $(v, u) = \overline{(u, v)}$. The simplest example of a complex inner product is defined on $\mathbb{C}$ itself by $(u, v) \stackrel{\text{def}}{=} u\overline{v}$, where if $v = a + ib$, then $\overline{v} = a - ib$. In this case, the corresponding norm of $v = a + ib$ is the modulus $|v| = \sqrt{a^2 + b^2}$ of the complex number v.

Exercise 2.5.7. For $a, b \in \ell_2$, define $\langle a, b \rangle = \sum_{n=1}^{\infty} a_n b_n$ (why does this series converge?). Show that $\langle\, ,\, \rangle$ is an inner product on ℓ_2.

Exercise 2.5.8. Verify the **Cauchy–Schwarz inequality**: $|\langle a, b \rangle| \leq \| a \|_2 \| b \|_2$, where $\| a \|_2 = \left[\sum_{n=1}^{\infty} a_n^2\right]^{1/2}$. Show that $a \to \| a \|_2$ is a norm. [Hint: consider the discriminant $B^2 - 4AC$ of the quadratic polynomial of λ, $Q(\lambda) = \| a + \lambda b \|_2^2 = A\lambda^2 + B\lambda + C \geq 0$. The quadratic formula $\lambda = \frac{B \pm \sqrt{B^2 - 4AC}}{2A}$ for the roots of $Q(\lambda)$ shows that, since $Q \geq 0$, (i) $B^2 - 4AC \leq 0$ and (ii) $B^2 - 4AC = 0$ if and only if Q has a (unique) root and this root equals $\frac{B}{A}$ as it is a root of $Q'(\lambda) = 2A\lambda + 2B = 0$. (See also Proposition 5.1.4.)]

Example 2.5.9. The Poisson distribution on $\mathbb{N} \cup \{0\}$ with mean 1 can be viewed as a probability distribution given by an ℓ_1-density a, where $a_n = e^{-1}(\frac{1}{n!})$. Is this density in ℓ_2?

6. DIFFERENTIATION UNDER THE INTEGRAL SIGN

In many circumstances functions $f(x)$ are transformed by integrating with a kernel $K(x, y)$, a function of $x, y \in \mathbb{R}$. For example,
(i) $K(x, y) = \frac{1}{\sqrt{2\pi}} e^{-\frac{(x-y)^2}{2}}$ or
(ii) $K(x, y) = e^{-xy}$.

The transformed function $Kf(x) = \int K(x,y)f(y)dy$, assuming the integral (with respect to Lebesgue measure dx on $\mathbb{R}$) exists. The kernel may be differentiable in x, and it is then natural to expect that Kf is differentiable and that
$$\frac{d}{dx}Kf(x) = \int \frac{\partial}{\partial x}K(x,y)f(y)dy.$$

When this is so, differentiation under the integral sign allows the computation of $(Kf)'$.

Theorem 2.6.1. Let $y \to K(x,y)f(y) \in L^1(\mathbb{R})$ for $a < x < b$. If $a < x_0 < b$, assume that there is an L^1-function $Y(y)$ with
$$\left|\frac{K(x_0+h,y) - K(x_0,y)}{h}\right| |f(y)| \leq Y(y),$$
for almost all $y \in \mathbb{R}$ and all h sufficiently small.

If $\frac{\partial}{\partial x}K(x,y)$ exists for all y at $x = x_0$, then $Kf(x) = \int K(x,y)f(y)dy$ is differentiable at x_0 and $(Kf)'(x_0) = \int \frac{\partial}{\partial x}K(x_0,y)f(y)dy$.

Proof. Let $a < x_0 < b$, and let $(h_n)_{n \geq 1}$ be a sequence with $a < x_0 + h_n < b$ for all n. Set $F_n(y) = \{\frac{K(x_0+h_n,y) - K(x_0,y)}{h_n}\}f(y)$. If $h_n \to 0$ as $n \to \infty$, then by hypothesis $F_n(y) \to \frac{\partial}{\partial x}K(x_0,y)f(y)$. Since $|F_n(y)| \leq Y(y)$, by dominated convergence (Theorem 2.1.38),
$$\int \frac{\partial}{\partial x}K(x_0,y)f(y)dy = \lim_{h_n \to 0} \frac{Kf(x_0+h_n) - Kf(x_0)}{h_n}.$$

Hence, $(Kf)'(x_0)$ exists and is this limit since the sequence $(h_n)_{n \geq 1}$ was arbitrary. □

7. Signed measures and the Radon–Nikodym theorem*

Let $(\Omega, \mathfrak{F}, \mu)$ be a σ-finite measure space. Exercise 2.4.2, indicates how to define a new measure ν by means of a non-negative function $X \in L^1(\Omega, \mathfrak{F}, \mu)$: set $\nu(A) = \int_A X d\mu \stackrel{\text{def}}{=} \int 1_A(\omega)X(\omega)d\omega$.

The measure ν is non-negative because X is non-negative. However, the set function $\nu(A) = \int_A X d\mu$ is still defined without this requirement of positivity, and it continues to satisfy the condition (M_2) of Definition 2.2.1.

Exercise 2.7.1. Let $X \in L^1(\Omega, \mathfrak{F}, \mu)$, and define $\nu(A) = \int_A X d\mu$ for all $A \in \mathfrak{F}$. Show that

(1) $\nu(A) \in \mathbb{R}$ for all $A \in \mathfrak{F}$ (i.e., ν is finite),
(2) if $A = \cup_{n=1}^\infty A_n$, where the A_n are pairwise disjoint sets in $\mathfrak{F}$, then $\nu(A) = \sum_{n=1}^\infty \nu(A_n)$ [Hint: make use of dominated convergence (Theorem 2.1.38) and the fact that $|X| \in L^1(\Omega, \mathfrak{F}, \mu)$.],
(3) show that ν is the difference of two (non-negative) measures. [Hint: $X = X^+ - X^-$.]

7. SIGNED MEASURES AND THE RADON–NIKODYM THEOREM

The set function thus defined by an L^1-function is an example of a finite signed measure.

Definition 2.7.2. *A real-valued function ν defined on a σ-algebra $\mathfrak{F}$ will be called a **finite signed measure** if whenever $A = \cup_{n=1}^{\infty} A_n$, with the A_n pairwise disjoint sets in $\mathfrak{F}$, then $\nu(A) = \sum_{n=1}^{\infty} \nu(A_n)$.*

Remarks.

(1) If ν is a finite signed measure, then $\nu(\emptyset) = 0$ as $\emptyset = \emptyset \cup \emptyset$.

(2) The word "measure" will be reserved to mean a non-negative measure in the sense of Definition 2.2.1. At times, this usage will be reinforced by also writing "non-negative measure".

(3) The term **signed measure** denotes a set function that, in addition to taking real values, may also take one of the values $\pm\infty$ (see Halmos [H1]). The results that follow for finite signed measures hold, with appropriate modification, for signed measures (see Halmos [H1]).

(4) A **σ-finite signed measure** is a signed measure such that Ω is a countable union of sets of finite measure.

(5) Complex-valued measures are also considered, but not in this book: they are σ-additive complex-valued functions; their real and imaginary parts are finite signed measures in the sense of Definition 2.7.2.

The finite signed measure ν determined by an L^1-function can be written as the difference of two finite measures ν_1 and ν_2 as indicated in Exercise 2.7.1 (3): let ν_1 be the measure determined by X^+ and ν_2 be the measure determined by X^-. Given any two finite measures ν_1 and ν_2, it is easy to see that their difference $\nu_1 - \nu_2$ is a finite signed measure whose value at $A \in \mathfrak{F}$ is $\nu_1(A) - \nu_2(A)$ (and it is a σ-finite signed measure if one of the two measures is σ-finite while the other is finite). It is a natural question to ask whether every finite signed measure can be written as the difference of two finite measures. This is in fact the case, as will be shown below.

As a first step in proving this result, one has the following characterization, for $X \in L^1$, of the set where $X \geq 0$, in terms of the measure ν defined by X.

Proposition 2.7.3. *Let $X \in L^1((\Omega, \mathfrak{F}, \mu))$ and $A \in \mathfrak{F}$. The following properties of A are equivalent, where ν is the measure defined by X:*

(1) $\nu(A) = \nu^+(A)$, where $\nu^+(A) = \int_A X^+ d\nu$;
(2) $\nu(A) = \int_A X^+ d\mu$;
(3) $\nu^-(A) = 0$, where $\nu^-(A) = \int_A X^- d\nu$;
(4) *if* $B \subset A, B \in \mathfrak{F}$, *then* $\nu(B) \geq 0$;
(5) $\mu(A \setminus \{X^+ \geq 0\}) = 0$.

Proof. (1) implies (2) by definition of ν^+. Since $\nu(A) = \nu^+(A) - \nu^-(A)$, it is obvious that (2) implies (3). If $B \subset A$, then $0 \leq \nu^-(B) = \int_B X^- d\mu \leq \int_A X^- d\mu$. Hence, if $\nu^-(A) = 0$, it follows that $\nu^-(B) = 0$ and so $\nu(B) = \nu^+(B) \geq 0$, i.e., (3) implies (4).

Assume (4) and let $B = A\setminus\{X^+ \geq 0\}$. Then $B = \cup_{n=1}^{\infty} B_n$, where $B_n = A \cap \{X^- \geq \frac{1}{n}\} = A \cap \{X \leq -\frac{1}{n}\}$. If $\mu(B) > 0$, then for some n, $\mu(B_n) > 0$. Now $\nu(B_n) = \int_{B_n} X d\mu \leq -\frac{1}{n}\mu(B_n) < 0$, which contradicts (4).

Finally, (5) implies (1). Since $A\setminus\{X^+ \geq 0\} = A \cap \{X^- > 0\}$ has μ-measure zero, it follows that $\int_A X d\mu = \int_A 1_{\{X^+ \geq 0\}} X d\mu = \int_A X^+ d\mu = \nu^+(A)$. □

Note that condition (4) in this proposition is formulated in terms of the finite signed measure ν alone. It motivates the following definition.

Definition 2.7.4. Let ν be a finite signed measure on a measurable space $(\Omega, \mathfrak{F})$. A set $A \in \mathfrak{F}$ is said to be ν-**positive** or **positive** if $\nu(B) \geq 0$ for all $\mathfrak{F}$-measurable subsets B of A. It is said to be ν-**negative** or **negative** if $\nu(B) \leq 0$ for all $\mathfrak{F}$-measurable subsets B of A.

Remark. Obviously, a set is ν-negative if and only it is $(-1)\nu$-positive.

Using this terminology, Proposition 2.7.3 implies that if $d\nu = X d\mu$ (i.e., if $\nu(A) = \int_A X d\mu$ for all $A \in \mathfrak{F}$), then a set is ν-positive if and only if it is a subset of $\{X \geq 0\}$ except for a set of μ-measure zero.

Given a subset $E \in \mathfrak{F}$ that is ν-positive, the restriction of ν to E is a finite (non-negative) measure as the next exercise shows.

Exercise 2.7.5. Let ν be a finite signed measure on a measurable space $(\Omega, \mathfrak{F})$ and $E \in \mathfrak{F}$ be a ν-positive set. Show that the set function $\eta(A) \stackrel{\text{def}}{=} \nu(A \cap E)$ is a finite measure on $(\Omega, \mathfrak{F})$.

This suggests that one way to decompose a finite signed measure ν into the difference of two finite measures is to look for a largest ν-positive set E and see if the difference between ν and its restriction to E is a finite measure. The first thing to check is whether in fact, given a finite signed measure ν, there are any ν-positive sets. Let $E \in \mathfrak{F}$, and assume that $A \subset E$. Then $\nu(E) = \nu(A) + \nu(E\setminus A)$. If the set E is ν-positive, then it is clear that $\nu(E) \geq \nu(A)$. This property characterizes ν-positive sets as stated in the next exercise.

Exercise 2.7.6. Show that a set $E \in \mathfrak{F}$ is ν-positive if and only if, for any subset $A \in \mathfrak{F}$ of E, one has $\nu(E) \geq \nu(A)$.

It is therefore of interest to investigate the set function $\overline{V}(E)$, where $\overline{V}(E) \stackrel{\text{def}}{=} \sup_{A \subset E} \nu(A)$ for $E \in \mathfrak{F}$. It is called the **upper variation of** ν (see Halmos [H1], p. 122 and Wheeden and Zygmund [W1], p. 164).

Exercise 2.7.7. Show that
(1) the upper variation of a finite signed measure ν is a measure [*Hint:* first show it is countably subadditive (Definition 1.4.10).],
(2) the upper variation of a (non-negative) measure μ coincides with μ,

7. SIGNED MEASURES AND THE RADON–NIKODYM THEOREM

(3) the upper variation of a finite signed measure equals zero if the finite signed measure is $-\mu$, with μ a finite (non-negative) measure.

In fact, not only is the upper variation of ν a measure, it is a finite measure, as the next proposition shows.

Proposition 2.7.8. *If ν is a finite signed measure, then $\overline{V}(\Omega) < \infty$.*

The key to proving this result is the following lemma.

Lemma. *Let ν be a finite signed measure, and assume that $A \in \mathfrak{F}$ has $\overline{V}(A) = +\infty$. If $\alpha > 0$, there is a subset B of A with (i) $\overline{V}(B) = +\infty$ and (ii) $\nu(B) \geq \alpha$.*

Proof of the lemma. Since $\overline{V}(A) = +\infty$, there is a subset B_1 of A with $\nu(B_1) > \alpha$. As $\overline{V}$ is a measure, either (i) $\overline{V}(B_1) = +\infty$ or (ii) $\overline{V}(B_1) < +\infty$. Assume that the second alternative holds (otherwise B_1 is a set with the desired property). Then $\overline{V}(A \backslash B_1) = +\infty$, and hence there is a subset B_2 of $A \backslash B_1$ with $\nu(B_2) \geq \alpha$. Again assume that the second alternative holds and continue in this way, always assuming the second alternative holds. Then there is a sequence of pairwise disjoint subsets $(B_k)_{k \geq 1}$ of A with $\nu(B_k) \geq \alpha$. Since ν is finite-valued, this is impossible, as $\nu(\cup_{k \geq 1} B_k) = \sum_{k \geq 1} \nu(B_k) = +\infty$.

It follows, therefore, that there is a subset B of A with the desired property. □

Proof of Proposition 2.7.8. If $\overline{V}(\Omega) = +\infty$, then by the lemma there is a subset A_1 with (i) $\overline{V}(A_1) = +\infty$ and (ii) $\nu(A_1) \geq 1$. Using the lemma once again, one gets a subset A_2 of A_1 with (i) $\overline{V}(A_2) = +\infty$ and $\nu(A_2) \geq \nu(A_1)+1$. Continuing in this way, one obtains a sequence $(A_k)_{k \geq 1}$ of subsets of E with $A_k \supset A_{k+1}$ and $\nu(A_{k+1}) \geq \nu(A_k) + 1$ for all $k \geq 1$. Let $A_\infty = \cap_{n=1}^\infty A_n$, and let $A = A_1 \backslash A_\infty = \cup_{k=1}^\infty A_k \backslash A_{k+1}$. Once again, since $\nu(A)$ is finite, this leads to a contradiction, as $\nu(A_k) = \nu(A_k \backslash A_{k+1}) + \nu(A_{k+1})$ implies that $\nu(A_k \backslash A_{k+1}) \leq -1$ and so $\nu(A) = \sum \nu(A_k \backslash A_{k+1}) = -\infty$. □

Proposition 2.7.9. *Let ν be a finite signed measure that takes on both negative and positive values. If $E_1 \in \mathfrak{F}$ and $\nu(E_1) > 0$, then there is a subset E of E_1 with $\nu(E) > 0$ that is ν-positive.*

Proof. The idea of the proof is to remove successively from E_1 sets of negative measure that are as large as possible. What is left over will then turn out to be the desired ν-positive set. Of course, if by luck every subset of E_1 has non-negative ν-measure, then one takes E to be E_1.

The concept of "as large as possible" amounts to the following: choose a measurable subset A_1 of E_1 (i.e., one belonging to $\mathfrak{F}$), with $\nu(A_1) \leq -\frac{1}{n_1}$ where $n_1 \geq 1$ is as small as possible. If $E_1 \backslash A_1$ is a ν-positive set, it has positive ν-measure as $0 < \nu(E_1) = \nu(A_1) + \nu(E_1 \backslash A_1)$, and one sets $E = E_1 \backslash A_1$. If $E_1 \backslash A_1$ is not a positive set, then again choose a subset A_2

of negative measure that is less than or equal to $-\frac{1}{n_2}$ where $n_2 \geq 1$ is as small as possible: note that $n_2 \geq n_1$. If $E_1 \setminus (A_1 \cup A_2)$ is not ν-positive, continue, by choosing as large as possible a subset A_3 of E_1 disjoint from $A_1 \cup A_2$ that has negative ν-measure, and so on. If this procedure stops at the ℓth stage, then $E_1 \setminus \left(\cup_{k=1}^{\ell} A_k \right)$ is a ν-positive set of positive ν-measure.

On the other hand, if it never stops, then one has a sequence $(A_k)_{k \geq 1}$ of pairwise disjoint subsets A_k of E_1 with $\nu(A_k) \leq -\frac{1}{n_k}$, where n_k is as small as possible. Let $A = \cup_{k=1}^{\infty} A_k$. Then $-\infty < \nu(A) = \sum_{k=1}^{\infty} \nu(A_k) \leq -\sum_{k=1}^{\infty} \frac{1}{n_k}$. This implies that the series $\sum_{k=1}^{\infty} \frac{1}{n_k}$ converges, and hence $\lim_k n_k = +\infty$.

Now suppose that $B \subset E_1$ is disjoint from A. If $\nu(B) < 0$, then there is an integer $k \geq 1$ with $\nu(B) \leq -\frac{1}{k}$. This means that at every stage of the above procedure, $\nu(A_k) \leq -\frac{1}{k}$ (i.e., $\frac{1}{n_\ell} \geq \frac{1}{k}$). This contradicts the fact that the series $\sum_{k=1}^{\infty} \frac{1}{n_k}$ converges. Hence, $\nu(B) \geq 0$, i.e., $E = E_1 \setminus A$ is a ν-positive set and, since $0 < \nu(E_1) = \nu(A) + \nu(E)$, it follows that $\nu(E) > 0$. □

Without loss of generality, for a finite signed measure one can assume $\nu(\Omega) > 0$ (if not, consider $-\nu$).

The basic idea of the above proof is now used to determine a measurable subset of Ω that is ν-positive and has a ν-negative complement, a so-called **Hahn decomposition** of Ω relative to ν. The argument uses the following observation.

Exercise 2.7.10. Let (A_n) be a sequence of pairwise disjoint ν-positive sets. Then $A = \cup_{n=1}^{\infty} A_n$ is also ν-positive.

Theorem 2.7.11. (Decompositions associated with a finite signed measure) Let ν be a finite signed measure on $(\Omega, \mathfrak{F})$.

(1) There is a set $E^+ \in \mathfrak{F}$ that is positive and whose complement $\Omega \setminus E^+ \stackrel{\text{def}}{=} E^-$ is negative (*this is a* **Hahn decomposition** *of Ω relative to ν*).

(2) If ν^+ denotes the restriction of ν to E^+ and ν^- denotes the restriction of $(-1)\nu$ to E^-, then $\nu^\pm$ are finite measures and $\nu = \nu^+ - \nu^-$ (*see the* **Jordan decomposition** *of ν below*).

Proof. Statement (2) is obvious given the existence of the sets E^+ and E^-: each set $A = A \cap E^+ \cup A \cap E^-$; by Exercise 2.7.5, $\nu^\pm$ are finite measures and $\nu(A) = \nu(A \cap E^+) + \nu(A \cap E^-) = \nu^+(A) - \nu^-(A)$.

It suffices to verify (1) when $\nu(\Omega) > 0$. To determine the set E^+, let A_1 be a measurable subset of Ω that is (i) ν-positive with positive ν-measure and (ii) as large as possible (i.e., $\nu(A_1) \geq \frac{1}{n_1}$, where $n_1 \geq 1$ is as small as possible). If $\Omega \setminus A_1$ is not ν-negative, let A_2 be a subset that is (i) ν-positive and (ii) as large as possible. If this procedure stops at the ℓth stage set $E^+ = \cup_{k=1}^{\ell} A_k$ and $E^- = \Omega \setminus E^+$. By Exercise 2.7.10, these sets have the desired properties.

7. SIGNED MEASURES AND THE RADON-NIKODYM THEOREM

If the procedure does not stop, then one has, as in the proof of Proposition 2.7.9, a sequence $(A_k)_{k\geq 1}$ of pairwise disjoint ν-positive sets A_k with $\nu(A_k) \geq \frac{1}{n_k}$ and n_k as defined above. Since $\sum_{k=1}^{\infty} \nu(A_k) = \nu(A) < +\infty$, the series $\sum_{k=1}^{\infty} \frac{1}{n_k}$ converges: as a result, $\lim_k n_k = +\infty$. If B is disjoint from A, then by an argument similar to the one used in the proof of Proposition 2.7.9, one has $\nu(B) \leq 0$. Set $E^+ = A$. By Exercise 2.7.10, it has the desired property. □

The measures $\nu^{\pm}$ are given by the upper and lower variations of ν, as shown in the next exercise. For this reason, the decomposition of ν is its so-called **Jordan decomposition** $\nu = \overline{V} - \underline{V}$.

Exercise 2.7.12. Show that

(1) the measure ν^+ equals $\overline{V}$.

Define the **lower variation** $\underline{V}$ of ν by setting $\underline{V}(E) = -\inf_{A \subset E} \nu(A) = \sup_{A \subset E} -\nu(A)$. Show that

(2) $\nu^- = \underline{V}$; and hence, conclude that

(3) $\nu = \overline{V} - \underline{V}$ (the **Jordan decomposition** of ν).

Note that the **total variation** V of ν is defined to be $\overline{V} + \underline{V}$ (see [H1], p. 122 and [W1], p. 164). In case $d\nu = X d\mu$ with $X \in L^1(\Omega, \mathfrak{F}, \mu)$, show that

(4) $d\overline{V} = X^+ d\mu$ (i.e., $\overline{V}(A) = \int_A X^+ d\mu$),

(5) $d\underline{V} = X^- d\mu$ (i.e., $\underline{V}(A) = \int_A X^- d\mu$), and

(6) $dV = |X| d\mu$ (i.e., $V(A) = \int_A |X| d\mu$).

Remark. These decompositions exist for σ-finite signed measures η. One decomposes the whole space Ω into a sequence of disjoint measurable sets A_n on each of which the restriction of η is a finite signed measure ν_n, i.e., a finite signed measure. On each of these sets, Hahn and Jordan decompositions for ν_n may be defined, and one merely puts them together in the obvious way to get a decomposition for the whole space. For example, if $A_n = A_n^+ \cup A_n^-$ is a Hahn decomposition for ν_n, then $A^+ = \cup_n A_n^+$ and $A^- = \cup_n A_n^-$ give a Hahn decomposition of Ω for η.

Definition 2.7.13. Let μ_1 and μ_2 be two measures defined on a measure space $(\Omega, \mathfrak{F})$. They are said to be **mutually singular** if there is no set $A \in \mathfrak{F}$ with both $\mu_1(A) > 0$ and $\mu_2(A) > 0$.

It follows from Theorem 2.7.11 that the two measures $\nu^+ = \overline{V}$ and $\nu^- = \underline{V}$ are mutually singular. This property characterizes the Jordan decomposition of a finite signed measure ν.

Exercise 2.7.14. Let $\nu = \nu_1 - \nu_2 = \mu_1 - \mu_2$ be a finite signed measure where $\nu_1, \nu_2, \mu_1,$ and μ_2 are finite measures. Then $\nu_1 = \mu_1$ and $\nu_2 = \mu_2$ if ν_1 and ν_2, as well as μ_1 and μ_2, are pairs of mutually singular measures.

Hence, not only is every finite signed measure the difference of two finite measures, but in addition the measures in this decomposition are unique and "live" on disjoint sets if they are mutually singular.

Hahn decompositions of the underlying set Ω into positive and negative parts are by no means unique, because a set of measure zero need not be empty (see the following exercise).

Exercise 2.7.15. Let ν denote the finite signed measure $\varepsilon_1 - \varepsilon_{-1}$ on $(\mathbb{R}, \mathfrak{B}(\mathbb{R}))$. Determine all the Hahn decompositions of $\mathbb{R}$ associated with ν.

The Radon–Nikodym theorem.

An essential feature of a finite signed measure ν given by $X \in L^1(\Omega, \mathfrak{F}, \mu)$ is that $\nu(A) = 0$ if $\mu(A) = 0$.

Definition 2.7.16. *Let μ be a measure on a measurable space $(\Omega, \mathfrak{F})$. A finite signed measure ν on $(\Omega, \mathfrak{F})$ is said to be* **absolutely continuous with respect to** *μ if $\mu(A) = 0$ implies that $\nu(A) = 0$. This is denoted by writing $\nu \ll \mu$.*

In other words, the finite signed measures determined by L^1-functions are all absolutely continuous with respect to the basic measure μ. The Radon–Nikodym theorem states that the converse is true.

Remark 2.7.17. Note that a finite signed measure is absolutely continuous with respect to μ if and only if $\nu^\pm$ are both absolutely continuous with respect to μ.

To prove the Radon–Nikodym theorem, it suffices to prove it for a (nonnegative) measure. If $X \in L^1(\mu)$ is non-negative, the measure $d\nu = X d\mu$ clearly has the property that if $F_{n,k} = \{\frac{k}{2^n} \leq X < \frac{k+1}{2^n}\}$, where $k \geq 0$, then $\frac{k}{2^n}\mu(A \cap F_{n,k}) \leq \nu(A \cap F_{n,k}) \leq \frac{k+1}{2^n}\mu(A \cap F_{n,k})$.

Given the connection between a Hahn decomposition for $\nu - \frac{k}{2^n}\mu$ and the sets $\{\frac{k}{2^n} \leq X\}, \{X < \frac{k}{2^n}\}$, the above suggests a way to approximate the measure ν by measures defined by simple functions s_n.

Let ν be any finite measure, and let D denote the set of positive dyadic rationals (i.e., $r \in D$ if and only if $r = \frac{k}{2^n}$ for some $k, n \in \mathbb{N}$). Assume that for each $\alpha \in D$ it is possible to determine a Hahn decomposition $\Omega = E_\alpha^+ \cup E_\alpha^-$ for $\xi = \nu - \alpha\mu$ such that if $\alpha_1 < \alpha_2$, then $E_{\alpha_1}^+ \supset E_{\alpha_2}^+$ (while it is not totally obvious that this monotone condition can be satisfied, it is not hard to find such a countable family of Hahn decompositions, as shown later in Proposition 2.7.28).

Since ν is a measure (i.e., it is non-negative), the Hahn decomposition corresponding to 0 can be assumed to be $\Omega = E_0^+$ and $\emptyset = E_0^-$.

The function s_n is constructed from the Hahn decompositions associated with D_n^+, the set of non-negative dyadic rationals of the form $\frac{k}{2^n}, 1 \leq k \leq 2^{2n}, n \geq 1$.

Let
$$F_{n,0} = E_{\frac{1}{2^n}}^-,$$
$$F_{n,k} = E_{\frac{k}{2^n}}^+ \setminus E_{\frac{k+1}{2^n}}^+ = E_{\frac{k}{2^n}}^+ \cap E_{\frac{k+1}{2^n}}^-, \text{ for } 1 \leq k < 2^{2n},$$
$$F_{n,2^{2n}} = E_{2^n}^+.$$

7. SIGNED MEASURES AND THE RADON–NIKODYM THEOREM

These sets are pairwise disjoint, and for any $A \in \mathfrak{F}$, it follows that, for $0 \le k < 2^{2n}$,

$$\frac{k}{2^n}\mu(A \cap F_{n,k}) \le \nu(A \cap F_{n,k}) \le \frac{k}{2^n}\mu(A \cap F_{n,k}) + \frac{1}{2^n}\mu(A \cap F_{n,k}).$$

Define
$$s_n(\omega) = \begin{cases} 0 & \text{if } \omega \in F_{n,0}, \\ \frac{k}{2^n} & \text{if } \omega \in F_{n,k}, \text{ for } 1 \le k < 2^{2n}, \\ 2^n & \text{if } \omega \in F_{n,2^{2n}}. \end{cases}$$

It follows that, for any $A \in \mathfrak{F}$, one has

(†) $$\int_{A \cap E_{2^n}^-} s_n \, d\mu \le \nu(A \cap E_{2^n}^-) \le \int_{A \cap E_{2^n}^-} s_n \, d\mu + \frac{1}{2^n}\mu(\Omega),$$

where $E_{2^n}^- = (E_{2^n}^+)^c = \cup_{0 \le k < 2^{2n}} F_{n,k}$.

Since $E_{\frac{k}{2^n}}^+ \supset E_{\frac{2k+1}{2^{n+1}}}^+ \supset E_{\frac{k+1}{2^n}}^+$, it follows that $s_n \le s_{n+1}$ for all $n \ge 1$. Let $X = \lim_n s_n$. It follows from (†) that $\int X_{A \cap E_{2^n}^-} \, d\mu = \nu(A \cap E_{2^n}^-)$ for any $A \in \mathfrak{F}$. Let E_f denote the union of all the sets $E_{2^n}^-, n \ge 1$. This is the set on which X is finite.

Lemma 2.7.18. *Let E_f denote $E_{+\infty}^- = \cup_n E_{2^n}^-$. Then $\mu(E_f) = \mu(\Omega)$, i.e., $\mu(E_f^c) = 0$.*

Proof. If $B \subset E_{2^n}^+$, then $\nu(B) \ge 2^n \mu(B)$. Since $\nu(B) < +\infty$, this implies that $\mu(B) = 0$ if $B \subset \cap_n E_{2^n}^+$. Hence, $\mu(E_f^c) = 0$. □

This completes most of the proof of the following theorem.

Theorem 2.7.19. (Radon–Nikodym theorem: measures) *Let μ be a finite measure on a measure space $(\Omega, \mathfrak{F})$ and ν be another finite measure that is absolutely continuous with respect to μ (i.e., $\nu \ll \mu$). Then there is a non-negative function $X \in L^1(\Omega, \mathfrak{F}, \mu)$ such that*

(*) $$\int_A X \, d\mu = \nu(A) \text{ for all } A \in \mathfrak{F}.$$

Furthermore, X is unique in the sense that if $Y \in \mathfrak{F}$ satisfies (*), then $X = Y$ μ-a.e. (i.e., $\mu(\{\omega \mid X(\omega) = Y(\omega)\}) = 0$).

Proof. Using the previous notations, $\nu(E_f^c) = 0$ since $\nu \ll \mu$ and so, for any set $A \in \mathfrak{F}$, one has

$$\nu(A) = \lim_n \nu(A \cap E_{2^n}^-) = \lim_n \int_{A \cap E_{2^n}^-} X \, d\mu = \int_A X \, d\mu.$$

Since ν is finite, this also implies that $X \in L^1(\mu)$.

Assume that Y satisfies (*), and let $A = \{Y \ge X - \frac{1}{n}\}$. Then $\int_A X \, d\mu = \int_A Y \, d\mu \ge \int_A X \, d\mu - \frac{1}{n}\mu(A)$. Hence, $\mu(A) = 0$ and so $Y \ge X$ μ-a.e. By symmetry, it follows that $X = Y$ μ-a.e. □

Corollary 2.7.20. (Radon–Nikodym Theorem: finite signed measures) Let μ be a finite measure on a measure space $(\Omega, \mathfrak{F})$ and ν be a finite signed measure that is absolutely continuous with respect to μ (i.e., $\nu \ll \mu$). Then there is a function $X \in L^1(\Omega, \mathfrak{F}, \mu)$ such that

(*) $$\int_A X\,d\mu = \nu(A) \quad \text{for all } A \in \mathfrak{F}.$$

Furthermore, X is unique in the sense that if $Y \in \mathfrak{F}$ satisfies (*), then $X = Y$ μ – a.e. (i.e., $\mu(\{\omega \mid X(\omega) = Y(\omega)\}) = 0$).

Proof. Let $\nu = \nu^+ - \nu^-$ denote the Jordan decomposition of ν. Then, as pointed out in Remark 2.7.17, both $\nu\pm$ are absolutely continuous. Let $d\nu^+ = X_1 d\mu$ and $d\nu^- = X_2 d\mu$. Then $X = X_1 - X_2 \in L^1(\mu)$ and satisfies (*). Uniqueness is proved as in the case of measures. □

Remark. The Radon–Nikodym theorem is usually stated when the basic measure μ is assumed to be σ-finite and ν is a finite signed measure. The extension from the case when μ is finite to the more general σ-finite case may be carried out by cutting Ω up into a countable collection of pairwise disjoint sets of finite μ-measure and applying Corollary 2.7.20 to each piece.

Exercise 2.7.21. Prove the Radon–Nikodym theorem when μ is σ-finite.

The Lebesgue decomposition of a finite signed measure with respect to a finite measure.

The basic argument just used to prove the Radon–Nikodym theorem only made use of the hypothesis of absolute continuity to guarantee that $\nu(A) = \nu(A) \cap E_f)$. In fact, if ν is any finite measure, one always has

(†) $$\nu(A \cap E_f) = \lim_n \nu(A \cap E_{2^n}^-) = \lim_n \int_{A \cap E_{2^n}^-} X\,d\mu = \int_A X\,d\mu$$

since $\mu(E_f^c) = 0$ by Lemma 2.7.18.

Let $\eta(A) = \nu(A \cap E_f^c)$. Then η is a finite measure that is singular with respect to μ, and so one has

$$\nu(A) = \eta(A) + \nu(A \cap E_f) = \eta(A) + \int_A X\,d\mu \text{ for any } \in \mathfrak{F}.$$

This proves the following result.

Theorem 2.7.22. (Lebesgue decomposition theorem: measures) Let μ be a finite measure on a measure space $(\Omega, \mathfrak{F})$ and ν be another finite measure. Then there is a unique decomposition of ν as a sum

(*) $$\nu = \xi + \eta$$

of an absolutely continuous measure ξ and a singular measure η.

7. SIGNED MEASURES AND THE RADON–NIKODYM THEOREM

In terms of the notations used earlier, $\eta(A) = \nu(A \cap E_f^c)$ and $\xi(A) = \int_A X d\mu$ with X a non-negative function $\in L^1(\Omega, \mathfrak{F}, \mu)$, i.e.,

$$\nu(A) = \nu(A \cap E_f) + \eta(A) = \int_A X d\mu + \eta(A) \text{ for any } A \in \mathfrak{F}.$$

The $L^1(\mu)$-function X is unique modulo a null function.

Proof. If $\xi(A) = \nu(A \cap E_f)$, then by (†) it follows that η and ξ are two mutually singular measures whose sum is ν with $\xi \ll \mu$ and η singular with respect to μ.

Assume that ν is the sum $\eta_1 + \xi_1$ of two mutually singular measures with $\xi_1 \ll \mu$ and η_1 singular with respect to μ.

Lemma 2.7.23. *Let $\nu = \eta_1 + \xi_1$ with η_1 singular with respect to μ. Let E be a set in $\mathfrak{F}$ on which $\nu - \alpha\mu$ is negative for some $0 < \alpha + \infty$. Then $\eta_1(B) = 0$ for all $B \subset E$.*

Proof. Since $\eta_1(B) \le \nu(B) \le \alpha\mu(B)$, the fact that η_1 and μ are singular implies that $\eta_1(B) = 0$. □

It follows from this lemma that for any $B \subset E_{2^n}^-$ one has $\eta_1(B) = 0$. Hence, for all $B \subset E_f$, it follows that $\eta_1(B) = 0$.

As a result, for any $A \in \mathfrak{F}$, it follows that $\nu(A \cap E_f) = \eta_1(A \cap E_f) + \xi_1(A \cap E_f) = \xi_1(A \cap E_f)$. Since $\xi_1 \ll \mu$ and $\mu(E_f^c) = 0$, it follows that $\xi_1(A) = \xi_1(A \cap E_f) + \xi_1(A \cap E_f^c) = \xi_1(A \cap E_f)$. Hence, $\xi_1 = \xi$ and so $\eta_1 = \eta$. The uniqueness of X follows in the usual way, or by appeal to the Radon–Nikodym theorem (Theorem 2.7.19). □

Exercise 2.7.24. Formulate and prove the Lebesgue decomposition theorem (i) when ν is a finite signed measure with μ finite and (ii) when μ is σ-finite and ν is a finite signed measure.

Exercise 2.7.25. Let $\mathbf{P}$ be a probability on $(\mathbb{R}, \mathfrak{B}(\mathbb{R}))$ that is neither absolutely continuous nor singular. Show that there is a unique probability $\mathbf{P}_s$ that is singular with respect to Lebesgue measure dx, a unique non-negative L^1-function f with $\int f(x) dx = 1$, and a unique $\lambda \in [0, 1]$ such that, for any Borel set A, one has

$$\mathbf{P}(A) = \lambda \int_A f(x) dx + (1 - \lambda) \mathbf{P}_s(A).$$

In the terminology of the next section, the probability determined by the L^1-function f is absolutely continuous with respect to Lebesgue measure. In other words, every probability that is neither absolutely continuous nor singular is a unique convex combination of a singular probability and an absolutely continuous one, i.e., if $\mathbf{P}_{ac}(A) = \int_A f(x) dx$, then $\mathbf{P}$ has a unique representation

$$\mathbf{P} = \lambda \mathbf{P}_{ac} + (1 - \lambda) \mathbf{P}_s$$

as a convex combination of an absolutely continuous probability and a singular probability.

Countable families of Hahn decompositions.

It remains to show that the basic assumption made above can be verified: namely, that for each $\alpha \in D$, the set of dyadic rationals, it is possible to determine a Hahn decomposition $\Omega = E_\alpha^+ \cup E_\alpha^-$ for $\xi = \nu - \alpha\mu$ such that if $\alpha_1 < \alpha_2$, then $E_{\alpha_1}^+ \supset E_{\alpha_2}^+$.

The first step is to see how Hahn decompositions of $\xi = \nu - \alpha\mu$ may be chosen to be "compatible" in the above sense for two values of α.

Exercise 2.7.26. Let E_α^+ denote the positive subset of a Hahn decomposition for $\nu - \alpha\mu$. Show that

(1) if $\alpha_1 < \alpha_2$, then $\mu(E_{\alpha_2}^+ \setminus E_{\alpha_1}^+) = 0$ [Hint: if $A \subset E_\alpha^+$, then $\nu(A) \geq \alpha\mu(A)$, and if $A \subset E_\alpha^-$, then $\nu(A) \leq \alpha\mu(A)$],
(2) if A is $(\nu - \alpha_1\mu)$-negative, it is $(\nu - \alpha_2\mu)$-negative,
(3) $E_{\alpha_1}^+ \cap E_{\alpha_2}^+$ is the positive set for a Hahn decomposition associated with $\nu - \alpha_2\mu$, and
(4) $E_{\alpha_1}^- \cup E_{\alpha_2}^-$ is the negative set for a Hahn decomposition associated with $\nu - \alpha_2\mu$.

Therefore, the original Hahn decomposition $E_{\alpha_2}^+$ and $E_{\alpha_2}^-$ for $\nu - \alpha_2\mu$, where $\alpha_1 < \alpha_2$, may be replaced by $E_{\alpha_1}^+ \cap E_{\alpha_2}^+$ and $E_{\alpha_1}^- \cup E_{\alpha_2}^-$. The modification amounts to observing that the union of the two $(\nu - \alpha_2\mu)$-negative sets $E_{\alpha_2}^-$ and $E_{\alpha_1}^-$ is also a $(\nu - \alpha_2\mu)$-negative set. This observation about the union of negative sets may be extended to countable unions, as indicated in the following exercise.

Exercise 2.7.27. Let ξ be a finite signed measure and $(B_k)_{k \geq 1}$ be a sequence of ξ-negative sets. Show that $\cup_{k \geq 1} B_k$ is a ξ-negative set. [Hint: express the union as a countable disjoint union of sets (see Remark 1.2.3).]

Proposition 2.7.28. Let D be a countable subset of $\mathbb{R}$. For each $\alpha \in D$, there is a Hahn decomposition of $\nu - \alpha\mu$ such that if E_α^+ is the positive set of this decomposition, then $\alpha_1 < \alpha_2$ implies $E_{\alpha_1}^+ \supset E_{\alpha_2}^+$ for $\alpha_1, \alpha_2 \in D$.

Proof. For each $\alpha \in D$, let E'_α be the positive set of some Hahn decomposition associated with $\nu - \alpha\mu$. Define E_α^+ to be $\cap_{\beta \in D, \beta \leq \alpha} E'_\beta$. Then E_α^+ is $(\nu - \alpha\mu)$-positive, as it is a subset of the $(\nu - \alpha\mu)$-positive set E'_α and its complement $\left(E_\alpha^+\right)^c = \cup_{\beta \in D, \beta \leq \alpha} \left(E'_\beta\right)^c$. Since $\left(E'_\beta\right)^c$ is $(\nu - \alpha\mu)$-negative by Exercise 2.7.26 (2), if $\beta \leq \alpha$, it follows from Exercise 2.7.27 that $\left(E'_\beta\right)^c$ is $(\nu - \alpha\mu)$-negative. Let E_α^- equal $\left(E'_\beta\right)^c$.

Then $E_\alpha^+ \cup E_\alpha^-$ is a Hahn decomposition of $\nu - \alpha\mu$. The "compatibility" property holds by construction. □

8. FINITE SIGNED MEASURES ON $\mathbb{R}$ AND FUNCTIONS OF BOUNDED VARIATION*

Functions of bounded variation.
Let $\nu = \nu^+ - \nu^-$ be a finite signed measure on $\mathfrak{B}(\mathbb{R})$, and define

$$\Gamma(x) = \begin{cases} \nu((0,x]) & \text{if } 0 < x, \\ 0 & \text{if } x = 0, \\ -\nu((x,0]) & \text{if } x < 0. \end{cases}$$

Then, if $-\infty < a < b < +\infty$, it follows that $\Gamma(b) - \Gamma(a) = \nu((a,b])$. Furthermore, while Γ is right continuous, it is not necessarily non-decreasing, as the finite signed measure ν is not necessarily a measure.

On the other hand, the measures $\nu^\pm$ define non-decreasing, right continuous functions $G^\pm$ by Theorem 2.2.2:

$$G^\pm(x) = \begin{cases} \nu^\pm((0,x]) & \text{if } 0 < x, \\ 0 & \text{if } x = 0, \\ -\nu^\pm((x,0]) & \text{if } x < 0. \end{cases}$$

It follows from these definitions that $\Gamma = G^+ - G^-$. This formula has the following proposition as an immediate consequence since $t_{i-1} < t_i$ implies that

$$|\Gamma(t_i) - \Gamma(t_{i-1})| \leq \{G^+(t_i) - G^+(t_{i-1})\} + \{G^-(t_i) - G^-(t_{i-1})\}$$
$$= \nu^+((t_{i-1}, t_i]) + \nu^-((t_{i-1}, t_i]).$$

Proposition 2.8.1. *Let $\pi : t_0 = a < t_1 < t_2 < \cdots < t_n = b$ be a finite partition of the bounded interval $[a, b]$. Then*

$$\sum_{i=1}^{n} |\Gamma(t_i) - \Gamma(t_{i-1})| \leq \{G^+(b) - G^+(a)\} + \{G^-(b) - G^-(a)\}$$
$$= \nu^+((a,b]) + \nu^-((a,b]).$$

Since the upper bound $\nu^+((a,b]) + \nu^-((a,b])$, while dependent upon the interval, does not depend upon the partition, the function Γ is of **bounded variation** on every finite interval $[a,b]$: in point of fact, in this instance, the upper bound $\nu^+((a,b]) + \nu^-((a,b]) \leq \nu^+(\mathbb{R}) + \nu^-(\mathbb{R}) < \infty$ since a finite signed measure is the difference of two finite measures.

Definition 2.8.2. *Let Φ be a function defined on $[a,b]$. It is said to have* **bounded variation** *or to be of bounded variation if there is a constant $M > 0$ such that, for any partition $\pi : t_0 = a < t_1 < t_2 < \cdots < t_n = b$ of $[a,b]$, one has*
$$\sum_{i=1}^{n} |\Phi(t_i) - \Phi(t_{i-1})| \stackrel{\text{def}}{=} V(\Phi, \pi) \leq M.$$
The least upper bound of $\{V(\Phi, \pi) \mid \pi \text{ a partition of } [a,b]\}$ is defined to be the **total variation** *of Φ on $[a,b]$ and will be denoted by $V(\Phi, [a,b])$.*

Remark. If π' refines π (i.e., if it contains all the partition points of π), then $V(\Phi, \pi) \leq V(\Phi, \pi')$.

It will be shown in what follows that every function of bounded variation is the difference of two non-decreasing functions and that in the case of the function Γ associated with a finite signed measure, its total variation over a finite interval $[a,b]$ is exactly $\nu^+((a,b]) + \nu^-((a,b])$ (see Proposition 2.8.11).

The argument used to establish Proposition 2.8.1 shows that if Φ is a function defined on $[a,b]$ and if Φ is the difference of two non-decreasing functions (defined on $[a,b]$), then Φ is of bounded variation. The converse is stated as the following result.

Proposition 2.8.3. *Let Φ be a function of bounded variation on $[a,b]$, and let $V(\Phi)(x) \stackrel{\text{def}}{=} V(\Phi, [a,x])$. Then $V(\Phi)$ is a finite non-decreasing function on $[a,b]$ and $V(\Phi) \pm \Phi$ is also a non-decreasing function on $[a,b]$. Hence, $\Phi = V(\Phi) - \{V(\Phi) - \Phi\} = \{V(\Phi) + \Phi\} - V(\Phi)$ is the difference of two non-decreasing functions.*

This result is an easy consequence of the following lemma.

Lemma 2.8.4. *Let Φ be a function of bounded variation on the finite interval $[a,b]$. If $a < x < x + h \leq b$, then*
$$V(\Phi, [a, x+h]) = V(\Phi, [a,x]) + V(\Phi, [x, x+h]).$$

Proof. To begin, if π_1 and π_2 are two partitions contained in $[a,b]$ and if the partition π_1 ends at the first point of π_2, then their union $\pi_1 \cup \pi_2$ is a partition such that $V(\Phi, \pi_1) + V(\Phi, \pi_2) = V(\Phi, \pi_1 \cup \pi_2)$. From this, it follows that $V(\Phi, [a, x+h]) \geq V(\Phi, [a,x]) + V(\Phi, [x, x+h])$. Since one may always add the point x to any partition of $[a, x+h]$, it follows that $V(\Phi, [a, x+h]) \leq V(\Phi, [a,x]) + V(\Phi, [x, x+h])$. □

Proof of Proposition 2.8.3. First observe that, when $h > 0$, every partition π of $[a,x]$ can be extended to give a partition π_h of $[a, x+h]$ simply by adding the point $x+h$. Hence, $V(\Phi, \pi) \leq V(\Phi, \pi_h)$, which implies that $V(\Phi)(x) \leq V(\Phi)(x+h)$, i.e., $V(\Phi)$ is a non-decreasing function of x.

It follows from Lemma 2.8.4 that $V(\Phi)(x+h) - V(\Phi)(x) = V(\Phi, [x, x+h]) \geq |\Phi(x+h) - \Phi(x)| \geq \pm\{\Phi(x+h) - \Phi(x)\}$, and so $V(\Phi)(x) \pm \Phi(x)$

are two non-decreasing functions of x on $[a, b]$, the sign choice being fixed independent of x. □

While every function Φ of bounded variation is a difference of finite, non-decreasing functions, say H_1 and H_2, this representation is not unique since not only can a constant be added to a non-decreasing function without changing this property, but also any non-decreasing function may be added. Note that if $\Phi = H_1 - H_2$, then $V(\Phi) \leq H_1 + H_2$, and if $H_1 = \frac{1}{2}\{V(\Phi) + \Phi\}$ and $H_2 = \frac{1}{2}\{V(\Phi) - \Phi\}$, then the representation of Φ given by $\Phi = H_1 - H_2$ is such that $H_1 + H_2 = V(\Phi)$. The converse is true since, given γ and δ, if two numbers α and β are such that $\alpha + \beta = \gamma$ and $\alpha - \beta = \delta$, then $\alpha = \frac{1}{2}\{\gamma + \delta\}$ and $\beta = \frac{1}{2}\{\gamma - \delta\}$. This proves the following proposition.

Proposition 2.8.5. *Let Φ be a function of bounded variation on $[a, b]$ and $V(\Phi)(x) = V(\Phi, [a, x])$ be its total variation on $[a, x]$. Then, the representation of Φ given by*

$$\Phi(x) = \frac{1}{2}\{V(\Phi)(x) + \Phi(x)\} - \frac{1}{2}\{V(\Phi)(x) - \Phi(x)\}$$

is the unique representation of Φ as the difference of two non-decreasing functions whose sum is the total variation $V(\Phi)$.

Functions of bounded variation determine finite signed measures when they are right continuous. Any non-decreasing function H on $[a, b]$ may always be "regularized" by looking at the corresponding right continuous, non-decreasing function $\tilde{H}$, where $\tilde{H}(x) \stackrel{\text{def}}{=} \inf\{H(u) \mid x < u\}$.

Exercise 2.8.6. Show that $\tilde{H}$ is right continuous on $[a, b]$.

Consequently, every function Φ of bounded variation on $[a, b]$ may also be "regularized" by taking right limits, and so determines a finite signed measure.

Given a function f on a closed finite interval $[a, b]$, let its "natural" extension to all of $\mathbb{R}$ (also denoted by f) be defined by setting $f(x) = f(a)$ if $x < a$ and $f(x) = f(b)$ if $b < x$. With this convention, to each right continuous function Φ of bounded variation on a finite interval, there corresponds a right continuous function on $\mathbb{R}$ that is of bounded variation on every finite interval.

Exercise 2.8.7. Let Φ be a real-valued function on $\mathbb{R}$ whose restriction to every finite interval is of bounded variation. Show that

(1) if $\check{\Phi}(y) \stackrel{\text{def}}{=} -\Phi(-y)$, then $V(\check{\Phi}, [a, b]) = V(\Phi, [-b, -a])$,
(2) there are non-decreasing functions H_1 and H_2 on $\mathbb{R}$ such that $\Phi = H_1 - H_2$ [Hint: first define these functions for $0 < x < +\infty$, then use (1) and imitate the formula in Theorem 2.2.2.],
(3) if Φ is right continuous, the functions H_i in (2) can be assumed to be right continuous and "standardized" to have value zero at 0.

Further, if the total variation $V(\Phi, [a, b]) \leq M$ for all finite intervals $[a, b]$ for some constant $M > 0$, show that

(4) Φ is a bounded function and H_1 and H_2 may be taken to be bounded.

It follows from this exercise and Theorem 2.2.2, that to each right continuous function Φ on $\mathbb{R}$ for which the total variation $V(\Phi, [a, b])$ is uniformly bounded by a constant $M > 0$ (i.e., independent of the interval), there corresponds a finite signed measure μ on $\mathfrak{B}(\mathbb{R})$ such that $\mu((a, b]) = \Phi(b) - \Phi(a)$ for each finite interval $(a, b]$.

Remark. If one wants to include σ-finite signed measures in the discussion, then at least one of the non-decreasing functions in a representation of Φ has to be bounded. When Φ is a right continuous function on $\mathbb{R}$ of bounded variation on every finite interval, it suffices that there is a constant $M > 0$ such that for every finite interval $[a, b]$ one has either (i) $V(\Phi, [a, b]) - \{\Phi(b) - \Phi(a)\} \leq M$ or (ii) $V(\Phi, [a, b]) + \{\Phi(b) - \Phi(a)\} \leq M$.

The finite signed measure μ has a total variation $|\mu| \stackrel{\text{def}}{=} \mu^+ + \mu^-$. This total variation measure (Exercise 2.7.12) is related to the total variation of the function Φ: in fact, $|\mu|((a, b]) = V(\Phi, [a, b])$ for any finite interval $[a, b]$. This is not entirely obvious and will require some additional discussion of the variation $V(\Phi, \pi)$ given by any partition.

Lemma 2.8.8. If $\pi : t_0 = a < t_1 < t_2 < \cdots < t_n = b$ is a partition of the finite interval $[a, b]$, let $V^+(\Phi, \pi) \stackrel{\text{def}}{=} \sum_{i=1}^{n} \{\Phi(t_i) - \Phi(t_{i-1})\} \vee 0$ and $V^-(\Phi, \pi) \stackrel{\text{def}}{=} (-1)\left[\sum_{i=1}^{n} \{\Phi(t_i) - \Phi(t_{i-1})\} \wedge 0\right] = \sum_{i=1}^{n} [(-1)\{\Phi(t_i) - \Phi(t_{i-1})\}] \vee 0$. Then

(1) $V(\Phi, \pi) = V^+(\Phi, \pi) + V^-(\Phi, \pi)$,
(2) $V^+(\Phi, \pi) \leq \overline{V}(\mu)((a, b])$, and
(3) $V^-(\Phi, \pi) \leq \underline{V}(\mu)([(a, b])$,

where $\overline{V}(\mu)((a, b])$ is the value of the upper variation of the finite signed measure μ for the set $(a, b]$ and $\underline{V}(\mu)((a, b])$ is the value of the lower variation (Exercise 2.7.12) of the finite signed measure μ for the same set.

Hence,

(4) $V(\Phi, [a, b]) \leq V(\mu)((a, b])$,

where $V(\mu)((a, b]) = |\mu|((a, b])$ is the value of the total variation of the finite signed measure μ for the set $(a, b]$.

Proof. Identity (1) is obvious, since for any real number α one has by Exercise 1.1.16 that $|\alpha| = \alpha \vee 0 - \{\alpha \wedge 0\} = \alpha \vee 0 + \{-\alpha\} \vee 0$.

Let A denote the union of the intervals $(t_{i-1}, t_i]$ for which $\Phi(t_i) - \Phi(t_{i-1}) > 0$. Then $\mu(A) = V^+(\Phi, \pi)$ and so (2) $V^+(\Phi, \pi) \leq \overline{V}((a, b]) = \mu^+((a, b])$. Similarly, (3) $V^-(\Phi, \pi) \leq \underline{V}((a, b]) = \mu^-((a, b])$, and so by (1) $V(\Phi, \pi) \leq V((a, b]) = |\mu|((a, b])$, i.e., (4) holds. □

8. SIGNED MEASURES ON $\mathbb{R}$ AND FUNCTIONS OF BOUNDED VARIATION

An almost immediate consequence of this lemma and Lemma 2.8.4 is the following corollary.

Corollary 2.8.9. *Let Φ be a right continuous function on $\mathbb{R}$ that is of bounded variation on each finite interval. Then, for any real number a,*
(1) *the function $V(\Phi)(x) = V(\Phi, [a, x])$ is right continuous for $x > a$, and*
(2) *the function $V(\Phi, [x, a])$ is right continuous for $x < a$.*

Proof. Let $h > 0$. Lemma 2.8.4 implies that $|V(\Phi)(x) - V(\Phi)(x + h)| = V(\Phi, [x, x+h]) \leq |\mu|((x, x+h])$. To complete the argument for (1), note that $|\mu|((x, x+h])$ tends to zero as h tends to zero.

The argument for (2) is the same since $|V(\Phi, [x, a]) - V(\Phi, [x+h, a])| = V(\Phi, [x, x+h])$ as long as $0 < h < a - x$. $\square$

Exercise 2.8.10. Use Hahn and Jordan decompositions of μ to show that if the finite signed measure $\mu = \mu_1 - \mu_2$, then for any measurable set A one has $\mu_1(A) \geq \mu^+(A)$ and $\mu_2(A) \geq \mu^-(A)$. [*Hint*: if E^+ is the positive set of a Hahn decomposition, show that for any measurable set A, one has $\mu^+(A) \leq \mu_1(A \cap E^+)$.]

Proposition 2.8.11. *Let Φ be right continuous on $\mathbb{R}$ with uniformly bounded variation on each finite interval $[a, b]$, i.e., there is a constant $M > 0$ such that $V(\Phi, [a, b]) \leq M$ for all finite intervals $[a, b]$. Let μ be the corresponding finite signed measure. Then, for any finite interval $[a, b]$, $V(\Phi, [a, b]) = |\mu|((a, b])$.*

Proof. One may assume $a \geq 0$: if not, consider the right continuous function $\Phi_a(x) \stackrel{\text{def}}{=} \Phi(x + a)$. Let μ' denote the corresponding finite signed measure. Then $\mu'((c - a, d - a]) = \mu((c, d])$, $V(\Phi_a, [c - a, d - a]) = V(\Phi, [c, d])$, $|\mu'|((c - a, d - a]) = |\mu|(c, d])$, and $V(\Phi_a, [c - a, d - a]) = V(\Phi, [c, d])$.

Since $V(\Phi)(x) = V([0, x])$ is right continuous in x for $x \geq 0$, the representation of Φ (for $x \geq 0$) as $\Phi = \frac{1}{2}\{V(\Phi) + \Phi\} - \frac{1}{2}\{V(\Phi) - \Phi\}$ determines two measures μ_1 and μ_2 on $\mathbb{R}$ for which $\mu_1((-\infty, 0]) = \mu_2((-\infty, 0]) = 0$. They are the measures such that $\mu_1((a, b]) = \frac{1}{2}\{V(\Phi) + \Phi\}(b) - \frac{1}{2}\{V(\Phi) + \Phi\}(a)$ if $a \geq 0$ and $\mu_1((a, b]) = 0$ if $b \leq 0$, and μ_2 is similarly defined by $\frac{1}{2}\{V(\Phi) - \Phi\}$. It follows from Exercise 2.8.10 that $|\mu|((0, x]) \leq \mu_1((0, x]) + \mu_2((0, x]) = V(\Phi)(x) = V(\Phi, [0, x])$. This combined with Lemma 2.8.8 (4) proves that $V(\Phi, [0, x]) = |\mu|((0, x])$ for all $x > 0$. Since by Lemmas 2.8.4 and 2.8.8, $V(\Phi, [0, b]) = V(\Phi, [0, a]) + V(\Phi, [a, b]) \leq |\mu|((0, a]) + |\mu|((a, b]) = |\mu|((0, b]) = V(\Phi, [0, b])$, it follows that $V(\Phi, [a, b]) = |\mu|((a, b])$ for all $0 \leq a < b < +\infty$. $\square$

Absolutely continuous functions. Absolutely continuous measures on $\mathbb{R}$.

Let Φ be a function on the finite interval $[a, b]$. If it is of bounded variation and right continuous, then, for any $\epsilon > 0$, it follows from the right continuity of $V(\Phi, [a, x])$ and its additivity (Lemma 2.8.4) that there exists a $\delta = \delta(x) > 0$ such that $V(\Phi, [x, x + h]) < \epsilon$ if $0 < h < \delta(x)$.

Definition 2.8.12. *A function Ψ on $[a, b]$ is said to be* **absolutely continuous** *on $[a, b]$ if for any $\epsilon > 0$ there exists a $\delta = \delta_\epsilon > 0$ such that for any finite collection of points $a \leq a_1 < b_1 \leq a_2 < b_2 < \cdots < b_{n-1} \leq a_n < b_n \leq b$ (equivalently for any finite collection of subintervals $[a_i, b_i], 1 \leq i \leq n$, of $[a, b]$ with non-overlapping interiors), it follows that*

$$\sum_{i=1}^n |\Psi(b_i - \Psi(a_i)| < \epsilon \quad \text{if} \quad \sum_{i=1}^n (b_i - a_i) < \delta.$$

Remark. The key point in this definition is that the intervals are not required to be contiguous. When $b_{i-1} = a_i, 2 \leq i \leq n$, then $\sum_{i=1}^n |\Psi(b_i - \Psi(a_i)| = V(\Psi, [a_1, b_n])$.

The first thing to observe is that if one takes $\epsilon = 1$, then $V(\Psi, [x, x+h]) < 1$ if $0 < h < \delta_1$. From the additivity property (Lemma 2.8.4), it follows that Ψ is of bounded variation on $[a, b]$ if Ψ is absolutely continuous on $[a, b]$. It also follows from the definition that $V(\Psi, [0, x]) = V(\Psi)(x)$ and $\Psi(x)$ are right continuous functions of x (even continuous functions). The significance of the non-contiguity of the intervals is shown by the following result

Proposition 2.8.13. *Let ν denote the measure on $\mathfrak{B}(\mathbb{R})$ defined by the "natural" right continuous extension of Ψ to $\mathbb{R}$: it equals $\Psi(a)$ for $x < a$ and $\Psi(b)$ for $b < x$. If $E \subset [a, b]$ is a Borel set of measure zero, then $\nu(E) = |\nu|(E) = 0$. Hence, ν and $|\nu|$ are absolutely continuous with respect to Lebesgue measure dx.*

Proof. Since the endpoint a has Lebesgue measure zero, one may assume $E \subset (a, b]$. Recall that $|E| = 0$ implies that for any $\delta > 0$ there is a sequence $(a_n, b_n])_n$ of intervals that can be assumed to be pairwise disjoint with (i) $E \subset \cup_{n=1}^\infty (a_n, b_n]$ and (ii) $|\cup_{n=1}^\infty (a_n, b_n]| = \sum_{n=1}^\infty (b_n - a_n) < \delta$.

Furthermore, one has $|\nu|(E) = 0$ if for any $\epsilon > 0$ this sequence can also be chosen so that (iii) $|\nu|(\cup_{n=1}^\infty (a_n, b_n]) = \sum_{n=1}^\infty |\nu|((a_n, b_n]) < \epsilon$.

Since $E \subset (a, b]$, all these intervals can be assumed to be subintervals of $(a, b]$. If $(a_n, b_n] \subset (a, b]$, then $|\Psi(b_n) - \Psi(a_n)| \leq V(\Psi, [a_n, b_n]) = |\nu|((a_n, b_n])$. Assume that $\delta = \delta_\epsilon$, where δ_ϵ implies that

$$\sum_{i=1}^n |\Psi(b_i - \Psi(a_i)| < \epsilon \quad \text{if} \quad \sum_{i=1}^n (b_i - a_i) < \delta_\epsilon.$$

Clearly, one wants to replace $\sum_{i=1}^n |\Psi(b_i) - \Psi(a_i)|$ in this inequality by $\sum_{i=1}^n V([a_i, b_i], \Psi) = \sum_{i=1}^n |\nu|((a_i, b_i])$. It is now shown that this gives a similar inequality that holds for $\delta = \delta_\epsilon$ but with ϵ replaced by 2ϵ. This suffices to prove the proposition.

Observe that for each interval $(a_n, b_n]$ one may find a partition π_n such that $V(\Psi, \pi_n) + \frac{\epsilon}{2^n} \geq V(\Psi, [a_n, b_n]) = |\nu|((a_n, b_n])$. Each partition π_n subdivides $[a_n, b_n]$ into subintervals with non-overlapping interiors. It follows that $\sum_{i=1}^n V(\Psi, \pi_n) < \epsilon$ since the sum of the lengths of the intervals involved equals $\sum_{i=1}^n (b_i - a_i) < \delta = \delta_\epsilon$. Hence, $\sum_{i=1}^n |\nu|((a_i, b_i]) = \sum_{i=1}^n V(\Psi, [a_i, b_i]) \leq \sum_{i=1}^n V(\Psi, \pi_n) + \sum_{i=1}^n \frac{\epsilon}{2^i} < 2\epsilon$. Since this inequality holds independent of n, it follows that $\sum_{n=1}^\infty |\nu|((a_n, b_n]) < 2\epsilon$. As a result, $|\nu|(E) = 0$ and so also $\nu(E) = 0$. □

From this, it follows that if Ψ is an absolutely continuous function on $[a, b]$, then there is a finite signed measure ν on $\mathfrak{B}(\mathbb{R})$ such that all the mass of $|\nu|$ and hence of ν is concentrated on $[a, b]$, which is absolutely continuous in the sense of Definition 2.7.16. Hence, by the Radon–Nikodym theorem (Theorem 2.7.20), there is an L^1-function ψ such that $\nu(A) = \int_A \psi(x) dx$ and $|\nu|(A) = \int_A |\psi(x)| dx$. As a result, it follows that one may assume that $\{x \mid |\psi(x)| \neq 0\} \subset [a, b]$ (since this is true up to a set of measure zero, one may assume it by modifying ψ if necessary on a set of Lebesgue measure zero); in other words, $\psi \in L^1([a, b], dx)$. This essentially proves the first part of the following result.

Theorem 2.8.14. *A function Ψ on $[a, b]$ is absolutely continuous if and only if there is a function $\psi \in L^1([a, b], dx)$ with $\Psi(x) = \int_a^x \psi(u) du, a \leq x \leq b$.*

Proof. If Ψ is absolutely continuous in the sense of Definition 2.8.12, the resulting finite signed measure ν is determined by the fact that $|\nu|(A) = 0$ if $A \subset (-\infty, a) \cup (b, +\infty)$ is a Borel subset and $\Psi(x) = \nu((a, x]) = \int_a^x \psi(u) du, a \leq x \leq b$, where the last inequality follows from the Radon–Nikodym theorem.

On the other hand, if ν is given by a function $\psi \in L^1([a, b], dx)$ in the sense that $\nu(A) = \int_A \psi(u) du$ for any $A \in \mathfrak{B}(\mathbb{R})$, then given $\epsilon > 0$, $|\nu|(A) < \epsilon$ if $|A| < \delta = \delta(\epsilon)$ as stated in the following exercise.

Exercise 2.8.15. (see Lemma 4.5.5) Let $\psi \in L^1(\mathbb{R})$, and define $|\nu|(A) = \int_A |\psi|(u) du$. Show that, for any $n > 0$, one has

(1) $|\nu|(A) \leq n|A \cap \{x \mid |\psi(x)| \leq n\}| + \int_{\{x \mid |\psi(x)| > n\}} |\psi(u)| du$, and

(2) $\int_{\{x \mid |\psi(x)| > n\}} |\psi(u)| du \to 0$ as $n \to \infty$ [Hint: use dominated convergence (Theorem 2.1.35)], and conclude that

(3) if $\epsilon > 0$, then there is a $\delta = \delta(\psi, \epsilon, n)$ such that $|\nu|(A) < \epsilon$ if $|A| < \delta$.

It follows from this that if $\Psi(x) = \int_a^x \psi(u) du$, then Ψ is absolutely continuous in the sense of Definition 2.8.12: observe that if $a_i < b_i$, then $|\Psi(b_i) - \Psi(a_i)| \leq V(\Psi, [a_i, b_i]) = |\nu|((a_i, b_i])$. Hence, if in Definition 2.8.12 $|\cup_{i=1}^n (a_i, b_i]| = \sum_{i=1}^n (b_i - a_i) < \delta(\psi, \epsilon, n)$, then $\sum_{i=1}^n |\Psi(b_i) - \Psi(a_i)| \leq \sum_{i=1}^n V(\Psi, (a_i, b_i]) = |\nu|(\cup_{i=1}^n (a_i, b_i]) < \epsilon$. □

In other words, the absolutely continuous functions on $[a, b]$ are the "indefinite" integrals in the sense of Lebesgue of the L^1-functions on $[a, b]$

(relative to Lebesgue measure on $[a,b]$). From calculus, one is familiar with the fact that for continuous functions ψ, the derivative of its indefinite integral coincides with ψ. This is true a.e. for the indefinite Lebesgue integral, but its proof is more difficult and is postponed until Chapter IV, where it is proved as a consequence of Lebesgue's differentiation theorem, which states that, if $f \in L^1(\mathbb{R})$, then a.e. $\frac{1}{2h}\int_{x-h}^{x+h} f(u)du \to f(x)$ as $h \to 0$.

9. ADDITIONAL EXERCISES*

Exercise 2.9.1. Define $G(x) = [x]$, where $[x] = n$ if $n \leq x < n+1$.
  (1) Show that every subset of $\mathbb{R}$ is μ^*-measurable, where μ is the measure for which $\mu((a,b]) = G(b) - G(a)$.
  (2) What are the μ-measurable functions?
  (3) Determine $L^1(\mathbb{R}, \mathfrak{P}(\mathbb{R}), \mu)$ (recall from (1) that $\mathfrak{P}(\mathbb{R})$ is the σ-field of μ^*-measurable sets).

Exercise 2.9.2. Show that
  (1) the union of a countable family of sets of measure zero is also of measure zero (here "measure" refers to any σ-finite measure, for example, one constructed from a right continuous, non-decreasing function $G: \mathbb{R} \to \mathbb{R}$),
  (2) E is a Lebesgue measurable set if and only if there is a Borel set B with $E \Delta B$ a set of Lebesgue measure zero where $E \Delta B \stackrel{\text{def}}{=} (E\backslash B) \cup (B\backslash E)$ (this set is called the **symmetric difference** of E and B. [Hint: see Exercise 2.9.12.]

Exercise 2.9.3. Let $E \subset \mathbb{R}$. Show that
  (1) the outer Lebesgue measure $\lambda^*(E)$ is the infimum of $\{|O| \mid E \subset O, O \text{ open}\}$ [Hint: if $E \subset \cup_{n=1}^\infty (a_n, b_n]$, replace each interval $(a_n, b_n]$ by $(a_n, b_n + \frac{\epsilon}{2^n})$.],
  (2) if E is bounded and Lebesgue measurable, then for any $\epsilon > 0$ there is a bounded open set $O \supset E$ with $|O\backslash E| < \frac{\epsilon}{2}$.

Assume that $E \subset [-N, N]$ is Lebesgue measurable. Show that

  (3) if $P \supset E^c \cap [-N, N] \stackrel{\text{def}}{=} E_N^c$ is open and $|P\backslash E_N^c| < \frac{\epsilon}{2}$, then $|E\backslash C| < \frac{\epsilon}{2}$ where $C = P^c \cap [-N, N] \subset E$.

The set C is closed and bounded, i.e., it is compact (Exercise 1.5.6). Conclude that

  (4) if E is a bounded Lebesgue measurable set and $\epsilon > 0$, then there exist a compact set C and an open set O with $C \subset E \subset O$ and $|O\backslash C| < \epsilon$, and
  (5) if E is a Lebesgue measurable set with $|E| < +\infty$, then there exist a compact set C and an open set O with $C \subset E \subset O$ and $|O\backslash C| < \epsilon$.

Finally, show that

(6) for any Lebesgue measurable set E, $|E| = \sup\{|C| \mid C \text{ compact } \subset E\} = \inf\{|O| \mid O \text{ open } \supset E\}$.

Exercise 2.9.4. (Differentiating the Fourier transform) Let $f \in L^1(\mathbb{R})$. Show that

(1) $\int \cos(tx)f(x)dx$ is defined for all $t \in \mathbb{R}$, and
(2) $\int \sin(tx)f(x)dx$ is defined for all $t \in \mathbb{R}$.

Assume that $x \to xf(x) \in L^1(\mathbb{R})$. Show that

(3) $\frac{d}{dt}\int \cos(tx)f(x)dx = -\int x\sin(tx)f(x)dx$, and
(4) $\frac{d}{dt}\int \sin(tx)f(x)dx = \int x\cos(tx)f(x)dx$.

Using complex notation with $e^{ixt} = \cos(xt) + i\sin(xt)$ (see the appendix to Chapter VI), this shows that the derivative of the Fourier transform $\hat{f}(t) \overset{\text{def}}{=} \int e^{-ixt}f(x)dx$ (see Körner [K3]) is given by

(5) $\frac{d}{dt}\int e^{ixt}f(x)dx = -\int ixe^{ixt}f(x)dx$.

In probability and statistics, when $f \geq 0$ and $\int f(x)dx = 1$ (i.e., f is a probability density function), a variant of this form of the Fourier transform is called the **characteristic function** of the corresponding probability. The value of the characteristic function at t is the value $\hat{f}(-t)$ at $-t$ of the Fourier transform.

Exercise 2.9.5. (Differentiating the Laplace transform) Let f be a Borel measurable function on $\mathbb{R}$ with $f(x) = 0$, $x < 0$. Assume that for some constants C and $\alpha > 0$, $|f(x)| \leq Ce^{\alpha x}$, $x \geq 0$. Let $\mathcal{L}f(s) = \int e^{-sx}f(x)dx, s > \alpha$. Show that

(1) $\mathcal{L}f$ is differentiable on $(\alpha, +\infty)$, and
(2) $(\mathcal{L}f)'(s) = -\int xe^{-sx}f(x)dx$.

Exercise 2.9.6. Let $n(x) \overset{\text{def}}{=} n_1(x) = \frac{1}{\sqrt{2\pi}}e^{-(\frac{x^2}{2})}$ and $n_t(x) \overset{\text{def}}{=} \frac{1}{\sqrt{t}}n(\frac{x}{\sqrt{t}}) = \frac{1}{\sqrt{2\pi t}}e^{-(\frac{x^2}{2t})}$. The main purpose of this exercise is to show that for any t and bounded measurable function f, the function $n_t \star f$ is smooth — it has derivatives of all orders and hence is a so-called **infinitely differentiable function** — where $n_t \star f(x) \overset{\text{def}}{=} \int n_t(x-y)f(y)dy$. This exercise is long and is divided into parts A,B, and C. The main reason for all the difficulty is that at this point convolution (Exercise 3.3.21) has not been defined and the relation (Theorem 4.2.5) between continuous functions and Lebesgue integrable functions has not been made clear (see also Exercise 4.2.10). This exercise follows from Proposition 4.2.29 once it is shown that all the derivatives of $n_t(x)$ are integrable [part C (3)].

Part A. Let $a > 0$. Show that, for all $r \geq 0$, one has

(1) $e^{-\frac{r^2}{2a}} \leq C_0(a)e^{-r}$ [Hint: complete the square $r^2 - 2ar$]

(2) $r^m \leq C(m)e^{\frac{r}{2}}$ if $m \geq 1$ [Hint: recall the power series for e^u] and
(3) $r^m e^{-r} \leq C(m)e^{-\frac{r}{2}}$ if $m \geq 1$.

Part B. Let f be a non-negative even function that takes its maximum value at $x = 0$ ("even" means that $f(x) = f(-x)$ for all $x \in \mathbb{R}$). Assume that $f \in L^1(\mathbb{R}) = L^1(\mathbb{R}, \mathfrak{F}, dx)$, where $\mathfrak{F}$ is the σ-algebra of Lebesgue measurable subsets of $\mathbb{R}$, is non-increasing on $[0, +\infty)$.

(1) Find a function $\varphi_N \in L^1(\mathbb{R})$ such that, if $|b| \leq N$, then $\varphi_N(x) \geq f(x-b)$ for all $x \in \mathbb{R}$. [Hint: "split" the graph of the function at $x = 0$ and "insert" a constant value C on $[-N, N]$ so that $\int \varphi_N(x)dx = 2CN + \int f(x)dx$.]

(2) Let $c > 0$. Find a function $\varphi_N \in L^1(\mathbb{R})$ such that, if $|b| \leq N$, $\varphi_N(x) \geq e^{-c|x-b|}$ for all $x \in \mathbb{R}$.

Show that

(3) $|n_t(x)| \leq C(t)e^{-|x|}$ [Hint: see part A], and use (2), together with dominated convergence, to show that $(n_t \star f)(x)$ is a continuous function of x if f is a bounded Borel function.

Part C. Recall that $n_t(x) = \frac{1}{\sqrt{2\pi t}} e^{-(\frac{x^2}{2t})}$. Use mathematical induction to show that

(1) $\frac{\partial^n}{\partial x^n} n_t(x) = H_n(x, t) n_t(x)$, for each $n \geq 1$, where for each fixed $t > 0$, $H_n(x, t)$ is a polynomial of degree n in x, the so-called **Hermite polynomial** of degree n when $t = 1$.

Show that

(2) $|H_n(x, t) n_t(x)| \leq C(t, n) e^{-\frac{|x|}{2}}$ [Hint: see part A].

Conclude that $\frac{\partial^n}{\partial x^n} n_t \in L^1(\mathbb{R})$.

Use the mean value theorem to show that

(3) if $|h| \leq 1$ then $\frac{1}{|h|} |n_t(x - y + h) - n_t(x - y)| \leq C(t, 1) e^{\frac{1}{2}} e^{-\frac{1}{2}|x-y|}$.

Make use of (3) to conclude that, if f is a bounded Borel (or, more generally, Lebesgue) measurable function,

(4) $(n_t \star f)'(x) = \int \frac{\partial}{\partial x} n_t(x-y) f(y) dy = \int H_1(x-y, t) n_t(x-y) f(y) dy$
and $x \to (n_t \star f)'(x)$ is a Borel function.

By a similar argument show that if f is a bounded Borel (or more generally Lebesgue) measurable function, then

(5) $\frac{\partial^n}{\partial x^n} \left[\int n_t(x-y) f(y) dy \right] = \int H_n(x-y, t) n_t(x-y) f(y) dy$, for each $n \geq 1$.

Conclude that $\int n_t(x-y) f(y) dy$ is a C^∞-function of x (i.e., it is smooth).

Exercise 2.9.7. Find a sequence $(\varphi_n)_{n \geq 1}$ of Riemann integrable functions φ_n on $[0, 1]$ that are uniformly bounded, converge everywhere to a function φ, and are such that

(1) the Riemann integral $\int_0^1 \varphi_n(x) dx = 1$, for all $n \geq 1$, and
(2) the Riemann integral $\int_0^1 \varphi(x) dx$ does not exist.

9. ADDITIONAL EXERCISES

Note that $\int \varphi(x)1_{[0,1]}(x)dx = \lim_n \int \varphi_n(x)1_{[0,1]}(x)dx = 1$, in view of the theorem of dominated convergence (2.1.38). [*Hint*: enumerate the dyadic rationals in $[0,1]$.]

Exercise 2.9.8. Let X be non-negative. Show that $E[X] = \int X d\mathbf{P} = \sup\{\int s d\mathbf{P} \mid 0 \leq s \leq X, s \text{ simple}\}$.

Remark. The integral $\int X d\mathbf{P}$ is often defined as $\sup\{\int s d\mathbf{P} \mid 0 \leq s \leq X, s \text{ simple}\}$, see Rudin [R4]. One then has to prove the additivity of the integral, which amounts to verifying the above exercise using the definition of the integral given in Definition 2.1.23.

Exercise 2.9.9. Let ε_a denote the Dirac measure at a or unit point mass at a (also denoted by δ_a). If $X \in L^1(\mathbb{R}, \mathfrak{B}(\mathbb{R}), \varepsilon_a)$, compute $\int X d\varepsilon_a$. For any measurable space $(\Omega, \mathfrak{F})$, define an analogous measure ε_{ω_0}, for any point $\omega_0 \in \Omega$, and compute $\int X d\varepsilon_{\omega_0}$.

Exercise 2.9.10. Let Y be a random variable on $(\Omega, \mathfrak{F}, \mathbf{P})$, and let $\sigma(Y)$ denote the smallest σ-algebra $\mathfrak{G}$ contained in $\mathfrak{F}$ such that Y is $\mathfrak{G}$-measurable (i.e., is a random variable on $(\Omega, \mathfrak{G}, \mathbf{P})$). Show that $A \in \mathfrak{G}$ if and only if there is a Borel set B with $A = Y^{-1}(B)$; equivalently, $1_A = 1_B \circ Y$.

Exercise 2.9.11. Let G be any non-decreasing, real-valued function on $\mathbb{R}$, and let μ be the corresponding σ-finite measure that Carathéodory's procedure (Theorem 1.4.13) produces. It is defined on the σ-algebra $\mathfrak{F}$ of μ^*-measurable subsets of $\mathbb{R}$. Show that

(1) if $E \in \mathfrak{F}$, then there are two Borel sets A and B with (i) $B \subset E \subset A$ and (ii) $\mu(A \setminus B) = 0$ (see Exercise 2.2.7), and
(2) $\mathfrak{F}$ is the smallest σ-algebra, containing $\mathfrak{B}(\mathbb{R})$ and all the sets that are subsets of a Borel set B with $\mu(B) = 0$; and
(3) consider μ restricted to the Borel σ-algebra, and show that if one starts the Carathéodory procedure with $\mathfrak{B}(\mathbb{R})$ and μ, in place of $\mathfrak{A}$ and μ, the resulting measure is the same, i.e., the same σ-algebra $\mathfrak{F}$ results and the measure on it is μ (see Exercise 1.5.4).

Exercise 2.9.12. If μ is a σ-finite measure on $\mathfrak{B}(\mathbb{R})$, let $\mathfrak{N}$ denote the collection of sets that are subsets of a Borel set B with $\mu(B) = 0$. Define $\mathfrak{F}$ to be the set of symmetric differences $A \triangle N$, where $A \in \mathfrak{B}(\mathbb{R})$ and $N \in \mathfrak{N}$. Show that

(1) $\mathfrak{F}$ is a σ-algebra, and
(2) the function ν defined by $\nu(A \triangle N) = \mu(A)$ is a σ-finite measure on $\mathfrak{F}$.

Remark. The σ-algebra determined in Exercise 2.9.12 is called the **completion** of $\mathfrak{B}(\mathbb{R})$ with respect to μ and is often denoted by $\overline{\mathfrak{B}(\mathbb{R})}^{\mu}$. Exercise 2.9.12 says that a σ-finite measure on $\mathfrak{B}(\mathbb{R})$ can be extended as a measure to the completion $\overline{\mathfrak{B}(\mathbb{R})}^{\mu}$. The intersection over all possible σ-finite (even finite) measures μ of the completions $\overline{\mathfrak{B}(\mathbb{R})}^\mu$ is an important σ-algebra. It is called the σ-algebra of **universally measurable sets**. Every σ-finite measure can be extended as a measure to the σ-algebra of universally measurable sets.

Exercise 2.9.13. This exercise shows that a non-decreasing function is not too discontinuous. In particular, the set of points at which it is discontinuous has Lebesgue measure zero, since it is a countable set.

Let H be a non-decreasing, finite-valued function defined on $\mathbb{R}$. Observe that

(1) for any $a \in \mathbb{R}$, $H(a)$ is an upper bound of the set $\{H(x) \mid x < a\}$ and is a lower bound of $\{H(x) \mid x > a\}$.

Define $H(a-)$ to be the least upper bound of $\{H(x) \mid x < a\}$ and $H(a+)$ to be the greatest lower bound of $\{H(x) \mid x > a\}$. Show that

(2) $\lim_{x \uparrow a} H(x) = H(a-)$, where $\lambda \stackrel{\text{def}}{=} \lim_{x \uparrow a} H(x)$ if for any $\epsilon > 0$ there is a $\delta > 0$ such that $|H(x) - \lambda| < \epsilon$ if $a - \delta < x < a$,

(3) $\lim_{x \downarrow a} H(x) = H(a+)$, where $\lambda \stackrel{\text{def}}{=} \lim_{x \downarrow a} H(x)$ if for any $\epsilon > 0$ there is a $\delta > 0$ such that $|H(x) - \lambda| < \epsilon$ if $a < x < a + \delta$,

(4) there are at most a countable number of points a with $H(a+) \ne H(a-)$ (note that the difference $H(a+) - H(a-)$ measures the size of the jump of H at a) [*Hint:* estimate the number of points a in $[-N, N]$ with the jump of H at a greater than $\frac{1}{n}$.],

(5) H is continuous at a if and only if $H(a-) = H(a+)$,

(6) if H_1 and H_2 are two non-decreasing functions on $\mathbb{R}$ that are right continuous, they coincide if $H_1(x) = H_2(x)$ for any point x at which both functions are continuous. [*Hint:* make use of Exercise 1.3.15.]

Exercise 2.9.14. Let F be a distribution function that is not continuous. Then there is a non-void, at most countable set of points x at which F is not continuous. Since F is right continuous, Exercise 2.9.13 implies that this happens if and only if $F(x) \ne F(x-)$

Let $G(x) \stackrel{\text{def}}{=} \sum_{a \le x}\{F(a) - F(a-)\}$ (see the following remark). Show that

(1) G is non-decreasing, right continuous, with $\lim_{x \downarrow -\infty} G(x) = 0$,
(2) $F(x) - G(x)$ is continuous and non-decreasing.

Let $G(+\infty) = \sum_{a \in \mathbb{R}}\{F(a) - F(a-)\}$ and let $F_d(x) \stackrel{\text{def}}{=} \frac{G(x)}{G(+\infty)}$. Show that if $G \ne F$, then

(3) there is a continuous distribution function F_c and a number $\lambda \in (0, 1)$ with $F = \lambda F_c + (1 - \lambda) F_d$, i.e., F is a convex combination of F_c and F_d. Equivalently, every probability that is neither continuous nor discrete is a unique convex combination of a continuous and a discrete probability.

Show that

(4) every continuous probability that is neither absolutely continuous nor singular is a unique convex combination of an absolutely continuous probability and a continuous singular probability.

A measure μ is said to be **continuous** if $\mu(\{x\}) = 0$ for all $x \in \mathbb{R}$; **singular** if there is a Borel set E with $|E| = 0$ and $\mu(E^c) = 0$; and **discrete** if there is a countable set D with $\mu(D^c) = 0$. Finally, show that

(5) given any probability $\mathbf{P}$, there are three unique non-negative measures $\mu_{ac}, \mu_{sc}, \mu_d$ are absolutely continuous, singular and continuous, and discrete, respectively, such that

$$\mathbf{P} = \mu_{ac} + \mu_{sc} + \mu_d;$$

and

(6) if a probability is neither continuous nor discrete, neither absolutely continuous nor singular, then there are three unique positive scalars $\lambda_1, \lambda_2, \lambda_3$ with $\lambda_1 + \lambda_2 + \lambda_3 = 1$ and three unique probabilities $\mathbf{P}_{ac}, \mathbf{P}_{cs}$ and $\mathbf{P}_d$ that are absolutely continuous, singular and continuous, and discrete, respectively, for which

$$\mathbf{P} = \lambda_1 \mathbf{P}_{ac} + \lambda_2 \mathbf{P}_{sc} + \lambda_3 \mathbf{P}_d.$$

Remark. Since all the terms are non-negative, the sum $\sum_{a \leq x} \{F(a) - F(a-)\}$ may be defined as $\sup_J \sum_{a \leq x, a \in J} \{F(a) - F(a-)\}$, where the supremum is taken over the finite subsets J of $(-\infty, x]$. Note that these finite sums are always bounded above by $F(x)$.

Exercise 2.9.15. Let ξ be an irrational number and let $G = \{m + n\xi \mid m, n \in \mathbb{Z}\}$. Show that

(1) G is an additive subgroup of $\mathbb{R}$, i.e., (G is closed under addition and subtraction),
(2) if $x = m + n\xi = m' + n'\xi$, then $n = n'$ and $m = m'$,
(3) if $n \neq 0$, there is a unique integer m_n such that $0 \leq m_n + n\xi < 1$ (in fact, $0 < m_n + n\xi < 1$),
(4) if $0 < k$, at least one of the k intervals $(\frac{\ell}{k}, \frac{\ell+1}{k}), 0 \leq \ell < k-1$, contains two distinct points of G (in fact, an infinite number),
(5) if $0 < k$, there is a point $\alpha \in G$ with $0 < \alpha < \frac{1}{k}$,
(6) if $-\infty < a < b < +\infty$, show that (a, b) contains a point of G [Hint: consider the subgroup $\mathbb{Z}\alpha$ and recall the use in Exercise 2.1.12 of $\mathbb{Z}\frac{1}{n}$.], and finally,
(7) if O is a non-void open subset of $\mathbb{R}$, then $O \cap G \neq \emptyset$, i.e., G is dense in $\mathbb{R}$.

Exercise 2.9.16. (a singular distribution function on $[0,1]$)
(See Wheeden and Zygmund [W1], pp. 34–35, where it is referred to as the Cantor–Lebesgue function.) Let
$$F_0(x) = x \text{ if } 0 \leq x \leq 1$$
and
$$F_1(x) = \begin{cases} \frac{3}{2}x & \text{if } 0 \leq x \leq \frac{1}{3}, \\ \frac{1}{2} & \text{if } \frac{1}{3} < x < \frac{2}{3}, \\ \frac{3}{2}x - \frac{1}{2} & \text{if } \frac{2}{3} \leq x \leq 1. \end{cases}$$

The graph of the function F_1 may be obtained from the graph of F_0 by viewing it as a diagonal of the rectangle defined by $(0,0)$ and $(1,1)$; divide the base of this rectangle into three equal pieces by two lines parallel to the y-axis and the vertical sides into two halves by a line parallel to the x-axis. This subdivides the original rectangle into 6 new rectangles. The graph of F_1 coincides with the line bisecting the vertical sides between the two lines that divide the base and the top into thirds. Outside these two lines, it is given by two diagonal lines in two of the 6 new rectangles: the line joining $(0,0)$ to $(\frac{1}{3}, \frac{1}{2})$ and the line joining $(\frac{2}{3}, \frac{1}{2})$ to $(1,1)$.

The graph of F_2 is obtained from the graph of F_1 by carrying out this construction for each of the new rectangles in which the graph of F_1 is given by a diagonal line. Continuing by induction on n, this defines a sequence of non-decreasing, continuous (even piecewise linear) functions F_n that map $[0,1]$ to $[0,1]$.

Show that

(1) for each $n \geq 1$, the maximum of $|F_{n-1}(x) - F_n(x)|, 0 \leq x \leq 1$ is $\frac{1}{6}(\frac{1}{2})^{n-1}$.

Conclude that

(2) the sequence of continuous functions converges uniformly on $[0,1]$ to a non-decreasing function F with $F(0) = 0$ and $F(1) = 1$.

Extend F to all of $\mathbb{R}$ by setting $F(x) = 0$ if $x < 0$ and $F(x) = 1$ if $X > 0$. Then F is a continuous distribution function. If **P** is the corresponding probability, show that

(3) $\mathbf{P}(C) = 1$, where C is the Cantor set (see Example 1.2.5), and
(4) **P** is singular with respect to Lebesgue measure (and any probability that has a probability density function relative to Lebesgue measure).

Exercise 2.9.17. (see Titchmarsh [T], pp. 229–230 and pp. 366–367) It follows from Exercise 1.2.8 that the intervals removed at the nth stage of the procedure that produce the Cantor set are of the form $(\sum_{i=1}^{n-1} \frac{b_i}{3^i} + \frac{1}{3^n}, \sum_{i=1}^{n-1} \frac{b_i}{3^i} + \frac{2}{3^n})$, where $b_i \in \{0, 2\}$. Let $c_i = b_i/2, 1 \leq i \leq n-1$. Show that

(1) if $\sum_{i=1}^{n-1} \frac{b_i}{3^i} + \frac{1}{3^n} \leq x \leq \sum_{i=1}^{n-1} \frac{b_i}{3^i} + \frac{2}{3^n}$, then $F(x) = \sum_{i=1}^{n-1} \frac{b_i}{2^i} + \frac{1}{2^n} = F_n(x)$, where $F_n(x)$ and $F(x)$ are defined in Exercise 2.9.16.

Each $x \in [0,1]$ has a triadic expansion of the form $x = 0.a_1 a_2 \ldots a_n \ldots$, where $a_i \in \{0,1,2\}$; in other words, $x = \sum_{i=1}^{\infty} \frac{a_i}{3^i}$. Show that

(2) $x \in C$ if and only if x has a triadic expansion in which only 0 or 2 occurs; e.g., $\frac{1}{3} = 0.1\ldots = 0.0222\ldots$, and so $\frac{1}{3} \in C$, and

(3) if $x \in C$, then $F(0.a_1 a_2 \ldots a_n \ldots) = 0.b_1 b_2 \ldots b_n \ldots$, where $a_i \in \{0,2\}$ and $b_i = \frac{1}{2} a_i$ (note that the expansion of $y = F(x) = 0.b_1 b_2 \ldots b_n \ldots$ is understood to be dyadic; i.e., $y = \sum_{i=1}^{\infty} \frac{b_i}{2^i}$.

Exercise 2.9.18. (characterization of a bounded Riemann integrable function on $[a,b]$) In the proof of Proposition 2.3.4, show that

(1) each partition π_n can be assumed to contain (for example) all the dyadic rationals on $[0,1]$ of the form $\frac{k}{2^n}, 0 \leq k \leq 2^n$, which implies that the "mesh" of π_n (which is defined as the maximum distance between adjacent partition points) is $\leq \frac{1}{2^n}$.

Let J denote the union of all the points in all the partitions π_n. Then $|J| = 0$. Show that

(2) if $f(x) = g(x)$ and $x \notin J$ (see Proposition 2.3.4 for the definition of f and g), then φ is continuous at x; and conversely,

(3) if φ is continuous at $x \notin J$, then $f(x) = g(x)$;

(4) conclude that φ is Riemann integrable if and only if it is continuous at almost every point of $[a,b]$ (i.e., φ is a.e. continuous on $[a,b]$).

Exercise 2.9.19. Let W be a random variable and assume that $W = X_1 - Y_1 = X_2 - Y_2$, where X_1, X_2, Y_1, Y_2 are all non-negative random variables and in addition Y_1 and Y_2 are integrable. Show that

(1) $E[X_1] - E[Y_1] = E[X_2] - E[Y_2]$.

Define $E[W]$ to be $E[X] - E[Y]$ if X and Y are non-negative and Y is integrable.

Now assume that $W = Z + X$ with Z non-negative and X integrable. Show that

(2) $E[W] = E[Z] + E[X]$.

CHAPTER III

INDEPENDENCE AND PRODUCT MEASURES

1. Random vectors and Borel sets in $\mathbb{R}^n$

Let $(\Omega, \mathfrak{F}, \mathbf{P})$ be a probability space and let $X : \Omega \to \mathbb{R}^2$ be a vector valued function, i.e., the values are vectors in $\mathbb{R}^2$. For each $\omega \in \Omega$, let $X(\omega) = (X_1(\omega), X_2(\omega))$, where $X_1(\omega)$ and $X_2(\omega)$ are the components of $X(\omega)$ with respect to the canonical basis of $\mathbb{R}^2$ consisting of $\mathbf{e}_1 = (1,0)$ and $\mathbf{e}_2 = (0,1)$.

It is a natural question to ask what condition X should satisfy in order to be a **measurable vector-valued function**, in other words, a **random vector**.

Suppose that X_1 and X_2 are random variables, i.e., they are measurable when $\mathbb{R}$ is equipped with the Borel σ-algebra $\mathfrak{B}(\mathbb{R})$. What happens if instead of the basis $\{\mathbf{e}_1, \mathbf{e}_2\}$ one uses another basis? The new components remain measurable, as they are linear combinations of X_1 and X_2. So this is a basis-free condition and is the one that will be imposed: X will be said to be **measurable** or a **random vector** if of its components X_1, X_2 with respect to the canonical basis are measurable, i.e., if they are random variables.

A (real-valued) random variable X has a distribution (on $\mathbb{R}$). Does a vector-valued random variable X (with values in $\mathbb{R}^2$, say) have a distribution on $\mathbb{R}^2$? To answer this question, it is necessary to introduce a σ-algebra on $\mathbb{R}^2$ — the Borel σ-algebra $\mathfrak{B}(\mathbb{R}^2)$ of $\mathbb{R}^2$.

Proposition 3.1.1. *Let $X : \Omega \to \mathbb{R}^2$ and let $\mathfrak{B}$ be the smallest σ-algebra on $\mathbb{R}^2$ containing all the sets $E_1 \times E_2$ where E_1 and $E_2 \in \mathfrak{B}(\mathbb{R})$. The following statements are equivalent:*
  (1) $B \in \mathfrak{B}$ *implies* $X^{-1}(B) \in \mathfrak{F}$;
  (2) *if* $X = X_1 \mathbf{e}_1 + X_2 \mathbf{e}_2$, *then* X_1 *and* X_2 *are random variables on* $(\Omega, \mathfrak{F}, \mathbf{P})$.

Proof. Assume (1). Let $E_1 \in \mathfrak{B}(\mathbb{R})$. Then $X_1^{-1}(E_1) \stackrel{\text{def}}{=} \{\omega \mid X_1(\omega) \in E_1\} = \{\omega \mid X(\omega) \in E_1 \times \mathbb{R}\}$. Since $E_1 \times \mathbb{R} \in \mathfrak{B}$, (1) implies $X_1^{-1}(E_1) \in \mathfrak{F}$, i.e., X_1 is a random variable. Similarly, (1) implies that X_2 is a random variable.

Assume (2) and let $E_1, E_2 \in \mathfrak{B}(\mathbb{R})$. Then $\{\omega \mid X(\omega) \in E_1 \times E_2\} = \{\omega \mid X_1(\omega) \in E_1\} \cap \{\omega \mid X_2(\omega) \in E_2\} \in \mathfrak{F}$ (i.e., $X^{-1}(E_1 \times E_2) \in \mathfrak{F}$).

Let $\mathfrak{G}$ be the collection of subsets E of $\mathbb{R}^2$ such that $X^{-1}(E) \in \mathfrak{F}$. Then, by Exercise 2.1.18 (with $\mathbb{R}$ replaced by $\mathbb{R}^2$), $\mathfrak{G}$ is a σ-algebra.

Hence, since $E_1 \times E_2 \in \mathfrak{G}$ for all $E_1, E_2 \in \mathfrak{B}(\mathbb{R})$, the σ-algebra $\mathfrak{G} \supset \mathfrak{B}$. $\square$

1. RANDOM VECTORS AND BOREL SETS IN $\mathbb{R}^N$

Definition 3.1.2. *The σ-algebra of Borel subsets of $\mathbb{R}^2$ (respectively $\mathbb{R}^n$), is the smallest σ- algebra containing all the open subsets (Definition 3.1.3) of $\mathbb{R}^2$ (respectively, $\mathbb{R}^n$). It will be denoted by $\mathfrak{B}(\mathbb{R}^2)$ (respectively, $\mathfrak{B}(\mathbb{R}^n)$).*

Definition 3.1.3. *A subset $O \subset \mathbb{R}^2$ is said to be an **open subset of $\mathbb{R}^2$** if $a \in O$ implies, that for some $r > 0$, O contains the open disc about a of radius r (i.e., $\{x \mid (x_1 - a_1)^2 + (x_2 - a_2)^2 < r^2\} \subset O$). Similarly, $O \subset \mathbb{R}^n$ is said to be an **open subset of $\mathbb{R}^n$** if $a \in O$ implies that there is some $r > 0$ such that O contains the open ball about a of radius r (i.e., $B(a; r) = \{x \mid \sum_{i=1}^n (x_i - a_i)^2 < r^2\}$, is a subset of 0).*

Exercise 3.1.4. On $\mathbb{R}^n$ define $\| x \|^2 = \sum_{i=1}^n x_i^2$ and $|x| = \max_{1 \leq i \leq n} |x_i|$.
  (1) Show that $\| \cdot \|$ and $|\cdot|$ are norms on $\mathbb{R}^n$ (see Definition 2.5.3 for the definition of a norm).
  (2) Given a norm on $\mathbb{R}^n$, define the open ball about x_0 of radius r to be the set of x such that the norm of $x - x_0$ is less than r. Sketch the open ball about $0 \in \mathbb{R}^2$ of radius 1 for each of the norms in (1).
  (3) Show that a set $O \subset \mathbb{R}^n$ is open if and only if for any $x_0 \in O$ there exists r such that $\{x \mid |x - x_0| < r\} \subset O$, where $|y|$ is defined above.
  (4) Show that an open ball is an open set.

Remark. This exercise shows that the open sets of $\mathbb{R}^n$ do not depend upon which of the two norms is used to define open balls. In fact, any norm on $\mathbb{R}^n$ determines the open sets in the same way. The collection of open sets is a topology (Exercise 1.3.11), and, as in $\mathbb{R}$, the closure of a set A is defined to be the smallest closed set that contains it.

Proposition 3.1.5. *The σ-algebra $\mathfrak{B}(\mathbb{R}^2)$ is the smallest σ-algebra $\mathfrak{B}$ containing all the sets $E_1 \times E_2, E_i \in \mathfrak{B}(\mathbb{R}), i = 1, 2$.*

To carry out the proof, it will be convenient to use the following lemma.

Lemma. *Every open set $O \subset \mathbb{R}^2$ is the union of those open balls with rational radii that are centered at points with rational coordinates and that lie inside O.*

Exercise 3.1.6. Prove this lemma. [Hints: use $|x| = \max\{|x_1|, |x_2|\}$ to define the open balls; show that if $a = (a_1, a_2)$ and $\{x \mid |x - a| < \epsilon\} \subset O$, then this ball contains a point $r = (r_1, r_2), r_i \in \mathbb{Q}$ (see Exercise 2.1.12), with $|a - r| < \frac{\epsilon}{2}$; find $\eta \in \mathbb{Q}$ with $a \in \{x \mid |x - r| < \eta\} \subset O$.]

Remark 3.1.7. The lemma says that there is a fixed, countable collection of open balls such that any open set is a union of a certain number of these balls (this implies that $\mathbb{R}^2$ satisfies what is known as the **second axiom of countability**).

Proof of Proposition 3.1.5. For convenience, use the norm $|\cdot|$, since an open ball is a "box". In view of the lemma, $\mathfrak{B}(\mathbb{R}^2)$ is the smallest σ-algebra of subsets of $\mathbb{R}^2$ that contains every open ball. Now $\{x \mid |x - a| < r\} = (a_1 - r, a_1 + r) \times (a_2 - r, a_2 + r)$ if $a = (a_1, a_2)$, and so $\mathfrak{B}(\mathbb{R}^2) \subset \mathfrak{B}$ since $\mathfrak{B}$ contains all of these "boxes".

To show that $\mathfrak{B} \subset \mathfrak{B}(\mathbb{R}^2)$, it suffices to show that if E_1 and $E_2 \in \mathfrak{B}(\mathbb{R})$, $E_1 \times \mathbb{R}$ and $\mathbb{R} \times E_2 \in \mathfrak{B}(\mathbb{R}^2)$, since then $E_1 \times E_2 = (E_1 \times \mathbb{R}) \cap (\mathbb{R} \times E_2) \in \mathfrak{B}(\mathbb{R}^2)$.

Let $\mathfrak{G}$ be the collection of subsets E of $\mathbb{R}$ such that $E \times \mathbb{R} \in \mathfrak{B}(\mathbb{R}^2)$, i.e., $\mathfrak{G} = \{E \subset \mathbb{R} \mid pr_1^{-1}(E) \in \mathfrak{B}(\mathbb{R}^2)\}$, where $pr_1(x_1, x_2) = x_1$. Then, by Exercise 2.1.18 (with a minor change), $\mathfrak{G}$ is a σ-algebra. Since it contains all open intervals (a, b), it contains $\mathfrak{B}(\mathbb{R})$. Therefore, $E_1 \in \mathfrak{B}(\mathbb{R})$ implies $E_1 \times \mathbb{R} \in \mathfrak{B}(\mathbb{R}^2)$. Similarly, $\mathbb{R} \times E_2 \in \mathfrak{B}(\mathbb{R}^2)$ if $E_2 \in \mathfrak{B}(\mathbb{R})$. □

Exercise 3.1.8. (1) Show that $\mathfrak{B}(\mathbb{R}^n)$ is the smallest σ-algebra $\mathfrak{B}$ containing all the sets $E_1 \times E_2 \times \cdots \times E_n$, $E_i \in \mathfrak{B}(\mathbb{R})$.

(2) Let $X : \Omega \to \mathbb{R}^n$. Show that if $X = (X_1, \ldots, X_n)$, then each component $X_i, i = 1, 2, \ldots, n$, is a random variable if and only if for all $E \in \mathfrak{B}(\mathbb{R}^n), X^{-1}(E) \in \mathfrak{F}$ (here Ω denotes a probability space $(\Omega, \mathfrak{F}, \mathbf{P})$).

Finally, one defines a **random vector** as follows.

Definition 3.1.9. Let $X : \Omega \to \mathbb{R}^n$ where $(\Omega, \mathfrak{F}, \mathbf{P})$ is a probability space. Then X is a **random vector** if it is measurable with respect to or relative to $\mathfrak{B}(\mathbb{R}^n)$, i.e., $E \in \mathfrak{B}(\mathbb{R}^n)$ implies $X^{-1}(E) \in \mathfrak{F}$.

Proposition 3.1.10. Let $(\Omega, \mathfrak{F}, \mathbf{P})$ be a probability space and $X : \Omega \to \mathbb{R}^n$ be a random vector. Then there is a unique probability $\mathbf{Q}$ on $\mathfrak{B}(\mathbb{R}^n)$ such that $E \in \mathfrak{B}(\mathbb{R}^n)$ implies $\mathbf{Q}(E) = \mathbf{P}(X^{-1}(E))$.

Proof. Use the argument that established Proposition 2.1.19. □

Definition 3.1.11. The probability $\mathbf{Q}$ in Proposition 3.1.10 is called the **distribution** or **law of the random vector** X.

Remark. A probability $\mathbf{Q}$ on $\mathbb{R}^n$ is determined by a distribution function F as in the case of $\mathbb{R}$: $F(x_1, x_2, \ldots, x_n) = \mathbf{Q}((-\infty, x_1] \times (-\infty, x_2] \times \cdots \times (-\infty, x_n]) = \mathbf{P}(X_1 \le x_1, X_2 \le x_2, \ldots, X_n \le x_n)$. The distribution function determines the values of $\mathbf{Q}$ on sets of the form $(a_1, b_1] \times (a_2, b_2] \times \cdots \times (a_n, b_n]$, which determines the probability $\mathbf{Q}$ on $\mathfrak{B}(\mathbb{R}^n)$ in view of the next exercise.

Exercise 3.1.12. Show that $\mathfrak{B}(\mathbb{R}^n)$ is the smallest σ-algebra of subsets of $\mathbb{R}^n$ that contains any one of the following collections $\mathfrak{C}_i$ of sets:

(1) $\mathfrak{C}_1$: the finite "boxes" or intervals of the form $(a_1, b_1] \times (a_2, b_2] \times \cdots \times (a_n, b_n]$;

(2) $\mathfrak{C}_2$: the finite "boxes" or intervals of the form $(a_1, b_1) \times (a_2, b_2) \times \cdots \times (a_n, b_n)$;

(3) $\mathfrak{C}_3$: the finite "boxes" or intervals of the form $[a_1, b_1) \times [a_2, b_2) \times \cdots \times [a_n, b_n)$;

(4) $\mathfrak{C}_4$: the finite "boxes" or intervals of the form $[a_1, b_1] \times [a_2, b_2] \times \cdots \times [a_n, b_n]$;
(5) $\mathfrak{C}_5$: the closed subsets of $\mathbb{R}^n$.

If $X = (X_1, \ldots, X_n)$, the "randomness" of a vector-valued function $X : \Omega \to \mathbb{R}^n$ is determined by that of its components X_i,. This leads one to ask what can be said about the distribution of X in terms of the distributions of its components? For example, can it be calculated knowing the distribution of each of the X_i's?

For example, given a two dimensional random vector X for which the so-called **marginal distributions** (i.e., the distributions of X_1, X_2) are known, does this enable one to compute or to say anything about the distribution X? More concretely, suppose that X measures two variables or parameters of an individual, say height and weight. Knowing that height and weight have normal distributions with given means and variances, can one say anything about the joint distribution of X? (More dramatic examples involving, say, smoking and lung cancer, etc., can be imagined.)

In general, one cannot determine the distribution of $X = (X_1, \ldots, X_n)$ from the distributions of its components X_i **unless they are independent**. Then, the distribution of X is the product of the distributions of the components, as will be shown in the next section.

2. INDEPENDENCE

One of the ways in which probability is distinguished from merely being a branch of analysis is the emphasis given to (and the importance of) the notion of independence.

Definition 3.2.1. *Let $(\Omega, \mathfrak{F}, \mathbf{P})$ be a probability space and $\mathfrak{C}_1, \mathfrak{C}_2$ be two subsets of $\mathfrak{F}$. They are said to be* **independent collections of events** *(relative to $\mathbf{P}$) if*

$$A_1 \in \mathfrak{C}_1, A_2 \in \mathfrak{C}_2 \text{ implies } \mathbf{P}(A_1 \cap A_2) = \mathbf{P}(A_1)\mathbf{P}(A_2).$$

Example 3.2.2. Let $A, B \in \mathfrak{F}$ be such that $\mathbf{P}(A \cap B) = \mathbf{P}(A)\mathbf{P}(B)$, in which case A and B are said to be **independent events**. Set $\mathfrak{F}_1 = \{\phi, A, A^c, \Omega\}$ and $\mathfrak{F}_2 = \{\phi, B, B^c, \Omega\}$. Then $\mathfrak{F}_1$ and $\mathfrak{F}_2$ are independent σ-algebras.

Proposition 3.2.3. *Let $\mathfrak{F}_1, \mathfrak{F}_2$ be independent sub σ-algebras of $\mathfrak{F}$ and let $X_1 \in \mathfrak{F}_1$, $X_2 \in \mathfrak{F}_2$ (i.e., X_1 is an $(\Omega, \mathfrak{F}_1, \mathbf{P})$ random variable and X_2 is an $(\Omega, \mathfrak{F}_2, \mathbf{P})$ random variable).*

Assume that X_1, X_2 are either both integrable or both non-negative. Then,

$$E[X_1 X_2] = E[X_1]E[X_2].$$

In particular, $X_1 X_2$ is integrable (in both cases) if each X_i is integrable.

Proof. Consider first the case where X_1 and X_2 are both non-negative. Let $X_2 = 1_{A_2}$, $A_2 \in \mathfrak{F}_2$, and $\Psi_1 = \{X_1 \geq 0, X_1 \in \mathfrak{F}_1 \mid E[X_1 1_{A_2}] = E[X_1]E[1_{A_2}] = E[X_1]\mathbf{P}(A_2)\}$.

Since the σ-algebras are independent, $1_{A_1} \in \Psi_1$ for all $A_1 \in \mathfrak{F}_1$. Hence, every simple function $s = \sum_{n=1}^{N} a_n 1_{A_n}, A_n \in \mathfrak{F}_1, 1 \leq n \leq N$, is in Ψ_1. The principle of monotone convergence and (RV_8) in Proposition 2.1.11 imply that Ψ_1 contains all the non-negative random variables $X_1 \in \mathfrak{F}_1$. Now fix a non-negative $X_1 \in \mathfrak{F}_1$ and look at $\Psi_2 = \{X_2 \in \mathfrak{F}_2, X_2 \geq 0 \mid E[X_1 X_2] = E[X_1]E[X_2]\}$. Since by what has been shown, Ψ_2 contains 1_{A_2} for all $A_2 \in \mathfrak{F}_2$, it follows (as above) that Ψ_2 contains all the non-negative $X_2 \in \mathfrak{F}_2$. This proves the result for X_1 and X_2 when they are both non-negative.

Now assume $X_1 = Y_1 - Z_1$, $X_2 = Y_2 - Z_2$, where Y_i and Z_i are non-negative, belong to $\mathfrak{F}_i$, and both are integrable (i.e., $E[Y_i], E[Z_i] < +\infty$). Then $Y_1 Y_2, Y_1 Z_2, Z_1 Y_2$, and $Z_1 Z_2$ are all integrable by what has been proved. Also, for example, $E[Y_1 Y_2] = E[Y_1]E[Y_2] < +\infty$. Hence, $X_1 X_2 = (Y_1 Y_2 + Z_1 Z_2) - (Y_1 Z_2 + Z_1 Y_2)$ is integrable and $E[X_1 X_2] = E[Y_1 Y_2 + Z_1 Z_2] - E[Y_1 Z_2 + Z_1 Y_2] = E[X_1]E[X_2]$, since for each of the products $Y_1 Y_2, Y_1 Z_2, Z_1 Y_2$, and $Z_1 Z_2$, the expectation of the product is the product of the expectations of the factors. □

Remark. In general, the product of two integrable random variables is not integrable. However, if each of their squares is integrable, then the product is integrable (see Proposition 4.1.6).

The next result shows that the marginal distributions of $X = (X_1, X_2)$ determine the law of X if X_1 and X_2 are independent random variables.

Proposition 3.2.4. *Let $\mathfrak{F}_1$ and $\mathfrak{F}_2$ be independent sub σ-algebras of $\mathfrak{F}$, where $(\Omega, \mathfrak{F}, \mathbf{P})$ is a probability space.*

Let $X_1 \in \mathfrak{F}_1$ and $X_2 \in \mathfrak{F}_2$ have distributions $\mathbf{Q}_1$ and $\mathbf{Q}_2$, respectively. Let $X = (X_1, X_2)$. The distribution $\mathbf{Q}$ of X is determined by $\mathbf{Q}_1$ and $\mathbf{Q}_2$ and is the unique probability $\mathbf{Q}$ on $\mathfrak{B}(\mathbb{R}^2)$ such that

(*) $\qquad \mathbf{Q}(E_1 \times E_2) = \mathbf{Q}_1(E_1)\mathbf{Q}_2(E_2) \quad \text{if } E_i \in \mathfrak{B}(\mathbb{R}).$

Proof. $X^{-1}(E_1 \times E_2) = X_1^{-1}(E_1) \cap X_2^{-1}(E_2)$. Let $A_i = X_i^{-1}(E_i)$. Then $\mathbf{P}(A_1 \cap A_2) = \mathbf{P}(A_1)\mathbf{P}(A_2)$, i.e., $\mathbf{Q}(E_1 \times E_2) = \mathbf{Q}_1(E_1)\mathbf{Q}_2(E_2)$. It remains to verify that there is at most one probability on $\mathfrak{B}(\mathbb{R}^2)$ satisfying (*). Let $\mathfrak{A}$ denote the collection of finite unions $\cup_{i=1}^{n} E_{1,i} \times E_{2,i}$, where $E_{1,i}$ and $E_{2,i} \in \mathfrak{B}(\mathbb{R})$. Then, in view of Exercise 3.2.5, $\mathfrak{A}$ is a Boolean algebra. Any two probabilities on $\mathfrak{B}(\mathbb{R}^2)$ that satisfy (*) agree on $\mathfrak{A}$. Since by Proposition 3.1.5 $\mathfrak{B}(\mathbb{R}^2)$ is the smallest σ-algebra containing $\mathfrak{A}$, it follows from the monotone class argument used in the proof of Theorem 1.4.13 that two probabilities on $\mathfrak{B}(\mathbb{R}^2)$ that agree on $\mathfrak{A}$ also agree on $\mathfrak{B}(\mathbb{R}^2)$. □

2. INDEPENDENCE

Exercise 3.2.5. This is an algebraic exercise that has to do with the relation between collections $\mathfrak{C}$ of sets that are closed under finite intersections and the smallest Boolean algebra $\mathfrak{A}$ containing $\mathfrak{C}$. It has three parts.

Part A. Let $\mathfrak{C}$ be a collection of subsets of Ω such that
  (1) $C_1, C_2 \in \mathfrak{C}$ implies $C_1 \cap C_2 \in \mathfrak{C}$, and
  (2) $C \in \mathfrak{C}$ implies C^c is the union of a finite number of sets from $\mathfrak{C}$.

Let $\mathfrak{A} = \{A \mid A = \cup_{i=1}^n C_i, C_i \in \mathfrak{C}, n \geq 1\}$. Show that
  (3) $\mathfrak{A}$ is a Boolean algebra of subsets of Ω.

Part B. Let $\mathfrak{E}$ be a collection of subsets of Ω that is closed under complements, i.e., if $E \in \mathfrak{E}$, then $E^c \in \mathfrak{E}$. Let $\mathfrak{C}$ be the collection of finite intersections of sets from $\mathfrak{E}$: $C \in \mathfrak{C}$ implies $C = \cap_{i=1}^n E_i$, $E_i \in \mathfrak{E}$.
Show that
  (1) if $C \in \mathfrak{C}$, then C^c is a finite disjoint union of sets from $\mathfrak{C}$. [Hint: use Remark 1.2.3.]

Now assume that $\mathfrak{C}$ has property (1). Show that
  (2) each $A \in \mathfrak{A}$ may be written as a finite disjoint union of sets from $\mathfrak{C}$.

Part C. Let $\mathfrak{C}$ be a collection of subsets of Ω such that
  (1) $C_1, C_2 \in \mathfrak{C}$ implies $C_1 \cap C_2 \in \mathfrak{C}$,

and let $\mathfrak{L} \supset \mathfrak{C}$ be a collection of subsets of Ω such that
  (2) $\Omega \in \mathfrak{L}$, and
  (3) if $A \subset B$ are sets in $\mathfrak{L}$, then $B \backslash A \in \mathfrak{L}$.

Let $\mathfrak{E} = \mathfrak{C} \cup \mathfrak{C}^c$, where $B \in \mathfrak{C}^c$ if and only if $B^c \in \mathfrak{C}$. Show that
  (4) $\mathfrak{E} \subset \mathfrak{L}$,
  (5) if $A, B_1, B_2, \ldots, B_p \in \mathfrak{C}$ then $A \cap (\cap_{i=1}^p B_i^c) \in \mathfrak{L}$. [Hints: $A \cap B_1^c \in \mathfrak{L}$ since it coincides with $A \backslash (A \cap B_1)$; since $A \cap B_1^c \in \mathfrak{L}$ for any $A, B_1 \in \mathfrak{C}$, then $A \cap B_1^c \cap B_2 \in \mathfrak{L}$, and so $(A \cap B_1^c) \backslash (A \cap B_1^c \cap B_2) = A \cap B_1^c \cap B_2^c \in \mathfrak{L}$.]

Let $\mathfrak{C}_0$ be the collection of finite intersections of sets from $\mathfrak{E}$. Then (5) implies that $\mathfrak{C}_0 \subset \mathfrak{L}$. It follows that $\mathfrak{C}_0$ satisfies part B (1), and so the collection of finite disjoint unions of sets from $\mathfrak{C}_0$ is a Boolean algebra by part B (2). Show that
  (6) if $\mathfrak{A}$ is the smallest Boolean algebra containing $\mathfrak{C}$, then $\mathfrak{A} \supset \mathfrak{C}_0$,
  (7) $\mathfrak{A}$ is the collection of finite disjoint unions of sets from $\mathfrak{C}_0$, and
  (8) if $E_1, E_2, \ldots, E_q$ are pairwise disjoint sets from $\mathfrak{C}_0$, then $\cup_{i=1}^q E_i \in \mathfrak{L}$. [Hint: if $E_1, E_2 \in \mathfrak{L}$ and $E_1 \cap E_2 = \emptyset$, then $E_1^c \supset E_2$ and so $E_1^c \cap E_2^c \in \mathfrak{L}$.]

Conclude that $\mathfrak{A} \subset \mathfrak{L}$.

The uniqueness question that was settled before this long exercise can be handled very easily using the next result, which is due to Dynkin [D3] (referred to as Dynkin's π-λ theorem by Billingsley [B1]). It follows from part C of the previous exercise that it is in effect another version of the monotone class theorem (1.4.16) and is therefore more appropriately referred to nowadays by that name. The advantage of Dynkin's version of this theorem is that its hypotheses are easier to apply than those of the earlier form in Exercise 1.4.16.

Proposition 3.2.6. (Dynkin's version of the monotone class theorem). *Let $\mathfrak{C}$ be a class of subsets of Ω that is closed under finite intersections. Let $\mathfrak{L} \supset \mathfrak{C}$ be a monotone class (Exercise 1.4.15) of subsets such that (i) $\Omega \in \mathfrak{L}$ and (ii) $A \subset B$ sets $\in \mathfrak{L}$ implies $B \backslash A \in \mathfrak{L}$. Then $\mathfrak{L}$ contains the smallest σ-algebra $\mathfrak{F}_0$ containing $\mathfrak{C}$.*

Proof. Let $\mathfrak{L}_0 \supset \mathfrak{C}$ be the smallest monotone class satisfying (i) and (ii). Then $\mathfrak{F}_0 \supset \mathfrak{L}_0$. Since $\mathfrak{L}_0$ is closed under complementation (in view of (i) and (ii)), it is closed under finite intersections if and only if it is closed under finite unions. When either of these two conditions is verified, $\mathfrak{L}_0$ is a σ-algebra since $\cup_{n=1}^{\infty} A_n$ is the monotone union $\cup_{N=1}^{\infty} (\cup_{n=1}^{N} A_n)$. Hence, $\mathfrak{L}_0 = \mathfrak{F}_0$ if $\mathfrak{L}_0$ is closed under finite intersections.

To show that $\mathfrak{L}_0$ is closed under finite intersections, first note that $\mathfrak{L}_1 = \{E \in \mathfrak{L}_0 \mid E \cap A \in \mathfrak{L}_0 \text{ for all } A \in \mathfrak{C}\}$ is a monotone class containing $\mathfrak{C}$ that satisfies (i) and (ii). Hence $\mathfrak{L}_1 = \mathfrak{L}_0$ (i.e., $E \in \mathfrak{L}_0, A \in \mathfrak{C}$ implies $E \cap A \in \mathfrak{L}_0$).

Now let $\mathfrak{L}_2 = \{E \in \mathfrak{L}_0 \mid E \cap F \in \mathfrak{L}_0 \text{ for all } F \in \mathfrak{L}_0\}$. Then $\mathfrak{L}_2$ is a monotone class containing $\mathfrak{C}$ (by the previous observation) and satisfies (i) and (ii). Hence, $\mathfrak{L}_2 = \mathfrak{L}_0$. □

Remarks. (1) One cannot replace condition (ii) by the weaker condition (ii): $\mathfrak{L}$ is closed under complements. Consider the class consisting of the unbounded intervals of $\mathbb{R}$ and the empty set: it is a monotone class, closed under complements, contains $\mathbb{R}$, and yet does not contain the smallest σ-algebra containing all the intervals $(-\infty, x], x \in \mathbb{R}$, which is $\mathfrak{B}(\mathbb{R})$.

(2) A monotone class $\mathfrak{L}$ satisfying the conditions of Proposition 3.2.6 is called a λ-system by Dynkin [D3] if, in addition, it contains $A_1 \cup A_2$ whenever $A_1, A_2 \in \mathfrak{L}$ and $A_1 \cap A_2 = \emptyset$.

Corollary 3.2.7. *Let Ω be a set and $\mathfrak{F}$ be a σ-algebra of subsets of Ω. Let $\mathbf{P}_1$ and $\mathbf{P}_2$ be two probabilities on $\mathfrak{F}$. If $\mathfrak{C} \subset \mathfrak{F}$ is closed under finite intersections and $\mathbf{P}_1(A) = \mathbf{P}_2(A)$ for all $A \in \mathfrak{C}$, then $\mathbf{P}_1$ and $\mathbf{P}_2$ agree on $\mathfrak{F}_0$, the smallest σ-algebra containing $\mathfrak{C}$.*

Proof. Let $\mathfrak{L} = \{E \in \mathfrak{F} \mid \mathbf{P}_1(E) = \mathbf{P}_2(E)\}$. Then $\mathfrak{L} \supset \mathfrak{C}$ is a monotone class satisfying (i) and (ii) of Proposition 3.2.6 and so $\mathfrak{L} \supset \mathfrak{F}_0$. □

Remark. This corollary gives the uniqueness of the probability $\mathbf{Q}$ in Proposition 3.2.4: let $\mathfrak{C} = \{E_1 \times E_2 \mid E_i \in \mathfrak{B}(\mathbb{R})\}$.

Let $(X_i)_{1 \leq i \leq n}$ be a finite collection of real-valued functions on a set Ω. Consider all the σ-algebras $\mathfrak{G}$ on Ω for which all the X_i are measurable.

2. INDEPENDENCE

Such σ-algebras exist, e.g., $\mathfrak{G} = \mathfrak{P}(\Omega)$, the σ-algebra of all subsets of Ω. Let $\sigma(\{X_i \mid 1 \leq i \leq n\})$ denote the smallest of these σ-algebras (the intersection of all the above σ-algebras). If, for each i, $E_i \in \mathfrak{B}(\mathbb{R})$, then $X_i^{-1}(E_i) = \{X_i \in E_i\} \in \sigma(\{X_i \mid 1 \leq i \leq n\})$ and so too is $\cap_{i=1}^n X_i^{-1}(E_i)$. Let $\mathfrak{C}$ be the collection of all sets of this form (it includes, for example, $\cap_{i=1}^{n-1} X_i^{-1}(E_i)$, by setting $E_n = \mathbb{R}$, and each $X_i^{-1}(E_i)$ by setting all the other $E_j = \mathbb{R}$). Then $\mathfrak{C}$ is closed under finite intersections, and $\sigma(\{X_i \mid 1 \leq i \leq n\})$ is the smallest σ-algebra of subsets of Ω containing $\mathfrak{C}$.

In particular, if $(\Omega, \mathfrak{F}, \mathbf{P})$ is a probability space and the X_i are random variables, then $\mathfrak{C} \subset \mathfrak{F}$ and hence $\sigma(\{X_i \mid 1 \leq i \leq n\}) \subset \mathfrak{F}$. The result in Proposition 3.2.4 about the distribution of $X = (X_1, X_2)$ when X_1 and X_2 are independent, — when they are measurable with respect to independent σ-algebras — has a corresponding version for n random variables, which is stated as the next proposition. Notice that here the form of the distribution of $X = (X_1, X_2, \ldots, X_n)$ is equivalent to a statement about independence.

Proposition 3.2.8. *Let $(X_i)_{1 \leq i \leq n}$ be a finite collection of real-valued random variables on $(\Omega, \mathfrak{G}, \mathbf{P})$. The following conditions are equivalent:*

(1) *if $\mathbf{Q}$ is the distribution of $X = (X_1, X_2, \ldots, X_n)$ and $\mathbf{Q}_i$ is the distribution of X_i, $1 \leq i \leq n$, then*

$$\mathbf{Q}(E_1 \times E_2 \times \cdots \times E_n) = \prod \mathbf{Q}_i(E_i);$$

(2) *if $E_i \in \mathfrak{B}(\mathbb{R})$, for $1 \leq i \leq n$, then*

$$\mathbf{P}\left[\cap_{i=1}^n \{X_i \in E_i\}\right] = \prod_{i=1}^n \mathbf{P}\left[X_i \in E_i\right]$$

(where $\prod_{i=1}^n p_i$ denotes the product of the numbers p_i); and
(3) *for every $F \subset \{1, 2, \ldots, n\}$ and $F' = \{1, 2, \ldots, n\} \setminus F$, the σ-algebras*

$$\mathfrak{F} = \sigma(\{X_i \mid i \in F\}) \quad \text{and} \quad \mathfrak{F}' = \sigma(\{X_i \mid i \in F'\})$$

are independent.

Proof. The equivalence of (1) and (2) is evident. Assume (2). Let $\mathfrak{C}$ be the collection of sets of the form $\cap_{i \in F} \{X_i \in E_i\}$ and $\mathfrak{C}'$ be the collection of sets of the form $\cap_{i \in F'} \{X_i \in E_i\}$. Hypothesis (2) implies $\mathbf{P}(A \cap A') = \mathbf{P}(A)\mathbf{P}(A')$ if $A \in \mathfrak{C}$ and $A' \in \mathfrak{C}'$ (to see this, note, for example, that (2) implies that $\mathbf{P}[\cap_{i \in F} \{X_i \in E_i\}] = \Pi_{i \in F} \mathbf{P}[X_i \in E_i]$: set the remaining $E_j = \mathbb{R}$).

Let $\mathfrak{L} = \{E \in \mathfrak{F} \mid \text{for all } A' \in \mathfrak{C}', \mathbf{P}(E \cap A') = \mathbf{P}(E)\mathbf{P}(A')\}$. Then $\mathfrak{L} \supset \mathfrak{C}$ is a monotone class satisfying (i) and (ii) of Proposition 3.2.6 and so $\mathfrak{L} \supset \sigma(\{X_i \mid i \in F\})$. Similarly, if $\mathfrak{L}_1 = \{E \in \mathfrak{F} \mid \text{for all } A \in \sigma(\{X_i \mid i \in F\}), \mathbf{P}(A \cap E) = \mathbf{P}(A)\mathbf{P}(E)\}$, then $\mathfrak{L}_1$ is a monotone class containing $\mathfrak{C}'$ and hence $\sigma(\{X_i \mid i \in F'\})$. This proves (3).

Assume (3). If $n = 2$, then (2) is obvious. Assume that the conclusion holds for $n = k$. If $n = k+1$, let $F = \{1, 2, \ldots, k\}$ and $F' = \{k+1\}$. Then $\mathbf{P}[(\cap_{i=1}^{k}\{X_i \in E_i\}) \cap \{X_{k+1} \in E_{k+1}\}] = \mathbf{P}[\cap_{i=1}^{k}\{X_i \in E_i\}]\mathbf{P}[X_{k+1} \in E_{k+1}]$. By the inductive assumption, this is $\prod_{i=1}^{k+1} \mathbf{P}[\cap_{i=1}^{k+1}\{X_i \in E_i\}]$. □

Definition 3.2.9. *A finite collection of random variables* $(X_i)_{1 \leq i \leq n}$ *is said to be* **independent** *if*

$$\mathbf{P}[\cap_{i=1}^{n}\{X_i \in E_i\}] = \prod_{i=1}^{n} \mathbf{P}[X_i \in E_i],$$

for any collection $(E_i)_{1 \leq i \leq n}$ *of Borel sets* E_i.

Remark. Proposition 3.2.8 shows that a finite collection of independent random variables is a particular case of a **family of independent random variables** (see Definition 3.4.3.). Definition 3.2.9 is the standard definition of a finite collection of independent random variables (see Loève [L2], Chung [C], pp. 49–50, Billingsley [B1], p. 16).

Corollary 3.2.10. *Let* $(X_i)_{1 \leq i \leq n}$ *be a finite collection of real-valued random variables on* $(\Omega, \mathfrak{F}, \mathbf{P})$ *that is independent, i.e., is such that, if* $E_i \in \mathfrak{B}(\mathbb{R})$, $1 \leq i \leq n$, *then*

$$\mathbf{P}[\cap_{i=1}^{n}\{X_i \in E_i\}] = \prod_{i=1}^{n} \mathbf{P}[X_i \in E_i].$$

Let $\varphi(x_1, \ldots, x_k)$ *and* $\psi(x_{k+1}, \ldots, x_n)$ *be Borel functions on* $\mathbb{R}^k, \mathbb{R}^{n-k}$, $i \leq k < n$. *Define* $\Phi = \varphi(X_1, \ldots, X_k)$ *and* $\Psi = \psi(X_{k+1}, \ldots, X_n)$. *Then for* $A, B \in \mathfrak{B}(\mathbb{R})$,

$$\mathbf{P}[\Phi \in A, \Psi \in B] = \mathbf{P}[\Phi \in A]\mathbf{P}[\Psi \in B].$$

In other words, X and Y are independent random variables.

Proof. Let $X = (X_1, \ldots, X_k)$ and $Y = (X_{k+1}, \ldots, X_n)$. Then $\Phi^{-1}(A) = X^{-1}(\varphi^{-1}(A))$. Since $\varphi^{-1}(A) \in \mathfrak{B}(\mathbb{R}^k)$ and X is a random vector relative to $\mathfrak{F}_0 = \sigma(\{X_i \mid 1 \leq i \leq k\})$, it follows that $\Phi^{-1}(A) \in \mathfrak{F}_0$. Similarly, $\Psi^{-1}(B) \in \mathfrak{F}_1 = \sigma(\{X_i \mid k+1 \leq i \leq\})$.

Since these two σ-algebras are independent, the result follows. □

Exercise 3.2.11. Let $\mathfrak{C}$ be a class of measurable subsets of a probability space $(\Omega, \mathfrak{F}, \mathbf{P})$.

If $\mathfrak{C}$ is independent of itself, show that $\mathbf{P}(A) = 0$ or 1 if $A \in \mathfrak{C}$. In particular, show that if $\mathfrak{T} \subset \mathfrak{F}$ is a σ-algebra that is independent of itself, then the probability of any event is either 0 or 1. (The probability space $(\Omega, \mathfrak{T}, \mathbf{P})$ is therefore an example of an **ergodic probability space**.)

Exercise 3.2.12. Let $\Omega = \{0, 1, 2, \ldots, n\}$ and $\mathfrak{F} = \mathfrak{P}(\Omega)$. Determine all the probabilities on $(\Omega, \mathfrak{F})$ that take only the values 0 and 1.

Exercise 3.2.13. Let $\mathfrak{C}$ be a class of measurable subsets of a probability space $(\Omega, \mathfrak{F}, \mathbf{P})$. Show that the class $\mathfrak{L}$ of sets A such that $\mathbf{P}(A \cap B) = \mathbf{P}(A)\mathbf{P}(B)$ for all $B \in \mathfrak{C}$ is a λ-system (see the remarks following Proposition 3.2.6 for the definition of a λ-system). Show, with an example, that $\mathfrak{L}$ need not be a σ-algebra (see Feller [F1], p. 127).

3. PRODUCT MEASURES

At this point, it is not evident that independent random variables exist, even pairs of independent random variables. Intuitively, in cases like coin tossing, they can be seen to exist, but in these notes no mathematical construct has yet been given that explains their existence. In the case of the real line, it follows from the first two chapters that for any distribution function F_1 (equivalently, the corresponding probability $\mathbf{P}_1$) on the real line there is at least one probability space and a random variable on it whose distribution function is the given one: one may take the probability space to be $(\mathbb{R}, \mathfrak{B}(\mathbb{R}), \mathbf{P}_1)$ and $X(x) = x$ for all $x \in \mathbb{R}$.

In view of the earlier discussion in §1 concerning random vectors, one could also take the following space $(\mathbb{R}^2, \mathfrak{F}_1^0, \mathbf{P}_1^0)$ and $X_1(x, y) = x$ for all $(x, y) \in \mathbb{R}^2$, where $\mathfrak{F}_1^0$ consists of the sets $A \times \mathbb{R}$, $A \in \mathfrak{B}(\mathbb{R})$, and $\mathbf{P}_1^0(A \times \mathbb{R}) = \mathbf{P}_1(A)$.

Suppose that $\mathbf{P}_2$ is another probability on the real line. Then it is the distribution of the random variable X_2 defined on $(\mathbb{R}^2, \mathfrak{F}_2^0, \mathbf{P}_2^0)$, where $X_2(x, y) = y$ for all $y \in \mathbb{R}$, $\mathbf{P}_2^0(\mathbb{R} \times B) = \mathbf{P}_2(B)$ for all $B \in \mathfrak{B}(\mathbb{R})$, and $\mathfrak{F}_2^0$ consists of the sets $\mathbb{R} \times B$ $B \in \mathfrak{B}(\mathbb{R})$.

The two σ-algebras $\mathfrak{F}_1^0$ and $\mathfrak{F}_2^0$ generate the σ-algebra $\mathfrak{B}(\mathbb{R}^2)$ in view of Proposition 3.1.5, i.e., $\mathfrak{B}(\mathbb{R}^2)$ is the smallest σ-algebra containing them both. So the question arises: is there a probability $\mathbf{Q}$ on $\mathfrak{B}(\mathbb{R}^2)$ that agrees with $\mathbf{P}_i^0$ on each $\mathfrak{F}_i^0$? If so, then given two probabilities on the real line, one would have one probability space and a pair of random variables on it whose distributions are the respective probabilities. Notice that Proposition 3.2.4 shows that there is at most one solution $\mathbf{Q}$ to the problem, if in addition one requires X_1 and X_2 to be independent random variables.

To motivate further the construction of the probability $\mathbf{Q}$ from $\mathbf{P}_1$ and $\mathbf{P}_2$, consider the problem of choosing a point (x, y) at random from the unit square $[0, 1] \times [0, 1]$. From Fig. 3.1, it is intuitively clear that if $\frac{1}{8} < x \leq \frac{1}{2}$ and $\frac{1}{4} < y \leq \frac{3}{4}$, then this probability should be $|(\frac{1}{8}, \frac{1}{2}]| \times |(\frac{1}{4}, \frac{3}{4}]| = \frac{3}{16}$.

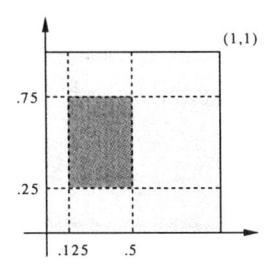

Fig. 3.1

What about the probability that the point lies in a more general set? For example, suppose that $x \in E_1$ and $y \in E_2$, where E_1 and E_2 are Borel subsets of [0,1]. The probability that x lies in E_1 is $|E_1|$ and that y lies in E_2 is $|E_2|$. This suggests that the probability that the point (x,y) lies in $E_1 \times E_2$ should be $|E_1| \times |E_2|$. What if $x \leq y$? Then the point (x,y) lies above the diagonal of the square and the probability should be $\frac{1}{2}$, the area of this portion of the square. For a general subset E of the square one expects the probability to be the area of the set E. This begs the question: which subsets of the unit square have an "area"? Quite general "boxes" $E_1 \times E_2$ with Borel sides E_1, E_2 should have an area equal to the product of the length of the sides, where length is determined by Lebesgue measure.

In general, the "area" of a Borel "box" $E_1 \times E_2$ is determined by two probabilities or positive measures, one for the first side and the other for the second. The problem is then how to extend this notion of "area" to the Borel subsets of the plane. Rather than proceeding with this specific example, consider the problem of defining the product of two probability spaces (or, more generally, two measure spaces) $(\Omega_1, \mathfrak{F}_1, \mathbf{P}_1)$ and $(\Omega_2, \mathfrak{F}_2, \mathbf{P}_2)$.

The first question to settle is how it is possible to define a "natural" σ-algebra $\mathfrak{F}$ on $\Omega_1 \times \Omega_2$ that contains all the "boxes" $E_1 \times E_2$, with $E_i \in \mathfrak{F}_i$, $i = 1, 2$.

The determination of this σ-algebra $\mathfrak{F}$ is not hard. Define two projections $pr_1 \colon \Omega_1 \times \Omega_2 \to \Omega_1$ and $pr_2 \colon \Omega_1 \times \Omega_2 \to \Omega_2$ by $pr_1(\omega_1, \omega_2) = \omega_1$ and $pr_2(\omega_1, \omega_2) = \omega_2$. Let $\mathfrak{F}$ be the smallest σ-algebra of subsets of $\Omega_1 \times \Omega_2$ that contains all the sets $pr_i^{-1}(E_i), E_i \in \mathfrak{F}_i$. Equivalently, let $\mathfrak{F}$ be the smallest σ-algebra of subsets of $\Omega_1 \times \Omega_2$ that contains all the sets $E_1 \times \Omega_2$ and $\Omega_1 \times E_2$. Note that $pr_1^{-1}(E_1) = E_1 \times \Omega_2$ and $pr_2^{-1}(E_2) = \Omega_1 \times E_2$. Fig. 3.2 is a diagram of $E_1 \times \Omega_2$.

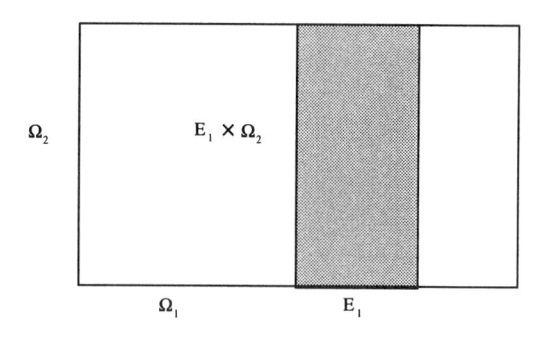

Fig. 3.2

The σ-algebra $\mathfrak{F}$ will be denoted by $\mathfrak{F}_1 \times \mathfrak{F}_2$ (it is often denoted by $\mathfrak{F}_1 \otimes \mathfrak{F}_2$). Note that in case $\Omega_i = \mathbb{R}$ and $\mathfrak{F}_i = \mathfrak{B}(\mathbb{R})$, then $\mathfrak{F}_1 \times \mathfrak{F}_2 = \mathfrak{B}(\mathbb{R}^2)$ by Proposition 3.1.5.

The next step is to make use of the two probabilities (or measures) to define a probability $\mathbf{P}$ (or measure) on $\mathfrak{F}_1 \times \mathfrak{F}_2$ such that

(*) $$\mathbf{P}(E_1 \times E_2) = \mathbf{P}_1(E_1)\mathbf{P}_2(E_2)$$

whenever $E_i \in \mathfrak{F}_i$, $i = 1, 2$.

The σ-algebra $\mathfrak{F}_1 \times \mathfrak{F}_2$ is generated by a Boolean algebra $\mathfrak{A}$. To describe it, let $\mathfrak{C}$ be the collection of sets of the form $E_1 \times E_2 \subset \Omega_1 \times \Omega_2$, where $E_i \in \mathfrak{F}_i, i = 1, 2$. If $\mathfrak{E}$ consists of the sets of the form $E_1 \times \Omega_2$ or $\Omega_1 \times E_2$ with $E_i \in \mathfrak{F}_i$, this collection is closed under complements and $\mathfrak{C}$ is the collection of finite intersections of sets from $\mathfrak{E}$. Hence $\mathfrak{C}$ satisfies the hypotheses of part B of Exercise 3.2.5. As a result, if $\mathfrak{A}$ is the collection of finite unions of sets from $\mathfrak{C}$, it follows that $\mathfrak{A}$ is a Boolean algebra and every set $A \in \mathfrak{A}$ can be written as a finite disjoint union of sets from $\mathfrak{C}$.

Exercise 3.3.1. Show that there is a unique finitely additive probability $\mathbf{P}$ on $\mathfrak{A}$ such that $\mathbf{P}(E_1 \times E_2) = \mathbf{P}_1(E_1)\mathbf{P}_2(E_2)$ for all sets $E_i \in \mathfrak{F}_i, i = 1, 2$. [Hints: First show that it suffices to show that if a "box" $E_1 \times E_2$ is a finite disjoint union of "boxes" $E_1^i \times E_2^i, 1 \leq i \leq n$, then

(**) $$\mathbf{P}(E_1 \times E_2) = \sum_{i=1}^{n} \mathbf{P}(E_1^i \times E_2^i).$$

The identity (**) may be proved in at least two ways.

(1) Use the "sides" of the boxes $E_1^i \times E_2^i$ to define an equivalence relation on E_1 and on E_2: set $x \sim y$ if $x \in E_1^i$ if and only if $y \in E_1^i$. The equivalence relation on E_1 splits it into a finite number of measurable sets

$A_1, A_2, \ldots, A_k, \ldots, A_K$, and similarly E_2 is split into a finite number of measurable sets $B_1, B_2, \ldots, B_\ell, \ldots, B_L$. Show that the sets $A_k \times B_\ell$ partition $E_1 \times E_2$, i.e., they are pairwise disjoint and their union is $E_1 \times E_2$. Make use of this partition to prove (**).

(2) The ideas involved in Lemma 3.3.3 may also be used to prove (**).]

If **P** is σ-additive on $\mathfrak{A}$, the Carathéodory procedure applies without change to the corresponding outer measure $\mathbf{P}^*$, which one defines by setting $\mathbf{P}^*(E) \stackrel{\text{def}}{=} \inf\{\sum_{n=1}^\infty \mathbf{P}(A_n) \mid E \subset \cup_{n=1}^\infty A_n, A_n \in \mathfrak{A}, \text{ for all } n \geq 1\}$. As a result, to obtain a probability **P** on $\mathfrak{F}$ satisfying (*), it suffices to show that **P** is σ-additive (see Definition 1.4.2) on $\mathfrak{A}$. Note that since $\mathfrak{F} = \mathfrak{F}_1 \times \mathfrak{F}_2$ is the smallest σ-algebra containing $\mathfrak{A}$, the outer measure $\mathbf{P}^*$ is a probability when restricted to $\mathfrak{F}$.

Exercise 3.3.2. Let **P** be a finitely additive probability on a Boolean algebra $\mathfrak{A}$. Show that the following conditions are equivalent:

(1) **P** is σ-additive on $\mathfrak{A}$;
(2) $(A_n)_{n\geq 1} \subset \mathfrak{A}$, $A_n \supset A_{n+1}$, and $\cap_{n=1}^\infty A_n = A$ imply $\mathbf{P}(A_n) \downarrow \mathbf{P}(A)$ if $A \in \mathfrak{A}$;
(3) $(A_n)_{n\geq 1} \subset \mathfrak{A}$, $A_n \supset A_{n+1}$, and $\cap_{n=1}^\infty A_n = \emptyset$ imply $\mathbf{P}(A_n) \downarrow 0$;
(4) if $(A_n)_{n\geq 1} \subset \mathfrak{A}$, $A_n \supset A_{n+1}$, and for all $n \geq 1$, $\mathbf{P}(A_n) \geq \delta$ for some $\delta > 0$, then $\cap_{n=1}^\infty A_n \neq \emptyset$; and
(5) $(A_n)_{n\geq 1} \subset \mathfrak{A}$, $A_n \subset A_{n+1}$, and $\cup_{n=1}^\infty A_n = A$ imply $\mathbf{P}(A_n) \uparrow \mathbf{P}(A)$ when $A \in \mathfrak{A}$.

The principle of monotone convergence implies that the probability **P** is σ-additive on $\mathfrak{A}$. To see this, the following lemma is useful.

Lemma 3.3.3. *If $A \in \mathfrak{A}$ and $\omega_1 \in \Omega_1$, then*

(1) *the function $\omega_2 \to 1_A(\omega_1, \omega_2)$ is in $\mathfrak{F}_2$, i.e., it is measurable relative to $\mathfrak{F}_2$,*
(2) *the function $\omega_1 \to \int 1_A(\omega_1, \omega_2) \mathbf{P}_2(d\omega_2)$ is in $\mathfrak{F}_1$, and*
(3) $\mathbf{P}(A) = \int [\int 1_A(\omega_1, \omega_2) \mathbf{P}_2(d\omega_2)] \mathbf{P}_1(d\omega_1)$.

Proof. Clearly, (1), (2), and (3) hold if $A = E_1 \times E_2$, $E_i \in \mathfrak{F}_i$. It is therefore true for finite disjoint unions of sets of the form $E_1 \times E_2$, $E_i \in \mathfrak{F}_i$, and hence for all sets in $\mathfrak{A}$. □

To prove Exercise 3.3.2 (4), and thus the σ-additivity of **P** on $\mathfrak{A}$, note that $1_{A_n} \uparrow 1_A$ if $A_n \subset A_{n+1}$ and $\cup_n A_n = A$. The principle of monotone convergence and Lemma 3.3.3 (2) imply that

$$Y_n(\omega_1) \stackrel{\text{def}}{=} \int 1_{A_n}(\omega_1, \omega_2) \mathbf{P}_2(d\omega_2) \uparrow Y(\omega_1) \stackrel{\text{def}}{=} \int 1_A(\omega_1, \omega_2) \mathbf{P}_2(d\omega_2)$$

and hence (again by monotone convergence and Lemma 3.3.3)

$$\mathbf{P}(A_n) = \int Y_n(\omega_1) \mathbf{P}_1(d\omega_1) \uparrow \int Y(\omega_1) \mathbf{P}_1(d\omega_1) = \mathbf{P}(A). \quad \square$$

This completes most of the proof of the following theorem.

3. PRODUCT MEASURES

Theorem 3.3.4. *Let $(\Omega_1, \mathfrak{F}_1, \mathbf{P}_1)$ and $(\Omega_2, \mathfrak{F}_2, \mathbf{P}_2)$ be two probability spaces. Then there is a smallest σ-algebra $\mathfrak{F}$ of subsets of $\Omega_1 \times \Omega_2$ and a unique probability $\mathbf{P}$ on $\mathfrak{F}$ such that*

(1) $E_1 \times E_2 \in \mathfrak{F}$ if $E_i \in \mathfrak{F}_i, i = 1, 2$, and
(2) $\mathbf{P}(E_1 \times E_2) = \mathbf{P}_1(E_1)\mathbf{P}_2(E_2)$ if $E_i \in \mathfrak{F}_i, i = 1, 2$.

Proof. It remains to show the uniqueness of $\mathbf{P}$. This follows from Corollary 3.2.7. □

Remarks. (1) The probability $\mathbf{P}$ is called the **product** of $\mathbf{P}_1$ and $\mathbf{P}_2$ and is denoted by $\mathbf{P}_1 \times \mathbf{P}_2$. When $\mathfrak{F}_1 \times \mathfrak{F}_2$ is denoted by $\mathfrak{F}_1 \otimes \mathfrak{F}_2$, it is usual to denote $\mathbf{P}_1 \times \mathbf{P}_2$ by $\mathbf{P}_1 \times \mathbf{P}_2$. The probability space $(\Omega_1 \times \Omega_2, \mathfrak{F}_1 \times \mathfrak{F}_2, \mathbf{P}_1 \times \mathbf{P}_2)$ is called the **product** of the probability spaces $(\Omega_1, \mathfrak{F}_1, \mathbf{P}_1)$ and $(\Omega_2, \mathfrak{F}_2, \mathbf{P}_2)$.
(2) If $\mathbf{P} = \mathbf{P}_1 \times \mathbf{P}_2$, an integral $\int X d\mathbf{P} = \int X d(\mathbf{P}_1 \times \mathbf{P}_2)$ will sometimes be denoted by $\int X(\omega) \mathbf{P}_1 \times \mathbf{P}_2(d\omega), \int X(\omega_1, \omega_2) \mathbf{P}_1 \times \mathbf{P}_2(d\omega_1, d\omega_2)$, or $\int \int X(\omega_1, \omega_2) \mathbf{P}_1(d\omega_1) \mathbf{P}_2(d\omega_2)$. This last notation is especially compatible with Fubini's theorem (Theorems 3.3.5 and 3.3.6), where the integral is computed as an iterated integral.
(3) Proposition 3.2.4 states in effect that the distribution $\mathbf{Q}$ of (X_1, X_2) is $\mathbf{Q}_1 \times \mathbf{Q}_2$ if $\mathbf{Q}_i$ is the distribution of X_i and the random variables X_1, X_2 are independent.
(4) If $\mathfrak{G}_1 = \{E_1 \times \Omega_2 \mid E_1 \in \mathfrak{F}_1\}$ and $\mathfrak{G}_2 = \{\Omega_1 \times E_2 \mid E_2 \in \mathfrak{F}_2\}$, then $\mathfrak{G}_1$ and $\mathfrak{G}_2$ are independent relative to $\mathbf{P} = \mathbf{P}_1 \times \mathbf{P}_2$.

The next theorem has to do with computing an integral on a product of probability spaces as an iterated integral.

Theorem 3.3.5. (Fubini's theorem for positive random variables functions)

Let $(\Omega_1 \times \Omega_2, \mathfrak{F}_1 \times \mathfrak{F}_2, \mathbf{P}_1 \times \mathbf{P}_2)$ be the product of the probability spaces $(\Omega_1, \mathfrak{F}_1, \mathbf{P}_1)$ and $(\Omega_2, \mathfrak{F}_2, \mathbf{P}_2)$. Let $\mathbf{P} = \mathbf{P}_1 \times \mathbf{P}_2$.
If X is a non-negative random variable on $(\Omega_1 \times \Omega_2, \mathfrak{F}_1 \times \mathfrak{F}_2, \mathbf{P})$, then

(1) *for all $\omega_1 \in \Omega_1$, the function $\omega_2 \to X(\omega_1, \omega_2) \in \mathfrak{F}_2$,*
(2) *the function $\omega_1 \to \int X(\omega_1, \omega_2) \mathbf{P}_2(d\omega_2) \in \mathfrak{F}_1$, and*
(3) $\int [\int X(\omega_1, \omega_2) \mathbf{P}_2(d\omega_2)] \mathbf{P}_1(d\omega_1) = \int X d\mathbf{P}$.

Proof. First, assume that X is the characteristic function of a set $A \in \mathfrak{F}$. If $A \in \mathfrak{A}$, then, by Lemma 3.3.3, $X = 1_A$ satisfies (1), (2), and (3). Let $\mathfrak{M} = \{A \in \mathfrak{F} \mid (1), (2),$ and (3) hold for $1_A\}$. Then $\mathfrak{M}$ is a monotone class: it is closed under increasing unions by the principle of monotone convergence and under decreasing intersections by the theorem of dominated convergence. Since it contains $\mathfrak{A}$, it follows from Exercise 1.4.16 that $\mathfrak{M} = \mathfrak{F}$.

Conditions (1), (2), and (3) hold for $X_1 + X_2$ and $\lambda X, \lambda \geq 0$, if they are satisfied by X_1, X_2, and X. Consequently, the result holds for any simple function.

If (1), (2), (3) are true for $X_n, n \geq 1$, with $0 \leq X_n \leq X$ and $\lim_n X_n = X$, then X satisfies these three conditions. This follows by monotone convergence. □

Corollary 3.3.6. (Fubini's theorem for integrable random variables) *Let $(\Omega_1 \times \Omega_2, \mathfrak{F}_1 \times \mathfrak{F}_2, \mathbf{P}_1 \times \mathbf{P}_2)$ be the product of the probability spaces $(\Omega_1, \mathfrak{F}_1, \mathbf{P}_1)$ and $(\Omega_2, \mathfrak{F}_2, \mathbf{P}_2)$. Let $\mathbf{P} = \mathbf{P}_1 \times \mathbf{P}_2$.*
If X is an integrable random variable on $(\Omega_1 \times \Omega_2, \mathfrak{F}_1 \times \mathfrak{F}_2, \mathbf{P})$, then
   (1) *for $\mathbf{P}_1$-almost all ω_1, the function $\omega_2 \to X(\omega_1, \omega_2) \in \mathfrak{F}_2$ is integrable,*
   (2) *there is an integrable random variable Y on $(\Omega_1, \mathfrak{F}_1, \mathbf{P}_1)$ such that $\mathbf{P}_1$-almost everywhere $\int X(\omega_1, \omega_2)\mathbf{P}_2(d\omega_2) = Y(\omega_1)$, and*
   (3) $\int Y(\omega_1)\mathbf{P}_1(d\omega_1) = \int X d\mathbf{P}$.

Proof. By Proposition 2.1.33, $X = X^+ - X^-$ is integrable if and only if X^+ and X^- are integrable. Fubini's theorem for positive functions (Theorem 3.3.5) applied to X^+ (respectively, X^-) implies that $\omega_2 \to X(\omega_1, \omega_2) \in \mathfrak{F}_2$ and, in view of Exercise 2.1.26, that, for $\mathbf{P}_1$-almost all ω_1 (i.e., the exceptional set has $\mathbf{P}_1$-measure zero), $\omega_2 \to X^\pm(\omega_1, \omega_2)$ is an integrable random variable on $(\Omega_2, \mathfrak{F}_2, \mathbf{P}_2)$. Hence, there is a set $N_1 \in \mathfrak{F}_1$ with $\mathbf{P}_1(N_1) = 0$ such that if $\omega_1 \notin N_1$, $\int X^+(\omega_1, \omega_2)\mathbf{P}_2(d\omega_2) < +\infty$ and $\int X^-(\omega_1, \omega_2)\mathbf{P}_2(d\omega_2) < +\infty$. As a result, $\int X(\omega_1, \omega_2)\mathbf{P}_2(d\omega_2) = \int X^+(\omega_1, \omega_2)\mathbf{P}_2(d\omega_2) - \int X^-(\omega_1, \omega_2)\mathbf{P}_2(d\omega_2)$ is defined for $\omega_1 \notin N_1$.

$$\text{Define } Y(\omega_1) = \begin{cases} 0 & \text{if } \omega_1 \in N_1, \\ \int X(\omega_1, \omega_2)\mathbf{P}_2(d\omega_2) & \text{if } \omega_1 \notin N_1. \end{cases}$$

Now, by Exercise 2.1.29,

$$\int_{\Omega_1 \setminus N_1} \left[\int X^+(\omega_1, \omega_2)\mathbf{P}_2(d\omega_2) \right] \mathbf{P}_1(d\omega_1) = \int X^+ d\mathbf{P} < +\infty,$$

and

$$\int_{\Omega_1 \setminus N_1} \left[\int X^-(\omega_1, \omega_2)\mathbf{P}_2(d\omega_2) \right] \mathbf{P}_1(d\omega_2) = \int X^- d\mathbf{P} < +\infty.$$

Hence, $\int X d\mathbf{P} = \int Y(\omega_1)\mathbf{P}_1(d\omega_1)$. □

Remarks 3.3.7. Condition (2) in Corollary 3.3.6 is a bit clumsy because one does not know that for integrable X, $\int X(\omega_1, \omega_2)\mathbf{P}_2(d\omega_2)$ is defined for all ω_1. All that can be said is that it is defined and finite outside a set of $\mathbf{P}_1$-probability zero. The formulation of Corollary 3.3.6 (2) can be tidied up if one uses the phrase "integrable random variable" to mean a function Y defined on a subset $\tilde{\Omega}_1$ of Ω where $Y : \tilde{\Omega}_1 \to \mathbb{R} \cup \{-\infty, +\infty\}$ and such that
   (a) there is a finite integrable random variable Z on $(\Omega, \mathfrak{F}, \mathbf{P})$ (i.e., $E[Z^+], E[Z^-] < +\infty$), with
   (b) $\mathbf{P}(\{Z = Y\}) = 1$; note that $\{Z = Y\} \subset \tilde{\Omega}_1$.

Then, it follows from Exercise 2.1.31 that $E[Z]$ depends only on Y. One sets $E[Y]$ equal to $E[Z]$.

With this terminological modification, Corollary 3.3.6 (2) and (3) can be replaced by

(2') the function $\omega_1 \to \int X(\omega_1, \omega_2)\mathbf{P}_2(d\omega_2)$ is an integrable random variable on $(\Omega_1, \mathfrak{F}_1, \mathbf{P}_1)$, and

(3') $\int [\int X(\omega_1, \omega_2)\mathbf{P}_2(d\omega_2)]\mathbf{P}_1(d\omega_1) = \int X d\mathbf{P}$.

This meaning of the phrase "integrable random variable" is clearly at variance with Convention 2.1.32 which requires an integrable random variable to be finite valued. That convention was introduced to make it easy to see why $L^1(\Omega, \mathfrak{F}, \mathbf{P})$ is a real vector space.

With this extended notion of integrable random variables, it can be seen that they form a real vector space provided (i) one views null functions as "equal to zero" and (ii) one defines $(X+Y)(\omega)$ to be $X(\omega)+Y(\omega)$ whenever both $X(\omega)$ and $Y(\omega)$ are finite and zero otherwise.

These somewhat clumsy devices can be avoided by passing to the vector space of equivalence classes of integrable functions, where X is equivalent to Y if X equals $Y+$ a null function. This formalizes the identification of the null functions with zero. In this book however, this formal step will not be taken.

Once the product of two probability spaces has been defined, arbitrary finite products $\Omega_1 \times \Omega_2 \times \cdots \times \Omega_n = (\Omega_1 \times \Omega_2 \times \cdots \times \Omega_n, \mathfrak{F}_1 \times \mathfrak{F}_2 \times \cdots \times \mathfrak{F}_n, \mathbf{P}_1 \times \mathbf{P}_2 \times \cdots \times \mathbf{P}_n)$ can be defined starting with $\Omega_1 \times \Omega_2$, then $(\Omega_1 \times \Omega_2) \times \Omega_3$, etc., One notes that there is no "associativity" problem, that is $\Omega_1 \times (\Omega_2 \times \Omega_3)$ and $(\Omega_1 \times \Omega_2) \times \Omega_3$ are the same and so the product $\Omega_1 \times \Omega_2 \times \cdots \times \Omega_n$ may be defined via the obvious formula: $\Omega_1 \times \Omega_2 \times \cdots \times \Omega_n \stackrel{\text{def}}{=} (\Omega_1 \times \Omega_2 \times \cdots \times \Omega_{n-1}) \times \Omega_n$.

The following result makes this statement more explicit.

Proposition 3.3.8. Let $(\Omega_i, \mathfrak{F}_i, \mathbf{P}_i)$, $1 \leq i \leq n$, be n probability spaces. Let $\mathfrak{F}_1 \times \mathfrak{F}_2 \times \cdots \times \mathfrak{F}_n$ denote the smallest σ-algebra on $\Omega_1 \times \Omega_2 \times \cdots \times \Omega_n$ that contains all the sets of the form $E_1 \times E_2 \times \cdots \times E_n$, $E_i \in \mathfrak{F}_i$, $1 \leq i \leq n$. Then there is a unique probability $\mathbf{P}$ on $\mathfrak{F}_1 \times \mathfrak{F}_2 \times \cdots \times \mathfrak{F}_n$ such that

(*) $\qquad \mathbf{P}(E_1 \times E_2 \times \cdots \times E_n) = \mathbf{P}_1(E_1)\mathbf{P}_2(E_2) \cdots \mathbf{P}_n(E_n)$

for all $E_i \in \mathfrak{F}_i$.

Furthermore, if $1 \leq k \leq n$, $\mathfrak{F}_1 \times \mathfrak{F}_2 \times \cdots \times \mathfrak{F}_n$ is the smallest σ-algebra on $\Omega_1 \times \Omega_2 \times \cdots \times \Omega_n$ containing the sets $A \times B$, where $A \in \mathfrak{F}_1 \times \mathfrak{F}_2 \times \cdots \times \mathfrak{F}_k$ and $B \in \mathfrak{F}_{k+1} \times \mathfrak{F}_{k+2} \times \cdots \times \mathfrak{F}_n$. Also, $\mathbf{P}$ is the unique probability on $\mathfrak{F}_1 \times \mathfrak{F}_2 \times \cdots \times \mathfrak{F}_n$ such that

(**) $\mathbf{P}(A \times B) = (\mathbf{P}_1 \times \mathbf{P}_2 \times \cdots \times \mathbf{P}_k)(A)(\mathbf{P}_{k+1} \times \mathbf{P}_{k+2} \times \cdots \times \mathbf{P}_n)(B)$

for all

$$A \in \mathfrak{F}_1 \times \mathfrak{F}_2 \times \cdots \times \mathfrak{F}_k \text{ and } B \in \mathfrak{F}_{k+1} \times \mathfrak{F}_{k+2} \times \cdots \times \mathfrak{F}_n.$$

Proof. Let $\mathfrak{C}$ denote the collection of sets of the form $E_1 \times E_2 \times \cdots \times E_n$, $E_i \in \mathfrak{F}_i, 1 \leq i \leq n$. It is closed under finite intersections, and so by Corollary 3.2.7 there is at most one probability $\mathbf{P}$ on $\mathfrak{F}_1 \times \mathfrak{F}_2 \times \cdots \times \mathfrak{F}_n$ satisfying (*).

For a similar reason, there is at most one probability $\mathbf{P}$ on $\mathfrak{F}_1 \times \mathfrak{F}_2 \times \cdots \times \mathfrak{F}_n$ satisfying (**) provided $\mathfrak{F}_1 \times \mathfrak{F}_2 \times \cdots \times \mathfrak{F}_n$ is also the smallest σ-algebra $\mathfrak{G}$ containing all the sets $A \times B$, $A \in \mathfrak{F}_1 \times \mathfrak{F}_2 \times \cdots \times \mathfrak{F}_k$, and $B \in \mathfrak{F}_{k+1} \times \mathfrak{F}_{k+2} \times \cdots \times \mathfrak{F}_n$. This will now be proved.

Since $E_1 \times E_2 \times \cdots \times E_k = A \in \mathfrak{F}_1 \times \mathfrak{F}_2 \times \cdots \times \mathfrak{F}_k$ and $E_{k+1} \times E_{k+2} \times \cdots \times E_n = B \in \mathfrak{F}_{k+1} \times \mathfrak{F}_{k+2} \times \cdots \times \mathfrak{F}_n$ when $E_i \in \mathfrak{F}_i$, $1 \leq i \leq n$, it follows that $\mathfrak{G} \supset \mathfrak{F}_1 \times \mathfrak{F}_2 \times \cdots \times \mathfrak{F}_n$. Fix $B = E_{k+1} \times E_{k+2} \times \cdots \times E_n, E_i \in \mathfrak{F}_i, k+1 \leq i \leq n$.

Then $\{A \in \mathfrak{F}_1 \times \mathfrak{F}_2 \times \cdots \times \mathfrak{F}_k \mid A \times B \in \mathfrak{F}_1 \times \mathfrak{F}_2 \times \cdots \times \mathfrak{F}_n\}$ is a σ-algebra that contains all sets of the form $E_1 \times E_2 \times \cdots \times E_k, E_j \in \mathfrak{F}_j, 1 \leq j \leq k$. It therefore equals $\mathfrak{F}_1 \times \mathfrak{F}_2 \times \cdots \times \mathfrak{F}_k$, and so for all $A \in \mathfrak{F}_1 \times \mathfrak{F}_2 \times \cdots \times \mathfrak{F}_k$ and $E_i \in \mathfrak{F}_i$, $k+1 < i < n$, one has $A \times (E_{k+1} \times E_{k+2} \times \cdots \times E_n) \in \mathfrak{F}_1 \times \mathfrak{F}_2 \times \cdots \times \mathfrak{F}_n$. Now fix $A \in \mathfrak{F}_1 \times \mathfrak{F}_2 \times \cdots \times \mathfrak{F}_k$. By an entirely similar argument, $A \times B \in \mathfrak{F}_1 \times \mathfrak{F}_2 \times \cdots \times \mathfrak{F}_n$ for all $B \in \mathfrak{F}_{k+1} \times \mathfrak{F}_{k+2} \times \cdots \times \mathfrak{F}_n$. Hence, $\mathfrak{F}_1 \times \mathfrak{F}_2 \times \cdots \times \mathfrak{F}_n \supset \mathfrak{G}$.

To complete the proof of this proposition, it remains to establish the existence of $\mathbf{P}$ satisfying (*). Proceeding by induction, when $n=2$ it restates the result proved earlier as Theorem 3.3.4. Assume that the proposition holds for $n=N$, i.e., that the product probability $\mathbf{P}_1 \times \mathbf{P}_2 \times \cdots \times \mathbf{P}_N$ exists on $\mathfrak{F}_1 \times \mathfrak{F}_2 \times \cdots \times \mathfrak{F}_N$ in this case.

Now assume $n = N+1$. The product probability $\mathbf{P}_1 \times \mathbf{P}_2 \times \cdots \times \mathbf{P}_N = \mathbf{Q}$ exists on $\mathfrak{F}_1 \times \mathfrak{F}_2 \times \cdots \times \mathfrak{F}_N$, and there is a unique probability $\mathbf{P}$ on $(\mathfrak{F}_1 \times \mathfrak{F}_2 \times \cdots \times \mathfrak{F}_N) \times \mathfrak{F}_{N+1}$ such that $\mathbf{P}(A \times B) = \mathbf{Q}(A)\mathbf{P}_{N+1}(B)$ whenever $A \in \mathfrak{F}_1 \times \mathfrak{F}_2 \times \cdots \times \mathfrak{F}_N$ and $B \in \mathfrak{F}_{N+1}$. Taking $A = E_1 \times E_2 \times \cdots \times E_n, E_i \in \mathfrak{F}_i, 1 \leq i \leq N$, it is clear that $\mathbf{P}$ satisfies (*). □

Definition 3.3.9. *The probability determined in Proposition 3.3.8 will be referred to as the* **product probability** *and will be denoted by* $\mathbf{P}_1 \times \mathbf{P}_2 \times \cdots \times \mathbf{P}_n$.

The probability space $(\Omega_1 \times \Omega_2 \times \cdots \times \Omega_n, \mathfrak{F}_1 \times \mathfrak{F}_2 \times \cdots \times \mathfrak{F}_n, \mathbf{P}_1 \times \mathbf{P}_2 \times \cdots \times \mathbf{P}_n)$ *will be called the* **product of the probability spaces** $(\Omega_i, \mathfrak{F}_i, \mathbf{P}_i), 1 \leq i \leq n$.

Remark. The second part of the previous proposition states that $(\Omega_1 \times \Omega_2 \times \cdots \times \Omega_n, \mathfrak{F}_1 \times \mathfrak{F}_2 \times \cdots \times \mathfrak{F}_n, \mathbf{P}_1 \times \mathbf{P}_2 \times \cdots \times \mathbf{P}_n) = (\tilde{\Omega}_1, \tilde{\mathfrak{F}}_1, \tilde{\mathbf{P}}_1) \times (\tilde{\Omega}_2, \tilde{\mathfrak{F}}_2, \tilde{\mathbf{P}}_2)$, where $\tilde{\Omega}_1 = \Omega_1 \times \Omega_2 \times \cdots \times \Omega_k, \tilde{\mathfrak{F}}_1 = \mathfrak{F}_1 \times \mathfrak{F}_2 \times \cdots \times \mathfrak{F}_k, \tilde{\mathbf{P}}_1 = \mathbf{P}_1 \times \mathbf{P}_2 \times \cdots \times \mathbf{P}_k$, and $\tilde{\Omega}_2, \tilde{\mathfrak{F}}_2$, and $\tilde{\mathbf{P}}_2$ are similarly defined using the spaces $(\Omega_i, \mathfrak{F}_i, \mathbf{P}_i), k+1 \leq i \leq n$.

The possibility of defining such a product allows one to solve the following problem. Let $\mathbf{Q}_1, \mathbf{Q}_2, \ldots, \mathbf{Q}_n$ be n probabilities on $\mathbb{R}$. Are there a probability space $(\Omega, \mathfrak{F}, \mathbf{P})$ and a collection of random variables $(X_i)_{1 \leq i \leq n}$

on Ω such that (1) they are independent and (2) $\mathbf{Q}_i$ is the distribution of $X_i, 1 \leq i \leq n$? Yes. It suffices to take $\Omega = \mathbb{R}^n, X_i(x_1,\ldots,x_n) = x_i$, and $\mathbf{P} = \mathbf{Q}_1 \times \mathbf{Q}_2 \times \cdots \times \mathbf{Q}_n$ (in view of Proposition 3.2.8).

These results on the products of probability spaces extend with very little additional work to products of σ-finite measure spaces. To begin with, given two σ-finite measure spaces $(\Omega_1, \mathfrak{F}_1, \mu_1)$ and $(\Omega_2, \mathfrak{F}_2, \mu_2)$, define the set function μ on "boxes" $E_1 \times E_2$ by setting $\mu(E_1 \times E_2) = \mu_1(E_1)\mu_2(E_2)$, where this product is taken to be zero if one of the terms $\mu_i(E_i)$ equals zero even if the other is $+\infty$. Note that in general the values of μ lie in $[0, +\infty]$. Then μ extends as a finitely additive function (recall that $(+\infty) + \alpha = \alpha + (+\infty) = +\infty$ for any real number α) to the smallest Boolean algebra $\mathfrak{A}$ containing the sets $E_1 \times E_2 \times \cdots \times E_n$ since by Exercise 3.2.5 each $A \in \mathfrak{A}$ is a finite disjoint union of "boxes" and Exercise 3.3.1 holds in this context. In addition, Lemma 3.3.3 is still valid and so μ is countably additive on $\mathfrak{A}$.

The Carathéodory procedure applies, and since $\mathfrak{F} = \mathfrak{F}_1 \times \mathfrak{F}_2$ is the smallest σ-algebra containing $\mathfrak{A}$, it follows that the restriction to $\mathfrak{F}$ of the outer measure μ^* is a σ-finite measure such that $\mu(E_1 \times E_2) = \mu_1(E_1)\mu_2(E_2)$ for any pair of sets $E_i \in \mathfrak{F}_i$.

This property defines a unique measure as the next exercise shows.

Exercise 3.3.10. Let $(\Omega_i, \mathfrak{F}_i, \mu_i), 1 \leq i \leq n$, be n σ-finite measure spaces. Assume that none of the measures μ_i is identically equal to zero. Let μ and μ' be two σ-finite measures defined on $\mathfrak{F}_1 \times \mathfrak{F}_2 \times \cdots \times \mathfrak{F}_n \stackrel{\text{def}}{=} \mathfrak{F}$ that agree on sets of the form $E_1 \times E_2 \times \cdots \times E_n$, where $E_i \in \mathfrak{F}_i$ and $\mu_i(E_i) < +\infty$, with common value $\prod_{i=1}^n \mu_i(E_i)$. Show that μ and μ' agree on $\mathfrak{F}$. [*Hints:* (i) if $E \in \mathfrak{F}$ and $\mu(E)$ and $\mu'(E)$ are finite and positive for some $E \in \mathfrak{F}$, use the basic idea of Exercise 2.2.4 — see the remark that follows the exercise — to show that if $A \in \mathfrak{F}, A \subset E$, then $\mu(A) = \mu'(A)$ (it will be slightly simpler to first show that E may be written as $E_1^0 \times E_2^0 \times \cdots \times E_n^0$ with $0 < \mu_i(E_i^0) < +\infty, 1 \leq i \leq n$); and (ii) show that $\Omega_1 \times \Omega_2 \times \cdots \times \Omega_n$ is a countable increasing union of sets E for which both $\mu(E)$ and $\mu'(E)$ are finite.]

One consequence of this is that the same measure results if, instead of defining μ on the Boolean algebra $\mathfrak{A}$ determined by arbitrary "boxes", one first defines it only on the Boolean ring $\mathfrak{R}$ generated by the "boxes" whose sides E_i have finite measure $\mu_i(E_i)$ and then applies the procedure of Carathéodory.

Using the uniqueness result of Exercise 3.3.10, the proof of Proposition 3.3.8 also proves the following result.

Proposition 3.3.11. *Let $(\Omega_i, \mathfrak{F}_i, \mu_i), 1 \leq i \leq n$, be σ-finite measure spaces. Then there is a unique σ-finite measure μ on $\mathfrak{F}_1 \times \mathfrak{F}_2 \times \cdots \times \mathfrak{F}_n$ such that if $E_i \in \mathfrak{F}_i$, $1 \leq i \leq n$, then*

(*) $$\mu(E_1 \times E_2 \times \cdots \times E_n) = \mu_1(E_1)\mu_2(E_2)\ldots\mu_n(E_n).$$

This measure μ, denoted by $\mu_1 \times \mu_2 \times \cdots \times \mu_n$, is also the unique measure μ on $\mathfrak{F}_1 \times \mathfrak{F}_2 \times \cdots \times \mathfrak{F}_n$ such that

(**) $\quad \mu(A \times B) = (\mu_1 \times \mu_2 \times \cdots \times \mu_k)(A)(\mu_{k+1} \times \mu_{k+1} \times \cdots \times \mu_n)(B)$

for all $A \in \mathfrak{F}_1 \times \mathfrak{F}_1 \times \cdots \times \mathfrak{F}_k$ and $B \in \mathfrak{F}_{k+1} \times \mathfrak{F}_{k+2} \times \cdots \times \mathfrak{F}_n$.

These observations applied to Lebesgue measure on $\mathfrak{B}(\mathbb{R})$ determine its n-fold product on $\mathfrak{B}(\mathbb{R}) \times \mathfrak{B}(\mathbb{R}) \times \cdots \times \mathfrak{B}(\mathbb{R}) = \mathfrak{B}(\mathbb{R}^n)$.

Definition 3.3.12. *The σ-finite measure defined by the product of n copies of $(\mathbb{R}, \mathfrak{B}(\mathbb{R}), dx)$ is called **Lebesgue measure on $\mathfrak{B}(\mathbb{R}^n)$**.*

Remark. If, in the above definition, the σ-algebra $\mathfrak{F}$ of Lebesgue measurable sets is used, instead of the σ-algebra $\mathfrak{B}(\mathbb{R})$, the resulting measure extends Lebesgue measure on $\mathfrak{B}(\mathbb{R}^n)$. However, it is not be the "largest" extension. This extension may be obtained by applying the Carathéodory procedure to the function λ defined on the Boolean ring $\mathfrak{R}$ of sets that are finite unions of "boxes" of the type $(a_1, b_1] \times (a_2, b_2] \times \cdots \times (a_n, b_n]$, where the endpoints of all the intervals are all finite, and for which $\lambda((a_1, b_1] \times (a_2, b_2] \times \cdots \times (a_n, b_n]) = \Pi_{i=1}^n (b_i - a_i)$ — the volume of the "box". Since the restriction of Lebesgue measure on the Borel sets of $\mathbb{R}^n$ to $\mathfrak{R}$ coincides with λ, it follows that λ is σ-additive on $\mathfrak{R}$. Hence, one may then apply the procedure of Carathéodory. The resulting measure, which will also be denoted by λ, is defined on what is called the **σ-algebra of Lebesgue measurable subsets of $\mathbb{R}^n$**. Since it agrees with Lebesgue measure on the "boxes" $(a_1, b_1] \times (a_2, b_2] \times \cdots \times (a_n, b_n]$, it agrees with Lebesgue measure on the Borel sets in view of Exercise 3.1.12. The measure λ defined on the σ-algebra of Lebesgue measurable subsets of $\mathbb{R}^n$ is called **Lebesgue measure on $\mathbb{R}^n$**.

Note that λ is not a product measure, because the σ-algebra of Lebesgue measurable subsets of $\mathbb{R}^n$ is not a product of σ-algebras, one for each copy of $\mathbb{R}$ (see Exercise 3.6.20).

There is another way to obtain Lebesgue measure on $\mathbb{R}^n$ from the Lebesgue measure on the Borel sets. It involves what is called **completing the measure** and is explained in the following exercise.

Exercise 3.3.13. Show that

(1) a subset N of $\mathbb{R}^n$ is Lebesgue measurable and has $|N| = 0$ if and only if for any $\epsilon > 0$ there is a sequence $(A_n)_{n \geq 1}$ of Borel sets with $N \subset \cup_{n=1}^\infty A_n$ and $\sum_{n=1}^\infty |A_n| < \epsilon$. A set with this property will be called a **set of Lebesgue measure zero** or a **null set**; and

(2) a set has Lebesgue measure zero if and only if it is a subset of a Borel set of Lebesgue measure zero.

Let $\mathfrak{F}$ be the collection of symmetric differences $A \Delta N \stackrel{\text{def}}{=} (A \backslash N) \cup (N \backslash A)$ (see Exercise 2.9.2), where $A \in \mathfrak{B}(\mathbb{R}^n)$ and N is a set of measure zero.

Show that

(3) if, for all $n \geq 1$, $A_n \in \mathfrak{B}(\mathbb{R}^n)$ and N_n has Lebesgue measure zero, then $\cup_{n=1}^{\infty}(A_n \Delta N_n) \in \mathfrak{F}$ [Hint: show $E \in \mathfrak{F}$ if and only if $E = (A \backslash N_0) \cup N_1$, where N_0 and N_1 are sets of Lebesgue measure zero and $A \in \mathfrak{B}(\mathbb{R}^n)$],
(4) $\mathfrak{F}$ is a σ-algebra,
(5) $\mathfrak{F}$ is the σ-algebra of Lebesgue measurable subsets of $\mathbb{R}^n$, and
(6) $\mathfrak{F}$ is the smallest σ-algebra containing $\mathfrak{B}(\mathbb{R}^n)$ and all the sets that are subsets of a Borel set of measure zero (see Exercise 2.9.12).

Definition 3.3.14. *Let $(\Omega, \mathfrak{F}, \mu)$ be a measure space. A subset of Ω is said to have **measure zero** (relative to μ) if, for any $\epsilon > 0$, there is a sequence $(A_n)_{n \geq 1} \subset \mathfrak{F}$ with (i) $E \subset \cup_{n=1}^{\infty} A_n$ and (ii) $\sum_{n=1}^{\infty} \mu(A_n) < \epsilon$. A σ-finite measure space $(\Omega, \mathfrak{F}, \mu)$ is said to be a **complete measure space** if every subset of Ω of measure zero is in $\mathfrak{F}$.*

Exercise 3.3.15. Show that the σ-finite measure space $(\mathbb{R}^n, \mathfrak{F}, dx)$ is complete, where $\mathfrak{F}$ is the σ-algebra of Lebesgue measurable sets. Show that if X is an $\mathfrak{F}$-measurable function, then there is a Borel function Y with $\{X \neq Y\}$ of Lebesgue measure zero.

Exercise 3.3.13 illustrates a procedure that can be applied to any σ-finite measure space to obtain a complete measure space. This complete measure space is called the **completion** of the original one (see Exercise 2.9.12 and the remark that follows).

Having now established the existence of products of measure spaces, the following Fubini theorems may be proved almost exactly as in the case of probability spaces (Theorem 3.3.5 and Corollary 3.3.6). The only slight difference that occurs is in the proof of Theorem 3.3.5. The argument involving the class $\mathfrak{M}$ needs to be handled a little carefully. One way is to proceed as in the proof of Theorem 2.2.2. The whole space $\Omega_1 \times \Omega_2 = \cup_{n=1}^{\infty} A_n$, $A_n = E_1^n \times E_2^n$, $\mu_i(E_i^n) < +\infty, i = 1, 2$. Let $\mathfrak{M}_n = \{A_n \cap E \mid E \in \mathfrak{F}$ such that (1), (2), and (3) of Theorem 3.3.16 hold for $1_{A \cap A_n}\}$. Then, for each n, $\mathfrak{M}_n$ is a monotone class that satisfies the hypotheses of Dynkin's result (Proposition 3.2.6). Since it contains the collection $\mathfrak{C}_n \stackrel{\text{def}}{=} \{A_n \cap (E_1 \times E_2) \mid E_i \in \mathfrak{F}_i, i = 1, 2\}$, which is closed under finite intersections, it follows from Proposition 3.2.6, that it contains $\mathfrak{F}_n \stackrel{\text{def}}{=} \{A_n \cap E \mid E \in \mathfrak{F}\}$. This shows that Fubini's theorem is true for any function $X = 1_{A_n \cap E}, E \in \mathfrak{F}$. It then follows from monotone convergence that Fubini's theorem holds for $X = 1_E, E \in \mathfrak{F}$. Once this is established, the rest of the proof goes through without change.

Theorem 3.3.16. (**Fubini's theorem for positive functions**)
Let $(\Omega_1 \times \Omega_2, \mathfrak{F}_1 \times \mathfrak{F}_2, \mu_1 \times \mu_2)$ be the product of the measure spaces $(\Omega_1, \mathfrak{F}_1, \mu_1)$ and $(\Omega_2, \mathfrak{F}_2, \mu_2)$. Let μ denote $\mu_1 \times \mu_2$. Let X be a non-

negative, measurable function. Then
   (1) for all $\omega_1 \in \Omega_1$, the function $\omega_2 \to X(\omega_1, \omega_2) \in \mathfrak{F}_2$,
   (2) the function $\omega_1 \to \int X(\omega_1, \omega_2) \mu_2(d\omega_2) \in \mathfrak{F}_1$, and
   (3) $\int [\int X(\omega_1, \omega_2) \mu_2(d\omega_2)] \mu_1(d\omega_1) = \int X d(\mu_1 \times \mu_2)$.

This theorem extends to L^1-functions exactly as in the case of probability measures.

Corollary 3.3.17. (Fubini's theorem for L^1-functions)
Let $(\Omega_1 \times \Omega_2, \mathfrak{F}_1 \times \mathfrak{F}_2, \mu_1 \times \mu_2)$ be the product of the measure spaces $(\Omega_1, \mathfrak{F}_1, \mu_1)$ and $(\Omega_2, \mathfrak{F}_2, \mu_2)$. Let μ denote $\mu_1 \times \mu_2$.
If X is an integrable function on $(\Omega_1 \times \Omega_2, \mathfrak{F}_1 \times \mathfrak{F}_2, \mu)$, then
   (1) for μ_1-almost all ω_1, the function $\omega_2 \to X(\omega_1, \omega_2)$ is an integrable function on $(\Omega_2, \mathfrak{F}_2, \mu_2)$,
   (2) there is an integrable function Y on $(\Omega_1, \mathfrak{F}_1, \mu_1)$ such that μ_1-almost everywhere $\int X(\omega_1, \omega_2) \mu_2(d\omega_2) = Y(\omega_1)$, and
   (3) $\int Y(\omega_1) \mu_1(d\omega_1) = \int X d\mu$.

Remark. Fubini's theorem is also valid for Lebesgue measurable functions on $\mathbb{R}^n$ (see Exercise 3.6.5). The statement is complicated by the fact that the σ-algebra of Lebesgue measurable subsets of $\mathbb{R}^n$ is not a product of σ-algebras.

An alternate way to construct the product of a finite number of σ-finite measures is illustrated in the next exercise for Lebesgue measures. This exercise therefore gives another way to construct the Lebesgue measure in $\mathbb{R}^n$ and is the n-dimensional version of Exercise 2.2.4.

Exercise 3.3.18. Let B_m be the closed box in $\mathbb{R}^n$ all of whose sides equal $[-m, m]$, i.e., $B_m = [-m, m] \times [-m, m] \times \cdots \times [-m, m]$. Show that
   (1) there is a unique probability $\mathbf{P}_m$ on $\mathfrak{B}(\mathbb{R}^n)$ such that, for any box $B = (a_1, b_1] \times (a_2, b_2] \times \cdots \times (a_n, b_n]$, $\mathbf{P}_m(B) = \frac{1}{(2m)^n} |B \cap B_m|$, where $|B \cap B_m|$ denotes the volume of the box $|B \cap B_m|$, i.e., $|B \cap B_m| = \prod_{i=1}^n |[-m, m] \cap (a_i, b_i]| = \prod_{i=1}^n (b_i \wedge m - a_i \vee (-m))$.
Define $\mu_m = (2m)^n \mathbf{P}_m, m \geq 1$.
Show that
   (2) if $A \in \mathfrak{B}(\mathbb{R}^n)$ is a subset of B_m, then $\mu_m(A) = \mu_{m+1}(A)$,
   (3) there is a unique σ-additive measure μ on $\mathfrak{B}(\mathbb{R}^n)$ such that $\mu(A) = \mu_m(A)$ if $A \subset B_m$, and
   (4) $\mu((a_1, b_1] \times (a_2, b_2] \times \cdots \times (a_n, b_n]) = |(a_1, b_1] \times (a_2, b_2] \times \cdots \times (a_n, b_n]|$ for any finite box $(a_1, b_1] \times (a_2, b_2] \times \cdots \times (a_n, b_n]$.
Conclude that μ equals Lebesgue measure λ on $\mathfrak{B}(\mathbb{R}^n)$.

Riemann and Lebesgue integration on $\mathbb{R}^n$.
Consider a bounded function φ on a finite closed "box" $B = [a_1, b_1] \times [a_2, b_2] \times \cdots \times [a_n, b_n]$. To define the Riemann integral of φ on B, one

3. PRODUCT MEASURES

subdivides each of the sides of the box B to get subboxes, and forms upper and lower sums. By definition, the function is Riemann integrable over the box B if and only if the supremum of the lower sums equals the infimum of the upper sums. In exactly the same way as in the case of $\mathbb{R}$ (see Chapter II §3), one shows that if the Riemann integral of φ exists, then there are two bounded Borel functions f and g with $f \leq \varphi \leq g$ and $\int f(x)dx = \int g(x)dx$. This implies as before that (i) φ is Lebesgue measurable — we extend it to have value zero outside the box B —, (ii) its Lebesgue integral exists (i.e., it is in L^1), and (iii) the two integrals agree.

Example 3.3.19. Let $D = \{(x,y) \mid 0 \leq x, 0 \leq y \leq x\}$, and let $f(x,y) = e^{-(\frac{x^2}{2})}$. Then $f 1_D$ is a Borel function, and Fubini's theorem shows that

$$\int_0^{+\infty} \left[\int_y^{+\infty} f(x,y)dx \right] dy = \int_0^{+\infty} \left[\int_0^x f(x,y)dy \right] dx = 1,$$

since both integrals coincide with the Lebesgue integral over $\mathbb{R}^2$ of $1_D f$ and the second one is computable. To see this, note that $F(y) = \int_y^{+\infty} f(x,y)dx = \int 1_D(x,y)f(x,y)dx$ (Lebesgue integral) by what was said earlier about improper Riemann integrals. Fubini's theorem states that F is Borel measurable and $\int F(y)dy$ is the Lebesgue integral $\int 1_D f dxdy$ of f on $\mathbb{R}^2$. Since F is non-negative, it follows from Remark 2.3.12 that the improper integral $\int_0^\infty F(y)dy = \int F(y)dy$, providing that the Riemann integral $\int_0^b F(y)dy$ exists for any $b > 0$. For this it suffices to show that F is continuous: if $y_1 < y_2, F(y_2) - F(y_1) = \int_{y_1}^{y_2} e^{-(\frac{x^2}{2})}dx$, which tends to zero as $y_1 \to y_2$ or vice versa. Hence, $\int_0^{+\infty}[\int_y^{+\infty} f(x,y)dx]dy = \int 1_D f dxdy$. Note that, although this is not needed for the above, in fact $\int_0^\infty F(y)dy < \infty$ as $\int_y^{+\infty} e^{-(\frac{x^2}{2})}dx \sim \frac{1}{y}e^{-(\frac{y^2}{2})}$ (see Exercise 3.6.5), as y tends to ∞.

To see that the other iterated integral equals $\int 1_D f dxdy$, one proceeds in a similar fashion.

Example 3.3.20. A random vector X (on $(\Omega, \mathfrak{F}, \mathbf{P})$) has a (non-singular) **multivariate normal distribution** with mean θ and (positive definite) covariance matrix Σ if the distribution $\mathbf{Q}$ of X has a density $f(x)$ with respect to Lebesgue measure $dx_1, \ldots, dx_n$ on $\mathbb{R}^n$ and

$$\mathbf{Q}(E) = \int_E f(x)dx_1 dx_2 \ldots dx_n \text{ for } E \in \mathfrak{B}(\mathbb{R}^n),$$

where

$$f(x) = \frac{1}{(2\pi|det\Sigma|)^{n/2}} e^{-\frac{1}{2}(x-\theta)^t \Sigma^{-1}(x-\theta)}.$$

Notice that to show that $\mathbf{Q}$ is indeed a probability, one must resort ultimately to computing an improper Riemann integral in $\mathbb{R}^n$. First, one observes that $\Sigma = ODO^*$, where O is an orthogonal matrix and D is a diagonal matrix with all its entries strictly positive. Assuming that the usual change-of-variable formula is valid, the integral reduces to computing the integral for the case where $\Sigma = D$. In other words, it reduces to showing that

$$\frac{1}{(2\pi|detD|)^{n/2}} \int \exp\{-\frac{1}{2}\sum_{i=1}^{n} d_i(x_i - \theta_i)^2\} dx_1 dx_2 \ldots dx_n = 1,$$

where the d_i are the diagonal entries of D.

One shows, as in one dimension, that the integral coincides with the improper Riemann integral

$$\int_{-\infty}^{+\infty} \int_{-\infty}^{+\infty} \cdots \int_{-\infty}^{+\infty} \exp\{-\frac{1}{2}\sum_{i=1}^{n} d_i(x_i - \theta_i)^2\} dx_1 dx_2 \cdots dx_n.$$

By making use of Fubini's theorem, this integral is easily seen to equal $(2\pi|detD|)^{n/2}$.

Convolution of measures.

Exercise 3.3.21. Let $\mathbf{P}_1 \times \mathbf{P}_2$ be a probability on $(\mathbb{R}^2, \mathfrak{B}(\mathbb{R}^2))$. The random variable $s : \mathbb{R}^2 \to \mathbb{R}$ defined by $s(x, y) = x + y$ has a distribution $\mathbf{Q}$. This distribution is called the **convolution** $\mathbf{P}_1 \star \mathbf{P}_2$ of $\mathbf{P}_1$ and $\mathbf{P}_2$.

If $A \in \mathfrak{B}(\mathbb{R})$, then $\mathbf{P}_1 \star \mathbf{P}_2(A) \stackrel{\text{def}}{=} \mathbf{P}_1 \times \mathbf{P}_2(\{(x, y) \mid x + y \in A\}) = \int 1_A(x + y)(\mathbf{P}_1 \times \mathbf{P}_2)(dx, dy) = \int \int 1_A(x + y) \mathbf{P}_1(dx) \mathbf{P}_2(dy)$ (see remark (2) following Theorem 3.3.4). It follows from Fubini's theorem, (Theorem 3.3.5), that

$$\mathbf{P}_1 \star \mathbf{P}_2(A) = \int 1_A(x + y)(\mathbf{P}_1 \times \mathbf{P}_2)(dx, dy)$$
$$= \int \left[\int 1_A(x + y) \mathbf{P}_2(dy) \right] \mathbf{P}_1(dx) = \int \mathbf{P}_2(A - x) \mathbf{P}_1(dx)$$
$$= \int \left[\int 1_A(x + y) \mathbf{P}_1(dx) \right] \mathbf{P}_2(dy) = \int \mathbf{P}_1(A - y) \mathbf{P}_2(dy)$$
$$= \mathbf{P}_2 \star \mathbf{P}_1(A),$$

where $A - u \stackrel{\text{def}}{=} \{a - u \mid a \in A\}$.

Assume that $\mathbf{P}_1(dx) = f_1(x)dx$. Show that

(1) $\mathbf{P}_1(A - y) = \int_A f_1(x - y)dx$ [Hint: Lebesgue measure is invariant under translation (Exercise 2.2.8).],
(2) $\mathbf{P}_1 \star \mathbf{P}_2(du) = g(u)du$, where $g(u) = f_1 \star \mathbf{P}_2(u) \stackrel{\text{def}}{=} \int f_1(u-y) \mathbf{P}_2(dy)$.

If, in addition, $\mathbf{P}_2(dy) = f_2(y)dy$, note that $g(u) = f_1 \star f_2(u)$, where $f_1 \star f_2(u) \stackrel{\text{def}}{=} \int f_1(u-y)f_2(y)dy$.

(3) If $f_1(x) = \frac{1}{\sqrt{2\pi t_1}}e^{-\frac{1}{2t_1}(x-m_1)^2}$ and $f_2(x) = \frac{1}{\sqrt{2\pi t_2}}e^{-\frac{1}{2t_2}(x-m_2)^2}$, show that $f_1 \star f_2(x) = \frac{1}{\sqrt{2\pi(t_1+t_2)}}e^{-\frac{1}{2(t_1+t_2)}(x-[m_1+m_2])^2}$.

Extend the definition of convolution to σ-finite measures, and show that

(4) $(\mu_1 + \mu_2) \star \nu = \mu_1 \star \nu + \mu_2 \star \nu$ for any three σ-finite measures μ_1, μ_2, and ν.

Compute

(5) $\varepsilon_a \star \varepsilon_b$ (see Exercise 1.5.2).

Let $0 \le p, q \le 1$, and $p + q = 1$. Compute

(6) $(q\varepsilon_0 + p\varepsilon_1) \star (q\varepsilon_0 + p\varepsilon_1) \stackrel{\text{def}}{=} (q\varepsilon_0 + p\varepsilon_1)^{\star 2}$ [Hint: use (4).],

(7) $(q\varepsilon_0 + p\varepsilon_1)^{\star n}$, the n-fold convolution of $q\varepsilon_0 + p\varepsilon_1$ with itself.

Finally, show that if X_1 and X_2 are two independent random variables, then

(8) the distribution of $X_1 + X_2$ is the convolution of the distributions of X_1 and X_2, and

(9) the distribution function $F_{X_1+X_2}$ is the convolution $F_{X_1} \star \mathbf{P}_{X_2}$ of the distribution function of X_1 and the distribution $\mathbf{P}_{X_2}$ of X_2 (see (2) for the definition of the convolution of a function and a probability).

Remark. While it is usual to denote the distribution function $F_{X_1} \star \mathbf{P}_{X_2}$ by $F_{X_1} \star F_{X_2}$, this notation will not be used in these notes. It is an unfortunate notation since it is standard mathematical usage to define the convolution of two functions by the formula $f_1 \star f_2(x) = \int f_1(x-y)f_2(y)dy$. The (mathematical) convolution $F_{X_1} \star F_{X_2}(x) = \int F_{X_1}(x-y)F_{X_2}(y)dy$, whereas $F_{X_1} \star \mathbf{P}_{X_2}(x) = \int F_{X_1}(x-y)d\mathbf{P}_{X_2}(y) = \int F_{X_1}(x-y)dF_{X_2}(y)$, using the notation $dF(y)$ for $d\mathbf{P}(y)$ when F is the distribution function associated with $\mathbf{P}$. For example, $H_a \star H_b \ne H_{a+b}$, where $H_a(x) = H(x-a)$ is the distribution of the point mass ε_a. In case $\mathbf{P}_{X_2}$ has a density f_{X_2} then, in fact, $F_{X_1} \star \mathbf{P}_{X_2}(x) = F_{X_1} \star f_{X_2}(x)$. Further, if f_{X_1} is the density of $\mathbf{P}_{X_1}$, the distribution of X_1, then it is true that $f_{X_1} \star f_{X_2}$ is the density $f_{X_1+X_2}$ of the distribution of $X_1 + X_2$ as shown in Exercise 3.3.21 (2).

Exercise 3.3.22. Let X_1 and X_2 be two independent Poisson random variables with means λ_1 and λ_2 (a Poisson random variable has a Poisson distribution; see Exercise 2.1.24 (2)). Show that $X_1 + X_2$ is a Poisson random variable with mean $\lambda_1 + \lambda_2$.

Exercise 3.3.23. Compute $H_a \star H_b$ for $a < b$ (where $H_c(x) = H(x-c)$ is the distribution of the point mass ε_c), and show why this convolution is not a distribution function.

Exercise 3.3.24. Let μ be any probability on $\mathbb{R}$ (i.e., on $\mathfrak{B}(\mathbb{R})$). A bounded Borel function ψ is said to **vanish at infinity** if, for any $\epsilon > 0$, there is an integer $N \geq 1$ such that $|x| > N$ implies that $|\psi(x)| < \epsilon$. Show that if ψ vanishes at infinity, then $\psi \star \mu$ also vanishes at infinity. [Hints: $\psi \star \mu(x) = \int \psi(x-y)\mu(dy) = \int_{\{|y| \leq M\}} \psi(x-y)\mu(dy) + \int_{\{|y| > M\}} \psi(x-y)\mu(dy)$; choose M with $\mu([-M, M]) > 1 - \epsilon$, and observe that one can force $|x - y|$ to be very large for all $y \in [-M, M]$ if $|x|$ is very large.]

4. Infinite Products

Let $(\Omega, \mathfrak{F}, \mathbf{P})$ be a probability space.

Definition 3.4.1. A **stochastic process** on $(\Omega, \mathfrak{F}, \mathbf{P})$ is a family $(X_\iota)_{\iota \in I}$ of random variables on $(\Omega, \mathfrak{F}, \mathbf{P})$.

Heuristically speaking, one may think of I as a set of "times" and X_ι as the observation (of some phenomenon) at time "ι". If one takes a finite set $F \subset I$, say $F = \{\iota_1, \ldots, \iota_n\}$ then the n-random variables X_{ι_k} give "information" about the states of the phenomenon or process "during F". One may calculate probabilities then for the events $E \in \sigma(\{X_\iota \mid \iota \in F\})$. The random vector $(X_{\iota_1}, \cdots, X_{\iota_n})$ associates with this σ-algebra and $\mathbf{P}$ a distribution $\mathbf{Q}$ on $\mathbb{R}^n$. This is called the **finite-dimensional joint distribution** of the stochastic process $(X_\iota)_{\iota \in I}$, corresponding to the finite set $F \subset I$.

Just as for a finite collection of random variables, there is a smallest σ-algebra relative to which they are all measurable, so too for any collection $(X_\iota)_{\iota \in I}$ of random variables (or stochastic process) there is a smallest σ-algebra for which the functions X_ι are all measurable. It will be denoted by $\sigma(\{X_\iota \mid \iota \in I\})$.

The events that the stochastic process $(X_\iota)_{\iota \in I}$ determines belong to $\mathfrak{G} = \sigma(\{X_\iota \mid \iota \in I\})$, and the probability $\mathbf{P}$ on $\mathfrak{G}$ is completely determined by knowing all the finite-dimensional joint distributions of the process. This fact is an immediate consequence of the following proposition.

Proposition 3.4.2. Let $(X_\iota)_{\iota \in I}$ be a stochastic process on $(\Omega, \mathfrak{F}, \mathbf{P})$. Let $\mathfrak{G}$ denote $\sigma(\{X_\iota \mid \iota \in I\})$ and $\mathfrak{G}_F = \sigma(\{X_\iota \mid \iota \in F\})$, where $F \subset I$ denotes an arbitrary finite subset. Then

(1) $\mathfrak{A} = \cup_{F \subset I} \mathfrak{G}_F$ is a Boolean algebra, and
(2) $\mathfrak{G}$ is the smallest σ-algebra containing $\mathfrak{A}$.

Proof. (1) If $A_1, A_2 \in \mathfrak{A}$, then there are finite sets $F_i \subset I$ with $A_i \in \mathfrak{G}_{F_i}$. Since $F = F_1 \cup F_2$ is finite, $A_1, A_2 \in \mathfrak{G}_F$ and so $A_1 \cup A_2 \in \mathfrak{A}$ and $A_1 \cap A_2 \in \mathfrak{A}$. It is clear that $A \in \mathfrak{A}$ implies $A^c \in \mathfrak{A}$.

(2) If $J \subset I$, then $\mathfrak{G}_J \subset \mathfrak{G}_I = \mathfrak{G}$ and so $\mathfrak{A} \subset \mathfrak{G}$. If $\mathfrak{H}$ is a σ-algebra $\supset \mathfrak{A}$, then each random variable X_ι is measurable with respect to $\mathfrak{H}$ and so $\mathfrak{H} \supset \sigma(\{X_\iota \mid \iota \in I\}) = \mathfrak{G}$. $\square$

Remark. If I_0 denotes an arbitrary at most countable subset of I, then $\mathfrak{G} = \cup_{I_0 \subset I} \mathfrak{G}_{I_0}$.

There is a special class of stochastic processes for which the finite-dimensional joint distributions are always finite products of the distributions of the individual random variables. This is the class of processes that are independent.

Definition 3.4.3. *A family $(X_\iota)_{\iota \in I}$ of random variables on $(\Omega, \mathfrak{F}, \mathbf{P})$ (i.e., a stochastic process) is said to be an **independent family of random variables** if, for any $J \subset I$, the σ-algebras $\mathfrak{F}_1 = \sigma(\{X_\iota \mid \iota \in J\})$ and $\mathfrak{F}_2 = \sigma(\{X_\iota \mid \iota \in I \setminus J\})$ are independent. A family $(E_\iota)_{\iota \in I}$ of events or sets is said to be an **independent family of events or sets** if the family $(1_{E_\iota})_{\iota \in I}$ of random variables 1_{E_ι} is independent (see Exercise 3.6.8).*

Remark. In view of Proposition 3.2.8, this definition agrees with Definition 3.2.9 in case I is finite.

Proposition 3.4.4. **(Kolmogorov's 0–1 law)** *Let $(X_n)_{n \geq 1}$ be a sequence of independent random variables on a probability space $(\Omega, \mathfrak{F}, \mathbf{P})$. Let $\mathfrak{T}_n = \sigma(\{X_i \mid i \geq n\})$ and $\mathfrak{T} = \cap_{n=1}^{\infty} \mathfrak{T}_n$. Then, for $\Lambda \in \mathfrak{T}$, it follows that either $\mathbf{P}(\Lambda) = 0$ or $\mathbf{P}(\Lambda) = 1$. In addition, if $Z \in \mathfrak{T}$, then Z is a.s. constant.*

Proof. Let $\mathfrak{F}_n = \sigma(\{X_i \mid 1 \leq i \leq n\})$ and $\mathfrak{A} = \cup_{n=1}^{\infty} \mathfrak{F}_n$. It follows from Definition 3.4.3 that $\mathfrak{F}_n$ and $\mathfrak{T}_{n+1}$ are independent for all $n \geq 1$ and, hence, $\mathfrak{T}$ is independent of each $\mathfrak{F}_n$, $n \geq 1$. Hence,

$$\mathbf{P}(A \cap \Lambda) = \mathbf{P}(A)\mathbf{P}(\Lambda) \quad \text{if } A \in \mathfrak{A} \text{ and } \Lambda \in \mathfrak{T}.$$

Let $\mathfrak{F}_\infty = \sigma(\{X_i \mid i \geq 1\})$, and let $\mathfrak{M} = \{B \in \mathfrak{F}_\infty \mid \mathbf{P}(B \cap \Lambda) = \mathbf{P}(B)\mathbf{P}(\Lambda) \text{ for all } \Lambda \in \mathfrak{T}\}$. Note that $\mathfrak{F}_\infty$ is the smallest σ-algebra containing $\mathfrak{A}$.

Exercise. Show that $\mathfrak{M}$ is a monotone class (Exercise 1.4.15).

Now $\mathfrak{A}$ is a Boolean algebra: since each $\mathfrak{F}_n$ is a σ-algebra, it contains Ω and A^c if $A \in \mathfrak{A}$; also, if $A_1, A_2 \in \mathfrak{A}$, then there is an integer n such that both A_1 and A_2 belong to $\mathfrak{F}_n$; hence, $A_1 \cup A_2 \in \mathfrak{A}$. Since $\mathfrak{A} \subset \mathfrak{M}$, it follows from Exercise 1.4.16 that $\mathfrak{M} \supset \mathfrak{F}_\infty$.

It follows, for any $B \in \mathfrak{F}_\infty$ and $\Lambda \in \mathfrak{T}$, that $\mathbf{P}(B \cap \Lambda) = \mathbf{P}(B)\mathbf{P}(\Lambda)$. In particular, if $\Lambda \in \mathfrak{T}$, then $\mathbf{P}(\Lambda) = \mathbf{P}(\Lambda)^2$, i.e., $\mathbf{P}(\Lambda)(1 - \mathbf{P}(\Lambda)) = 0$. Hence, either $\mathbf{P}(\Lambda) = 0$ or $\mathbf{P}(\Lambda) = 1$.

If $Z \in \mathfrak{T}$ and $\lambda \in \mathbb{R}$, then $\mathbf{P}[Z < \lambda]$ equals 0 or 1. If it is always 0, then $Z = +\infty$ a.s. If it is always 1, then $Z = -\infty$ a.s. If neither is the case, there is a unique $\lambda_0 \in \mathbb{R}$ with $\mathbf{P}[Z < \lambda_0] = 0$ and $\mathbf{P}[Z < \lambda_0 + \delta] = 1$ for any $\delta > 0$. This implies $Z = \lambda_0$ a.s. as $\mathbf{P}[\lambda_0 \leq Z < \lambda_0 + \frac{1}{n}] = 1$ for all $n \geq 1$. $\square$

Remark. One may also prove this result by making use of Dynkin's version of the monotone class theorem (Proposition 3.2.6) if one prefers to argue from the independence of the σ-algebra $\mathfrak{T}$ and the class $\mathfrak{C}$ of sets of the form $\cap_{i=1}^n X_i^{-1}(E_i)$, $E_i \in \mathfrak{B}(\mathbb{R})$.

Exercise 3.4.5. Let $(X_\iota)_{\iota \in I}$ be a family of independent variables defined on a probability space $(\Omega, \mathfrak{F}, \mathbf{P})$. For each ι, let $\psi_\iota : \mathbb{R} \to \mathbb{R}$ be a Borel function. Let $Y_\iota = \psi_\iota \circ X_\iota$. Show that $(Y_\iota)_{\iota \in I}$ is again an independent family. In particular, show that if $(E_\iota)_{\iota \in I}$ is a family of independent sets (i.e., the family of random variables $(1_{E_\iota})_{\iota \in I}$ is independent), then the family $(E_\iota^c)_{\iota \in I}$ of sets E_ι^c is also independent (see Exercise 3.6.7).

The question of independence involves only the finite-dimensional joint distributions, as the next result shows.

Proposition 3.4.6. *A family $(X_\iota)_{\iota \in I}$ of random variables is independent if and only if, for every finite $F \subset I$, the family $(X_\iota)_{\iota \in F}$ is independent.*

Proof. If $(X_\iota)_{\iota \in I}$ is independent and $I' \subset I$, then $(X_\iota)_{\iota \in I'}$ is independent. Hence, for any finite $F \subset I$, the family $(X_\iota)_{\iota \in F}$ is independent.

Conversely, if $J \subset I$, let $\mathfrak{C}_1$ be the class of sets of the form $\cap_{\iota \in F_1} X_\iota^{-1}(E_\iota)$, where F_1 runs over the collection of finite subsets of J, and the sets E_ι are Borel subsets of $\mathbb{R}$. Let $\mathfrak{C}_2$ be the class of sets of the form $\cap_{\iota \in F_2} X_\iota^{-1}(E_\iota)$, where F_2 runs over the collection of finite subsets of $I \setminus J$, and the sets E_ι are as before.

Since the union of two finite sets is finite, each of these classes $\mathfrak{C}_i$ is closed under finite intersections. Furthermore, if $A_1 \in \mathfrak{C}_1$ and $A_2 \in \mathfrak{C}_2$, $\mathbf{P}(A_1 \cap A_2) = \mathbf{P}(A_1)\mathbf{P}(A_2)$, since by hypothesis every finite subfamily $(X_\iota)_{\iota \in F}$ is independent. The σ-algebra $\sigma(\{X_\iota \mid \iota \in J\})$ is the smallest σ-algebra containing $\mathfrak{C}_1$. The σ-algebra $\sigma(\{X_\iota \mid \iota \in I \setminus J\})$ is the smallest σ-algebra containing $\mathfrak{C}_2$. It therefore follows, as in the proof of Proposition 3.2.8, that these two σ-algebras are independent. $\square$

As in the case of finite families of distributions (or probabilities) $\mathbf{Q}_n$ on $\mathbb{R}$, the problem arises of constructing a sequence of independent random variables $(X_n)_{n \geq 1}$ on a probability space with a prescribed distribution $\mathbf{Q}_n$ for each X_n. For example, $Q_n(dx)$ could be $\frac{1}{2}\{\epsilon_0(dx) + \epsilon_1(dx)\}$ for each n, and then one wants to find an independent sequence of random variables X_n with equally likely values of 0 or 1 (intuitively, such a space exists in view of the possibilities for tossing a fair coin).

To show that there are probability spaces $(\Omega, \mathfrak{F}, \mathbf{P})$ with such sequences of random variables on them, one now proceeds to construct the (countably) infinite product of a sequence of probability spaces.

To begin with, if $\Omega_1, \Omega_2, \ldots, \Omega_n, \ldots$ is a sequence of sets, what is the set $\times_{n=1}^\infty \Omega_n$? If one notices that $\Omega_1 \times \ldots \times \Omega_n$ (as a set) can be viewed as the set of functions $\omega : \{1, \ldots, n\}$ with $\omega(i) \in \Omega_i$, $1 \leq i \leq n$, then it is clear how to define the set $\times_{n=1}^\infty \Omega_n$.

4. INFINITE PRODUCTS

Definition 3.4.7. *The (infinite) product set $\times_{n=1}^{\infty} \Omega_n$ is the set of functions $\omega : \mathbb{N} \to \cup_{n=1}^{\infty} \Omega_n$ such that $\omega(n) \in \Omega_n$ for each $n \in \mathbb{N}$.*

On the infinite product $\times_{n=1}^{\infty} \Omega_n$ there is a "natural" sequence $(X_n)_{n \geq 1}$ of functions (the coordinate projections) where $X_n(\omega) = \omega(n)$ for each n (i.e., $X_n(\omega)$ is the evaluation of the function ω in $\times_{n=1}^{\infty} \Omega_n$ at n). If, for example, $\Omega_n = \mathbb{R}$ for each n, then $\times_{n=1}^{\infty} \Omega_n$ is the set of all real-valued sequences $a = (a_n)_{n \geq 1}$, and $X_n(a) = a_n$ is the nth term of the sequence a. If one plots the sequences in the plane, one obtains a "picture" of this product space, as shown in Fig.3.3.

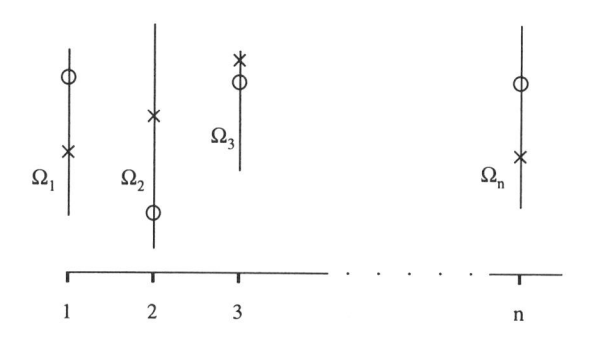

Fig. 3.3

In general, one may think schematically as follows: arrange the sets $\Omega_1, \Omega_2, \ldots, \Omega_n, \ldots$ in a sequence, viewing Ω_n as over the integer n; choosing a point in each set Ω_n over n determines a point $\Omega \in \times_{n=1}^{\infty} \Omega_n$. In Fig. 3.3, two points in $\times_{n=1}^{\infty} \Omega_n$ are indicated: the values of one point are indicated by the locations of an "×" and the other by an "o".

Suppose now that, for each $n \geq 1$, a probability space $(\Omega_n, \mathfrak{F}_n, \mathbf{P}_n)$ is given. Form the set $\times_{n=1}^{\infty} \Omega_n = \Omega$, and consider the functions $X_n : \Omega \to \Omega_n$ given by $X_n(\omega) = \omega(n)$. On the set Ω_n there is a σ-algebra $\mathfrak{F}_n$.

Definition 3.4.8. *The smallest σ-algebra $\mathfrak{F}$ on Ω such that each $X_n : \Omega \to \Omega_n, n \geq 1$, is measurable with respect to $\mathfrak{F}$ and $\mathfrak{F}_n$ will be called the **product of the σ-algebras** $\mathfrak{F}_n, n \geq 1$. It will be denoted by $\times_{n=1}^{\infty} \mathfrak{F}_n$ (often denoted by $\otimes_{n=1}^{\infty} \mathfrak{F}_n$). Note that $\times_{n=1}^{\infty} \mathfrak{F}_n = \sigma(\{X_n \mid n \geq 1\})$.*

For each n, the finite product $(\Omega_1 \times \Omega_2 \times \cdots \times \Omega_n, \mathfrak{F}_1 \times \mathfrak{F}_2 \times \cdots \mathfrak{F}_n, \mathbf{P}_1 \times \mathbf{P}_2 \times \cdots \times \mathbf{P}_n)$ is defined. The aim is to construct a probability $\mathbf{P}$ on $\times_{n=1}^{\infty} \mathfrak{F}_n$ such that for each $n \geq 1$, $\mathbf{P}_1 \times \mathbf{P}_2 \times \cdots \times \mathbf{P}_n$ is the (finite-dimensional) joint distribution of $(X_1, X_2, \cdots, X_n)$, in other words, $\mathbf{P}(E_1 \times \cdots \times E_n \times \Omega_{n+1} \times \cdots) = \prod_{i=1}^{n} \mathbf{P}_i(E_i)$.

Heuristically, the finite product corresponds to forgetting about the remaining factors. Formally, for each n, one may define $pr_n : \Omega \to \Omega_1 \times \Omega_2 \times \cdots \times \Omega_n$ by setting $pr_n(\omega) = (\omega(1), \omega(2), \ldots, \omega(n))$. Then, if

$\tilde{X}_k(\omega_1 \times \omega_2 \times \cdots \times \omega_n) = \omega_k$, $1 \leq k \leq n$, it follows that $X_k = \tilde{X}_k \circ pr_n$, $1 \leq k \leq n$. In other words, Fig. 3.4 is commutative.

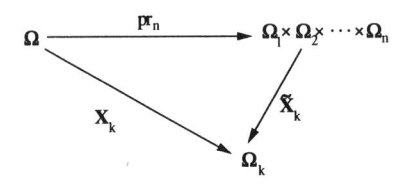

Fig. 3.4

The function pr_n, which "forgets" the tail end of a point in Ω, allows one to set up a correspondence between the sets in $\times_{n=1}^\infty \mathfrak{F}_n$, which are determined by the coordinate functions $X_1, X_2 \cdots, X_n$, and the sets in $\mathfrak{F}_1 \times \mathfrak{F}_2 \times \cdots \times \mathfrak{F}_n$. More precisely, if $\mathfrak{G}_n = \sigma(\{X_i \mid 1 \leq i \leq n\})$, one has the following result.

Proposition 3.4.9. *The following statements about $\Lambda \subset \Omega$ are equivalent:*

(1) $\Lambda \in \mathfrak{G}_n$;
(2) $\Lambda = pr_n^{-1}(A)$, $A \in \mathfrak{F}_1 \times \mathfrak{F}_2 \times \cdots \times \mathfrak{F}_n$.

Furthermore, the map $A \to pr_n^{-1}(A)$ maps $\mathfrak{F}_1 \times \mathfrak{F}_2 \times \cdots \times \mathfrak{F}_n$ in a one-to-one way onto $\mathfrak{G}_n$. In addition,

(3) $\Lambda = A \times (\times_{k=n+1}^\infty \Omega_k)$, $A \in \mathfrak{F}_1 \times \mathfrak{F}_2 \times \cdots \times \mathfrak{F}_n$.

Proof. To verify that (1) is equivalent to (2), first note that $\cap_{i=1}^n X_i^{-1}(E_i) = pr_n^{-1}(E_1 \times E_2 \times \cdots \times E_n)$ since $\omega \in \cap_{i=1}^n X_i^{-1}(E_i)$ if and only if, for all i, $1 \leq i \leq n$, $X_i(\omega) = \omega(i) \in E_i$. Hence, the proposition holds for $\Lambda = \cap_{i=1}^n X_i^{-1}(E_i)$.

Let $\mathfrak{F} = \{A \in \mathfrak{F}_1 \times \mathfrak{F}_2 \times \cdots \times \mathfrak{F}_n \mid pr_n^{-1}(A) \in \mathfrak{G}_n\}$. Then $\mathfrak{F}$ is a σ-algebra (see Exercise 2.1.18). Since it contains all the sets $E_1 \times E_2 \times \cdots \times E_n$, $E_i \in \mathfrak{F}_i$, $1 \leq i \leq n$, it follows that $\mathfrak{F} = \mathfrak{F}_1 \times \mathfrak{F}_2 \times \cdots \times \mathfrak{F}_n$, and so $A \in \mathfrak{F}_1 \times \mathfrak{F}_2 \times \cdots \times \mathfrak{F}_n$ implies that $pr_n^{-1}(A) \in \mathfrak{G}_n$. Conversely, let $\mathfrak{G}_n' = \{\Lambda \in \mathfrak{G}_n \mid \Lambda = pr_n^{-1}(A) \text{ for some } A \in \mathfrak{F}_1 \times \mathfrak{F}_2 \times \cdots \times \mathfrak{F}_n\}$. It is also a σ-algebra, and as it contains all the sets $\cap_{i=1}^n X_1^{-1}(E_i)$, it coincides with $\mathfrak{G}_n$. This shows that (1) is equivalent to (2).

To see that Λ and A determine each other, observe that

(a) if $(x_1, x_2, \ldots, x_n) = x \in \times_{i=1}^n \Omega_i$, then there is a function $\omega \in \Omega$ with $\omega(i) = x_i, 1 \leq i \leq n$ (i.e., $x = pr_n(\omega)$ for some $\omega \in \Omega$), and
(b) if $\omega \in \Lambda \in \mathfrak{G}_n$ and $pr_n(\omega') = pr_n(\omega)$ (i.e., if $\omega(i) = \omega'(i), 1 \leq i \leq n$), then $\omega' \in \Lambda$.

The first statement is clear enough: pick any function $\omega_0 \in \times_{n=1}^{\infty} \Omega_n$ and define $\omega(i) = x_i, 1 \leq i \leq n$ and $\omega(i) = \omega_0, i > n$. The second one is true because (i) it is true for any set Λ of the form $\cap_{i=1}^{n} X_i^{-1}(E_i)$; if it is true for Λ_1 and for $\Lambda_0 \subset \Lambda_1$ it is true for $\Lambda_1 \setminus \Lambda_0$; and finally, the class of sets in $\mathfrak{G}$ for which it is true is a monotone class (Exercise 1.4.15); hence, by Proposition 3.2.6 it is true for every set in $\mathfrak{G}$ as the class $\mathfrak{C}$ of sets of the form $\cap_{i=1}^{n} X_i^{-1}(E_i)$ is closed under finite intersections. Let $pr_n^{-1}(A) = \Lambda$. Then $A = pr_n(\Lambda)$: if $x \in A \subset \times_{i=1}^{n} \Omega_i$, then by (a) above, there is a function $\omega \in \Omega$ such that $x_i = \omega(i), 1 \leq i \leq n$; then $\omega \in \Lambda$ and $pr_n(\omega) = x$; since $pr_n(\Lambda) \subset A$ in any case, $A = pr_n(\Lambda)$. Hence, $A_1 = A_2$ if $pr_n^{-1}(A_1) = pr_n^{-1}(A_2)$.

On the other hand, if $pr_n(\Lambda_1) = pr_n(\Lambda_2) = A$, then $\Lambda_1 = \Lambda_2$: if $\omega_1 \in \Lambda_1$ and $x = pr_n(\omega_1) = pr_n(\omega_2)$ with $\omega_2 \in \Lambda_2$, then by (b) above, $\omega_1 \in \Lambda_2$ (i.e., $\Lambda_1 \subset \Lambda_2$); by symmetry, $\Lambda_1 = \Lambda_2$.

This shows that A and Λ determine each other. Formula (3) follows from the above discussion. $\square$

The main goal here is to construct a probability $\mathbf{P}$ on $\times_{n=1}^{\infty} \mathfrak{F}_n$. Using Proposition 3.4.9, one determines $\mathbf{P}$ on the Boolean algebra $\mathfrak{A} = \cup_{n=1}^{\infty} \mathfrak{G}_n$ as follows.

Corollary 3.4.10. *If $\Lambda \in \mathfrak{A} = \cup_{n=1}^{\infty} \mathfrak{G}_n$, define $\mathbf{P}(A) = (\mathbf{P}_1 \times \mathbf{P}_2 \times \cdots \times \mathbf{P}_n)(A)$, where $\Lambda = A \times (\times_{k=n+1}^{\infty} \Omega_k)$. Then $\mathbf{P}$ is a finitely additive probability on $\mathfrak{A}$.*

Proof. Since $\Lambda \in \mathfrak{A}$, for some n, $\Lambda \in \mathfrak{G}_n$. By Proposition 3.4.9, $\Lambda = A \times (\times_{k=n+1}^{\infty} \Omega_n)$ for a unique $A \in \mathfrak{F}_1 \times \mathfrak{F}_2 \times \cdots \times \mathfrak{F}_n$. Since $\Lambda \in \mathfrak{G}_{n+1}$, one has $\Lambda = A \times \Omega_{n+1} \times (\times_{k=n+2}^{\infty} \Omega_k)$.

In order to see that the formula for $\mathbf{P}$ defines a function on $\mathfrak{A}$, it suffices to show that, for any $A \in \mathfrak{F}_1 \times \mathfrak{F}_2 \times \cdots \times \mathfrak{F}_n$, $(\mathbf{P}_1 \times \mathbf{P}_2 \times \cdots \times \mathbf{P}_n)(A) = (\mathbf{P}_1 \times \mathbf{P}_2 \times \cdots \times \mathbf{P}_n \times \mathbf{P}_{n+1})(A \times \Omega_{n+1})$. However, by Proposition 3.3.8, this last probability is $((\mathbf{P}_1 \times \mathbf{P}_2 \times \cdots \times \mathbf{P}_n) \times \mathbf{P}_{n+1})(A \times \Omega_{n+1}) = (\mathbf{P}_1 \times \mathbf{P}_2 \times \cdots \times \mathbf{P}_n)(A)\mathbf{P}_{n+1}(\Omega_{n+1}) = (\mathbf{P}_1 \times \mathbf{P}_2 \times \cdots \times \mathbf{P}_n)(A)$.

The fact that $\mathbf{P}$ is a finitely additive probability on $\mathfrak{A}$ follows from the fact that $\mathbf{P}$ restricted to each $\mathfrak{G}_n$ is a probability. $\square$

Exercise 3.4.11. Verify this last statement.

At this stage, one has $(\times_{n=1}^{\infty} \Omega_n, \times_{n=1}^{\infty} \mathfrak{F}_n)$ defined and $\mathbf{P}$ defined on $\mathfrak{A} = \cup_{n=1}^{\infty} \mathfrak{G}_n$, which is a Boolean algebra with the property that $\times_{n=1}^{\infty} \mathfrak{F}_n = \mathfrak{F}$ is the smallest σ-algebra containing $\mathfrak{A}$. As a result, there is at most one way to define $\mathbf{P}$ on $\mathfrak{F}$, and this is possible if and only if $\mathbf{P}$ is σ-additive on $\mathfrak{A}$.

Comment. The reader may avoid the following technicalities and skip to the statement of Theorem 3.4.14.

The σ-additivity of $\mathbf{P}$ on $\mathfrak{A}$ is proved using a technique due to Jessen (see Anderson and Jessen[1]) and to Kakutani.[2] The argument given here can be found in Halmos [H1], pp. 157–158. To see more closely how this technique works, it is worthwhile first to consider two probability spaces $(\tilde{\Omega}_1, \tilde{\mathfrak{F}}_1, \tilde{\mathbf{P}}_1)$, $(\tilde{\Omega}_2, \tilde{\mathfrak{F}}_2, \tilde{\mathbf{P}}_2)$ and Fubini's theorem. Let $E \in \tilde{\mathfrak{F}}_1 \times \tilde{\mathfrak{F}}_2$ and, for each $\omega_1 \in \tilde{\Omega}_1$, let $E(\omega_1)$ be the "slice" or section of E over ω_1, i.e., $E(\omega_1) = \{\omega_2 \mid (\omega_1, \omega_2) \in E\}$ (see Fig. 3.5).

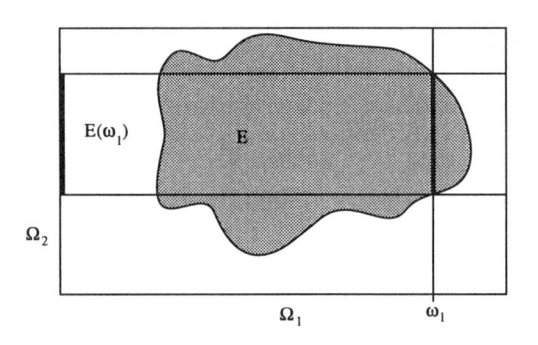

Fig. 3.5

Fubini's theorem (Theorem 3.3.5 (1)) states that

(i) $E(\omega_1) \in \tilde{\mathfrak{F}}_2$ for each $\omega_1 \in \tilde{\Omega}_1$,
(ii) $\omega_1 \to \tilde{\mathbf{P}}_2(E(\omega_1))$ is a random variable on $(\tilde{\Omega}_1, \tilde{\mathfrak{F}}_1, \tilde{\mathbf{P}}_1)$, and
(iii) $(\tilde{\mathbf{P}}_1 \times \tilde{\mathbf{P}}_2)(E) = \int \tilde{\mathbf{P}}_2(E(\omega_1)) \tilde{\mathbf{P}}_1(d\omega_1)$.

Lemma 3.4.12. *If* $(\tilde{\mathbf{P}}_1 \times \tilde{\mathbf{P}}_2)(E) \geq \delta > 0$, *then* $\tilde{\mathbf{P}}_1(\{\omega_1 \mid \tilde{\mathbf{P}}_2(E(\omega_1)) \geq \frac{\delta}{2}\}) \geq \frac{\delta}{2}$.

Proof. Let $B = \{\omega_1 \mid \tilde{\mathbf{P}}_2(E(\omega_1)) \geq \frac{\delta}{2}\} \in \mathfrak{F}_1$. Then

$$0 < \delta \leq \int \tilde{\mathbf{P}}_2(E(\omega_1)) \mathbf{P}_1(d\omega_1)$$
$$= \int_B \tilde{\mathbf{P}}_2(E(\omega_1)) \mathbf{P}_1(d\omega_1) + \int_{B^c} \tilde{\mathbf{P}}_2(E(\omega_1)) \mathbf{P}_1(d\omega_1)$$
$$\leq \mathbf{P}_1(B) + \frac{\delta}{2} \mathbf{P}_1(B^c) \leq \mathbf{P}_1(B) + \frac{\delta}{2}. \quad \square$$

To prove the σ-additivity of $\mathbf{P}$ on A, one uses a variant of this lemma over and over again to construct a function $\omega^0 \in \cap_{n=1}^{\infty} C_n$ if $(C_n)_{n \geq 1} \subset \mathfrak{A}$

---
[1] *Some limit theorems on integrals in an abstract set*, Det Kgl. Danske Videnskabernes Selskab Mat-Fys. Medd. **Bind XXII Nr. 14** (1946), 3–29.
[2] *Note on Infinite Product Measure Spaces I*, Proc. Imp. Acad. Tokyo, **vol XIX** (1943), 148–151.

is decreasing and $\mathbf{P}(C_n) \geq \delta > 0$ for all $n \geq 1$. It follows from this that $\mathbf{P}$ is σ-additive on $\mathfrak{A}$ by Exercise 3.3.2 (4): if not, then there is a decreasing sequence $(C_n)_{n\geq 1}$ of sets $C_n \in \mathfrak{A}$ and a positive number δ such that (i) $\cap_{n=1}^{\infty} C_n = \emptyset$ and (ii) $\lim_n \mathbf{P}(C_n) = \delta > 0$. If, as will now be shown in the context of the infinite product, (ii) implies that (i) is false, then by Exercise 3.3.2 (4), $\mathbf{P}$ is σ-additive on $\mathfrak{A}$.

The argument goes as follows. After relabeling (if necessary), one may assume that

$$C_n = A_n \times (\times_{k=n+1}^{\infty} \Omega_k) \text{ with } A_n \in \mathfrak{F}_1 \times \mathfrak{F}_2 \times \cdots \times \mathfrak{F}_n.$$

Exercise. Explain this statement. [For example, if $C_1 \in \mathfrak{G}_{n_1}$, one may redefine C_n to be Ω for $1 \leq n < n_1$ and C_{n_1} to be C_1. Explain how to continue: show that there is a strictly increasing sequence $(k_n)_n$ such that $C_n \in \mathfrak{G}_{k_n}$.]

Having arranged the labeling of the sets C_n, the problem of showing that $\cap_{n=1}^{\infty} C_n \neq \emptyset$ can be reformulated. It amounts to proving the following lemma.

Lemma 3.4.13. *Assume that, for each $n \geq 1$, one has a set $A_n \in \mathfrak{F}_1 \times \mathfrak{F}_2 \times \cdots \times \mathfrak{F}_n$ such that, for all $n \geq 1$,*

(1) $A_n \times \Omega_{n+1} \supset A_{n+1}$, and
(2) $(\mathbf{P}_1 \times \mathbf{P}_2 \times \cdots \times \mathbf{P}_n)(A_n) \geq \delta > 0.$

Then there is a function $\omega^0 \in \times_{k=1}^{\infty} \Omega_k$ such that $(\omega^0(1), \omega^0(2), \ldots, \omega^0(n)) \in A_n$ for all $n \geq 1$.

To determine a value for $\omega^0(1)$, let $A_n(\omega_1)$ be the slice of A_n over ω_1 in $\Omega_1 \times (\Omega_2 \times \cdots \times \Omega_n)$ when $n \geq 2$, (i.e., in Lemma 3.4.12, take $\tilde{\Omega}_1 = \Omega_1$ and $\tilde{\Omega}_2 = \Omega_2 \times \cdots \times \Omega_n$). Since $(\mathbf{P}_1 \times \mathbf{P}_2 \times \cdots \times \mathbf{P}_n)(A_n) \geq \delta$ for all $n \geq 2$, it follows from Lemma 3.4.12 that $\mathbf{P}_1(B_n) \geq \frac{\delta}{2}$ for all $n \geq 2$, where $B_n = \{\omega_1 \mid (\mathbf{P}_2 \times \cdots \times \mathbf{P}_n)(A_n(\omega_1)) \geq \frac{\delta}{2}\}$.

Since $A_1 \times \Omega_2 \supset A_2$, if $A_2(\omega_1) \neq \emptyset$, then $\omega_1 \in A_1 : (\omega_1, \omega_2) \in A_2$ implies $\omega_1 \in A_1$. Hence, $A_1 \supset B_2$.

Further, $A_n \times \Omega_{n+1} \supset A_{n+1}$ implies that $A_n(\omega_1) \times \Omega_{n+1} \supset A_{n+1}(\omega_1)$: if $(\omega_2, \ldots, \omega_{n+1}) \in A_{n+1}(\omega_1)$, then $(\omega_1, \omega_2, \ldots, \omega_{n+1}) \in A_{n+1}$ and so $(\omega_1, \ldots, \omega_n) \in A_n$ since $(\omega_1, \omega_2, \ldots, \omega_{n+1}) \in A_n \times \Omega_{n+1}$. This implies that $(\omega_2, \ldots, \omega_{n+1}) \in A_n(\omega_1) \times \Omega_{n+1}$. Hence, $(\mathbf{P}_2 \times \cdots \times \mathbf{P}_{n+1})(A_{n+1}(\omega_1)) \leq (\mathbf{P}_2 \times \cdots \times \mathbf{P}_n)(A_n(\omega_1))$ and so $B_n \supset B_{n+1}$ for all $n \geq 2$.

Since $\mathbf{P}_1$ is a probability and $A_1 \supset B_n \supset B_{n+1}$ for all $n \geq 2$, the fact that $\mathbf{P}_1(B_n) \geq \frac{\delta}{2}$ for all $n \geq 2$ implies that $\mathbf{P}_1(\cap_{n\geq 2} B_n) \geq \frac{\delta}{2}$. Hence, there is a point $\omega_1^0 \in \cap_{n\geq 2} B_n$. Note that $\omega_1^0 \in A_1$ and $A_n(\omega_1^0) \neq \emptyset$ for all $n \geq 2$.

This establishes the following variant of Lemma 3.4.12.

Lemma 3.4.12*. *Assume that*

$$(\mathbf{P}_1 \times \mathbf{P}_2 \times \cdots \times \mathbf{P}_n)(A_n) \geq \delta \text{ for all } n \geq 2.$$

One now "forgets" about Ω_1 and replaces the sets A_n by their slices $A_n(\omega_1^0) \in \mathfrak{F}_2 \times \mathfrak{F}_3 \times \cdots \times \mathfrak{F}_n$. Since $(\mathbf{P}_2 \times \mathbf{P}_3 \times \cdots \times \mathbf{P}_n)(A_n(\omega_1^0)) \geq \frac{\delta}{2}$ for all $n \geq 2$, it follows from the above lemma that there is a point $\omega_2^0 \in A_2(\omega_1^0)$ such that $(\mathbf{P}_3 \times \cdots \times \mathbf{P}_n)(A_n(\omega_1^0, \omega_2^0)) \geq \frac{\delta}{2^2}$ for all $n \geq 3$, where $A_n(\omega_1^0, \omega_2^0) = A_n(\omega_1^0)(\omega_2^0)$. Using this lemma, it follows, by induction on m, that for each $m \geq 2$, if $n \geq m$ there are $\omega_1^0, \omega_2^0, \ldots, \omega_{m-1}^0$ such that

(a) $(\omega_1^0, \omega_2^0, \ldots, \omega_{m-1}^0) \in A_{m-1}$, and
(b) $(\mathbf{P}_m \times \mathbf{P}_{m+1} \times \cdots \times \mathbf{P}_n)(A_n(\omega_1^0, \omega_2^0, \ldots, \omega_{m-1}^0)) \geq \frac{\delta}{2^{m-1}}$.

To go from the case $m = k$ to the case $m = k+1$, one replaces the sets $A_n(\omega_1^0, \omega_2^0, \ldots, \omega_{k-1}^0)$ by their slices $A_n(\omega_1^0, \omega_2^0, \ldots, \omega_k^0)$, where $A_n(\omega_1^0, \omega_2^0, \ldots, \omega_k^0) = A_n(\omega_1^0, \omega_2^0, \ldots, \omega_{k-1}^0)(\omega_k^0)$, and then makes use of the lemma.

This proves Lemma 3.4.13 and hence the σ-additivity of $\mathbf{P}$ on $\mathfrak{A}$.

This completes the proof of the following result.

Theorem 3.4.14. *Let $((\Omega_n, \mathfrak{F}_n, \mathbf{P}_n))_{n \geq 1}$ denote a sequence of probability spaces. Then there is a unique probability $\mathbf{P}$ on the σ-algebra $\times_{n=1}^\infty \mathfrak{F}_n$ on $\times_{n=1}^\infty \Omega_n$ such that, for all $\Lambda \in \mathfrak{G}_n = \sigma(\{X_i \mid 1 \leq i \leq n\})$, $\mathbf{P}(\Lambda) = (\mathbf{P}_1 \times \cdots \times \mathbf{P}_n)(A)$ if $\Lambda = A \times (\times_{k=n+1}^\infty \Omega_k)$. It will be denoted by $\times_{n=1}^\infty \mathbf{P}_n$.*

Definition 3.4.15. *The (infinite)* **product of the probability spaces** *$((\Omega_n, \mathfrak{F}_n, \mathbf{P}_n))_{n \geq 1}$ is defined to be the probability space constructed by Theorem 3.4.14. It will be denoted by $(\times_{k=1}^\infty \Omega_k, \times_{k=1}^\infty \mathfrak{F}_k, \times_{n=1}^\infty \mathbf{P}_k)$.*

As a result, given a sequence of probabilities $(\mathbf{Q}_n)_{n \geq 1}$ on $\mathbb{R}$, one can find a probability space $(\Omega, \mathfrak{F}, \mathbf{P})$ and a sequence of random variables $(X_n)_{n \geq 1}$ such that (1) the sequence is independent and (2) the distribution of each X_n is $\mathbf{Q}_n$.

Exercise 3.4.16. Let $(\Omega_n, \mathfrak{F}_n, \mathbf{P}_n) = (\mathbb{R}, \mathfrak{B}(\mathbb{R}), \mathbf{Q}_n), n \geq 1$. Show that the infinite product of the spaces with the random variables $X_n(\omega) = \omega(n)$ gives a probability space with the desired properties.

Remarks. (1) The first proof of this theorem is due to Von Neumann.[3] (2) A corollary of the theorem is its extension to arbitrary products as stated below. The reason is that if $(X_\iota)_{\iota \in \mathbf{I}}$ is any family of functions, the σ-additivity of a set function $\mathbf{P}$ on the Boolean algebra $\mathfrak{A} = \cup_{\text{finite } F \subset \mathbf{I}} \sigma(\{X_\iota \mid \iota \in F\})$ may be determined by using only a countable number of the variables (see the remark following Proposition 3.4.2).

Corollary 3.4.17. *Let $(\Omega_\iota, \mathfrak{F}_\iota, \mathbf{P}_\iota), \iota \in \mathbf{I}$, be any family of probability spaces. Let Ω be the product set $\times_{\iota \in I} \Omega_\iota$ (i.e., the set of functions $\omega : I \to \cup_{\iota \in I} \Omega_\iota$ with $\omega(\iota) \in \Omega_\iota$, for all $\iota \in I$) and let $\times_{\iota \in I} \mathfrak{F}_\iota$ be the σ-algebra $\sigma(\{X_\iota \mid \iota \in I\})$, where $X_\iota(\omega) = \omega(\iota)$ for all $\omega \in \Omega$, $\iota \in I$. Then there is a*

---
[3] *Functional operators, Vol. I: Measure and Integration*, Annals of Mathematics Studies no.21, Princeton University Press, Princeton, N.J., 1950.

unique probability $\mathbf{P}$ on $\times_{\iota \in I} \mathfrak{F}_\iota$ such that

(*) $$\mathbf{P}(\cap_{\iota \in I}\{X_\iota \in E_\iota\}) = \prod_{\iota \in I} \mathbf{P}_\iota(E_\iota)$$

if $E_\iota \in \mathfrak{F}_\iota$, for all ι, and $E_\iota = \Omega_\iota$ for all but a finite number of indices.

5. SOME REMARKS ON MARKOV CHAINS*

This starred section of the book may be omitted without any loss of continuity in the succeeding chapters. It is even more technical than the proof of σ-additivity for the product measure. In fact, it deals with an important generalization of the concept of a product measure, namely a so-called Markov Chain, one simple example of which may be constructed from a series of real-valued i.i.d. random variables $(X_n)_{n \geq 1}$ by considering the partial sums of the random series $\sum_{n=0}^{\infty} X_n$, where X_0 is an arbitrary random variable.

In Neveu [N1], the above results on infinite products are obtained as consequences of a more general result due to I. Tulcea (see [N1], p. 162; also in Doob [D1], pp. 613–615). This more general result also applies to Markov chains. What follows is a presentation of the result for Markov chains (see Revuz [R1], Theorem 2.8).

Let $(X_n)_{n \geq 0}$ be a real-valued stochastic process defined on a probability space $(\Omega, \mathfrak{F}, \mathbf{P})$, where the parameter n is to be thought of as the time at which the nth observation of some phenomenon is made. In general, one might suppose that the $(n+1)$st observation depends upon all the previous ones (the complete past). Since these observations are known only probabilistically (i.e., via their distributions $\mathbf{P}_n, n \geq 0,$), this suggests that to compute $\mathbf{P}_{n+1}(B) = \mathbf{P}[X_{n+1} \in B]$, one should average over the past using the joint distribution $\mathbf{Q}_n$ of $(X_0, X_1, \ldots, X_n)$. Intuitively, there is a transition that occurs from the observation X_n to the observation X_{n+1} that in principle involves $X_0, X_1, \ldots, X_n$. The process is said to be a **Markov chain** if in fact this transition depends only upon X_n. Furthermore, if the way the transition occurs does not depend upon the particular time, the chain is said to be **homogeneous**.

In this case, there is a function $\mathbf{N}(x, B)$ of position x and Borel set B that describes the transition from being in "state" x at one time to being in B at the next time with probability $\mathbf{N}(x, B)$. Using the so-called transition function, one has (for a homogeneous chain) $\mathbf{P}[X_{n+1} \in B] = E[\mathbf{N}(X_n, B)]$, the average over the possible positions of X_n of the probability of then being in B after time increases by one unit. For this formula to make sense, the transition function $\mathbf{N}$ has to satisfy certain conditions, which are spelled out in the following definition. Note that in the above heuristic discussion, $\mathbf{N}$ is a transition function from $(\mathbb{R}, \mathfrak{B}(\mathbb{R}))$ to $(\mathbb{R}, \mathfrak{B}(\mathbb{R}))$ in the terminology to follow.

Definition 3.5.1. Let $(\Omega_0, \mathfrak{F}_0)$ and $(\Omega_1, \mathfrak{F}_1)$ be measurable spaces (i.e., each one consists of a set Ω and a σ-algebra $\mathfrak{F}$ e.g., $\Omega = \mathbb{R}^n$, $\mathfrak{F} = \mathfrak{B}(\mathbb{R}^n)$). A **transition function** (or **Markovian kernel**) $\mathbf{N}$ from Ω_0 to Ω_1 is a function $\mathbf{N} : \Omega_0 \times \mathfrak{F}_1 \to \mathbb{R}^+$ such that
  (1) for each $\omega_0 \in \Omega_0$, $A \to \mathbf{N}(\omega_0, A)$ is a probability measure on $\mathfrak{F}_1$ (a probability because $\mathbf{N}$ is to be Markovian),
  (2) for each $A \in \mathfrak{F}_1$, the function $\omega_0 \to \mathbf{N}(\omega_0, A) \in \mathfrak{F}_0$ (see Revuz [R1]).

Exercise 3.5.2. If $X \in \mathfrak{F}_1$ is non-negative, show that $\mathbf{N}X \in \mathfrak{F}_0$, where $\mathbf{N}X(\omega_0) \stackrel{\text{def}}{=} \int \mathbf{N}(\omega_0, d\omega_1) X(\omega_1)$.

Examples.
  (1) $\Omega_i = \mathbb{R}$, $\mathfrak{F}_i = \mathfrak{B}(\mathbb{R})$, and $\mathbf{N}(x, A) = \frac{1}{(2\pi)^{1/2}} \int_A e^{-\frac{(y-x)^2}{2}} dy$,
  (2) $\Omega_i = \mathbb{Z}$, $\mathfrak{F}_i = $, all subsets of $\mathbb{Z}$, and $\mathbf{N}(n, A) = \frac{1}{2}|A \cap \{n-1, n+1\}|$.
These two examples of kernels have extensions to higher dimensions.
  (3) $\Omega_i = \mathbb{R}^n$, $\mathfrak{F}_i = \mathfrak{B}(\mathbb{R}^n)$, and $\mathbf{N}(x, A) = \frac{1}{(2\pi)^{n/2}} \int_A e^{-\frac{\|y-x\|^2}{2}} dy$,
  (4) $\Omega_i = \mathbb{Z}^d$, $\mathfrak{F}_i = $ all subsets of $\mathbb{Z}^d$, and $\mathbf{N}(\mathbf{n}, A) = \frac{1}{2d}|A \cap \{\mathbf{n} \pm \mathbf{e}_i, 1 \le i \le d\}|$, where $\mathbf{e}_i$ is the canonical basis vector of $\mathbb{R}^n$ that has a 1 in the ith position and zero elsewhere, and $\mathbf{n} = (n_1, n_2, \ldots, n_d)$, i.e., $2^d \mathbf{N}(\mathbf{n}, A)$ is the number of so-called **nearest neighbours** of $\mathbf{n}$ that are in A.
  (5) Ω_i is a countable set, say $\mathbb{N}$, and $\mathfrak{F}_i$ is the σ-algebra of all subsets of Ω_i. Let $(\pi(i,j))_{i,j \ge 1}$ be an infinite stochastic matrix (i.e., $\pi(i,j) \ge 0$ for all i,j and $\sum_{j=1}^\infty \pi(i,j) = 1$ for all i). Define $\mathbf{N}(i, A) = \sum_{j \in A} \pi(i, j)$.

The first four examples are all **convolution kernels**, i.e., they have the form $\mathbf{N}(x, A) = \mu(A - x)$, where μ is a probability on the abelian group $\mathbb{R}^n$ or $\mathbb{Z}^d$. For example, in (1), the probability is the unit normal $n(x)dx$ since $\frac{1}{(2\pi)^{1/2}} \int 1_{A-x}(u) e^{-(\frac{u^2}{2})} du = \frac{1}{(2\pi)^{1/2}} \int 1_A(u+x) e^{-(\frac{u^2}{2})} du = \frac{1}{(2\pi)^{1/2}} \int 1_A(y) e^{-\frac{(y-x)^2}{2}} dy$.

Knowing $\mathbf{N}$ and X_n, or rather its distribution $\mathbf{P}_n$, one may immediately compute $\mathbf{P}_{n+1}$ as $\mathbf{P}_{n+1}(B) = E[\mathbf{N}(X_n, B)]$. The following lemma, applied to $\varphi(x) = \mathbf{N}(x, B)$, implies that

(1) $$\mathbf{P}_{n+1}(B) = \int \mathbf{P}_n(dx) \mathbf{N}(x, B).$$

Lemma 3.5.3. Let X be a random variable on $(\Omega, \mathfrak{F}, \mathbf{P})$ with distribution Q. Let $\varphi : \mathbb{R} \to \mathbb{R}^+$ be a non-negative Borel function. Then

$$E[\varphi \circ X] = \int \varphi(x) Q(dx).$$

Proof. Let $\varphi \ge 0$ be Borel and $s_n \uparrow \varphi$ be a sequence of Borel simple functions. Then $\int \varphi(x) Q(dx) = \lim_{n \to \infty} \int s_n(x) Q(dx)$. Also, $s_n \circ X \uparrow \varphi \circ X$

5. SOME REMARKS ON MARKOV CHAINS

and so $E[\varphi \circ X] = \lim_{n\to\infty} \int (s_n \circ X)(\omega)P(\omega)$. It therefore suffices to verify the formula for simple functions.

Let $s = \sum_{n=1}^{N} a_n 1_{A_n}$ be a Borel simple function. Then $\sum_{n=1}^{N} a_n 1_{\{X \in A_n\}} = s \circ X$, and since by definition $\mathbf{Q}(A_n) = \mathbf{P}[X \in A_n]$, $E[s \circ X] = \sum_{n=1}^{N} a_n \mathbf{P}[X \in A_n] = \int s(x) Q(dx)$. □

Remark. The integral $\int \mathbf{P}_n(dx)\mathbf{N}(x, B)$ equals $\int \varphi(x)\mathbf{P}_n(dx)$, where $\varphi(x) = \mathbf{N}(x, B)$. This different order in the expression for the integral is used because in effect two integrations occur in a specified order: first one computes $\mathbf{N}(x, B) = \int_B \mathbf{N}(x, dy)$; then, to compute $\mathbf{P}_{n+1}$, one integrates with respect to $\mathbf{P}_n$. There is a definite order here, suggested by writing the integral in this new way which looks "backwards". This usage will be common in what follows, (see Proposition 3.5.4).

Now suppose that one wants to determine the joint distribution of X_n and X_{n+1}. To compute $\mathbf{P}[X_n \in A, X_{n+1} \in B] = \mathbf{P}[(X_n, X_{n+1}) \in A \times B]$, note that if $\mathbf{P}[X_n \in A] = \mathbf{P}_n(A) \neq 0$, then formula (1) using the conditional probability $\mathbf{P}_n[\cdot \mid A]$ in place of $\mathbf{P}_n$ gives

$$\mathbf{P}[X_{n+1} \in B \mid X_n \in A] = \int \mathbf{P}_n(dx \mid A)\mathbf{N}(x, B).$$

Since

$$\mathbf{P}[X_n \in A, X_{n+1} \in B] = \mathbf{P}[X_n \in A]\mathbf{P}[X_{n+1} \in B \mid X_n \in A],$$

and

$$\int \mathbf{P}_n(dx|A)\psi(x) = \frac{1}{\mathbf{P}_n(A)} \int_A \mathbf{P}_n(dx)\psi(x)$$

it follows that

(2) $$\mathbf{P}[X_n \in A, X_{n+1} \in B] = \int_A \mathbf{P}_n(dx)\mathbf{N}(x, B).$$

To summarize, given a real-valued homogeneous Markov chain $(X_n)_{n \geq 0}$ on a probability space $(\Omega, \mathfrak{F}, \mathbf{P})$ with transition kernel $\mathbf{N}$, the distributions $\mathbf{P}_n$, are related by the transition kernel: $\mathbf{P}_{n+1}(B) = \int \mathbf{P}_n(dx)\mathbf{N}(x, B)$ and the joint distributions $\mathbf{P}_{n,n+1}$ of (X_n, X_{n+1}) are given by $\mathbf{P}_{n,n+1}(A \times B) = \int_A \mathbf{P}_n(dx)\mathbf{N}(x, B)$.

The goal now is to obtain the formulas for the joint distributions $\mathbf{Q}_n$ of $(X_0, X_1, \ldots, X_n)$ and to relate them to the distributions of the X_k and $\mathbf{N}$.

First, however, it is necessary to show that the right-hand side of formula (2) determines a probability on $\mathfrak{B}(\mathbb{R}^2)$: the reason is that ultimately one wants to show the existence of the Markov chain, i.e., of the basic probability space $(\Omega, \mathfrak{F}, \mathbf{P})$ and of the sequence of random variables X_n with the above properties. Of course, when the Markov chain exists, it is clear that (2) defines a probability, namely the distribution of (X_n, X_{n+1}). In fact, the right-hand side of (2) always determines a probability as the next result shows.

122   III. INDEPENDENCE AND PRODUCT MEASURES

Proposition 3.5.4. *Let $\mathbf{P}_0$ be a probability on $(\Omega_0, \mathfrak{F}_0)$ and let $\mathbf{N}$ be a Markovian kernel from Ω_0 to Ω_1. Then there is a unique probability $\mathbf{P}$ on $(\Omega_0 \times \Omega_1, \mathfrak{F}_0 \times \mathfrak{F}_1)$ such that*

(*)
$$\mathbf{P}(E_0 \times E_1) = \int_{E_0} \left[\int \mathbf{N}(\omega_0, d\omega_1) 1_{E_1}(\omega_1) \right] \mathbf{P}_0(d\omega_0)$$
$$\stackrel{\text{def}}{=} \int_{E_0} \mathbf{P}_0(d\omega_0) \left[\int \mathbf{N}(\omega_0, d\omega_1) 1_{E_1}(\omega_1) \right],$$

where $\int \mathbf{N}(\omega_0, d\omega_1) 1_{E_1}(\omega_1) \stackrel{\text{def}}{=} \mathbf{N}(\omega_0, E_1)$ and $E_i \in \mathfrak{F}_i, i = 1, 2$.

Proof. Let $\mathfrak{A}$ be the Boolean algebra of finite disjoint unions of sets of the form $E_0 \times E_1$, $E_i \in \mathfrak{F}_i$. The formula (*) determines $\mathbf{P}$ on $\mathfrak{A}$. To verify that $\mathbf{P}$ is σ-additive on $\mathfrak{A}$, one first proves a lemma for Markov kernels that corresponds to Lemma 3.3.3. Then the argument used for the σ-additivity in the case of the product of two measures applies. □

Interpretation of Proposition 3.5.4 in the case of Example (4) for $d = 2$. Consider an initial probability distribution $\mathbf{P}_0$ on $\mathbb{Z}^2 = \Omega$ together with the given transition kernel $\mathbf{N}$. Now $\mathbf{N}(\omega, A)$ can be interpreted as giving the probability for the transition or motion of a random particle starting at $\omega = (n_1, n_2) = \mathbf{n}$ to a point in $A \subset \mathbb{Z}^2$ after one jump. On $\Omega \times \Omega = \mathbb{Z}^2 \times \mathbb{Z}^2$ define $X_0(\omega_0, \omega_1) = \omega_0$ and $X_1(\omega_0, \omega_1) = \omega_1$. One may view X_0 as giving the initial position of the random particle and X_1 as giving its position after one jump, with the result distributed according to the transition kernel $\mathbf{N}$ and the initial distribution $\mathbf{P}_0$ of position. The probability $\mathbf{P}$, which is the joint distribution of (X_0, X_1), also determines the distribution $\mathbf{P}_0$ of X_0 since $\mathbf{P}[X_0 \in A] = \mathbf{P}(A \times \mathbb{Z}^2) = \int_A \mathbf{N}(\omega_0, \mathbb{Z}^2) \mathbf{P}_0(d\omega_0) = \mathbf{P}_0(A)$, and the distribution of X_1 since $\mathbf{P}[X_1 \in A] = \mathbf{P}(\mathbb{Z}^2 \times A) = \int_{\mathbb{Z}^2} \mathbf{N}(\mathbf{n}, A) \mathbf{P}_0(d\mathbf{n})$. For example, if $\mathbf{P}_0 = \frac{1}{2}\varepsilon_0 + \frac{1}{3}\varepsilon_{\mathbf{e}_1} + \frac{1}{6}\varepsilon_{\mathbf{e}_2}$, where $\varepsilon_{\mathbf{n}}$ is a unit or point mass at the point $\mathbf{n}$ (see Exercise 2.9.9) and $A = \{\mathbf{0}, \mathbf{e}_1, \mathbf{e}_2, \mathbf{e}_1 + \mathbf{e}_2\}$, then $\mathbf{P}[X_1 \in A] = \frac{1}{2}\mathbf{N}(\mathbf{0}, A) + \frac{1}{3}\mathbf{N}(\mathbf{e}_1, A) + \frac{1}{6}\mathbf{N}(\mathbf{e}_2, A) = (\frac{1}{2} + \frac{1}{3} + \frac{1}{6})\frac{1}{2} = \frac{1}{2}$, as each point in A has two nearest neighbours in A. If $B = \{\mathbf{0}, \mathbf{e}_1, \mathbf{e}_2, -\mathbf{e}_1 + \mathbf{e}_2\}$, then $\mathbf{P}[X_1 \in B] = \frac{1}{2}\mathbf{N}(\mathbf{0}, B) + \frac{1}{3}\mathbf{N}(\mathbf{e}_1, B) + \frac{1}{6}\mathbf{N}(\mathbf{e}_2, B) = \frac{1}{2}\frac{1}{2} + \frac{1}{3}\frac{1}{4} + \frac{1}{6}\frac{1}{2} = \frac{5}{12}$. Also, if $C = \{\mathbf{0}, -e_1, \mathbf{e}_2, -\mathbf{e}_1 + \mathbf{e}_2\}$, then $\mathbf{P}(C \times B) = \frac{1}{2}\mathbf{N}(\mathbf{0}, B) + \frac{1}{6}\mathbf{N}(\mathbf{e}_2, B) = \frac{1}{2}\frac{1}{2} + \frac{1}{6}\frac{1}{2} = \frac{1}{3}$.

Turning to the computation of the joint distributions $\mathbf{Q}_n$, observe that $\mathbf{Q}_0 = \mathbf{P}_0$ and $\mathbf{Q}_1$ is determined by $\mathbf{P}_0$ and $\mathbf{N}$ as

$$\mathbf{Q}_1(E_0 \times E_1) = \int_{E_0} \mathbf{P}_0(dx_0) \mathbf{N}(x_0, E_1).$$

Notice that

$$\mathbf{Q}_2(\mathbb{R} \times E_1 \times E_2) = \mathbf{P}_{1,2}(E_1 \times E_2) = \int_{E_1} \mathbf{P}_1(dx_1) \mathbf{N}(x_1, E_2).$$

Now
$$\mathbf{P}_1(E_1) = \int \mathbf{P}_0(dx_0)\mathbf{N}(x_0, E_1) = \int \mathbf{P}_0(dx_0)\left[\int_{E_1} \mathbf{N}(x_0, dx_1)\right]$$

i.e., $\mathbf{P}_1(dx_1) = \mathbf{P}_0(dx_0)\mathbf{N}(x_0, dx_1)$, and so

(3) $\quad \mathbf{Q}_2(\mathbb{R} \times E_1 \times E_2) = \int \mathbf{P}_0(dx_0)\left[\int_{E_1} \mathbf{N}(x_0, dx_1)\mathbf{N}(x_1, E_2)\right].$

Assume that $\mathbf{P}[X_0 \in E_0] = \mathbf{P}_0(E_0) \neq 0$. Then one may replace $\mathbf{P}_0$ in (3) by the conditional probability $\mathbf{P}_0[\,\cdot\,|E_0]$ and obtain

$$\mathbf{P}[X_1 \in E_1, X_2 \in E_2 \mid X_0 \in E_0]$$
$$= \int \mathbf{P}_0(dx_0|E_0)\left[\int_{E_1} \mathbf{N}(x_0, dx_1)\mathbf{N}(x_1, E_2)\right]$$
$$= \frac{1}{\mathbf{P}_0(E_0)} \int_{E_0} \mathbf{P}_0(dx_0)\left[\int_{E_1} \mathbf{N}(x_0, dx_1)\mathbf{N}(x_1, E_2)\right].$$

As a result,
$$Q_2(E_0 \times E_1 \times E_2) = \mathbf{P}[X_0 \in E_0, X_1 \in E_1, X_2 \in E_2]$$
$$= \int_{E_0} \mathbf{P}_0(dx_0)\left[\int_{E_1} \mathbf{N}(x_0, dx_1)\mathbf{N}(x_1, E_2)\right].$$

In this formula, the term $\int_{E_1} \mathbf{N}(x_0, dx_1)\mathbf{N}(x_1, E_2)$ in the brackets determines a kernel $\mathbf{M}_2(x_0, E)$ from $(\mathbb{R}, \mathfrak{B}(\mathbb{R}))$ to $(\mathbb{R}^2, \mathfrak{B}(\mathbb{R}))$, where

$$\mathbf{M}_2(x_0, E) \stackrel{\text{def}}{=} \int \mathbf{N}(x_0, dx_1)\left[\mathbf{N}(x_1, dx_2)1_E(x_1, x_2)\right].$$

Exercise 3.5.5. Verify this statement. [*Hint*: consider the collection of Borel sets $E \subset \mathbb{R}^2$ such that $\mathbf{N}(x_1, dx_2)1_E(x_1, x_2)$ is a Borel function of x_1.]

The kernel $\mathbf{M}_2$ will also be denoted by the following formal expression:

$$\mathbf{M}_2(x_0, dx) = \mathbf{M}_2(x_0, dx_1, dx_2) = \mathbf{N}(x_0, dx_1)\mathbf{N}(x_1, dx_2),$$

where dx stands for $dx_1 dx_2$ and the integer 2 indicates the dimension of $\mathbb{R}^2$.

In a similar way, the probability $\mathbf{Q}_2$ will also be denoted by the formal expression

$$\mathbf{Q}_2(dx_0, dx_1, dx_2) = \mathbf{P}_0(dx_0)\mathbf{N}(x_0, dx_1)\mathbf{N}(x_1, dx_2) = \mathbf{P}_0(dx_0)\mathbf{M}_2(x_0, dx).$$

These formal formulas for kernels and probabilities have to be interpreted for integration purposes as requiring integration first on x_2, then on x_1, and finally on x_0.

Since $\mathbf{Q}_2(dx_0, dx_1, dx_2) = \mathbf{P}_0(dx_0)\mathbf{M}_2(x_0, dx)$, it follows that $\mathbf{Q}_2$ is the probability determined by Proposition 3.5.4 from $\mathbf{P}_0$ and the kernel $\mathbf{M}_2$. It is also determined in the same way by $\mathbf{Q}_1$ and $\mathbf{N}$ since $\mathbf{Q}_2(dx_0, dx_1, dx_2) = \mathbf{Q}_1(dx_0, dx_1)\mathbf{N}(x_1, dx_2)$ as $\mathbf{Q}_1(dx_0, dx_1) = \mathbf{P}_0(dx_0)\mathbf{N}(x_0, dx_1)$. The kernel $\mathbf{N}(x_1, dx_2)$ is here viewed as a kernel from $(\mathbb{R}^2, \mathfrak{B}(\mathbb{R}^2))$ to $(\mathbb{R}, \mathfrak{B}(\mathbb{R}))$: as a function of (x_0, x_1), the probability $\mathbf{N}(x_1, A)$ of $A \in \mathfrak{B}(\mathbb{R})$ is Borel measurable.

The statement about how to use formulas like

$$\mathbf{Q}_2(dx_0, dx_1, dx_2) = \mathbf{Q}_1(dx_0, dx_1)\mathbf{N}(x_1, dx_2)$$

for integration follows from the next formal result, which can be viewed as a generalization of Fubini's theorem.

Proposition 3.5.6. (**An extension of Fubini's theorem**). *Let $\mathbf{P}_0$ be a probability on $(\Omega_0, \mathfrak{F}_0)$ and $\mathbf{N}$ be a Markov kernel from $(\Omega_0, \mathfrak{F}_0)$ to $(\Omega_1, \mathfrak{F}_1)$. Let $\mathbf{P}$ be the probability on $(\Omega_0 \times \Omega_1, \mathfrak{F}_0 \times \mathfrak{F}_1)$ such that*

$$\mathbf{P}(E_0 \times E_1) = \int_{E_0} \mathbf{P}_0(d\omega_0) \left[\int \mathbf{N}(\omega_0, d\omega_1) 1_{E_1}(\omega_1) \right]$$
$$\stackrel{def}{=} \int \mathbf{P}_0(d\omega_0) \left[\int \mathbf{N}(\omega_0, d\omega_1) 1_{E_0 \times E_1}(\omega_0, \omega_1) \right],$$

where $E_i \in \mathfrak{F}_i$, $i = 0$ or 1. Let X be a non-negative random variable on $(\Omega_0 \times \Omega_1, \mathfrak{F}_0 \times \mathfrak{F}_1, \mathbf{P})$. Then

(1) *for all $\omega_0 \in \Omega_0$, the function $\omega_1 \to X(\omega_0, \omega_1) \in \mathfrak{F}_1$,*
(2) *the function $\omega_0 \to \int \mathbf{N}(\omega_0, d\omega_1) X(\omega_0, \omega_1) \in \mathfrak{F}_0$, and*
(3) $E[X] = \int \mathbf{P}_0(d\omega_0) \left[\int \mathbf{N}(\omega_0, d\omega_1) X(\omega_0, \omega_1) \right].$

Proof. The proof of this proposition is essentially the same as the proof of the Fubini theorem (Theorem 3.3.5). One reduces to the case of $X = 1_E, E \in \mathfrak{F}_0 \times \mathfrak{F}_1$, and then uses the same argument. Note that (2) extends Exercise 3.5.2. □

Remark. If $E \in \mathfrak{F}_0 \times \mathfrak{F}_1$, and $E(\omega_0)$ is the slice of E over $\omega_0 \in \Omega_0$, this result applied to $X = 1_E$ shows that

$$\mathbf{P}(E) = \int [\mathbf{N}(\omega_0, E(\omega_0))] \mathbf{P}_0(d\omega_0).$$

Then, one has the following result.

5. SOME REMARKS ON MARKOV CHAINS

Lemma 3.5.7. *If* $\mathbf{P}(E) \geq \delta$, *then* $\mathbf{P}_0(\{\omega_0 \mid \mathbf{N}(\omega_0, E(\omega_0)) \geq \frac{\delta}{2}\}) \geq \frac{\delta}{2}$.

Proof. Repeat the proof of Lemma 3.4.12. □

It should be clear by now what to expect for the formula for $\mathbf{Q}_n$, namely

$$\mathbf{Q}_n(dx_0, dx_1, \ldots, dx_n) = \mathbf{Q}_{n-1}(dx_0, dx_1, \ldots, dx_{n-1})\mathbf{N}(x_{n-1}, dx_n).$$

Since

$$\mathbf{Q}_{n-1}(dx_0, dx_1, \ldots, dx_{n-1}) = \mathbf{Q}_{n-2}(dx_0, dx_1, \ldots, dx_{n-2})\mathbf{N}(x_{n-2}, dx_{n-1}),$$

it follows that

$$\mathbf{Q}_n(dx_0, dx_1, \ldots, dx_n)$$
$$= \mathbf{Q}_{n-2}(dx_0, dx_1, \ldots, dx_{n-2})\mathbf{N}(x_{n-2}dx_{n-1})\mathbf{N}(x_{n-1}, dx_n)$$
$$\vdots$$
$$= \mathbf{P}(dx_0)\mathbf{N}(x_0, dx_1)\mathbf{N}(x_1, dx_2)\cdots\mathbf{N}(x_{n-1}, dx_n).$$

This is formalized as the following result.

Proposition 3.5.8. *For any* $n \geq 1$, *there is a unique probability* $\mathbf{Q} = \mathbf{Q}_n$ *on* $(\mathbb{R} \times \mathbb{R}^n, \mathfrak{B}(\mathbb{R}^{n+1}))$ *such that*

$$\mathbf{Q}_n(dx_0, dx_1, \ldots, dx_n) = \mathbf{Q}_{n-1}(dx_0, dx_1, \ldots, dx_{n-1})\mathbf{N}(x_{n-1}, dx_n),$$

i.e., such that

$$\mathbf{Q}_n(E_0 \times E_1 \times \cdots \times E_n)$$
$$= \int_{E_0 \times \cdots \times E_{n-1}} \mathbf{Q}_{n-1}(dx_0, dx_1, \ldots, dx_{n-1})\mathbf{N}(x_{n-1}, E_n)$$
$$\vdots$$
$$= \int_{E_0} \mathbf{P}_0(dx_0)\left[\int_{E_1} \mathbf{N}(x_0, dx_1)\left[\int_{E_2} \mathbf{N}(x_1, dx_2) \cdots \right.\right.$$
(†) $$\left.\left. \cdots \left[\int_{E_n} \mathbf{N}(x_{n-1}, dx_n)\right] \cdots \right]\right],$$

where $E_i \in \mathfrak{B}(\mathbb{R})$, $0 \leq i \leq n$. *Furthermore, if* X *is a non-negative* random variable on $(\mathbb{R}^{n+1}, \mathfrak{B}(\mathbb{R}^{n+1}), \mathbf{Q}_n)$, then

$$E[X] = \int \mathbf{Q}_n(dx_0, dx_1, \ldots, dx_n) X(x_0, x_1, \ldots, x_n)$$
$$= \int \mathbf{Q}_{n-1}(dx_0, dx_1, \ldots, dx_{n-1}) \left[\int \mathbf{N}(x_{n-1}, dx_n) X(x_0, x_1, \ldots, x_n) \right]$$
$$\vdots$$

(††)
$$= \int \mathbf{P}_0(dx_0) \left[\int \mathbf{N}(x_0, dx_1) \left[\int \mathbf{N}(x_0, dx_2) \cdots \right. \right.$$
$$\left. \left. \cdots \left[\int \mathbf{N}(x_{n-1}, dx_n) X(x_0, x_1, \ldots, x_n) \right] \cdots \right] \right].$$

Proof. There is at most one such probability since $\mathfrak{B}(\mathbb{R}^{n+1})$ is the smallest σ-algebra containing all the sets $E_0 \times E_1 \times \cdots \times E_n$.

For $n = 1$ this result follows from Propositions 3.5.4 and 3.5.6. Assume the result to be true for $n = k$. Let $\mathbf{Q}_k = \tilde{\mathbf{P}}_0$ be the probability on $(\mathbb{R} \times \mathbb{R}^k, \mathfrak{B}(\mathbb{R}^{k+1}))$ such that (†) holds. Define $\tilde{\mathbf{N}}((x_0, x_1, \ldots, x_k), dx_{k+1}) = \mathbf{N}(x_k, dx_{k+1})$. Then $\tilde{\mathbf{N}}$ is a Markov kernel from $\Omega_0 = (\mathbb{R} \times \mathbb{R}^k, \mathfrak{B}(\mathbb{R}^{k+1}))$ to $\Omega_1 = (\mathbb{R}, \mathfrak{B}(\mathbb{R}))$. Let $\mathbf{Q}_{k+1}$ be the unique probability on $(\mathbb{R} \times \mathbb{R}^k \times \mathbb{R}, \mathfrak{B}(\mathbb{R}^{k+1}) \times \mathfrak{B}(\mathbb{R})) = (\mathbb{R} \times \mathbb{R}^{k+1}, \mathfrak{B}(\mathbb{R}^{k+2}))$ such that for $\tilde{E}_0 \in \mathfrak{B}(\mathbb{R}^{k+1})$ and $E_{k+1} \in \mathfrak{B}(\mathbb{R})$,

(4)
$$\mathbf{Q}_{k+1}(\tilde{E}_0 \times E_{k+1})$$
$$= \int_{\tilde{E}_0} \mathbf{Q}_k(dx_0, dx_1, \ldots, dx_k) \left[\int_{E_{k+1}} \tilde{\mathbf{N}}((x_0, x_1, \ldots, x_k), dx_{k+1}) \right]$$
$$= \int_{\tilde{E}_0} \mathbf{Q}_k(dx_0, dx_1, \ldots, dx_k) \left[\int_{E_{k+1}} \mathbf{N}(x_k, dx_{k+1}) \right]$$
$$= \int_{\tilde{E}_0} \mathbf{Q}_{k+1}(dx_0, dx_1, \ldots, dx_k) \mathbf{N}(x_k, E_{k+1}).$$

By using (††) for the case $n = k$ and the function $X(x_0, x_1, \ldots, x_k) = 1_{\tilde{E}_0}(x_0, x_1, \ldots, x_k) \mathbf{N}(x_k, E_{k+1})$, when $\tilde{E}_0 = E_0 \times E_1 \times \cdots \times E_k$, it follows that

$$\int_{\tilde{E}_0} \mathbf{Q}_k(dx_0, dx_1, \ldots, dx_k) \mathbf{N}(x_k, E_{k+1})$$
$$= \int_{E_0} \mathbf{P}_0(dx_0) \left[\int_{E_1} \mathbf{N}(x_0, dx_1) \left[\int_{E_2} \mathbf{N}(x_1, dx_2) \cdots \right. \right.$$
$$\left. \left. \cdots \left[\int_{E_{k+1}} \mathbf{N}(x_k, dx_{k+1}) \right] \cdots \right] \right],$$

5. SOME REMARKS ON MARKOV CHAINS

which verifies the first part (†) of the result for $n = k+1$.

To show that the "Fubini" formula (††) holds for $\mathbf{Q}_{k+1}$, it suffices to apply Proposition 3.5.6 to $\mathbf{Q}_k = \tilde{\mathbf{P}}_0$ and $\tilde{\mathbf{N}}$ and to use the fact that it holds for $\mathbf{Q}_k$. □

Remarks 3.5.9. (1) If $E \in \mathfrak{B}(\mathbb{R}^{n+1})$, $\mathbf{Q}_n(E)$ is the probability that the **trajectory** or **path** $(X_0(\omega), X_1(\omega), \ldots, X_n(\omega)) = (x_0, x_1, \ldots, x_n)$ of the random particle during the time set $\{0, 1, \ldots, n\}$ lies in E.

(2) Formula (4) in the proof of Proposition 3.5.8 implies that if $\tilde{E}_0 \in \mathfrak{B}(\mathbb{R}^{k+1})$, then

$$E[1_{\tilde{E}_0}(x)\psi(x_{k+1})] = \int_{\tilde{E}_0} \mathbf{Q}_{k+1}(dx_0, dx_1, \ldots, dx_{k+1})\psi(x_{k+1})$$

$$= \int_{\tilde{E}_0} \mathbf{Q}_k(dx_0, dx_1, \ldots, dx_k)\mathbf{N}(x_k, \psi)$$

$$= E[1_{\tilde{E}_0}(x)\mathbf{N}(x_k, \psi)]$$

for any non-negative Borel function ψ, where $\mathbf{N}(x, \psi) \stackrel{\text{def}}{=} \int \mathbf{N}(x, dy)\psi(y)$ and the expectation is taken relative to $\mathbf{Q}_{k+1}$. This observation will be explained in Chapter V in terms of conditional expectations and the **Markov property** (see Remark 5.2.11 and Exercise 5.2.12).

If the Markov kernel $\mathbf{N}(x, dy)$ does not depend on x (i.e., $\mathbf{N}(x, dy) = \mathbf{Q}(dy)$ for all x), the probability $\mathbf{Q}_n = \mathbf{P}_0 \times \mathbf{Q} \times \cdots \times \mathbf{Q}$. These probabilities are the finite-dimensional joint distributions of an independent process $(X_n)_{n\geq 0}$, with $\mathbf{P}_0$ the distribution of X_0 and the remaining X_n all identically distributed with common distribution $\mathbf{Q}$. This raises the problem as to whether for a general Markovian kernel $\mathbf{N}$ there is a stochastic process with the $\mathbf{Q}_n$ as the finite-dimensional joint distribution corresponding to $\{0, 1, \ldots, n\}$. Such processes exist and are called **Markov chains with transition probability N and initial distribution $\mathbf{P}_0$**. To simplify the following discussion, one assumes the Markov chain to be real-valued.

To construct such a process, one uses the probabilities $\mathbf{Q}_n$ to construct a probability on the Boolean algebra $\mathfrak{A} = \cup_{n=0}^{\infty} \mathfrak{G}_n$ of subsets of $\Omega = \times_{n=0}^{\infty} \Omega_n$, $\Omega_n = \mathbb{R}$ for all $n \geq 0$, where $\mathfrak{G}_n = \sigma(\{X_i \mid 0 \leq i \leq n\})$ and $X_i : \Omega \to \mathbb{R}$ is defined by $X_i(\omega) = \omega(i)$ for all $\omega \in \Omega$. These notations were used before when constructing the infinite product space $(\times_{n=1}^{\infty} \Omega_n, \times_{n=1}^{\infty} \mathfrak{F}_n, \times_{n=1}^{\infty} \mathbf{P}_n)$. Here $\Omega_n = \mathbb{R}$ and $\mathfrak{F}_n = \mathfrak{B}(\mathbb{R})$ for all $n \geq 0$, but the sequence $(\mathbf{P}_n)_{n \geq 1}$ of probabilities $\mathbf{P}_n$ is not given. In the construction of the infinite product space, it was shown in Proposition 3.4.9 that $\Lambda \in \mathfrak{G}_n$ if and only if $\Lambda = A \times (X_{k=n+1}^{\infty} \Omega_k)$, where A is uniquely determined by Λ and vice versa. Define $\mathbf{Q}$ on $\mathfrak{A}$ by setting $\mathbf{Q}(\Lambda) = \mathbf{Q}_n(A)$ (for all $n \geq 0$).

This defines a function on $\mathfrak{A}$ provided, for all $n \geq 0$ and $A \in \mathfrak{B}(\mathbb{R}^{n+1})$, $\mathbf{Q}_n(A) = \mathbf{Q}_{n+1}(A \times \mathbb{R})$. In view of formula (†) in Proposition 3.5.8, this is clear because $\int_{\mathbb{R}} \mathbf{N}(x_{n-1}, dx_n) = \mathbf{N}(x_{n-1}, \mathbb{R}) = 1$ since $\mathbf{N}$ is a Markovian kernel. To extend $\mathbf{Q}$ from $\mathfrak{A} = \bigcup_{n=0}^{\infty} \mathfrak{E}_n$ to $\times_{n=0}^{\infty} \mathfrak{F}_n$, $\mathfrak{F}_n = \mathfrak{B}(\mathbb{R})$ for all $n \geq 0$, one has therefore to prove that $\mathbf{Q}$ is σ-additive on A. In view of Lemma 3.5.7, the pattern of the argument used for the case of infinite products can be followed to show that if $(C_n)_{n\geq 0} \subset \mathfrak{A}$, $C_n \supset C_{n+1}$, and $\mathbf{Q}(C_n) \geq 2\delta > 0$ for all $n \geq 0$, then $\cap_{n=0}^{\infty} C_n \neq \emptyset$. As before, this implies that $\mathbf{Q}$ is σ-additive on $\mathfrak{A}$ (see Exercise 3.3.2 (4)).

As before one may assume $C_n \in \mathfrak{E}_n$ for all $n \geq 0$, i.e., $C_n = A'_n \times (\times_{k=n+1}^{\infty} \Omega_k)$, $A'_n \in \mathfrak{B}(\mathbb{R}^{n+1})$. This allows one to transpose the problem to the analogue of Lemma 3.4.13.

Before doing this, it will be helpful to define several kernels. Given the Markov kernel $\mathbf{N}$, define the Markov kernel $\mathbf{M}_n$ from $(\mathbb{R}, \mathfrak{B}(\mathbb{R}))$ to $(\mathbb{R}^n, \mathfrak{B}(\mathbb{R}^n))$ by setting

$$\mathbf{M}_n(x_0, dx_1, dx_2, \ldots, dx_n) = \mathbf{N}(x_0, dx_1)\mathbf{N}(x_1, dx_2) \cdots \mathbf{N}(x_{n-1}, dx_n).$$

Note that the subscript n on $\mathbf{M}_n$ indicates that the probability corresponding to x_0 is defined on the Borel subsets of $\mathbb{R}^n$.

In keeping with the convention established earlier, this formula is to be read as follows: if $E \in \mathfrak{B}(\mathbb{R}^n)$, then

$$\mathbf{M}_n(x_0, E) = \int \mathbf{N}(x_0, dx_1) \left[\int \mathbf{N}(x_1, dx_2) \cdots \right.$$
$$\left. \left[\int \mathbf{N}(x_{n-1}, dx_n) 1_E(x_1, x_2, \ldots, x_n) \right] \cdots \right].$$

In particular, if $E = E_1 \times E_2 \times \cdots \times E_n$, then

$$\mathbf{M}_n(x_0, E_1 \times E_2 \times \cdots \times E_n) = \int_{E_1} \mathbf{N}(x_0, dx_1) \left[\int_{E_2} \mathbf{N}(x_1, dx_2) \cdots \right.$$
$$\left. \left[\int_{E_n} \mathbf{N}(x_{n-1}, dx_n) \right] \cdots \right].$$

Hence, the earlier formula for $\mathbf{Q}_2$ extends, i.e.,

(*) $\qquad \mathbf{Q}_n(dx_0, dx_1, \ldots, dx_n) = \mathbf{P}_0(dx_0)\mathbf{M}_n(x_0, dx_1, \ldots, dx_n).$

For a fixed x_0, the probability $\mathbf{N}(x_0, dx_1)\mathbf{M}_{n-1}(x_1, dx_2, dx_3, \ldots, dx_n)$ is the probability defined by Proposition 3.5.4 with $\mathbf{P}_0(d\omega_0) = \mathbf{N}(x_0, d\omega_0)$ and $\mathbf{N}(\omega_0, d\omega_1) = \mathbf{M}_{n-1}(\omega_0, d\omega_1)$, i.e., $\omega_0 = x_1$ and $d\omega_1 = (dx_2, dx_3, \ldots, dx_n)$. It follows from Proposition 3.5.4 that the value of this probability for $E = (E_1 \times \tilde{E}_2)$ with $E_1 \subset \mathbb{R}$ and $\tilde{E}_2 \subset \mathbb{R}^{n-1}$ is $\int_{E_1} \mathbf{N}(x_0, dx_1)\mathbf{M}_{n-1}(x_1, \tilde{E}_2)$.

If $\tilde{E}_2 = E_2 \times \cdots \times E_n$, then

$$\int_{E_1} \mathbf{N}(x_0, dx_1)\mathbf{M}_{n-1}(x_1, E_2 \times \cdots \times E_n)$$
$$= \int_{E_1} \mathbf{N}(x_0, dx_1) \left[\int_{E_2} \mathbf{N}(x_1, dx_2) \cdots \left[\int_{E_n} \mathbf{N}(x_{n-1}, dx_n) \right] \cdots \right].$$

In other words, the kernel $\mathbf{M}_n(x_0, dx) = \mathbf{N}(x_0, dx_1)\mathbf{M}_{n-1}(x_1, d\tilde{x})$, where $dx = (dx_1, dx_2, \ldots, dx_n)$ and $d\tilde{x}$ denotes $(dx_2, dx_3, \ldots, dx_n)$ since two measures that agree on all the Borel sets $E_1 \times E_2 \times \cdots \times E_n$ are identical.

Returning to the proof that $\mathbf{Q}$ is σ-additive on $\mathfrak{A}$, as a first step one applies Lemma 3.5.7 to the probabilities $\mathbf{Q}_n$ using the decomposition given by formula (*). If $B_n = \{x_0 \mid M_n(x_0, A'_n(x_0)) \geq \delta\}$, then $\mathbf{P}_0(B_n) \geq \delta$, for all $n \geq 1$. As in the infinite product case, one has $B_n \supset B_{n+1}$ as $A'_n \times \mathbb{R} \supset A'_{n+1}$ for all $n \geq 1$ implies $A'_n(x_0) \times \mathbb{R} \supset A'_{n+1}(x_0)$ and $\mathbf{M}_n(x_0, A'_n(x_0)) = \mathbf{M}_{n+1}(x_0, A'_n(x_0) \times \mathbb{R}) \geq \mathbf{M}_{n+1}(x_0, A'_{n+1}(x_0))$. Hence $\mathbf{P}_0(\cap_{n=1}^\infty B_n) \geq \delta$, and so there is a point $x_0^0 \in A'_0$ such that $M_n(x_0^0, A'_n(x_0^0)) \geq \delta$ for all $n \geq 1$.

To continue the induction and simplify the notation, let A_n now denote $A'_n(x_0^0)$ for all $n \geq 1$. The analogue of Lemma 3.4.13 is the following lemma.

Lemma 3.5.10. *Assume that, for each $n \geq 1$, one has a set $A_n \in \mathfrak{B}(\mathbb{R}^n)$ such that, for all $n \geq 0$,*

(1) $A_n \times \mathbb{R} \supset A_{n+1}$, and
(2) $\mathbf{M}_n(x_0, A_n) \geq \delta > 0$.

Then there is a sequence $(x_k^0)_{k \geq 1}$ of real numbers such that, for all $n \geq 1$, one has $(x_1^0, x_2^0, \ldots, x_n^0) \in A_n$.

Once this lemma is proved, it follows that $\mathbf{Q}$ is σ-additive on $\mathfrak{A}$ and hence extends uniquely to $\mathfrak{F} = \times_{n=0}^\infty \mathfrak{F}_n$; the point $(x_0^0, x_1^0, \ldots, x_n^0, \ldots) \in \cap_{n=0}^\infty C_n$.

Since $\mathbf{M}_n(x_0, dx) = \mathbf{N}(x_0, dx_1)\mathbf{M}_{n-1}(x_1, d\tilde{x})$, it follows from Lemma 3.5.7 that if $B_n = \{x_1 \mid \mathbf{M}_{n-1}(x_1, A_n(x_1)) \geq \frac{\delta}{2}\}$, then $\mathbf{N}(x_0, B_n) \geq \frac{\delta}{2}$. Just as before, it follows that $A_n(x_1) \times \mathbb{R} \supset A_{n+1}(x_1)$ and so $B_n \supset B_{n+1}$ since $\mathbf{M}_{n-1}(x_1, A_n(x_1)) = \mathbf{M}_n(x_1, A_n(x_1) \times \mathbb{R}) \geq \mathbf{M}_n(x_1, A_{n+1}(x_1))$. Hence, $\mathbf{N}(x_0, \cap_{n=1}^\infty B_n) \geq \frac{\delta}{2}$, and so there is a point $x_1^0 \in A_1$ such that $\mathbf{M}_{n-1}(x_1^0, A_n(x_1^0)) \geq \frac{\delta}{2}$ for all $n \geq 2$.

This proves the following variant of Lemma 3.5.10.

Lemma 3.5.10*. *Assume that*

$$\mathbf{M}_n(x_0, A_n) \geq \delta \quad \text{for all } n \geq 2.$$

Then there is a point $x_1^0 \in A_1$ such that

$$\mathbf{M}_{n-1}(x_1^0, A_n(x_1^0)) \geq \frac{\delta}{2} \quad \text{for all } n \geq 2.$$

One now "forgets" about the first copy of $\mathbb{R}$ and replaces the sets A_n by their slices $A_n(x_1^0) \in \mathfrak{B}(\mathbb{R}^{n-1})$. Since $\mathbf{M}_{n-1}(x_1^0, A_n(x_1^0)) \geq \frac{\delta}{2}$ for all $n \geq 2$, and $\mathbf{M}_{n-1}(x_1, dx_2, \ldots, dx_n) = \mathbf{N}(x_1, dx_2)\mathbf{M}_{n-2}(x_2, dx_3, \ldots, dx_n)$, it follows from the above lemma that there is a point $x_2^0 \in A_2(x_1^0)$ (i.e., $(x_1^0, x_2^0) \in A_2$), such that $\mathbf{M}_{n-2}(x_2^0, A_n(x_1^0, x_2^0)) \geq \frac{\delta}{2^2}$ for all $n \geq 3$, where $A_n(x_1^0, x_2^0) = A_n(x_1^0)(x_2^0)$. Using this lemma, it follows by induction on m that, for each $m \geq 2$, if $n \geq m+1$, there are $x_1^0, x_2^0, \ldots, x_{m-1}^0$ such that

(a) $(x_1^0, x_2^0, \ldots, x_{m-1}^0) \in A_{m-1}$ and
(b) $\mathbf{M}_{n-(m-1)}(x_{m-1}^0, A_n(x_1^0, x_2^0, \ldots, x_{m-1}^0)) \geq \frac{\delta}{2^{m-1}}$.

To go from the case $m = k$ to the case $m = k+1$, one replaces the sets $A_n(x_1^0, x_2^0, \ldots, x_{k-1}^0)$ by their slices $A_n(x_1^0, x_2^0, \ldots, x_k^0)$, where $A_n(x_1^0, x_2^0, \ldots, x_k^0) = A_n(x_1^0, x_2^0, \ldots, x_{k-1}^0)(x_k^0)$, and then makes use of the lemma.

This proves Lemma 3.5.10 and hence the σ-additivity of $\mathbf{Q}$ on $\mathfrak{A}$, which completes the proof of the following result.

Theorem 3.5.11. *Let $\mathbf{P}_0$ be a probability on $\mathfrak{B}(\mathbb{R})$ and N be a Markov kernel from $(\mathbb{R}, \mathfrak{B}(\mathbb{R}))$ to $(\mathbb{R}, \mathfrak{B}(\mathbb{R}))$. Let Ω be the set of real-valued functions ω on $\mathbb{N} \cup \{0\}$, and define $X_n : \Omega \to \mathbb{R}$ by $X_n(\omega) = \omega(n)$ for all $n \geq 0$. Denote by $\mathfrak{F}$ the countable product of the Borel σ-algebras, i.e., $\mathfrak{F} = \sigma(\{X_k \mid k \geq 0\})$. Then there is a unique probability $\mathbf{Q}$ on $\mathfrak{F}$ such that (for all n) the finite-dimensional joint distribution of $(X_0, X_1, \ldots, X_n)$ is the probability $\mathbf{Q}_n$ on $\mathfrak{B}(\mathbb{R}^{n+1})$ for which*

$$\mathbf{Q}_n(E_0 \times E_1 \times \cdots \times E_n)$$
$$= \int_{E_0 \times \cdots \times E_{n-1}} \mathbf{Q}_{n-1}(dx_0, dx_1, \ldots, dx_{n-1}) N(x_{n-1}, E_n)$$
$$\vdots$$
$$= \int_{E_0} \mathbf{P}_0(dx_0) \int_{E_1} N(x_0, dx_1) \cdots \int_{E_n} N(x_{n-1}, dx_n).$$

Remarks. $(\Omega, \mathfrak{F}, \mathbf{Q})$ together with the family $(X_n)_{n \geq 0}$ of random variables X_n is a stochastic process that has been constructed from its finite-dimensional joint distributions. In general, to construct stochastic processes from their finite-dimensional distributions, one is obliged to use a different technique: namely Kolmogorov's extension theorem (Kolmogorov [K2], p. 31, also Billingsley [B1], p. 510 and Loève [L2], p. 94), which makes use of local compactness and hence requires a topology. The case of a real-valued process is discussed in [K2] and is well worth reading: there is a change in terminology, with "field" indicating a Boolean algebra and "Borel field" indicating a σ-field. In Loève, the theorem is called a consistency theorem. The issue is the following: given an arbitrary product of copies of the real numbers indexed by a set I, and for each finite subset F of I a probability $\mathbf{P}_F$ on the σ-algebra $\mathfrak{F}_F$ generated by the coordinate functions $X_\iota, \iota \in F$, such that these probabilities are "consistent", (i.e., $\mathbf{P}_{F_1}$ and $\mathbf{P}_{F_2}$ agree with $\mathbf{P}_{F_1 \cap F_2}$ on $\mathfrak{F}_{F_1 \cap F_2}$), is there a probability on the σ-algebra $\mathfrak{F}$ generated by all the coordinate functions? Clearly, $\mathfrak{A} \stackrel{\text{def}}{=} \cup_{F \subset I} \mathfrak{F}_F$ is a Boolean algebra, and the consistency of the probabilities shows that there is a unique finitely additive probability $\mathbf{P}$ on $\mathfrak{A}$ that agrees with each $\mathbf{P}_F$ on each $\mathfrak{F}_F$. The problem is, as always, to show that $\mathbf{P}$ is σ-additive on $\mathfrak{A}$. The proof of Kolmogorov uses the compactness of closed and bounded subsets of $\mathbb{R}$ and makes use of Cantor's diagonal procedure (see the discussion in the appendix of Chapter VI on Helly's selection principle).

6. ADDITIONAL EXERCISES*

Definition 3.6.1. *Let E be any subset of $\mathbb{R}^n$. A subset U of E is said to be an **open subset** of E or **open relative to** E if there is an open set O of $\mathbb{R}^n$ with $U = O \cap E$, i.e., U is open in the topological subspace E of $\mathbb{R}^n$.*

Definition 3.6.2. *The σ-algebra of **Borel subsets of E** is defined to be the smallest σ-algebra containing all the open subsets of E. It will be denoted by $\mathfrak{B}(E)$.*

Exercise 3.6.3. Show that A is a Borel subset of E if and only if there is a Borel subset F of $\mathbb{R}^n$ with $A = F \cap E$.

Exercise 3.6.4. (Fubini's theorem for Lebesgue integrable functions on $\mathbb{R}^n$) Let X be a Lebesgue measurable function on $\mathbb{R}^n$ that is integrable. By Exercise 3.3.15, there is a Borel function Y on $\mathbb{R}^n$ (which can be assumed to be finite valued) with $\Lambda = \{X \neq Y\}$ of Lebesgue measure zero on $\mathbb{R}^n$. Recall that by Exercise 2.1.31, Y is integrable. Let $n = p + q$, and if $x \in \mathbb{R}^n$, let $x = (x_1, x_2)$, with $x_1 \in \mathbb{R}^p$ and $x_2 \in \mathbb{R}^q$.

Show that

(1) the slice $\Lambda(x_1)$ of Λ over $x_1 \in \mathbb{R}^p$ has Lebesgue measure zero (in $\mathbb{R}^q$) for almost all x_1, where $x_2 \in \Lambda(x_1)$ if and only if $(x_1, x_2) \in \Lambda$,
(2) if $|\Lambda(x_1)| = 0$, then $x_2 \to X(x_1, x_2)$ is Lebesgue measurable on $\mathbb{R}^p$,
(3) for almost all x_1 with $|\Lambda(x_1)| = 0$, $x_2 \to X(x_1, x_2)$ is Lebesgue integrable with $\int X(x_1, x_2) dx_2 = \int Y(x_1, x_2) dx_2$ (recall that by Corollary 3.3.17, $|\{x_1 \mid Y(x_1, \cdot) \notin L^1\}| = 0$ as $Y \in L^1$), and
(4) $\int X dx = \int [\int X(x_1, x_2) dx_2] dx_1$, where the first integral is taken to be zero if x_1 is such that $|\Lambda(x_1)| = 0$ and $x_2 \to X(x_1, x_2)$ is not integrable. [*Hint:* make use of Fubini's theorem for Borel function.]

Note that the "arbitrary" definition of $\int X(x_1, x_2) dx_2 = 0$ on the bad set in (4) amounts to modifying the original function X to have value zero on a set Γ of Lebesgue measure zero (which does not affect its integral on $\mathbb{R}^n$): let $\Gamma = (N_0 \cup N_1) \times \mathbb{R}^q$, where $N_0 = \{x_1 \mid \Lambda(x_1)$ does not have Lebesgue measure zero on $\mathbb{R}^q\}$ and $N_1 = \{x_1 \mid |\Lambda(x_1)| = 0$ and $x_2 \to X(x_1, x_2)$ is not Lebesgue integrable on $\mathbb{R}^q\}$.

Exercise 3.6.5. Show that $\int_y^{+\infty} e^{-(\frac{x^2}{2})} dx \sim \frac{1}{y} e^{-(\frac{y^2}{2})}$ as $y \to +\infty$ (i.e., the ratio tends to 1 as $y \to +\infty$). [*Hint:* show that the ratio of the derivatives of the two expressions tends to 1 as $y \to +\infty$.]

Exercise 3.6.6. Let $(E_\iota)_{\iota \in I}$ be a family of events E_ι. Show that the σ-algebra $\sigma\{1_{E_\iota} \mid \iota \in I\}$ is the smallest σ-algebra $\mathfrak{F}$ that contains one of the following collections of sets:

(1) all the sets $E_\iota, \iota \in I$;
(2) all the sets $E_\iota^c, \iota \in I$;
(3) all the sets $A_\iota, \iota \in I$, where, for each ι, $A_\iota = E_\iota$ or E_ι^c;
(4) all the sets $\cap_{\iota \in F} E_\iota$, where F is any finite subset of I;

(5) all the sets $\cap_{\iota \in F} E_\iota^c$, where F is any finite subset of I;
(6) all the sets $\cap_{\iota \in F} A_\iota$, where F is any finite subset of I and, for each ι, $A_\iota = E_\iota$ or E_ι^c.

[Hint: note that $1_E \in \mathfrak{F}$ if and only if $E \in \mathfrak{F}$; equivalently, if and only if $E^c \in \mathfrak{F}$.]

Exercise 3.6.7. Let $(E_i)_{1 \leq i \leq n}$ be a finite collection of sets. Show that it is independent in the sense of Definition 3.4.3 if and only if one of the equivalent conditions is verified:

(1) if $A_i = E_i, E_i^c$, or Ω, then $\mathbf{P}(\cap_{i=1}^n A_i) = \Pi_{i=1}^n \mathbf{P}(A_i)$;
(2) if $A_i = E_i$ or Ω, then $\mathbf{P}(\cap_{i=1}^n A_i) = \Pi_{i=1}^n \mathbf{P}(A_i)$;
(3) for any $F \subset \{1, 2, \ldots, n\}$, $\mathbf{P}(\cap_{i \in F} A_i) = \Pi_{i \in F} \mathbf{P}(A_i)$, where $A_i = E_i$ for all $i \in F$.

Remark. Condition (3) is the usual definition of a collection of independent events (see Billingsley [B1], p. 48).

Exercise 3.6.8. Show that for a family $(E_\iota)_{\iota \in I}$ of events E_ι, the following are equivalent:

(1) the family of random variables $(1_{E_\iota})_{\iota \in I}$ is independent (Definition 3.4.3);
(2) for any finite subset $F \subset I$, $\mathbf{P}(\cap_{\iota \in F} E_\iota) = \Pi_{\iota \in F} \mathbf{P}(E_\iota)$.

Exercise 3.6.9. Let $\mathfrak{C}_i, 1 \leq i \leq n$, be n classes of events each of which is closed under finite intersections. Show that the following are equivalent:

(1) $\mathbf{P}(\cap_{i=1}^n A_i) = \Pi_{i=1}^n \mathbf{P}(A_i)$ if, for each i, either $A_i \in \mathfrak{C}_i$ or $A_i = \Omega$;
(2) $\mathbf{P}(\cap_{i \in F} A_i) = \Pi_{i \in F} \mathbf{P}(A_i), A_i \in \mathfrak{C}_i$, for any $F \subset \{1, 2, \ldots, n\}$;
(3) for any $F \subset \{1, 2, \ldots, n\}$ and $F' = \{1, 2, \ldots, n\} \backslash F$, the σ-algebras $\sigma(\cup_{i \in F} \mathfrak{C}_i)$ and $\sigma(\cup_{i \in F'} \mathfrak{C}_i)$ are independent, where, for example, $\sigma(\cup_{i \in F} \mathfrak{C}_i)$ is the smallest σ-algebra that contains all the sets in the classes $\mathfrak{C}_i, i \in F$.

Remark. Condition (1) is the definition (see Definition 3.6.10) in Billingsley ([B1], p. 50) of n independent classes of events (without the requirement that the classes be closed under finite intersections). Note that Billingsley's Theorem 4.2 ([B1], p. 50) is an immediate consequence of condition (2).

Definition 3.6.10. Let $(\mathfrak{C}'_\lambda)_{\lambda \in \Lambda}$ be a family of classes of events $\mathfrak{C}'_\lambda$. It is said to be an **independent family of classes of events** if $\mathbf{P}(\cap_{\lambda \in F} A_\lambda) = \Pi_{i \in F} \mathbf{P}(A_\lambda), A_\lambda \in \mathfrak{C}'_\lambda$, for any finite subset F of Λ.

Exercise 3.6.11. Let $(\mathfrak{C}'_\lambda)_{\lambda \in \Lambda}$ be a family of classes of events $\mathfrak{C}'_\lambda$. Show that the following conditions are equivalent, where $\mathfrak{C}_\lambda$ is the class of sets that are finite intersections of sets from $\mathfrak{C}'_\lambda$:

(1) the family $(\mathfrak{C}_\lambda)_{\lambda \in \Lambda}$ of classes $\mathfrak{C}_\lambda$ is an independent family of classes;
(2) for each $\Gamma \subset \Lambda$, the σ-algebras $\sigma(\cup_{\lambda \in \Gamma} \mathfrak{C}_\lambda)$ and $\sigma(\cup_{\lambda \in \Lambda \backslash \Gamma} \mathfrak{C}_\lambda)$ are independent.

Remark. Note that Exercise 3.6.6 implies that $\sigma(\cup_{\lambda\in\Gamma}\mathfrak{C}_\lambda) = \sigma(\cup_{\lambda\in\Gamma}\mathfrak{C}'_\lambda)$ and $\sigma(\cup_{\lambda\in\Lambda\setminus\Gamma}\mathfrak{C}_\lambda) = \sigma(\cup_{\lambda\in\Lambda\setminus\Gamma}\mathfrak{C}'_\lambda)$. Hence, one might think that condition (2) is equivalent to (1) for the classes $\mathfrak{C}'_\lambda$. This is not the case, as one may have the original family $(\mathfrak{C}'_\lambda)_{\lambda\in\Lambda}$ of classes independent and the family $(\mathfrak{C}_\lambda)_{\lambda\in\Lambda}$ derived from it by taking finite intersections of the sets in each class not independent. A counterexample is easily made: Feller [F1], p. 127 contains an example of three events A_1, A_2, A_3 any two of which are independent but for which $\mathbf{P}(A_1 \cap A_2 \cap A_3) \neq \mathbf{P}(A_1)\mathbf{P}(A_2)\mathbf{P}(A_3)$; take $\mathfrak{C}'_1 = \{A_1, A_2\}$ and $\mathfrak{C}'_2 = \{A_3\}$; then these two classes are independent, but the classes $\mathfrak{C}_1 = \{A_1 \cap A_2, A_1, A_2\}$ and $\mathfrak{C}_2 = \mathfrak{C}'_2 = \{A_3\}$ are not.

Exercise 3.6.12. Let $(E_\iota)_{\iota\in I}$ be a family of independent events E_ι. Let $I = \cup_{\lambda\in\Lambda} J_\lambda$, where the sets J_λ are pairwise disjoint (e.g., if $I = \mathbb{N}$ and $\Lambda = \{1,2\}$, one could have $J_1 = \{1,3,\ldots,2n+1,\ldots\}$ the set of odd natural numbers and $J_2 = \{2,4,\ldots,2n,\ldots\}$ the set of even natural numbers). Define $\mathfrak{C}'_\lambda$ to be the family $(E_\iota)_{\iota\in J_\lambda}$ of events corresponding to the index ι in J_λ. Let $\mathfrak{F}_\lambda$ be the σ-algebra $\sigma(\mathfrak{C}'_\lambda) = \sigma(1_{E_\iota} \mid \iota \in J_\lambda)$ (by Exercise 3.6.6 this is the smallest σ-algebra that contains all the sets $E_\iota, \iota \in J_\lambda$). Show that the family $(\mathfrak{F}_\lambda)_{\lambda\in\Lambda}$ is an independent family of collections of sets, i.e., an independent family of σ-algebras. [*Hint:* verify the independence of the family of classes $(\mathfrak{C}_\lambda)_{\lambda\in\Lambda}$, where $\mathfrak{C}_\lambda$ is derived from $\mathfrak{C}'_\lambda$ by taking finite intersections of the sets in $\mathfrak{C}'_\lambda$.]

Exercise 3.6.13. Let X, Y_1, and Y_2 be three independent, real-valued random variables on a probability space $(\Omega, \mathfrak{F}, \mathbf{P})$. Assume that Y_1 and Y_2 have the same distribution. Show that

(1) for all $a \in \mathbb{R}$, $\mathbf{P}[X + Y_1 \leq a] = \mathbf{P}[X + Y_2 \leq a]$.

Assume that all three random variables are strictly positive and that the distribution of X is the same as that of $\frac{1}{Y_1+X}$. Show that

(2) if $a > 0$, $\mathbf{P}[X \leq a] = \mathbf{P}[Y_3 \leq a]$, where $Y_3 = \frac{1}{Y_2 + \frac{1}{Y_1+X}}$. [*Hint:* use transformations of x, y_1, and y_2.]

The next series of exercises is devoted to proving the monotone class theorem for functions. The notations used in the statement of the theorem will be used in these exercises without additional comment.

Theorem 3.6.14. (*Monotone class theorem for functions*). *Let $\mathcal{H}$ be a vector space of bounded functions on a set Ω. Assume that the constant function 1 belongs to $\mathcal{H}$ and that if $(f_n) \subset \mathcal{H}$ is a sequence of functions that converges uniformly to a function f, then $f \in \mathcal{H}$, (i.e., $\mathcal{H}$ is uniformly closed). In addition, assume that $\mathcal{H}$ satisfies the following monotone condition:*

if (f_n) is a sequence of functions in $\mathcal{H}$, and M is a constant such that $f_n \leq f_{n+1} \leq M$ for all n, then $\lim_n f_n \in \mathcal{H}$.

Let $\mathcal{C} \subset \mathcal{H}$ be closed under multiplication of functions. Then $\mathcal{H}$ contains every bounded function that is measurable relative to the σ-field $\sigma(\mathcal{C}) = \sigma(\{f \mid f \in \mathcal{C}\})$.

To begin with, the notions of **uniform convergence** and **uniform closure** are defined.

Definition 3.6.15. A sequence $(f_n)_{n\geq 1}$ of real-valued functions f_n on a set Ω **converges uniformly** to a real-valued function f if, for any $\epsilon > 0$, there is an integer $N = N(\epsilon)$ such that $|f_n(\omega) - f(\omega)| < \epsilon$ for all $\omega \in \Omega$ if $n \geq N$. This will be denoted by writing $f_n \xrightarrow{u} f$.

A set $\mathcal{S}$ of real-valued functions is said to be **uniformly closed** if $f \in \mathcal{S}$ whenever there is a sequence of functions in $\mathcal{S}$ that converges uniformly to f. The **uniform closure** of a set $\mathcal{S}$ of functions is the intersection of all the uniformly closed sets of functions that contains $\mathcal{S}$. It is the smallest uniformly closed set containing $\mathcal{S}$.

Exercise 3.6.16. Let $\mathcal{B}$ denote the uniform closure of $\mathcal{L}$, the linear subspace of the vector space $\mathcal{F}_b(\Omega)$ of all bounded functions on Ω that is generated by $\mathcal{C}$ and the constant function 1 (hence all constant functions): note that $h \in \mathcal{L}$ if and only if $h = (\lambda_1 f_1 - \lambda_2 f_2) + \mu$, where $f_i \in \mathcal{C}$ and $\lambda_i, \mu \in \mathbb{R}$. Show that

(1) $\mathcal{B}$ is a linear subspace of $\mathcal{F}_b(\Omega)$ and is closed under multiplication (this amounts to saying that $\mathcal{B}$ is an algebra, and since it is also a Banach space, it is a so-called **Banach algebra**).

Make use of the Weierstrass approximation theorem (Theorem 4.3.15) to show that

(2) if $f \in \mathcal{B}$, then $|f| \in \mathcal{B}$ [Hint: show that for any $M > 0$, the function $x \to |x|$ can be uniformly approximated on $[-M, M]$ by a polynomial.],

(3) conclude that if $f, g \in \mathcal{B}$, then $f \wedge g$ and $f \vee g$ are in $\mathcal{B}$. [Hint: for two real numbers a, b, recall that $\max\{a,b\} \stackrel{\text{def}}{=} a \vee b = \frac{1}{2}\{a+b+|a-b|\}$].

Exercise 3.6.17. A set $\mathcal{M}$ of bounded functions will be called a **monotone class of functions** if it satisfies the following monotone conditions for sequences (f_n) of functions f_n in $\mathcal{M}$:

(1) if there is a constant M such that $f_n \leq f_{n+1} \leq M$ for all n, then $\lim_n f_n \in \mathcal{M}$, and

(2) if there is a constant m such that $m \leq f_{n+1} \leq f_n$ for all n, then $\lim_n f_n \in \mathcal{M}$.

If, in addition, $\mathcal{M}$ is uniformly closed, then it will be called a **closed monotone class**. Show that

(3) given any collection of bounded functions, there is a smallest closed monotone class that contains it.

6. ADDITIONAL EXERCISES   135

Let $\mathcal{M}_0$ be the smallest closed monotone class that contains $\mathcal{B}$. Show that
- (4) if $f \in \mathcal{B}$ and $g \in \mathcal{M}_0$, then $f + g \in \mathcal{M}_0$ [Hint: imitate the first hint for Exercise 1.4.16 by considering $\{g \in \mathcal{M}_0 \mid f + g \in \mathcal{M}_0\}$.],
- (5) $\mathcal{M}_0$ is closed under addition (i.e., if $g_1, g_2 \in \mathcal{M}_0$, then $g_1 + g_2 \in \mathcal{M}_0$) [Hint: imitate the second hint for Exercise 1.4.16.] and scalar multiplication,
- (6) if $g_1, g_2 \in \mathcal{M}_0$, then $g_1 \vee g_2 \in \mathcal{M}_0$ [Hint: copy the proof for addition],
- (7) if $g_1, g_2 \in \mathcal{M}_0$, then $g_1 \wedge g_2 \in \mathcal{M}_0$,
- (8) if $g_1, g_2 \in \mathcal{M}_0$ and are non-negative, then $g_1 g_2 \in \mathcal{M}_0$,
- (9) $\mathcal{M}_0$ is closed under multiplication and so is a Banach algebra. [Hint: every function in $\mathcal{M}_0$ is a difference of non-negative functions in $\mathcal{M}_0$.]

Exercise 3.6.18. Let φ be a non negative function in $\mathcal{M}_0$. Show that
- (1) $1_A \in \mathcal{M}_0$, where $A = \{\varphi > 0\}$ [Hint: the function $(n\varphi) \wedge 1 \in \mathcal{M}_0$; take a limit.], and
- (2) $\mathfrak{F} \stackrel{\text{def}}{=} \{A \mid 1_A \in \mathcal{M}_0\}$ is a σ-field.

With the aid of these exercises, one may now prove the monotone class theorem for functions.

Proof of Theorem 3.6.14. The monotone hypothesis satisfied by $\mathcal{H}$ ensures that $\mathcal{H} \supset \mathcal{M}_0$. Furthermore, if $f \in \mathcal{C}$ and $\lambda \in \mathbb{R}$, then $\{f < \lambda\} = \{\varphi > 0\}$, where $\varphi = \lambda - f \wedge \lambda \in \mathcal{M}_0$. It follows from Exercise 3.6.18 that $A = \{f < \lambda\} \in \mathfrak{F}$. The following exercise shows that $\mathfrak{F} \supset \sigma(\mathcal{C})$.

Exercise 3.6.19. The σ-field $\sigma(\mathcal{C}) = \sigma(\{f \mid f \in \mathcal{C}\})$ is the smallest σ-field that contains all the sets $\{f < \lambda\}, f \in \mathcal{C}, \lambda \in \mathbb{R}$.

To conclude the proof, it suffices to observe that if a monotone class of (bounded) functions is also a vector space and contains all the functions $1_A, A \in \mathfrak{F}$, then it contains all the bounded $\mathfrak{F}$-measurable functions since it necessarily contains every $\mathfrak{F}$-simple function. $\square$

Exercise 3.6.20. Let $\mathfrak{F}(2)$ denote the σ-algebra of Lebesgue measurable subsets of $\mathbb{R}^2$. Show that
- (1) if $\mathfrak{F}(2) = \mathfrak{G} \times \mathfrak{G}$, where $\mathfrak{G}$ is a σ-algebra of subsets of $\mathbb{R}$, then $\mathfrak{G} = \mathfrak{F}(1)$ [Hint: use Fubini's theorem to study the slices of a set in $\mathfrak{F}$.], and
- (2) $\mathfrak{F}(2) \neq \mathfrak{F}(1) \times \mathfrak{F}(1)$.

Exercise 3.6.21. (Integration by parts) Let μ and ν be two probabilities on $\mathfrak{B}(\mathbb{R})$ with distribution functions F and G. Let $\mu(dx)$ and $\nu(dx)$ also be denoted by $dF(x)$ and $dG(x)$. Let $B = (a, b] \times (a, b]$, and set $B^+ = \{(x, y) \in B \mid x < y\}, B^- = \{(x, y) \in B \mid x \geq y\}$.
- (1) Use Fubini's theorem to express $(\mu \times \nu)(B^-)$ as two distinct integrals.

(2) Let $F(x-) = \lim_{y \uparrow x} F(y)$. Show that $\mu((a, c)) = F(c-) - F(a)$.

(3) Use (1) and (2) to show that

$$(\mu \times \nu)(B) = \{F(b) - F(a)\}\{G(b) - G(a)\}$$
$$= \int_{(a,b]} \{F(u-) - F(a)\} dG(u) + \int_{(a,b]} \{G(u) - G(a)\} dF(u).$$

(4) Deduce that

$$\{F(b)G(b) - F(a)G(a)\} = \int_{(a,b]} F(u-) dG(u) + \int_{(a,b]} G(u) dF(u).$$

These results apply not only to probabilities but also to any two σ-finite measures on $\mathbb{R}$.

Now assume that F is a distribution function with $F(0-) = 0$. Show that

(5) $\int_{[0,n]} \{1 - F(x)\} dx = n\{1 - F(n)\} + \int_{[0,n]} x dF(x)$; and

if $\int x dF(x) < +\infty$, then

(6) $\int_0^{+\infty} \{1 - F(x)\} dx = \int x dF(x)$. [Hint: $n\mathbf{P}[X > n] \le \int_{\{X > n\}} X d\mathbf{P}$.]

Finally, use integration of parts to show that

(7) if $X \in L^p(\Omega, \mathfrak{F}, \mathbf{P})$, then $E[|X|^p] = \int_0^{+\infty} px^{p-1} \mathbf{P}[|X| > x] dx$.

Remark. This exercise on integration by parts is based on an article by E. Hewitt.[4] For another more direct proof of (6) and (7) see Exercise 4.7.1.

---

[4] *American Math. Monthly* **67** (1960), 419–422.

CHAPTER IV

CONVERGENCE OF RANDOM VARIABLES AND MEASURABLE FUNCTIONS

1. NORMS FOR RANDOM VARIABLES AND MEASURABLE FUNCTIONS

In what follows, the results will be stated and proved for probability spaces. Their extension to general σ-finite measure spaces are to be taken for granted unless commented upon. The probabilist's notation $E[X]$ for the integral $\int X dP$ or $\int X d\mu$ will be used frequently.

Let $(\Omega, \mathfrak{F}, \mathbf{P})$ be a probability space and L be the vector space of finite random variables on $(\Omega, \mathfrak{F}, \mathbf{P})$. Depending on the context, L will also denote the set of finite measurable functions on a σ-finite measure space $(\Omega, \mathfrak{F}, \mu)$.

Proposition 4.1.1. *The set L^1 of integrable random variables is a linear subspace of L. If $X \in L^1$, the function $X \to E[|X|] \stackrel{\text{def}}{=} \| X \|_1$ is a norm on L^1, i.e.,*

(1) $\| \lambda X \|_1 = |\lambda| \| X \|_1$ *if* $\lambda \in \mathbb{R}$, $X \in L^1$;
(2) $\| X + Y \|_1 \leq \| X \|_1 + \| Y \|_1$ *if* $X, Y \in L^1$;
(3) $\| X \|_1 = 0$ *implies X is a null function, i.e., a.s. equal to 0 and hence will be viewed as the same as the zero function.*

Proof. The first statement repeats Proposition 2.1.33 (1). Statements (1) and (2) follow from the properties of the expectation $E[X]$ and the fact that $|\lambda X| = |\lambda| |X|$ and $|X + Y| \leq |X| + |Y|$. In order that $\| \cdot \|_1$ be a norm, it must have the property that $\| X \|_1 = 0$ implies $X = 0$ (see Definition 2.5.3). Now $E[|X|] = 0 \Rightarrow P(\{|X| > 0\}) = 0$ (see Exercise 2.1.28 (4)), and so $\| X \|_1 = 0$ implies that $\mathbf{P}$-almost surely, $X = 0$. $\square$

Remark 4.1.2. In order to be entirely logical, one should not distinguish between integrable random variables that differ only on a set of measure zero. Then L^1 would be taken to be the vector space whose points are equivalence classes of integrable functions, and $\| \cdot \|_1$ would then be a norm on this vector space. However, it is common mathematical practice to ignore the equivalence classes and to use the actual functions themselves as the elements of L^1. Then one considers $\| \cdot \|_1$ as a norm with the reservation that $\| X \|_1 = 0$ does not imply $X(\omega) = 0$ for all ω but only for $\mathbf{P}$-almost all ω. Such a function is a null function. By "abus de langage" $\| \cdot \|_1$ will be called a norm on L^1.

From the statistical point of view, the actual random variable X is far less important than its distribution. This raises several questions for ran-

dom variables:
(1) is it possible to determine when X is in L^1 in terms of its distribution **Q**?
(2) if so, can $\| X \|_1$ be computed from **Q**?
(3) given $X, Y \in L^1$ with distributions **Q** and **R**, is it possible to compute the distribution of $X + Y$ from **Q** and **R**?

Remark. The concept of "distribution" is not used this way in general measure theory, although it makes sense. It is in effect what is sometimes called the **image measure** of the underlying measure under the measurable function, i.e., given a measurable function f on a measure space $(\Omega, \mathfrak{F}, \mu)$, the function $B \to \mu(f^{-1}(B))$ is a Borel measure that may be called the image measure or "push forward" of μ by f (see the remark following Proposition 2.1.19). There is a connection between the probabilist's concept of distribution function and what analysts call the "distribution function" of a non-negative measurable function (see Exercise 4.7.3). When the measure is finite, then, up to a scaling constant, the image measure is the distribution of the measurable function relative to the normalized measure. In Wheeden and Zygmund [W1], the "distribution function" is defined for finite measure spaces $(\Omega, \mathfrak{F}, \mu)$ as $\mu(\{f > \lambda\})$. In the case of a probability, this function is $1 - F$, where F is the distribution function of the random variable f.

The first two questions are answered by making use of the following lemma, which extends Lemma 3.5.3 to integrable functions.

Lemma 4.1.3. *Let X be a random variable on $(\Omega, \mathfrak{F}, \mathbf{P})$ with distribution* **Q**. *Let $\varphi : \mathbb{R} \to \mathbb{R}$ be either a non-negative Borel function or such that $\varphi \circ X$ is integrable. Then*

$$E[\varphi \circ X] = \int \varphi(x) \mathbf{Q}(dx).$$

In addition, if $\varphi : \mathbb{R} \to \mathbb{R}$ is a Borel function, then $\varphi \circ X \in L^1(\Omega, \mathfrak{F}, \mathbf{P})$ if and only if $\varphi \in L^1(\mathbb{R}, \mathfrak{B}(\mathbb{R}), \mathbf{Q})$.

Further, if $(\Omega, \mathfrak{F}, \mathbf{P})$ is a probability space and $X : \Omega \to \mathbb{R}^n$ is a random vector, then for any Borel function φ on $\mathbb{R}^n$ that is either non-negative or such that $\varphi \circ X \in L^1(\Omega, \mathfrak{F}, \mathbf{P})$, it follows that

$$E[\varphi \circ X] = \int \varphi(x) \mathbf{Q}(dx),$$

*where **Q** is the distribution (Definition 3.1.11) of X.*

Proof. Lemma 3.5.3 proves the formula for non-negative Borel functions. Since $\varphi = \varphi^+ - \varphi^-$, $\varphi \in L^1(\mathbb{R}, \mathfrak{B}(\mathbb{R}), \mathbf{Q}) = L^1(\mathbf{Q})$ if and only if $\varphi^+, \varphi^- \in L^1(\mathbf{Q})$. In this case $\int \varphi d\mathbf{Q} = \int \varphi^+ d\mathbf{Q} - \int \varphi^- d\mathbf{Q}$. Since $(\varphi \circ X)^\pm = \varphi^\pm \circ X$, the lemma follows from Lemma 3.5.3.

The extension of the result to random vectors has exactly the same proof as in the random variable case. □

1. NORMS FOR RANDOM VARIABLES AND MEASURABLE FUNCTIONS

Remarks. (1) $E[\varphi(\lambda X)] = \int \varphi(\lambda x) \mathbf{Q}(dx)$: let $\psi(x) = \varphi(\lambda x)$.

(2) The above result is essentially formal in the sense that if $X : \Omega \to E$ is measurable, with $(E, \mathfrak{E})$ a measurable space, then it extends immediately to measurable functions $\varphi : E \to \mathbb{R}$. The distribution $\mathbf{Q}$ of X is again defined on $\mathfrak{E}$ by $\mathbf{Q}(B) = \mathbf{P}(X^{-1}(B))$, and one has as before that $E[\varphi \circ X] = \int \varphi(x) \mathbf{Q}(dx)$ whenever φ is non-negative or $\varphi \circ X \in L^1(\Omega, \mathfrak{F}, \mathbf{P})$.

(3) For example, the image (see the above remarks) of Lebesgue measure dx on $(0, +\infty)$ under the map $x \to x^{\frac{1}{p}} = y$ ($1 < p < \infty$) is the measure $\nu(dy) = py^{p-1}dy$ (i.e., $|\{x \mid x^{\frac{1}{p}} \in A\}| = \int_A py^{p-1}dy$): this holds for all Borel sets $A \subset (0, +\infty)$ because $0 < a < x^{\frac{1}{p}} < b$ if and only if $a^p < y < b^p$ and $\int_a^b py^{p-1}dy = b^p - a^p$. A similar "change-of-variable" result holds whenever $y = \psi(x)$ and ψ is a strictly increasing, continuously differentiable function.

Corollary 4.1.4. *A random variable X with distribution $\mathbf{Q}$ is in L^1 if and only if $\int |x|\mathbf{Q}(dx) < +\infty$, i.e., if and only if $|x| \in L^1(\mathbb{R}, \mathfrak{B}(\mathbb{R}), \mathbf{Q})$. Also, $\| X \|_1 = \int |x|\mathbf{Q}(dx)$ and $E[X] = \int x \mathbf{Q}(dx)$.*

This corollary answers questions (1) and (2) posed earlier.

The answer to the third question is affirmative, provided X and Y are independent. The resulting distribution is the **convolution** of $\mathbf{Q}$ and $\mathbf{R}$ (see Exercise 3.3.21).

Example 4.1.5. Let X be a random variable whose distribution $\mathbf{Q}$ has a density f. If $f(x) = 1_{(-\infty, -1] \cup [1, +\infty)}(x) \frac{1}{2x^2}$, then X is not in L^1 and hence has no expectation $E[X]$ (see Feller [F1] for comments on random variables without expectations).

If one thinks of a random variable X as a vector with as many coordinates as there are points in Ω, then the norm $\| X \|_1$ is the analogue of the norm $|x| = |x_1| + |x_2| + |x_3|$ for the vector $x = (x_1, x_2, x_3) \in \mathbb{R}^3$. The usual Euclidean norm $\| x \| = (x_1^2 + x_2^2 + x_3^2)^{\frac{1}{2}}$ by analogy suggests consideration of $(\int X^2(\omega) P(d\omega))^{\frac{1}{2}} = E[X^2]^{\frac{1}{2}}$. Note that this is finite if and only if $\int x^2 \mathbf{Q}(dx) < \infty$.

Proposition 4.1.6. *Let L^2 be the set of $X \in L$ such that $E[X^2] < \infty$. Then L^2 is a linear subspace of L, and the function $X \to \| X \|_2 \stackrel{\text{def}}{=} E[X^2]^{1/2}$ is a norm (in the sense discussed in Remark 4.1.2) on L^2. For $X, Y \in L^2$, $XY \in L^1$ and*

$$|E[XY]| \le \| X \|_2 \| Y \|_2 \qquad \text{— the inequality of Cauchy–Schwarz.}$$

Proof. Since $2ab \le a^2 + b^2$, if $X, Y \in L^2$, then $XY \in L^1$. Consequently, $X, Y \in L^2$ implies that $X + Y \in L^2$. Let $\lambda \in \mathbb{R}$. Clearly, $X \in L^2$ implies $\lambda X \in L^2$ and so L^2 is a linear subspace of L. Also, if $X, Y \in L^2$, the function $\lambda \to \| X + \lambda Y \|_2^2 = \| X \|_2^2 + 2\lambda E[XY] + \lambda^2 \| Y \|_2^2$ is non-negative and so its discriminant $b^2 - 4ac = 4E[XY]^2 - 4 \parallel X \parallel_2^2 \parallel Y \parallel_2^2$ is negative, in other words $E[XY] \le \parallel X \parallel_2 \parallel Y \parallel_2$.

Since $\parallel X+Y \parallel^2 \le (\parallel X \parallel + \parallel Y \parallel)^2$ by the Cauchy–Schwarz inequality, it follows that $X \to \parallel X \parallel_2$ is a norm with the proviso as before that $\parallel X \parallel_2 = 0$ implies $X = 0$, P-a.s. □

Definition 4.1.7. *Let X be a random variable with distribution $\mathbf{Q}$. Let $1 \le p < \infty$. If $\int |x|^p \mathbf{Q}(dx) < \infty$ it is said to have a **pth absolute moment** (about zero). When this is the case, $\int x^p \mathbf{Q}(dx)$ exists and is called its **pth moment** (about zero).*

The next result is true only for finite measures.

Proposition 4.1.8. *Let $X \in L^2$. Then $X \in L^1$.*

Proof. $|X| = |X| 1_{\{|X| \le 1\}} + |X| 1_{\{|X| > 1\}} \le 1 + |X|^2 1_{\{|X| > 1\}}$. Since $1 \in L^1$, it follows that $X \in L^1$. □

Important remark. Proposition 4.1.8 is false for measure spaces that are σ-finite. Note that $1 \in L^1(\mu)$ if and only if μ is a bounded measure, i.e., $\mu(\Omega) < \infty$. Lebesgue measure on $\mathbb{R}$ gives standard counterexamples, as the next exercise shows.

Exercise 4.1.9. Let $f(x) = \frac{1}{x^r}, 1 \le x < +\infty, r \in \mathbb{R}, f(x) = 0$ otherwise. Show that $f \in L^1(\mathbb{R})$ if and only if $r > 1$. Determine a function $f \in L^2(\mathbb{R})$, with $f \notin L^1(\mathbb{R})$.

In the case of probability spaces, the L^2-norm is used to define two well-known statistical quantities.

Definition 4.1.10. *Let $X \in L^2$. The second moment of $X - E[X]$ is called the **variance** of X and is denoted by $\sigma^2(X) = \sigma^2$. The **standard deviation** of X is $\sigma(X) = \sqrt{\sigma^2} = \sigma$.*

Remark. The variance of a random variable is a measure of the concentration or "spread" of its distribution around the mean. For example, in the case of a normal random variable with distribution $n_t(x)dx$ (Exercise 2.9.6), as the variance $t = \sigma^2$ tends to zero, the density function concentrates as a "spike" around zero. The larger the variance, the more spread out is the distribution. In the extreme case when the variance of a random variable is zero, this indicates that it is a.s. equal to its mean value and so there is zero "spread".

Exercise 4.1.11. Let $X \in L^2$. Show that
(1) $\sigma^2(X) = E[X^2] - E[X]^2 = \parallel X \parallel_2^2 - m^2$, where $m = E[X]$ is the mean or expectation of X,
(2) $\sigma^2(X - a) = \sigma^2(X)$ for any $a \in \mathbb{R}$, and
(3) if X has a binomial distribution $b(n, p)$, $E[X] = np$ and $\sigma^2(X) = np(1-p) = npq$.

1. NORMS FOR RANDOM VARIABLES AND MEASURABLE FUNCTIONS

[*Hints*: the binomial distribution $b(n,p)$ is the distribution of the number S of "successes" or "heads" obtained when tossing a coin n times with the probability of success equal to p. It follows from Exercise 3.3.21 or directly, since $\binom{n}{k}$ is the number of ways to choose k objects from among n, that $\mathbf{P}[S=k] = \binom{n}{k}p^k(1-p)^{n-k}$. The law or distribution of S is

$$\mathbf{Q} = \sum_{k=0}^{n}\binom{n}{k}p^k(1-p)^{n-k}\varepsilon_k.$$

Note that all the mass is concentrated on the set $\{0, 1, 2, \ldots, k, \ldots, n\}$. Consequently,

$$E[f(S)] = \int f d\mathbf{Q} = \sum_{k=0}^{n}\binom{n}{k}p^k(1-p)^{n-k}f(k),$$

for any $\mathbf{Q}$-integrable function f. Scaling S by a positive constant $\lambda > 0$, the distribution of λS is

$$\mathbf{R} = \sum_{k=0}^{n}\binom{n}{k}p^k(1-p)^{n-k}\varepsilon_{k\lambda}.$$

The mass is now concentrated on the set $\{0, \lambda, 2\lambda, \ldots, k\lambda, \ldots, n\lambda\}$, and

$$E[f(\lambda S)] = \int f d\mathbf{R} = \sum_{k=0}^{n}\binom{n}{k}p^k(1-p)^{n-k}f(k\lambda) = \int f(\lambda x)\mathbf{Q}(dx),$$

for any $\mathbf{Q}$-integrable function f (see the remarks following Lemma 4.1.3).]

Definition 4.1.12. *Let $1 \leq p < \infty$. Define L^p to be the set of finite random variables X or measurable functions X for which $E[|X|^p] < +\infty$.*

Remark. In the case of a random variable X, it follows from Lemma 4.1.3 that $X \in L^p$ if and only if X possesses a pth absolute moment.

Remark. Since $(a+b)^p \leq 2^p(a^p + b^p)$ if $p > 0$, $a, b \geq 0$, it is clear that L^p is a linear subspace of L for any $p > 0$.

Exercise 4.1.13. *If $(x_1, x_2) \in \mathbb{R}^2$, let $\|x\|_p \stackrel{\text{def}}{=} (|x_1|^p + |x_2|^p)^{\frac{1}{p}}$. This exercise shows the dependence of $\|x\|_p$ on p. For $p = 1, 2, 3, 4$, sketch the set of points x in the plane for which $(|x_1|^p + |x_2|^p)^{\frac{1}{p}} = 1$. What is the limiting position of this set as $p \to +\infty$?*

If $X \in L^p$, one may define the analogue of the norm $\|x\|_p$, $x \in \mathbb{R}^n$, by setting $\|X\|_p = (\int |X|^p d\mathbf{P})^{\frac{1}{p}} = (E[|X|^p])^{\frac{1}{p}}$. Then it turns out that $X \to (E[|X|^p])^{\frac{1}{p}}$ is a norm.

In order to show that $X \to (E[|X|^p])^{\frac{1}{p}}$ is a norm, it will be necessary to discuss the analogue for L^p of the Cauchy–Schwarz inequality for L^2. This is called **Hölder's inequality**.

To begin, if $1 < p < +\infty$, let $q \in (1, +\infty)$ be the unique number such that $\frac{1}{p} + \frac{1}{q} = 1$. Note that $q = \frac{p}{p-1}$. Then p and q are said to be **conjugate indices**.

Proposition 4.1.14. *Let p and q be conjugate indices and $a, b \geq 0$. Then $G(a,b) \stackrel{\text{def}}{=} a^{1/p}b^{1/q} \leq \frac{1}{p}a + \frac{1}{q}b \stackrel{\text{def}}{=} A(a,b)$. In other words, the **geometric mean** is less than or equal to the **arithmetic mean**.*

Proof. The inequality is obvious if either a or b equals 0. Assume a and b are both positive. If $t = \frac{a}{b}$ and $\lambda = \frac{1}{p}$, the inequality is equivalent to $\lambda t + (1 - \lambda) \geq t^\lambda$, i.e., $\phi(t) = \lambda t - t^\lambda + (1 - \lambda) \geq 0$. Since, for any $\lambda \in (0, 1)$, $\phi'(1) = 0$, and $\phi''(t) > 0$ for all $t > 0$, it follows that $\phi(1) = 0$ is the minimum value of $\phi(t)$ on $(0, +\infty)$. □

Remarks. (1) For a proof using a Lagrange multiplier, see Exercise 4.7.4.
(2) $G(a,b) \leq A(a,b)$ is equivalent to a special case of **Young's inequality**: $cd \leq \frac{c^p}{p} + \frac{d^q}{q}$ if $c, d \geq 0$ (see Exercise 4.7.5).

Proposition 4.1.15. (Hölder's inequality). *Let $X \in L^p$, $Y \in L^q$, where p and $q \in (1, \infty)$ are conjugate indices. Then $XY \in L^1$ and*

$$E[\,|XY|\,] \leq \|X\|_p \|Y\|_q.$$

Proof. Let $U = |X|^p, V = |Y|^q$. Then both functions belong to L^1.
Hence, by Proposition 4.1.14,

$$\left[\frac{U}{E[U]}\right]^{\frac{1}{p}} \left[\frac{V}{E[V]}\right]^{\frac{1}{q}} \leq \left(\frac{1}{p}\right)\frac{U}{E[U]} + \left(\frac{1}{q}\right)\frac{V}{E[V]}.$$

Integrate both sides. The right-hand side has integral 1 as $\frac{1}{p} + \frac{1}{q} = 1$. Therefore,

$$\int \left[\frac{U}{E[U]}\right]^{\frac{1}{p}}(\omega) \left[\frac{V}{E[V]}\right]^{\frac{1}{q}}(\omega) P(d\omega) \leq 1.$$

Hence, $\int |X(\omega)Y(\omega)| P(d\omega) \leq \|X\|_p \|Y\|_q$ as $E[U]^{\frac{1}{p}} = \|X\|_p$ and $E[V]^{\frac{1}{q}} = \|Y\|_q$. Also, $XY \in L^1$. □

Hölder's inequality implies the following inequality, due to Minkowski, which in effect shows that $X \to \|X\|_p$ is a norm on L^p.

Proposition 4.1.16. (Minkowski's inequality)
Let $X, Y \in L^p$. Then $X + Y \in L^p$ and

$$\|X + Y\|_p \leq \|X\|_p + \|Y\|_p.$$

Proof. It has been pointed out (following Definition 4.1.12) that L^p is a linear subspace of L. Now,

$$E[|X+Y|^p] = \int |X+Y|^p dP \leq \int |X||X+Y|^{p-1}dP + \int |Y||X+Y|^{p-1}dP.$$

1. NORMS FOR RANDOM VARIABLES AND MEASURABLE FUNCTIONS 143

As $Z \in L^p$ implies $Z^{p-1} \in L^q$ if q and p are conjugate, since $\| Z^{p-1} \|_q = (E[|Z|^p])^{\frac{1}{q}} = \| Z \|_p^{\frac{p}{q}}$ as $p + q = pq$, it follows by Hölder's inequality that

$$E[|X+Y|^p] \le \| X \|_p \| |X+Y|^{p-1} \|_q + \| Y \|_p \| |X+Y|^{p-1} \|_q.$$

Therefore,

$$\| X+Y \|_p^p \le (\| X+Y \|_p^{\frac{p}{q}})[\| X \|_p + \| Y \|_p] < +\infty,$$

and so

$$\| X+Y \|_p^{p(1-\frac{1}{q})} \le \| X \|_p + \| Y \|_p.$$

Since $p(1 - \frac{1}{q}) = 1$, this completes the proof. □

Exercise 4.1.17. Let $1 \le p_1 < p_2 < \infty$. Show that
(1) for a finite measure space $(\Omega, \mathfrak{F}, \mu)$, $L^{p_2} \subset L^{p_1}$ [*Hint*: make use of the argument used to prove Proposition 4.1.8],
(2) this statement is false for $(\mathbb{R}, \mathfrak{B}(\mathbb{R}), dx)$. [*Hint*: see Exercise 4.1.9.]

Let $(\Omega, \mathfrak{B}(\mathbb{R}), \mathbf{P})$ be the probability space corresponding to the uniform distribution on $[0, 1]$ (see Example 1.3.9 (1)). Since the probability has no mass outside $[0,1]$, one might as well take Ω to be $[0, 1]$. The resulting measure space corresponds to Lebesgue measure on $[0,1]$, i.e., $([0, 1], \mathfrak{B}([0, 1]), dx)$. Let $f(x) = \frac{1}{x^r}, 0 < x \le 1$ — no need to define f at zero (why?). Show that

(3) $f \in L^1([0, 1])$ if and only if $r < 1$,
(4) $L^{p_1}([0, 1]) \not\subset L^{p_2}([0, 1])$ for $1 \le p_1 < p_2$.

Consider the spaces ℓ_p (also denoted by ℓ^p): for $1 \le p < \infty$, a sequence of real numbers $a = (a_n)_{n \ge 1}$ is said to be in ℓ_p if $\sum_{n=1}^{\infty} |a_n|^p < +\infty$ and $\|a\|_p \overset{\text{def}}{=} (\sum_{n=1}^{\infty} |a_n|^p)^{\frac{1}{p}}$: see Exercises 2.5.4 and 2.5.5 for $p = 1, 2$. If $p = \infty$, then ℓ_∞ is the set of bounded sequences and $\|a\|_\infty \overset{\text{def}}{=} \sup_n |a_n|$. Show that

(5) $1 \le p_1 < p_2 \le \infty$ implies that $\ell_{p_1} \subset \ell_{p_2}$.

Finally, if $1 \le p_1 < p_2 < \infty$, for Lebesgue measure on $\mathbb{R}$, show that

(6) $L^{p_1}(\mathbb{R}) \not\subset L^{p_2}(\mathbb{R})$ and $L^{p_2}(\mathbb{R}) \not\subset L^{p_1}(\mathbb{R})$.

Exercise 4.1.18. Use Hölder's inequality to show that, on a probability space,
(1) if $1 < p < +\infty$ and $X \in L^p$, then $X \in L^1$ and $\| X \|_1 \le \| X \|_p$ [*Hint*: $1 \in L^q$.],
(2) if $1 < p_1 < p_2 < \infty$ and $X \in L^{p_2}$, then $X \in L^{p_1}$ and $\| X \|_{p_1} \le \| X \|_{p_2}$. [*Hint*: use (1).]

Remark. What modification is needed to the above exercise in order that it be valid for a bounded measure?

The first result (1) in Exercise 4.1.18 is a special case of a more general inequality known as Jensen's inequality. This inequality involves convex functions. Recall that a function $\varphi : (a,b) \to \mathbb{R}$ is **convex** if for any $a < x < y < b$ and $t \in [0,1]$, $\varphi(tx + (1-t)y) \leq t\varphi(x) + (1-t)\varphi(y)$.

Proposition 4.1.19. (**Jensen's inequality**) *Assume that X is an integrable random variable-valued in (a,b). If φ is a convex function on (a,b) and $\varphi(X)$ is integrable, then*

$$\varphi(E[X]) \leq E[\varphi(X)].$$

Proof. This result is an easy consequence of the fact that any convex function φ is the supremum of a countable number of affine functions (see Exercise 4.7.25).

Let $L_n(x) = a_n x + b_n$ for $n \geq 1$ and assume that $\varphi(x) = \sup_n L_n(x)$. Since X is integrable, $E[L_n(X)] = a_n E[X] + b_n = L_n(E[X])$, it follows that $E[L_n(X)] \leq E[\varphi(X)]$ for all $n \geq 1$. The result follows. □

Remarks. (1) The function $t \to t^p$ is convex if $p > 1$ since its second derivative is non-negative. Hence, by Jensen's inequality, $\bigl(E[|X|]\bigr)^p \leq E[|X|^p]$ if $p > 1$ as long as $X \in L^1$. Note that the inequality is trivial if $X \notin L^p$.

(2) Jensen's inequality is only valid on probability spaces (see Exercise 4.7.6).

(3) The hypothesis that $\varphi(X)$ is integrable is not necessary once one has sorted out how to define the integral for the sum of a non-negative random variable and an integral one (see Exercise 2.9.19).

Exercise 4.1.12 asks for the limiting position as $p \to +\infty$ of $\{x = (x_1, x_2) \mid (|x_1|^p + |x_2|^p)^{1/p} = 1\}$. It is $\{x \mid \max\{|x_1|,|x_2|\} = 1\}$. This set is the unit sphere for the norm on $\mathbb{R}^2$ defined by $\| x \|_\infty = \max\{|x_1|,|x_2|\} =$ largest of the components. By analogy, one makes the following definition.

Definition 4.1.20. $X \in L^\infty$ *if $X \in L$ and there is a number $C > 0$ such that $P(\{|X| > C\}) = 0$. The smallest number with this property is called the* **essential supremum** *of X and is denoted by $\| X \|_\infty$.*

Remark. Note that $X \in L^\infty$ need not imply that X is bounded. However, modulo a null function, X is bounded. If $X \in L^\infty$, it is also said to be **essentially bounded**.

Exercise 4.1.21. Let $(\Omega, \mathfrak{F}, \mu)$ be a σ-finite measure space and let X, Y be measurable functions. Show that

(1) $\mu(\{|X+Y| > C+D\}) = 0$ if $\mu(\{|X| > C\}) = 0$ and $\mu(\{|Y| > D\}) = 0$ [*Hint:* consider $\Omega \setminus (E \cup F)$, where $E = \{|X| > C\}$ and $F = \{|Y| > D\}$.],

1. NORMS FOR RANDOM VARIABLES AND MEASURABLE FUNCTIONS   145

(2) L^∞ is a linear subspace of L,
(3) $X \to \| X \|_\infty$ is a norm on L^∞ with the usual proviso that $\| X \|_\infty = 0$ implies $X = 0, \mu$-a.e.

In the case of a probability space $(\Omega, \mathfrak{F}, \mathbf{P})$, show that

(4) $L^\infty \subset L^p$ if $1 \leq p < \infty$ and in general $L^\infty \neq \cap_{1 \leq p < \infty} L^p(\Omega, \mathfrak{F}, \mathbf{P})$,
[Hint: show that $X(x) = x$ is in every $L^p(\mathbb{R}, \mathfrak{B}(\mathbb{R}), \frac{1}{\sqrt{2\pi}} e^{-\frac{x^2}{2}} dx)$],
(5) $\| X \|_p \leq \| X \|_\infty$ if $X \in L^\infty$.

Remark. In the case of a probability space $(\Omega, \mathfrak{F}, \mathbf{P})$, $\sup_{1 \leq p < \infty} \| X \|_p = \| X \|_\infty$ if $X \in L^\infty$. This is not true in general for a σ-finite measure space. However, if the measure is finite, then $\| X \|_\infty = \lim_{p \to \infty} \| X \|_p$. It suffices to observe the effect on the L^p-norms of normalizing the total mass (see Exercise 4.7.11).

Exercise 4.1.22. Let X be a random variable with distribution $\mathbf{Q}$. Show that

(1) $X \in L^\infty$ if and only if $\mathbf{Q}([-M, M]) = 1$ for some $M > 0$,
(2) if $X \in L^\infty$ and $M = \|X\|_\infty$, then $\mathbf{Q}(\{x \mid M - \epsilon \leq |x| \leq M\}) > 0$ for any $\epsilon > 0$ that is less than M.

Use this property (2) together with Chebychev's inequality (Proposition 4.3.6) and (Exercise 1.1.18) to show that on a probability space, $\| X \|_\infty = \sup_{1 \leq p < \infty} \| X \|_p$.

Definition 4.1.23. Let $1 \leq p \leq +\infty$. A sequence $(X_n)_{n \geq 1}$ of r.v. in L^p **converges to 0 in L^p or in L^p-norm** if $\lim_{n \to \infty} \| X_n \|_p = 0$. This is often denoted by writing $X_n \xrightarrow{L^p} 0$.

Important remark. Assume that the measure space $(\Omega, \mathfrak{F}, \mu)$ is finite, i.e., $\mu(\Omega) < \infty$. If $p_1 < p_2$ and $(X_n)_{n \geq 1}$ is a sequence in L^{p_2}, it is also a sequence in L^{p_1}. In view of Exercise 4.1.18, if it converges to zero in L^{p_2}, it converges to zero in L^{p_1}, but not conversely. In particular, if all the random variables are in L^2 (i.e., $(X_n)_{n \geq 1} \subset L^2$), to say that it converges to 0 in L^2 is a stronger statement than saying that it converges to 0 in L^1. The higher the p, the stronger the condition "converges to 0 in L^p" becomes.

Example 4.1.24. Let $\Omega = [0,1]$, $\mathfrak{F} = \mathfrak{B}([0,1])$, and $\mathbf{P} = dx$. In other words, $\mathbf{P}$ is Lebesgue measure on $[0,1]$. Let $X_n = \sqrt{n} 1_{[0,1/n]}$. Then $(X_n)_{n \geq 1}$ is in L^∞ and $\| X_n \|_2 = 1$, for all n. However, $\| X_n \|_p = n^{1/2 - 1/p}$ and so $X_n \xrightarrow{L^p} 0$ if $1 \leq p < 2$. Clearly, X_n does not converge in L^2 to zero. Replacing $\sqrt{n}$ by a_n, one can get similar examples that fail to converge in L^{p_2}, for $1 < p_2 < +\infty$, but converge in L^{p_1}, $1 < p_1 < p_2$.

Important remark. The situation in the above example involving the vector space $L^2([0,1])$ is completely different from the situation on $\mathbb{R}^n$. There, because the dimension is finite, all norms give the same topology.

To sum up, for any measure space $(\Omega, \mathfrak{F}, \mu)$ or probability space $(\Omega, \mathfrak{F}, \mathbf{P})$, a large family of normed linear spaces is associated: namely, the spaces $L^p(\Omega, \mathfrak{F}, \mu)$ or $L^p(\Omega, \mathfrak{F}, \mathbf{P}), 1 \leq p \leq \infty$.

Definition 4.1.25. *A* **metric space** *is a set E together with a function $d: E \times E \to \mathbb{R}^+$ called a* **metric** *such that, for all $x, y, z \in E$,*

(1) $d(x, y) = d(y, x)$,
(2) $d(x, z) \leq d(x, y) + d(y, z)$ *(the* **triangle inequality***), and*
(3) $d(x, y) = 0$ *if and only if $x = y$.*

A sequence $(x_n)_{n \geq 1}$ in E **converges to** x if for any $\varepsilon > 0$ there is an integer $n(\epsilon)$ such that $d(x_n, x) < \varepsilon$ when $n \geq n(\varepsilon)$ (Definition 1.1.5).

A sequence $(x_n)_{n \geq 1}$ in E is said to be a **Cauchy sequence** if for any $\varepsilon > 0$, there is an integer $N(\epsilon)$ such that $d(x_m, x_n) < \varepsilon$ when $n, m \geq N(\varepsilon)$.

A metric space is said to be a **complete metric space** if every Cauchy sequence converges.

Remarks 4.1.26. (1) A norm $\|\cdot\|$ on a vector space V (see Definition 2.5.3) determines a metric d by setting $d(x, y) = \|x - y\|$. The simplest example is $V = \mathbb{R}$, with the norm being the absolute value and $d(a, b) = |a - b|$. It turns out (see Exercise 4.7.12) that $\mathbb{R}$ is complete under the metric given by the absolute value, as also are all the vector spaces $\mathbb{R}^n$ with either of the norms of Exercise 3.1.4 (or in fact any norm). A vector space with a norm is called a **normed vector space**. A normed vector space that is complete in the associated metric is called a **Banach space**.

(2) A metric has associated with it a topology: a set E in a metric space is said to be **open** if for any $a \in E$ there is an $\epsilon > 0$ such that $\{x \mid d(x, a) < \epsilon\}$ is a subset of E; the **open ball about** a **of radius** ϵ is defined to be $\{x \mid d(x, a) < \epsilon\}$; it is an open set in view of the triangle inequality. Hence, a set E is open if and only if for any $a \in E$ there is an open ball about a that is contained in E.

The next two results imply that the normed spaces $L^p(\Omega, \mathfrak{F}, \mu)$ and $L^p(\Omega, \mathfrak{F}, \mathbf{P}), 1 \leq p \leq \infty$, are all complete and so are Banach spaces. The first result is a general fact from functional analysis.

Proposition 4.1.27. *Let V be a normed real or complex vector space. The following are equivalent:*

(1) *it is complete (as a metric space);*
(2) *a series $\sum_{n=1}^{\infty} x_n$ converges in V provided $\sum_{n=1}^{\infty} \|x_n\|$ converges in $\mathbb{R}$ (such a series is said to be* **absolutely summable** *or* **normally summable***).*

Proof. Assume (1) and that $\sum_{n=1}^{\infty} \|x_n\| < +\infty$. The sequence $(y_n)_{n \geq 1}$ of partial sums $y_n = \sum_{k=1}^{n} x_k$ is a Cauchy sequence: $\|y_n - y_{n+m}\| \leq \sum_{k=n+1}^{n+m} \|x_k\|$, which tends to zero as n tends to infinity. Hence, the series converges to a vector x, which is denoted by $\sum_{n=1}^{\infty} x_n$.

1. NORMS FOR RANDOM VARIABLES AND MEASURABLE FUNCTIONS 147

Assume (2). Let $(y_n)_{n\geq 1}$ be a Cauchy sequence. One cannot immediately assume that the y_n are the partial sums of an absolutely summable series. However, by using the Cauchy property, one can determine a sequence $(n_k)_{k\geq 1}$ of integers $n_k \geq 1$ such that, for all $k \geq 1$,

$$\| y_m - y_n \| < \frac{1}{2^k} \quad \text{if } n, m \geq n_k.$$

One may also assume that $n_k \geq k$ for all $k \geq 1$. This ensures that the y_{n_k} are the partial sums of an absolutely summable series: define $x_1 = y_{n_1}$, $x_{k+1} = y_{n_{k+1}} - y_{n_k}$, and consider the series $\sum_{k=1}^{\infty} x_k$; then y_{n_k} is its kth partial sum. Since $\sum_{k=1}^{\infty} \| x_k \| \leq 1$, (2) implies that the series $\sum_{k=1}^{\infty} x_k$ converges to a limit y. Hence, $y = \lim_k y_{n_k}$.

The proof concludes with the observation that if a Cauchy sequence has a convergent subsequence, then it converges: let $\varepsilon > 0$ and let ℓ be such that $\frac{1}{2^\ell} < \frac{\varepsilon}{2}$; then $\| y_{n_k} - y \| < \frac{\varepsilon}{2}$ for all $k \geq \ell$; and $\| y_n - y \| \leq \| y_n - y_{n_\ell} \| + \| y_{n_\ell} - y \| < \varepsilon$ if $n \geq n_\ell$. □

To prove that all the L^p-spaces are Banach spaces, it therefore suffices to show that every sequence in L^p that is absolutely summable also converges.

Proposition 4.1.28. (Riesz–Fischer theorem) *A series $\sum_{n=1}^{\infty} X_n$ in the normed linear space $L^p, 1 \leq p \leq \infty$, converges if it is absolutely summable. Hence, $L^p, 1 \leq p \leq \infty$, is complete.*

Proof. First, consider the case where $1 \leq p < \infty$. Let $C = \sum_{n=1}^{\infty} \| X_n \|_p$. Let $Z_N = \sum_{n=1}^{N} |X_n|$. Since $\| X \|_p = \| |X| \|_p$ for any measurable function X, Minkowski's inequality (Proposition 4.1.16) implies that $\| Z_N \|_p \leq \sum_{n=1}^{N} \| X_n \|_p \leq \sum_{n=1}^{\infty} \| X_n \|_p = C$. Hence, it follows from monotone convergence that if $Z = \lim_N Z_N$, then $E[Z^p] \leq C^p < \infty$. As a result, Z is finite a.e.

If E denotes $\{Z = +\infty\}$, the series $\sum_{n=1}^{\infty} X_n(\omega)$ is absolutely convergent on $\Omega \setminus E$.

Define $Y(\omega) = \begin{cases} \sum_{n=1}^{\infty} X_n(\omega) & \text{if } \omega \notin E, \\ 0 & \text{if } \omega \in E. \end{cases}$

Then, if $Y_N(\omega) = \sum_{n=1}^{N} X_n(\omega)$, $|Y_N(\omega)| \leq Z_N(\omega) \leq Z(\omega)$, and so $|Y| \leq Z \in L^p$. Consequently, $Y \in L^p$.

Now $|Y - Y_N| \leq 2Z \in L^p$, since $|Y - Y_N|^p \leq 2^p Z^p \in L^1$. Since $|Y - Y_N|^p \to 0$ a.e., it follows from the theorem of dominated convergence (Theorem 2.1.38) that $E[|Y - Y_N|^p] \to 0$. In other words, $\| Y - Y_N \|_p \to 0$ and so the series converges in L^p to Y.

The case when $p = \infty$ is simpler. Assume that $C = \sum_{n=1}^{\infty} \| X_n \|_\infty < +\infty$. For each $n \geq 1$, let E_n be the set where $|X_n| > \| X_n \|_\infty$ and set $E = \cup_{n=1}^{\infty} E_n$. This is a set of measure zero, and on $\Omega \setminus E$ the series $\sum_{n=1}^{\infty} X_n$ converges uniformly: $\omega \notin E$ implies $|\sum_{n=N+1}^{\infty} X_n(\omega)| \leq \sum_{n=N+1}^{\infty} \| X_n \|_\infty$

Let $Y(\omega) = \begin{cases} \sum_{n=1}^{\infty} X_n(\omega), & \text{if } \omega \notin E \\ 0 & \text{if } \omega \in E. \end{cases}$

Then, if $Y_N = \sum_{n=1}^{N} X_n$, $\| Y - Y_N \|_\infty \le \sum_{n=N+1}^{\infty} \| X_n \|_\infty$. This is because E is a set of measure zero and on $\Omega \backslash E$ the inequality holds pointwise. Consequently, the series converges in L^∞ to Y. $\square$

In probability, there is a famous lemma, the Borel–Cantelli lemma, the first part of which is related to the proof of the completeness of L^1. Recall that $\| 1_E \|_1 = \mathbf{P}(E)$. Given a sequence $(E_n)_{n \ge 1}$ of events in $\mathfrak{F}$, let the set $\{\omega | E_n \text{ occurs infinitely often}\} = \{\omega | \omega \text{ belongs to an infinite number of sets } E_n\}$ be denoted by $\{E_n \text{ i.o.}\}$. Then $\{E_n \text{ i.o.}\} = \cap_{m=1}^{\infty}(\cup_{n=m}^{\infty} E_n)$, equivalently, $1_{\{E_n \text{ i.o.}\}} = 1_{\cap_{m=1}^{\infty}(\cup_{n=m}^{\infty} E_n)} = \limsup_n 1_{E_n}$.

Proposition 4.1.29. (Borel–Cantelli lemma) *Let $(E_n)_{n \ge 1}$ be a sequence of events in $\mathfrak{F}$.*

(1) *If $\sum_{n=1}^{\infty} \mathbf{P}(E_n) < +\infty$, then $\mathbf{P}(\{E_n \text{ i.o.}\}) = 0$.*
(2) *Conversely, if the events E_n are independent, then $\mathbf{P}(\{E_n \text{ i.o.}\}) = 0$ implies that $\sum_{n=1}^{\infty} \mathbf{P}(E_n) < +\infty$.*

Proof. (1) $\mathbf{P}(E_n) = \| 1_{E_n} \|_1$ and so by the proof of Proposition 4.1.28 for $p=1$, $Y = \sum_{n=1}^{\infty} 1_{E_n} \in L^1$ if $\sum_{n=1}^{\infty} \mathbf{P}(E_n) < +\infty$. Since $Y(\omega) < +\infty$ with probability one, $\{\omega | \text{ the sequence } (1_{E_n}(\omega))_n \text{ has an infinite number of ones}\} = \{\omega | E_n \text{ occurs infinitely often}\}$ has probability zero.

(2) Let $\Lambda = \{E_n \text{ i.o.}\}$. Then, as observed above, $1_\Lambda = \limsup_n 1_{E_n}$. If $\mathbf{P}(\Lambda) = 0$ then, as $m \to \infty$, $\mathbf{P}(\cup_{n=m}^{\infty} E_n) \downarrow 0$ and so $\mathbf{P}(\cap_{n=m}^{\infty} E_n^c) \uparrow 1$. Now the sets $(E_n^c)_{n \ge 1}$ are independent since the sets $(E_n)_{n \ge 1}$ are independent (see Exercises 3.4.5 and 3.6.7). Hence,

$$\mathbf{P}(\cap_{n=m}^{N} E_n^c) = \prod_{n=m}^{N} \mathbf{P}(E_n^c) = \prod_{n=m}^{N} [1 - \mathbf{P}(E_n)]$$

$$\le \prod_{n=m}^{N} \exp\{-\mathbf{P}(E_n)\} = \exp\{-\sum_{n=m}^{N} \mathbf{P}(E_n)\}, \text{ as } 1 - x \le e^{-x} \text{ if } x \ge 0.$$

Let $N \to +\infty$. This gives the inequality

$$\mathbf{P}(\cap_{n=m}^{\infty} E_n^c) \le \exp\{-\sum_{n=m}^{\infty} \mathbf{P}(E_n)\}.$$

Now $\mathbf{P}(\cap_{n=m}^{\infty} E_n^c) \uparrow 1$ as $m \to +\infty$ and so there is at least one m with $\mathbf{P}(\cap_{n=m}^{\infty} E_n^c) \ge \frac{1}{2}$ (say). The fact that $\frac{1}{2} \le \exp\{-\sum_{n=m}^{\infty} \mathbf{P}(E_n)\}$ implies that $\sum_{n=m}^{\infty} \mathbf{P}(E_n) < +\infty$. $\square$

2. Continuous functions and $L^{p}*$

This section can be omitted without loss of continuity. While it begins with some extra results about continuous functions that are outlined in exercises, its main purpose is to introduce Fourier series and to discuss the use of convolution in summing them (see Féjer's theorem 4.2.27). In addition, it is shown that convolution and differentiation commute, thereby extending Theorem 2.6.1.

It was shown in Proposition 2.1.15 that every continuous function on $\mathbb{R}$ is a Borel function, i.e., is measurable with respect to $\mathfrak{B}(\mathbb{R})$. There are many Borel functions that are not continuous, for example, the characteristic function 1_A of any Borel set different from $\mathbb{R}$ or the empty set $\emptyset$. However, in the L^p-spaces, for $1 \leq p < \infty$, the continuous functions can be used to approximate in L^p-distance any measurable function in L^p. Since such functions are limits of simple functions (if they are non-negative (Proposition 2.1.11), (RV_8)), it is more or less clear that the first thing to do, if one wants to verify this, is to approximate the characteristic function 1_A of a Borel set by a continuous function.

Exercise 4.2.2 sets the stage for this. To have an idea of what is going on, imagine approximating the function $1_{[0,1]}$ by the continuous piecewise linear functions φ_n, where

$$\varphi_n(x) = \begin{cases} 0 & \text{if } x < -\frac{1}{n}, \\ nx + 1 & \text{if } -\frac{1}{n} \leq x \leq 0, \\ 1 & \text{if } 0 < x < 1, \\ -nx + n + 1 & \text{if } 1 \leq x \leq 1 + \frac{1}{n}, \\ 0 & \frac{1}{n} < x. \end{cases}$$

The area of the region between the graphs of $1_{[0,1]}$ and φ_n tends to zero as n tends to infinity.

First, recall that in Exercise 2.9.3(4), it is shown that if E is a bounded Lebesgue measurable set and $\epsilon > 0$, then there exist a compact set C and a bounded open set O with $C \subset E \subset O$ and $|O \backslash C| < \epsilon$. Since, as explained in the next exercise, 1_E is continuous on $C \cup O^c$, it follows that the measurable function 1_E is continuous except on a set of Lebesgue measure less than ϵ. This a special case of Lusin's Theorem (see later). Recall that in Exercise 2.3.7, it was shown that a function $f : \mathbb{R} \to \mathbb{R}$ is Lebesgue measurable if, for any $\epsilon > 0$, it is continuous except on a set of measure less than ϵ.

Exercise 4.2.1. (Simple case of Lusin's theorem) Let $C, E,$ and O be as above. Show that

(1) the characteristic function 1_E of E is continuous (on $C \cup O^c$) when restricted to the closed set $C \cup O^c$ (Definition 2.1.13). [Hints: if $x \in C$, some small open interval about x lies inside O; if $x \in O^c$ some small open interval about x lies in C^c.]

Note that the Lebesgue measure of $\mathbb{R}\setminus\{C \cup O^c\} = O\setminus C$ is less than ϵ.

Now let $s = \sum_{n=1}^{N} a_n 1_{A_n}$ be a simple Lebesgue measurable function with $|A_n| < \infty$ for each n. Make use of Exercise 2.9.3(5) to show that

(2) if $\epsilon > 0$ there is a closed subset A of $\mathbb{R}$ such that (i) the restriction of s to A is continuous (on A); and (ii) $|\mathbb{R}\setminus A| < \epsilon$.

Since every non-negative measurable function can be approximated by simple functions, it is natural to ask whether (2) holds for any Lebesgue measurable function on $\mathbb{R}$. The answer is affirmative as stated in a preliminary version of the following theorem.

Lusin's theorem. (see Exercise 4.7.9) *A real-valued function on $\mathbb{R}$ is Lebesgue measurable if and only if, for any $\epsilon > 0$, there is a closed set A such that (i) the restriction of f to A is continuous on A, and (ii) $|\mathbb{R}\setminus A| < \epsilon$.*

While Lusin's theorem shows that measurable functions are not far from being continuous functions, there is another more quantitative way in which continuous functions are close to the functions in L^p. This involves actually approximating the characteristic function of a bounded Lebesgue measurable set by a continuous function defined on all of $\mathbb{R}$.

Exercise 4.2.2. Let C, E, and O be as above. Show that

(1) there exists a continuous function φ on $\mathbb{R}$, $0 \leq \varphi \leq 1$, such that $\varphi(x) = 0$ for $x \notin O, \varphi(x) = 1$ for $x \in C$ [Hint: consider $\psi(x) = \text{dist}(x, C)$ where $\text{dist}(x, C) = \inf\{|x - y| \mid y \in C\}$ (see Exercise 4.7.17); use results of that exercise to show that for some $R > 0$ one may take $\varphi = \frac{1}{R}(\psi \wedge R)$];

(2) $|1_E(x) - \varphi(x)| = 0$ for $x \in C \cup O^c$ and ≤ 1 for $x \in O\setminus C$. In other words, $|1_E - \varphi| \leq 1_{O\setminus C}$.

Proposition 4.2.3. *Let E be a Lebesgue measurable set with $|E| < +\infty$. For each p, $1 \leq p < \infty$, and $\epsilon > 0$, there is a continuous function $\varphi, 0 \leq \varphi \leq 1$, such that*

$$\| 1_E - \varphi \|_p < \epsilon \text{ and } \{x \mid \varphi(x) \neq 0\} \text{ is bounded.}$$

Proof. Let $E_n = E \cap [-n, n]$. Then, $1_{E_n} \uparrow 1_E$ and so, by dominated convergence, $\| 1_E - 1_{E_n} \|_p \to 0$. Choose an n such that $\| 1_E - 1_{E_n} \|_p < \frac{\epsilon}{2}$. It follows from Exercise 4.2.2 that there is a continuous function φ with $\{x \mid \varphi(x) \neq 0\} \subset [-(n+1), (n+1)]$ and such that $|1_{E_n} - \varphi| \leq 1_{O\setminus C}$, where $|O\setminus C| < (\frac{\epsilon}{2})^p$. Hence, $\| 1_{E_n} - \varphi \|_p < \frac{\epsilon}{2}$. □

Definition 4.2.4. *A continuous function φ on $\mathbb{R}$ is said to have* **compact support** *if the closure of $\{x \mid \varphi(x) \neq 0\}$ is compact. Similarly, if O is an open subset of $\mathbb{R}$, a continuous function φ on O is said to have* **compact support** *if the closure of $\{x \mid \varphi(x) \neq 0\}$ is a compact subset of O.*

Let $\mathcal{C}_c(\mathbb{R})$ denote the set of continuous functions on $\mathbb{R}$ with compact support. Similarly, if O is an open set, let $\mathcal{C}_c(O)$ denote the set of continuous functions on O with compact support. It is a linear subspace of the space of all real-valued functions on $\mathbb{R}$ because $\{x \mid \varphi(x) + \psi(x) \neq 0\} \subset \{x \mid \varphi(x) \neq 0\} \cup \{x \mid \psi(x) \neq 0\}$ and the union of two compact sets is compact (by part A of Exercise 1.5.6). In addition, for all p, $1 \leq p \leq +\infty$, $\mathcal{C}_c(\mathbb{R}) \subset L^p(\mathbb{R})$.

Theorem 4.2.5. *If $1 \leq p < +\infty$, then $\mathcal{C}_c(\mathbb{R})$ is a dense subspace of $L^p(\mathbb{R})$.*

Proof. It follows from Proposition 4.2.3 that the closure of $\mathcal{C}_c(\mathbb{R})$ in $L^p(\mathbb{R})$, which is a linear subspace, contains all the simple functions and their differences.

Note that $X \in L^p$, if and only if $|X| \in L^p$, which holds if and only if X^+ and $X^- \in L^p$. As a result, every function $X \in L^p$ is the limit in L^p of a sequence of differences of simple functions provided every positive function $X \in L^p$ is the limit in L^p of a sequence of simple functions.

Hence, to prove the theorem, it is enough to verify the last statement. Let $0 \leq X \in L^p$ and $(s_n)_{n \geq 1}$ be a sequence of simple functions that increases to X. It follows from dominated convergence (Theorem 2.1.38) that $E[|X - s_n|^p] \to 0$, as $|X - s_n|^p \leq X^p$. □

If $E \subset \mathbb{R}$, let $\mathcal{C}(E)$ denote the set of continuous functions $f : E \to \mathbb{R}$ (see Definition 2.1.13).

Corollary 4.2.6. $\mathcal{C}([a,b])$ *is dense in* $L^p([a,b])$, $1 \leq p < +\infty$.

Proof. $L^p([a,b])$ is the L^p-space associated with Lebesgue measure on $[a,b]$. Recall that this measure can be viewed as the restriction of Lebesgue measure to the Lebesgue measurable subsets of $[a,b]$ or as the measure obtained from the right continuous function $G(x) = a, x < a, G(x) = x, a \leq x < b$, and $G(x) = b, b \leq x$.

Every function $f \in L^p([a,b])$ can be viewed as a Lebesgue measurable function on $\mathbb{R}$ that vanishes outside $[a,b]$. If $\varphi \in \mathcal{C}_c(\mathbb{R})$ and $\| f - \varphi \|_p < \epsilon$, then
$$\int 1_{[a,b]}(x)|f(x) - \varphi(x)|^p dx < \int |f(x) - \varphi(x)|^p dx < \epsilon^p.$$
Since $\varphi|_{[a,b]} \in \mathcal{C}([a,b])$, the result follows. □

Exercise 4.2.7. Make use of the proof of Proposition 4.2.3 to show that the space $\mathcal{C}_c((a,b))$ is dense in $L^p([a,b])$. In particular, every function in L^p can be approximated in L^p by continuous functions ψ on $[a,b]$ for which $\psi(a) = \psi(b)$.

Remark. Note that $\varphi \in \mathcal{C}_c((a,b))$ if and only if it is continuous on $[a,b]$ and vanishes near the endpoints a and b.

These density results depend on a relation between the topology of $\mathbb{R}$ and the measure dx which is stated in Exercise 2.9.3. A measure μ on $\mathfrak{B}(\mathbb{R})$ for which this exercise holds is called a **regular Borel measure**. All the measures determined by a non-decreasing, right continuous function G as in Chapter II are regular. The concept of regularity makes sense in a fairly general setting; in particular on $\mathbb{R}^n$. It is not hard to verify Exercise 2.9.3(4) for Lebesgue measure on $\mathbb{R}^n$ (see Exercise 4.2.9). As a result, the following holds since, in addition, the proof suggested for Exercise 4.2.2 holds in $\mathbb{R}^n$.

Theorem 4.2.8. *The set $C_c(\mathbb{R}^n)$ of continuous functions on $\mathbb{R}^n$ with compact support is dense in $L^p, 1 \leq p < \infty$.*

Remark. The concept of compact support for a function on $\mathbb{R}^n$ is defined as in Definition 4.2.4. It requires that the concept of compactness be defined in $\mathbb{R}^n$ (one copies Definition 1.4.9). For this to have any meaning, one needs to know that, as before on $\mathbb{R}$, a set is compact if and only if it is closed and bounded, i.e., lies in some ball centered at the origin (see Marsden [M1]).

Exercise 4.2.9. Let E be a bounded Lebesgue measurable subset of $\mathbb{R}^n$ with $|E| < +\infty$. Let $\epsilon > 0$. Show that
  (1) there exists a bounded open set $O \supset E$ with $|O \setminus E| < \frac{\epsilon}{2}$. [Hint: show that the outer measure λ^* that may be used to define Lebesgue measure starting from the Boolean ring of "half-open boxes" $(a_1, b_1] \times (a_2, b_2] \times \cdots \times (a_n, b_n]$ can be computed by using open boxes; replace intervals of the form $(a_i, b_i]$ by open intervals $(a_i, b_i + \varepsilon)$ for small ε (depending on the box).]

Assume $E \subset [-N, N] \times [-N, N] \times \cdots \times [-N, N]$. Show that
  (2) if $P \supset E^c \cap [-N, N] \times [-N, N] \times \cdots \times [-N, N] \stackrel{\text{def}}{=} E_N^c$ is open and $|P \setminus E_N^c| < \frac{\epsilon}{2}$, then $|E \setminus C| < \frac{\epsilon}{2}$, where $C = P^c \cap [-N, N] \times [-N, N] \times \cdots \times [-N, N]$.

Conclude that
  (3) if E is a bounded Lebesgue measurable set, then there is a bounded open set O and a compact set C with $C \subset E \subset O$ and $|O \setminus C| < \epsilon$ (recall that a compact set is one that is closed and bounded).

The above density results are false for $p = \infty$. This is a consequence of either of the next two exercises.

Exercise 4.2.10. If f is a function on $\mathbb{R}$, let f_h denote the translate of f by h, i.e., $f_h(x) = f(x + h)$ for all $x \in \mathbb{R}$. Show that
  (1) if $\varphi \in C_c(\mathbb{R})$, show that $\| \varphi_h - \varphi \|_p \to 0$ as $|h| \to 0$ for $1 \leq p \leq \infty$,
  (2) if $f \in L^p(\mathbb{R})$ and $1 \leq p < \infty$, show that $\| f_h - f \|_p \to 0$ as $|h| \to 0$ [Hint: use the density result (Theorem 4.2.5) and relate $\| f_h - f \|_p$ to $\| \varphi_h - \varphi \|_p$ if φ approximates f; recall that Lebesgue measure is translation invariant (Exercise 2.2.8).], and
  (3) if $p = +\infty$, it is false that $\| f_h - f \|_p \to 0$ as $|h| \to 0$. [Hint: consider $f = 1_{[0,1]}$.]

Extend these results to $L^p(\mathbb{R}^n)$.

2. CONTINUOUS FUNCTIONS AND L^p

Remark. The definition of f_h in effect shifts the graph of f to the left if $h > 0$. To shift to the right, one either uses $-h$ or redefines the translation of f by h by setting $f_h(x) = f(x-h)$. The definition given in the exercise is compatible with the definition of the translation of a measure given in Proposition 6.5.4(8).

Exercise 4.2.11. Let φ denote a continuous bounded function on $\mathbb{R}$. Its **uniform norm** $\|\varphi\|_u$ is defined to be $\sup_{x \in \mathbb{R}} |\varphi(x)|$. Show that

(1) if φ_n is a sequence of bounded continuous functions that converges uniformly (see Definition 3.6.15) to a function φ, then φ is bounded and continuous, i.e., $\mathcal{C}_b(\mathbb{R})$, the vector space of continuous bounded functions on $\mathbb{R}$, is uniformly closed (Definition 3.6.15). [*Hint*: $|\varphi(x_0) - \varphi(x)| \le |\varphi(x_0) - \varphi_n(x_0)| + |\varphi_n(x_0) - \varphi_n(x)| + |\varphi_n(x) - \varphi(x)|.$]

Comment. This property of continuous functions holds in general: given any set $E \subset \mathbb{R}^n$ (say), the vector space $\mathcal{C}_b(E)$ of continuous bounded functions on E is uniformly closed: the above hint applies.

Continuing the exercise, show that

(2) if φ is a bounded continuous function, then $\|\varphi\|_u \ge \|\varphi\|_\infty$ (its L^∞-norm relative to Lebesgue measure), and
(3) $\|\varphi\|_u \le \|\varphi\|_\infty$. [*Hint*: if $M = \|\varphi\|_u$, then $\{x \mid \varphi(x) > M - \epsilon\}$ is a non-empty open set].

Conclude that

(4) the vector space $\mathcal{C}_b(\mathbb{R})$ of essentially bounded, Lebesgue measurable functions on $\mathbb{R}$ is not dense in $L^\infty(\mathbb{R})$, .

Remark. Since $\|\varphi\|_u = \|\varphi\|_\infty$ if φ is a bounded continuous function, it is usual to denote the uniform norm by $\|\varphi\|_\infty$.

For L^2, these density results, combined with the Stone–Weierstrass theorem, prove that $f \in L^2([0, 2\pi])$ is zero if f is orthogonal to all the trigonometric polynomials, where the inner product (Definition 2.5.6) is defined by $\langle f, g \rangle = \int_0^{2\pi} f(x)g(x)\,dx$.

Theorem 4.2.12. *Let $f \in L^2([0, 2\pi])$. Assume that*

$$\int_0^{2\pi} f(x) \cos nx\, dx = \int_0^{2\pi} f(x) \sin nx\, dx = 0, \quad \text{for all } n \ge 0.$$

Then $f = 0$, i.e., $f = 0$ a.e.

Proof. A **trigonometric polynomial** p is by definition a finite linear combination of the functions $\cos nx, \sin mx, n \ge 0, m \ge 1$ on $[0, 2\pi]$. It follows from the addition formulas for sines and cosines that the collection $\mathcal{P}$ of trigonometric polynomials is an algebra of functions (i.e., it is a linear subspace closed under (pointwise) multiplication). By the Stone–Weierstrass theorem, (see Marsden [M1], p. 120, Rudin [R4]), every continuous function φ on $[0, 2\pi]$ for which $\varphi(0) = \varphi(2\pi)$ is a uniform limit (Definition 3.6.15) of trigonometric polynomials.

Let $\epsilon > 0$ and $f \in L^2([0, 2\pi])$. By Exercise 4.2.7, there is a continuous function φ with $\varphi(0) = \varphi(2\pi)$ such that $\| f - \varphi \|_2 < \frac{\epsilon}{2}$. From the above observations, if $\eta > 0$, there is a trigonometric polynomial p with $\| \varphi - p \|_\infty < \eta$. This implies that $\int_0^{2\pi} |\varphi(x) - p(x)|^2 dx < 2\pi\eta^2$. Hence, if $\frac{\epsilon}{2} < \sqrt{2\pi}\eta$,

$$\| f - p \|_2 < \| f - \varphi \|_2 + \| \varphi - p \|_2 < \epsilon.$$

The assumption that f is orthogonal to all trigonometric polynomials p implies that $\| f - p \|_2^2 = \| f \|_2^2 + 2\langle f, p\rangle + \| p \|_2^2 = \| f \|_2^2 + \| p \|_2^2 < \epsilon^2$. Therefore, $\| f \|_2 < \epsilon$, for all $\epsilon > 0$ and so $f = 0$. □

The vector space $L^2([0, 2\pi])$ is complete in the L^2-norm in view of the Riesz–Fischer theorem (Theorem 4.1.28). As a result, it is a so-called **Hilbert space**: a complete normed vector space V whose norm is defined by an inner product (see Exercise 2.5.8 and Proposition 5.1.8).

Definition 4.2.13. *Let V be a Hilbert space. An **orthogonal system** in V is a sequence $(\psi_n)_{n \geq 1}$ of vectors such that any two distinct vectors are orthogonal. If, in addition, the vectors are all of length one, the system is called an **orthonormal system**. An orthogonal system is said to be **complete** if the only vector perpendicular to all the vectors ψ_n is the zero vector.*

In the Hilbert space $L^2([0, 2\pi])$, the functions $1, \cos nx, \sin nx, n \geq 1$ form an orthogonal system.

Exercise 4.2.14. Verify this statement. [Hint: make use of the addition formulas for sines and cosines (see the appendix of Chapter VI) to express (for example) $\sin mx \cos nx$ as a linear combination of sines and cosines.] Also verify that the length in $L^2([0, 2\pi])$ of 1 is $\sqrt{2\pi}$ and that for each of the other functions their length is $\sqrt{\pi}$.

In other words, Theorem 4.2.12 says that the functions $1, \cos nx, \sin nx$, $n \geq 1$ form a complete orthogonal system. Whenever this happens in a Hilbert space, one may expand any vector in a unique way in terms of the complete orthogonal system. For the above system in $L^2([0, 2\pi])$, this expansion for a function f is called its **Fourier series**.

This expansion will now be discussed by first explaining how it works in an abstract Hilbert space.

Let $(\phi_n)_{n \geq 1}$ be an orthonormal system in a Hilbert space V. Then it is necessarily a linearly independent set: if $\sum_{i=1}^n a_i \phi_i = 0$, then $0 = \langle \sum_{i=1}^n a_i \phi_i, \phi_k \rangle = \sum_{i=1}^n \langle a_i \phi_i, \phi_k \rangle = a_k$ if $1 \leq k \leq n$.

Let L_N be the linear subspace of V with basis $\phi_1, \phi_2, \ldots, \phi_N$. It is a closed subspace: if $\sum_{n=1}^N a_n(k) \phi_n = x_k \to x \in V$, then $\langle x_k, y \rangle \to \langle x, y \rangle$ for any $y \in V$; taking $y = \phi_j$, $1 \leq j \leq N$, shows that each sequence $(a_j(k))_{k \geq 1}$ converges to a limit a_j; hence, $\| x_k - \sum_{n=1}^{N} a_n \phi_n \|^2 = \sum_{n=1}^{N} \{a_n(k) - a_n\}^2$ tends to 0; as the limit is unique, $x = \sum_{n=1}^{N} a_n \phi_n \in L_N$.

Lemma 4.2.15. *If $u \in V$, then there exists a unique $\ell = \ell_N \in L_N$ such that $u - \ell$ is perpendicular to L_N. More specifically,*

$$\ell_N = \sum_{n=1}^{N} \langle u, \phi_n \rangle \phi_n \quad \text{and} \quad \| \ell_N \|^2 = \sum_{n=1}^{N} \langle u, \phi_n \rangle^2 \leq \| u \|^2.$$

Proof. Assume that it is possible to write $u = \ell + w$ with $\ell = \sum_{i=1}^{N} a_i \phi_i$ in L_N and w perpendicular to L_N. Since $0 = \langle u - \ell, \phi_n \rangle$ for $1 \leq n \leq N$, it follows that $\langle u, \phi_n \rangle = \langle \ell, \phi_n \rangle = a_n$. This shows that there is at most one $\ell \in L_N$ with $w = u - \ell$ perpendicular to L_N.

Define ℓ to be $\sum_{i=1}^{N} \langle u, \phi_n \rangle \phi_n$. Then $\langle u - \ell, \phi_n \rangle = 0$ for $1 \leq n \leq N$, i.e., $w = u - \ell$ is perpendicular to L_N.

Since $\langle \ell, w \rangle = 0$, it follows that

$$\| u \|^2 = \| \ell \|^2 + \| w \|^2 \geq \| \ell \|^2 = \sum_{n=1}^{n} \langle u, \phi_n \rangle^2. \quad \square$$

As an immediate corollary, one has the following result.

Corollary 4.2.16. *Let $(\phi_n)_{n \geq 1}$ be an orthonormal system in a Hilbert space V. If $u \in V$, then*

$$\sum_{n=1}^{\infty} \langle u, \phi_n \rangle^2 \leq \| u \|^2 \quad \textbf{(Bessel's inequality)}.$$

Bessell's inequality implies that the series $\sum_{n=1}^{\infty} \langle u, \phi_n \rangle \phi_n$ is absolutely summable and so converges in the Hilbert space V by Proposition 4.1.27.

The significance of the completeness of an orthonormal system is that in this case the above series converges to u and so gives an expansion of u in terms of the complete orthonormal system.

Proposition 4.2.17. *Let $(\phi_n)_{n \geq 1}$ be a complete orthonormal system in a Hilbert space V. If $u \in V$, then*

$$u = \sum_{n=1}^{\infty} \langle u, \phi_n \rangle \phi_n \quad \text{and}$$

$$\sum_{n=1}^{\infty} \langle u, \phi_n \rangle^2 = \| u \|^2 \quad \textbf{(Parseval's equality)}.$$

Proof. Let $\ell = \sum_{n=1}^{\infty} \langle u, \phi_n \rangle \phi_n$ and $\ell_N = \sum_{n=1}^{N} \langle u, \phi_n \rangle \phi_n$. Then since $\ell = \lim_{N \to \infty} \ell_N$, it follows that

$$\langle \ell, \phi_k \rangle = \lim_{N \to \infty} \langle \ell_N, \phi_k \rangle = \lim_{N \to \infty} \langle \sum_{n=1}^{N} \langle u, \phi_n \rangle \phi_n, \phi_k \rangle = \langle u, \phi_k \rangle.$$

As a result, $\langle u - \ell, \phi_k \rangle = 0$ for all $k \geq 1$. Completeness of the orthonormal system implies that $u = \ell$.

It remains to compute $\| \ell \|$. Since $\ell_N \to \ell$, it follows that $\| \ell_N \| \to \| \ell \|$. Now $\| \ell_N \|^2 = \sum_{n=1}^{N} \langle u, \phi_n \rangle^2$ by Lemma 4.2.15 and so $\| \ell \|^2 = \sum_{n=1}^{\infty} \langle u, \phi_n \rangle^2$. □

Corollary 4.2.18. (Parseval's theorem) *An orthonormal system is complete if and only if for any $u \in V$ one has*

(*) $$\sum_{n=1}^{\infty} \langle u, \phi_n \rangle^2 = \| u \|^2 .$$

Proof. It remains to show that the condition (*) implies completeness. The condition itself shows that $\| u \|^2 - \| \ell_N \|^2 \to 0$ for any $u \in V$. Since $u = \ell_N + w_N$ with w_N perpendicular to L_N, it follows that $\| w_N \|^2 \to 0$ and so $\ell_N \to u$.

If $u \in V$ and u is orthogonal to each ϕ_n, then $\ell_N = 0$ for all N and so $u = 0$, i.e., the system is complete. □

In the case of $L^2([0, 2\pi])$, the functions $1, \cos nx, \sin nx, n \geq 1$ form an orthogonal system. If in a Hilbert space one has an orthogonal system $(\psi_n)_{n \geq 1}$, then by normalizing the vectors ψ_n, one obtains an orthonormal system $(\phi_n)_{n \geq 1}$. Assuming the orthogonal system to be complete, and since $\phi_n = \frac{1}{\|\psi_n\|} \psi_n$, the expansion of u can be written as

$$u = \sum_{n=1}^{\infty} \langle u, \phi_n \rangle \phi_n = \sum_{n=1}^{\infty} \frac{\langle u, \psi_n \rangle}{\| \psi_n \|^2} \psi_n \quad \text{and}$$

$$\| u \|^2 = \sum_{n=1}^{\infty} \langle u, \phi_n \rangle^2 = \sum_{n=1}^{\infty} \frac{\langle u, \psi_n \rangle^2}{\| \psi_n \|^2} .$$

As a result, for any $f \in L^2([0, 2\pi])$, in view of Exercise 4.2.14, one has the following Fourier expansion of f:

$$f(x) = \frac{a_0}{2\pi} + \frac{1}{\pi} \sum_{n=1}^{\infty} \{a_n \cos nx + b_n \sin nx\},$$

where

$$a_n = \int_0^{2\pi} f(t)\cos ntdt, \text{ for all } n \geq 0, \text{ and}$$

$$b_n = \int_0^{2\pi} f(t)\sin ntdt, \text{ for all } n \geq 1.$$

This expansion is to be understood as in L^2. It does not mean that for all $x \in [0, 2\pi]$ the series sums to $f(x)$. The expansion could be written more appropriately as

$$f(\cdot) = \frac{a_0}{2\pi} + \frac{1}{\pi}\sum_{n=1}^{\infty}\{a_n \cos n(\cdot) + b_n \sin n(\cdot)\},$$

where the equality is read as convergence in L^2 of the partial sums and, for example, $\cos n(x) \stackrel{\text{def}}{=} \cos nx$.

The study of other types of convergence for this type of series is the subject of classical Fourier analysis (see Körner [K3]). For example, it was an open question for many years as to whether the series converged a.e. to f when $f \in L^2$. It was solved in the affirmative in 1966 by Lennart Carleson.[5]

Convolution and Fourier series.

Convolution was defined earlier (Exercise 3.3.21) primarily to discuss the distribution of sums of independent random variables. The definition of convolution for functions in $L^1(\mathbb{R})$ was implicit in part of that discussion.

Definition 4.2.19. *If f and g are in $L^1(\mathbb{R})$, the **convolution** of f with g is the function denoted by $f \star g$, where*

$$(f \star g)(x) = \int f(x-y)g(y)dy.$$

Notice that, for a specific $x \in \mathbb{R}$, there is no a priori reason why the product $f(x-y)g(y)$ should be integrable. However, Fubini's theorem implies that, for a.e. x, it is in $L^1(\mathbb{R})$.

Proposition 4.2.20. *If f and g are in $L^1(\mathbb{R})$, then*

$$f \star g \in L^1(\mathbb{R}) \quad \text{and} \quad \|f \star g\|_1 \leq \|f\|_1 \|g\|_1 .$$

Proof. One may assume that f and g are finite Borel functions — modifying f and g if necessary by adding null functions. Then $\Phi(x,y) = f(x-y)g(y)$ is a Borel function on $\mathbb{R}^2$: the function $(x,y) \to x-y$ is Borel and so

---
[5] Acta Math. **116**, (1966) pp. 135–157.

$f(x-y)$ is a Borel function (Proposition 2.1.17); $(x,y) \to g(y)$ is Borel for the same reason, and the product of measurable functions is measurable (Proposition 2.1.11) as $(u,v) \to uv$ is measurable (even continuous).

The function $\Phi \in L^1(\mathbb{R}^2)$: using Fubini's theorem (Theorem 3.3.16) one has

$$E[|\Phi|] = \int |g(y)| \left[\int |f(x-y)| dx \right] dy = \| f \|_1 \int |g(y)| dy = \| f \|_1 \| g \|_1,$$

as the invariance of Lebesgue measure under translation (Exercise 2.2.8) implies that $\int |f(x-y)| dx = \int |f(x)| dx$. Since Φ is integrable, by Fubini's theorem (Theorem 3.3.17), $\int f(x-y)g(y) dy$ is integrable for almost all x. Therefore, the function $f \star g$ is defined as a function a.e. (set it equal to zero (say), when the integral is not defined).

Furthermore,

$$\int |(f \star g)(x)| dx \leq \int \left[\int |f(x-y)g(y)| dy \right] dx = \| \Phi \|_1 = \| f \|_1 \| g \|_1. \quad \square$$

In principle, the convolution product $f \star g$ depends upon the order. However, $f \star g = g \star f$ for any two functions in L^1. This is explained in the next exercise, using the fact that for any Lebesgue measurable set A one has $|A| = |-A|$, where $-A = \{-x \mid x \in A\}$: in other words, Lebesgue measure is invariant under the transformation $x \to -x$.

Exercise 4.2.21. Show that

(1) if f is a Borel (respectively, Lebesgue measurable) function, the function $\check{f}(x) \stackrel{\text{def}}{=} f(-x)$ has the same property,
(2) if $f = 1_A$, then $\check{f} = 1_{-A}$, where $-A = \{-x \mid x \in A\}$,
(3) $|A| = |-A|$ for any Lebesgue measurable set A [Hint: first verify this for intervals.],
(4) if $f \in L^1(\mathbb{R})$, then $\check{f} \in L^1(\mathbb{R})$ and $\int f(x) dx = \int \check{f}(x) dx$ [Hint: first verify this for simple functions.],
(5) if $f, g \in L^1(\mathbb{R})$, then $(f \star g)(x) = \int \check{f}(y-x)g(y) dy = \int \check{f}(u)g(x+u) du$ [Hint: use translation invariance.], and finally,
(6) $\int \check{f}(u)g(x+u) du = \int f(u)g(x-u) du = (g \star f)(x)$. [Hint: use (4).]

Note that in (5) and (6), one assumes that x is such that $y \to f(x-y)g(y)$ is in L^1.

Conclude that

(7) if $f, g \in L^1(\mathbb{R})$, then $f \star g = g \star f$.

Remark 4.2.22. All of these observations about convolution carry over automatically to functions in $L^1(\mathbb{R}^n)$.

In classical Fourier analysis, the interval $[0, 2\pi]$ plays a privileged role. The functions f on $\mathbb{R}$ that are of interest are all 2π-periodic (i.e., $f(x+2\pi) =$

2. CONTINUOUS FUNCTIONS AND L^p

$f(x)$ for all x). These functions may therefore be identified with functions on $[0, 2\pi]$ that have the same value at 0 and 2π. By an "abus de langage", a 2π-periodic function f on $\mathbb{R}$ will be said to be a 2π-**periodic integrable function** or simply to be **integrable** if its restriction to $[0, 2\pi]$ is integrable: strictly speaking, such a function is locally integrable in the sense of Remark 4.6.8 Note that such functions have associated with them a Fourier series since the integrals that define the coefficients a_n and b_n are defined.

Since the interval $[0, 2\pi]$ can be identified with the unit circle by the map $t \to e^{it} = \cos t + i \sin t$, the above amounts to considering functions on the circle S^1 (also called a **one dimensional-torus** and denoted by $\mathbb{T}$). The laws of exponents for the complex exponential (equivalently, the addition formulas for sin and cos) imply that S^1 is a group, even a compact commutative group: $e^{is}e^{it} = e^{i(s+t)} = e^{it}e^{is}$ (see the appendix to Chapter VI).

The uniform distribution on $[0, 2\pi]$ (i.e., $\mu(dx) = \frac{1}{2\pi}1_{(0, 2\pi]}(x)dx$) has an image m on S^1 under the exponential map $t \to e^{it}$. This function may be viewed as a random vector on the probability space $([0, 2\pi], \mathfrak{B}([0, 2\pi]), \mu)$. The distribution m of the exponential map is then a probability on the circle: it is the uniform distribution on the circle, i.e. $m(\alpha) = \frac{1}{2\pi}|\alpha|$, where $|\alpha|$ is arc length for any arc α on the circle.

As a result, if f is 2π-periodic on $\mathbb{R}$ and $\phi(e^{it}) \stackrel{\text{def}}{=} f(t)$, for integrable f, it follows that

$$\frac{1}{2\pi} \int_0^{2\pi} f(t)dt = \frac{1}{2\pi} \int_0^{2\pi} \phi(e^{it})dt = \int \phi dm.$$

In this way $L^1([0, 2\pi])$ can be viewed as $L^1(dm)$.

It turns out that because S^1 is a group, convolution can be defined for functions on S^1 that are in $L^1(m)$ (see Exercise 4.7.26). However, it can be directly defined using 2π-periodic functions on $\mathbb{R}$: the important thing to observe is that if f is 2π-periodic, then by the translation invariance of Lebesgue measure, $\int_a^{a+2\pi} f(x)dx = \int_0^{2\pi} f(u)du$ for any $a \in \mathbb{R}$.

Proposition 4.2.23. *Let f and g be two 2π-periodic integrable functions on $\mathbb{R}$. Then*

$$\int_0^{2\pi} f(x-y)g(y)dy = \int_0^{2\pi} g(x-u)f(u)du$$

is an integrable 2π-periodic function of x.

Proof. First, observe that as in Proposition 4.2.20 the functions f and g may be assumed to be Borel and that

$$\int_0^{2\pi} \left[\int_0^{2\pi} |f(x-y)||g(y)|dy \right] dx = \int_0^{2\pi} \left[\int_0^{2\pi} |f(x-y)|dx \right] |g(y)|dy$$

$$= \left(\int_0^{2\pi} |f(x)|dx \right) \left(\int_0^{2\pi} |g(y)|dy \right)$$

since $\int_0^{2\pi} |f(x-y)| dx = \int_{-y}^{2\pi-y} |f(u)| du$, which equals $\int_0^{2\pi} |f(u)| du$ by 2π-periodicity. Hence, for almost all $x \in [0, 2\pi]$, the functions $y \to f(x-y)g(y)$ and $y \to g(x-y)f(y)$ are in $L^1([0, 2\pi])$.

Let $x \in [0, 2\pi]$ be such that $y \to f(x-y)g(y)$ and $y \to g(x-y)f(y)$ are in $L^1([0, 2\pi])$. The integral $\int_0^{2\pi} f(x-y)g(y)dy = \int 1_{[0,2\pi]}(y) f(x-y)g(y)dy$. Following the line of reasoning in Exercise 4.2.21, it follows that

$$\int 1_{[0,2\pi]}(y) f(x-y)g(y)dy = \int 1_{[-x,2\pi-x]}(u) \check{f}(u) g(x+u) du$$

$$= \int 1_{[x-2\pi,x]}(u) f(u) g(x-u) du$$

$$= \int_0^{2\pi} f(u) g(x-u) du,$$

since $\int_a^{a+2\pi} h(u) du = \int_0^{2\pi} h(x) dx$ for any $a \in \mathbb{R}$ as $u \to h(u) = f(u)g(x-u)$ is 2π-periodic.

The 2π-periodicity of $\int_0^{2\pi} f(x-y)g(y)dy$ is obvious. The integrability of $x \to \int_0^{2\pi} f(x-y)g(y)dy$ follows by Fubini's theorem from the initial computation, which showed that $\Phi(x,y) = f(x-y)g(y)$ is in $L^2([0,2\pi] \times [0,2\pi])$. □

This suggests that one could reasonably denote $\frac{1}{2\pi} \int_0^{2\pi} f(x-y)g(y)dy$ by $(f \star_{2\pi} g)(x)$ if f and g are 2π-periodic functions on $\mathbb{R}$. In [K3], Körner denotes this convolution by $\frac{1}{2\pi} \int_\mathbb{T} f(y) g(x-y) dy$.

In particular, if K is 2π-periodic and non-negative, and $\frac{1}{2\pi} \int_0^{2\pi} K(x) dx = 1$, then $1_{[0,2\pi]}(x) K(x) dx$ determines a probability μ on the circle S^1, and the convolution $(f \star_{2\pi} K)$ could be denoted by $\phi \star_{S^1} \mu$ if $\phi(e^{ix}) = f(x)$.

In classical Fourier analysis, there is a very well-known sequence $(\mu_n)_{n \geq 1}$ of probabilities μ_n of this type which are collectively referred to as the **Féjer kernel**. The nth probability μ_n is given by the function $K_n(x) \stackrel{\text{def}}{=} \left(\frac{1}{n+1}\right) \left(\frac{\sin(n+\frac{1}{2})x}{\sin \frac{x}{2}}\right)^2$ (note that it is not obvious from this formula that $\frac{1}{2\pi} \int_0^{2\pi} K_n(x) dx = 1$ — see Lemma 4.2.26). The significance of this kernel is that for any $n \geq 1$ and 2π-periodic function f, one has

(†) $\quad (f \star_{2\pi} K_n)(x) = \dfrac{1}{n+1} \sum_{\ell=0}^n \left(\dfrac{a_0}{2\pi} + \dfrac{1}{\pi} \sum_{k=1}^\ell \{a_k \cos kx + b_k \sin kx\} \right),$

where the a_k and b_k are the Fourier coefficients of f, and the summation on k is set equal to zero when $\ell = 0$. The expressions

$$s_\ell = \frac{a_0}{2\pi} + \sum_{k=1}^\ell \{a_k \cos kx + b_k \sin kx\}$$

2. CONTINUOUS FUNCTIONS AND L^p

are the partial sums s_ℓ of the Fourier series of f and so (†) states that the average of the first $n+1$ partial sums (see Exercise 4.3.14) is given by a convolution kernel.

To verify (†), it is useful to relate the Fourier coefficients of f to convolution operators. The Fourier coefficients of $f \in L^1([0, 2\pi])$ can be expressed in terms of the complex exponentials e^{ikx} since $\cos kx = \frac{e^{ikx} + e^{-ikx}}{2}$ and $\sin kx = \frac{e^{ikx} - e^{-ikx}}{2i}$. One has the following lemma.

Lemma 4.2.24. *If $f \in L^1([0, 2\pi])$ and $k \geq 1$, then*

$$\frac{1}{2\pi} \int_0^{2\pi} f(u) \{ e^{-iku} e^{ikx} + e^{iku} e^{-ikx} \} du = \frac{1}{\pi} \{ a_k \cos kx + b_k \sin kx \}.$$

In other words,

$$\frac{1}{\pi} \{ a_k \cos kx + b_k \sin kx \} = (f \star_{2\pi} \{ e^{ik} + e^{-ik} \})(x) = 2(f \star_{2\pi} \cos k)(x),$$

where $e^{ik}(x) \stackrel{\text{def}}{=} e^{ikx}$ and $\cos k(x) \stackrel{\text{def}}{=} \cos kx$.

Proof. If $k \geq 1$, then

$$\frac{1}{2\pi} \left(\int_0^{2\pi} f(u) e^{-iku} du \right) e^{ikx}$$

$$= \frac{1}{2\pi} \left(\int_0^{2\pi} f(u) \{ \cos ku - i \sin ku \} \{ \cos kx + i \sin kx \} du \right)$$

$$= \frac{1}{2\pi} \left(\int_0^{2\pi} f(u) \{ \cos ku \cos kx + \sin ku \sin kx \} du \right)$$

$$+ \frac{i}{2\pi} \left(\int_0^{2\pi} f(u) \{ \cos ku \sin kx - \sin ku \cos kx \} du \right)$$

$$= \frac{1}{2\pi} \{ a_k \cos kx + b_k \sin kx \}$$

$$+ \frac{i}{2\pi} \left(\int_0^{2\pi} f(u) \{ \cos ku \sin kx - \sin ku \cos kx \} du \right).$$

The result follows since f real implies that $\frac{1}{2\pi} \left(\int_0^{2\pi} f(u) e^{iku} du \right) e^{-ikx}$ is the complex conjugate of $\frac{1}{2\pi} \left(\int_0^{2\pi} f(u) e^{-iku} du \right) e^{ikx}$. □

Hence,

$$s_\ell = \frac{a_0}{2\pi} + \frac{1}{\pi}\sum_{k=1}^{\ell}\{a_k \cos kx + b_k \sin kx\}$$

$$= \sum_{k=-\ell}^{\ell} \frac{1}{2\pi} \int_0^{2\pi} f(u)e^{ik(x-u)} du$$

$$= (f \star_{2\pi} \{\sum_{k=-\ell}^{\ell} e^{ik}\})(x), \text{ and so}$$

$$\sigma_n = \frac{1}{n+1}\sum_{\ell=0}^{n} s_\ell$$

$$= (f \star_{2\pi} \frac{1}{n+1}\{\sum_{\ell=0}^{n}(\sum_{k=-\ell}^{\ell} e^{ik})\})(x)$$

$$= f \star_{2\pi} K_n(x),$$

in view of (5) in the next exercise. This completes the proof of (†).

Exercise 4.2.25. Show that
(1) $\sum_{k=0}^{n} z^k = \frac{1-z^{n+1}}{1-z}$ for any complex number $z \neq 1$.

Make use of (1) to show that

(2) $D_\ell(x) \stackrel{\text{def}}{=} \sum_{k=-\ell}^{\ell} e^{ikx} = \frac{\sin(\ell+\frac{1}{2})x}{\sin \frac{x}{2}}$ ($D_\ell(x)$ is called **Dirichlet's kernel**,

(3) $\sum_{\ell=0}^{n} e^{i(\ell+\frac{1}{2})x} = \frac{\sin(n+1)x + i\{1-\cos(n+1)x\}}{2\sin\frac{x}{2}}$, and conclude that

(4) $\sum_{\ell=0}^{n} \sin(\ell + \frac{1}{2})x = \frac{1-\cos(n+1)x}{2\sin\frac{x}{2}}$.

Finally, show that

(5) $\frac{1}{n+1}\sum_{\ell=0}^{n}(\sum_{k=-\ell}^{\ell} e^{ikx}) = \frac{1}{n+1}\sum_{\ell=0}^{n} D_\ell(x) = \left(\frac{\sin\frac{(n+1)x}{2}}{\sin\frac{x}{2}}\right)^2$.

Every continuous 2π-periodic function is in $L^1([0,1])$ and hence has a Fourier series. However, this is not enough to ensure that the partial sums $s_n(x) = \frac{a_0}{2\pi} + \sum_{k=1}^{n}\{a_n \cos kx + b_n \sin kx\}$ converge to $f(x)$. For classical counterexamples, see Körner [K3] pp. 67–73.

While the partial sums of the Fourier series of a continuous function need not converge to $f(x)$, in fact their averages $\sigma_n = \frac{1}{n+1}\sum_{\ell=0}^{n} s_\ell$ do so (i.e., the **Cesaro means** (see (Exercise 4.3.14), also [T] p. 411) of the Fourier series converge to $f(x)$).

The expression (†) states that the nth Cesaro mean is obtained from f by taking its convolution with the nth Féjer kernel $K_n(x)$. Féjer's theorem (Theorem 4.2.27) states that, if f is continuous, these convolutions converge uniformly to $f(x)$. This theorem is a formal consequence of the following lemma.

2. CONTINUOUS FUNCTIONS AND L^p

Lemma 4.2.26. *The Féjer kernel* $(K_n)_{n \geq 1}$ *has the following properties:*
 (1) $K_n(x) \geq 0$;
 (2) $\frac{1}{2\pi} \int_0^{2\pi} K_n(x)dx = 1$; *and*
 (3) *if* $\delta > 0$ *and* $\epsilon > 0$, *then* $K_n(x) < \epsilon$ *for all* $x \in [\delta, 2\pi - \delta]$ *if* n *is sufficiently large; equivalently,*
 (4) *if* $\delta > 0$ *and* $\epsilon > 0$, *then* $K_n(x) < \epsilon$ *for all* $x \in [-\pi, -\delta] \cup [\delta, \pi]$ *if* n *is sufficiently large.*

Proof. Property (1) is obvious. If $x \in [\delta, 2\pi - \delta]$, then $\sin \frac{x}{2} \geq \sin \frac{\delta}{2} = m > 0$, and so in this range $K_n(x) \leq \frac{1}{n+1}(\frac{1}{m})$, which is less than ϵ for $n + 1 > \frac{1}{m\epsilon}$. This proves (3) and hence (4).

Since $K_n(x) = \frac{1}{n+1} \sum_{\ell=0}^{n} (\sum_{k=-\ell}^{\ell} e^{ikx})$ and $\int_0^{2\pi} e^{ikx} dx = 0$ for all $k \neq 0$, property (2) follows immediately. □

Theorem 4.2.27. (Féjer) *If* f *is a continuous* 2π-*periodic function on* $\mathbb{R}$, *then, for* $0 \leq x \leq 2\pi$,

$$(f \star_{2\pi} K_n)(x) = \frac{1}{2\pi} \int_0^{2\pi} f(x-y) K_n(y) dy \xrightarrow{u} f(x) \text{ as } n \to \infty,$$

i.e., the Cesaro means $f \star_{2\pi} K_n$ *converge uniformly to* f.
 More generally, if f *is* 2π-*periodic and integrable, then*

$$(f \star_{2\pi} K_n)(x_0) \to \frac{1}{2}\{f(x_0+) + f(x_0-)\} \text{ as } n \to \infty$$

at any point x_0 *where both limits exist.*

Proof. If f is continuous, it is uniformly continuous on $[-\pi, \pi]$ by Property 2.3.8 (3). Let $\epsilon > 0$ and $\delta > 0$ be such that $|f(x) - f(y)| < \epsilon$ if $|x - y| < \delta$, $x, y \in [-\pi, \pi]$. Then

$$|(f \star_{2\pi} K_n)(x) - f(x)| = \left| \frac{1}{2\pi} \int_{-\pi}^{\pi} \{f(x-y) - f(x)\} K_n(y) dy \right|$$

$$\leq \frac{1}{2\pi} \int_{-\pi}^{\pi} |f(x-y) - f(x)| K_n(y) dy.$$

Now

$$\frac{1}{2\pi} \int_{-\pi}^{\pi} |f(x-y) - f(x)| K_n(y) dy = \frac{1}{2\pi} \int_{-\pi}^{-\delta} |f(x-y) - f(x)| K_n(y) dy$$

$$+ \frac{1}{2\pi} \int_{-\delta}^{\delta} |f(x-y) - f(x)| K_n(y) dy$$

$$+ \frac{1}{2\pi} \int_{\delta}^{\pi} |f(x-y) - f(x)| K_n(y) dy$$

$$\leq \frac{1}{2} \epsilon \| f \|_{\infty} + \epsilon + \frac{1}{2} \epsilon \| f \|_{\infty}$$

if n is large enough to imply that $K_n(x) < \epsilon$ on $[\delta, 2\pi - \delta]$.

The proof for the more general case follows in essentially the same way once one observes that the Féjer kernel is symmetric, i.e., it is an even function of x. As a result, $\int_0^\pi K_n(x)dx = \int_{-\pi}^0 K_n(x)dx = \pi$, and so one may "average" on each side of a point x_0 by splitting the integral over $[-\delta, \delta]$ into two integrals, one over $[-\delta, 0]$ and the other over $[0, \delta]$.

If $a, b \in \mathbb{R}$, then one has

(1)
$$\left| \frac{1}{2\pi} \int_0^\pi f(x_0 - y) K_n(y) dy - \frac{1}{2} a \right| = \left| \frac{1}{2\pi} \int_0^\pi \{f(x_0 - y) - a\} K_n(y) dy \right|$$
$$\leq \frac{1}{2\pi} \int_0^\pi |f(x_0 - y) - a| K_n(y) dy$$
$$= \frac{1}{2\pi} \int_0^\delta |f(x_0 - y) - a| K_n(y) dy$$
$$+ \frac{1}{2\pi} \int_\delta^\pi |f(x_0 - y) - a| K_n(y) dy$$
$$= I_1 + I_2$$

and

(2)
$$\left| \frac{1}{2\pi} \int_{-\pi}^0 f(x_0 - y) K_n(y) dy - \frac{1}{2} b \right| = \left| \frac{1}{2\pi} \int_{-\pi}^0 \{f(x_0 - y) - b\} K_n(y) dy \right|$$
$$\leq \frac{1}{2\pi} \int_{-\pi}^0 |f(x_0 - y) - b| K_n(y) dy$$
$$= \frac{1}{2\pi} \int_{-\delta}^0 |f(x_0 - y) - b| K_n(y) dy$$
$$+ \frac{1}{2\pi} \int_{-\pi}^{-\delta} |f(x_0 - y) - b| K_n(y) dy$$
$$= J_1 + J_2.$$

Assume that $f(x_0-)$ exists, and let $a = f(x_0-)$ in (1). Then, for small $\delta > 0$, it follows that $|f(x_0 - y) - f(x_0-)| < \epsilon$ if $0 < y < \delta$ and so $I_1 < \epsilon$ for small $\delta > 0$. Fix one such δ. Then, the second integral I_2 in (1) is dominated by $\epsilon\{|a| + \frac{1}{\pi} \int_0^{2\pi} |f(u)|du\}$ as long as $n \geq n(\delta, \epsilon)$.

If $f(x_0+)$ exists, let $b = f(x_0+)$ in (2). Then, for the same reasons, $J_1 + J_2 \leq \epsilon\{1 + |b| + \frac{1}{\pi} \int_0^{2\pi} |f(u)|du\}$ for a small $\delta > 0$ and $n \geq n(\delta, \epsilon)$ sufficiently large.

This shows that $2(f \star_{2\pi} K_n)(x_0) \to f(x_0+) + f(x_0-)$ as $n \to \infty$, provided both limits exist. □

Convolution and differentiation.

Let $f \in L^p(\mathbb{R}^n), 1 \leq p \leq \infty$, and let $k \in C_c(\mathbb{R}^n)$. Then, for all $x \in \mathbb{R}^n$, the function $y \to f(y)k(x - y)$ is in L^p, as it is dominated by $\|k\|_\infty |f(y)|$,

2. CONTINUOUS FUNCTIONS AND L^p

and by dominated convergence the convolution $(f \star k)(x)$ is a continuous function of x on $\mathbb{R}^n$.

If, in addition, $k \in C_c^1(\mathbb{R}^n)$, then it will now be shown that $f \star k$ is C^1 and its partial derivatives $\frac{\partial}{\partial x_i}(f \star k)$ are given by convolution with the partial derivatives $\frac{\partial}{\partial x_i} k$.

Let e_i be the canonical basis vector of $\mathbb{R}^n$ all of whose components are zero, except for a 1 in the ith position. The mean-value theorem implies that
$$\frac{k(x + he_i) - k(x)}{h} = \frac{\partial}{\partial x_i} k(x + \theta h e_i),$$
where $0 \leq \theta \leq 1$ and depends upon x, h, and i.

Since $\frac{\partial}{\partial x_i} k \in C_c(\mathbb{R}^n)$, there is a constant M with $|\frac{\partial}{\partial x_i} k(x)| \leq M$ for all i and x. Hence, if $B_N \stackrel{\text{def}}{=} \{x \mid \|x\| < N\}$ contains the support of k, i.e. $B_N \supset \overline{\{x \mid k(x) \neq 0\}}$, then
$$\left| \frac{k(x + he_i) - k(x)}{h} \right| \leq M 1_{B_N}, \text{ for all } i, \ 1 \leq i \leq n, \text{ and } x \in \mathbb{R}^n.$$

It follows from this, by dominated convergence, as in the proof of Theorem 2.6.1, that
$$\frac{(f \star k)(x + he_i) - (f \star k)(x)}{h} = \int f(y) \frac{k(x + h - ye_i) - k(x - y)}{h} dy$$
converges to $\int f(y) \frac{\partial}{\partial x_i} k(x - y) dy$ as $h \to 0$. This proves the following important result about differentiating under the integral sign.

Theorem 4.2.28. Let $f \in L^p(\mathbb{R}^n)$, $1 \leq p \leq \infty$, and $k \in C_c^1(\mathbb{R}^n)$. Then
$$f \star k \in C^1(\mathbb{R}^n) \text{ and } \frac{\partial}{\partial x_i}(f \star k) = f \star \frac{\partial}{\partial x_i} k, \text{ for all } i, \ 1 \leq i \leq n.$$

For bounded measurable functions f this result extends to convolution by functions k that are C^1 and integrable, but not necessarily of compact support. First, here is the one-dimensional version, which automatically solves Exercise 2.9.6 once one shows that all the derivatives of the Gaussian density $n(x) = \frac{1}{\sqrt{2\pi}} e^{-\frac{x^2}{2}}$ exist and are integrable.

Proposition 4.2.29. Let $f \in L^\infty(\mathbb{R})$ and $k \in C^1(\mathbb{R}) \cap L^1(\mathbb{R})$. Assume that $k' \in C(\mathbb{R}) \cap L^1(\mathbb{R})$. Then
$$(f \star k)' = f \star k'.$$

Proof. Let $\phi_N \in C_c^1(\mathbb{R})$ with $0 \leq \phi \leq 1$ be such that $\phi_N(x) = 1$ if $|x| \leq N$ and $\phi_N(x) = 0$ if $|x| > N + 1$. One can also assume that the derivative ϕ_N' is uniformly bounded, independent of N, by a positive constant M.

Since $f \in L^\infty$ and k is continuous, it follows from dominated convergence that $f \star k$ is continuous: if $x_n \to x$, then $f(y)k(x_n - y) \to f(y)k(x - y)$; and $f(y)k(x - y)$ is integrable in y for all x.

Also, $f \star k\phi_N \xrightarrow{pt} f \star k$ (i.e., pointwise) as $N \to \infty$ since $\int f(x - y)k(y)\phi_N(y)dy$ converges to $\int f(x - y)k(y)dy$ as $N \to \infty$, again by dominated convergence.

Furthermore, by Theorem 4.2.28,

(†) $$(f \star k\phi_N)' = f \star (k\phi_N)' = f \star k'\phi_N + f \star k\phi_N',$$

and hence

$$(f \star k\phi_N)' \xrightarrow{pt} f \star k',$$

since $|(f \star k\phi_N')(x)| \le M \| f \|_\infty \int_{\{|y|>N\}} |k(y)|dy \to 0$ as $n \to \infty$ and $f \star k\phi_N' \xrightarrow{pt} f \star k'$. Note that, for any u, (†) implies

(††) $$|(f \star k\phi_N)'(u)| \le (|f| \star |k'|)(u) + 2M \| f \|_\infty \text{ for large } N,$$

with $(|f| \star |k'|)(u)$ continuous.

Now

$$(f \star k)(x) = \lim_{N \to \infty} (f \star k\phi_N)(x)$$
$$= \lim_{N \to \infty} \left[(f \star k\phi_N)(0) + \int_0^x (f \star k\phi_N)'(u)du \right]$$
$$= (f \star k)(0) + \int_0^x (f \star k')(u)du,$$

where the last limit holds by dominated convergence in view of (††) since $(|f| \star |k'|)(u)$ is continuous. The continuity of k' ensures that $f \star k'$ is continuous and so $\frac{d}{dx} \int_0^x (f \star k')(u)du = (f \star k')(x)$. □

Corollary 4.2.30. *Let $f \in L^\infty(\mathbb{R}^n)$ and $k \in \mathcal{C}^1(\mathbb{R}^n) \cap L^1(\mathbb{R}^n)$. Assume that each partial derivative $\frac{\partial}{\partial x_i} k$ belongs to $L^1(\mathbb{R}^n)$. Then, for each i, $1 \le i \le n$,*

$$\frac{\partial}{\partial x_i}(f \star k) = f \star \frac{\partial}{\partial x_i} k.$$

Proof. Let function $\theta_N(x) \stackrel{\text{def}}{=} \phi_N(\|x\|)$, where ϕ_N is the cutoff function used above. The norm of its gradient $\nabla \theta_N(x) = \phi_N'(\|x\|)\frac{1}{\|x\|}x$ is then bounded by a positive constant M independent of N.

As a result, (†) and (††) hold with the derivative replaced in turn by each of the n partial derivatives $\frac{\partial}{\partial x_i}$.

Let $x \in \mathbb{R}^n$ and let $\mathbf{e}_i$ be a basis vector. By integrating along the line segment from x to $x + h\mathbf{e}_i$, one has, as before,

$$(f \star k)(x + he_i) = \lim_{N \to \infty} (f \star k\phi_N)(x + he_i)$$
$$= \lim_{N \to \infty} (f \star k\phi_N)(x) + \int_0^1 \langle \nabla(f \star k\phi_N)(x + the_i), he_i \rangle dt$$
$$= \lim_{N \to \infty} (f \star k\phi_N)(x) + \int_0^1 h \frac{\partial}{\partial x_i}(f \star k\phi_N)(x + the_i) dt$$
$$= (f \star k)(x) + \int_0^1 h(f \star \frac{\partial}{\partial x_i} k)(x + the_i) dt.$$

Since the partial derivative $\frac{\partial}{\partial x_i} k$ is continuous by hypothesis, it follows that $\frac{\partial}{\partial x_i}(f \star k)(x) = (f \star \frac{\partial}{\partial x_i} k)(x)$. □

3. Pointwise Convergence and Convergence in Measure or Probability

Let $(X_n)_{n \geq 1}$ be a sequence of random variables on $(\Omega, \mathfrak{F}, \mathbf{P})$ or measurable functions on a σ-finite measure space $(\Omega, \mathfrak{F}, \mu)$. Then, for each $\omega \in \Omega, (X_n(\omega))_{n \geq 1}$ is a sequence of real numbers and ω can be viewed as a parameter, a "random" parameter in the case of a probability space. One can then consider the set of parameter values for which the sequence $(X_n(\omega))_{n \geq 1}$ converges.

Definition 4.3.1. *A sequence* $(X_n)_{n \geq 1}$ *of random variables* **converges P-a.s.** *to a random variable* X *(or* **P**-*a.s. to* X*) if* $\mathbf{P}(\{\omega \mid \lim_{n \to \infty} X_n(\omega) = X(\omega)\}) = 1$. *This will be denoted by writing* $X_n \xrightarrow{\mathbf{P}\text{-}a.s.} X$ *or* $X_n \xrightarrow{a.s.} X$ *if the context determines* $\mathbf{P}$. *A sequence* $(X_n)_{n \geq 1}$ *of measurable functions* **converges** *to a measurable function* X *a.e.* (**almost everywhere**) *on a σ-finite measure space* $(\Omega, \mathfrak{F}, \mu)$ *if* $\mu(\Lambda) = 0$, *where* $\Lambda = \{\omega \mid X_n(\omega)$ *does not tend to* $X(\omega)$ *as* $n \to \infty\}$. *This will be denoted by writing* $X_n \xrightarrow{\mu\text{-}a.e.} X$ *or* $X_n \xrightarrow{a.e.} X$ *if μ is defined by the context.*

On a probability space (and also for any bounded positive measure) the following result holds.

Proposition 4.3.2. *If* $X_n \xrightarrow{a.s.} 0$ *then, for any* $\epsilon > 0$, $\mathbf{P}(\{\omega \mid |X_n(\omega)| > \epsilon\}) \to 0$ *as* $n \to \infty$. *The converse is false, as shown by Example 4.3.3.*

Proof. Let $\Lambda = \{\omega \mid X_n(\omega) \to 0 \text{ as } n \to +\infty\}$ and let $\epsilon > 0$. If $\Gamma(\epsilon, N) = \{\omega \mid |X_n(\omega)| < \epsilon \text{ for all } n \geq N\}$, then $\Gamma(\epsilon, N) \subset \Gamma(\epsilon, N + 1)$ and $\Lambda \subset \cup_{N=1}^{\infty} \Gamma(\epsilon, N)$, since for each $\omega \in \Lambda$ there is a first time $N = N(\omega)$ after which $|X_n(\omega)|$ is never $\geq \epsilon$.

Let $\delta > 0$. Since $\mathbf{P}(\Lambda) = 1$, there is an N_ϵ such that $\mathbf{P}(\Gamma(\epsilon, N)) > 1 - \delta$ if $N \geq N_\epsilon$. Now $\{\omega \mid |X_n(\omega)| \geq \epsilon\} \subset \Gamma(\epsilon, N)^c$ if $n \geq N$. Hence, $n \geq N_\epsilon$ implies that $\mathbf{P}(\{\omega \mid |X_n(\omega)| > \epsilon\}) \leq \delta$. □

Remark. The basic idea of the above proof is that $\mathbf{P}(\Gamma(\epsilon, N))$ is close to 1 if ϵ is small and N is large enough (in terms of ϵ). By modifying it to let ϵ tend to 0 appropriately, one obtains a set Λ with $\mathbf{P}(\Lambda) < \epsilon$ such that, on Λ, the sequence converges uniformly to zero. This result, known as **Egororov's theorem**, is used to prove Lusin's theorem (Exercise 4.7.9) (see also Exercise 4.2.1). See Exercise 4.7.8 for detailed hints on Egorov's theorem.

Example 4.3.3. Let $\Omega = [0,1], \mathfrak{F} = \mathfrak{B}(\mathbb{R})$, and $\mathbf{P}(dx) = dx$. For each n, divide $[0, 1]$ into 2^n closed intervals $I_n(k)$ of equal length. Enumerate the random variables $1_{I_n(k)} = 1_{[k/2^n, (k+1)/2^n]}$ in a sequence by first listing the intervals $I_2(k)$, then the intervals $I_3(k)$, and so on. Let the sequence of random variables that results be denoted by $(X_n)_{n \geq 1}$. Then, if $0 < \epsilon < 1$, $\mathbf{P}[|X_n| \geq \epsilon]) < \frac{1}{2^\ell}$ if $n \geq \sum_{i=0}^{\ell-1} 2^i$. Hence, this probability tends to zero as $n \to \infty$. On the other hand, $\mathbf{P}(\{x \mid \lim_{n \to \infty} X_n(x) \text{ exists}\}) = 0$ as one sees by considering what happens for x irrational.

Proposition 4.3.2 motivates the next definition.

Definition 4.3.4. *A sequence* $(X_n)_{n \geq 1}$ *of random variables* **converges in probability** *to a random variable X if, for any $\epsilon > 0$,*

$$\mathbf{P}\big[|X_n - X| \geq \epsilon\big] \to 0 \text{ as } n \to \infty.$$

This will be denoted by writing $X_n \xrightarrow{pr} X$.

A sequence $(X_n)_{n \geq 1}$ *of measurable functions* **converges in measure** *to a measurable function X if for any $\epsilon > 0$*

$$\mu(\{|X_n - X| \geq \epsilon\}) \to 0 \text{ as } n \to \infty.$$

This will be denoted by writing $X_n \xrightarrow{m} X$.

Proposition 4.3.2 states that convergence a.s. to zero implies convergence to zero in probability, and Example 4.3.3 shows that the converse is false.

Remark. On a general measure space, the result that corresponds to Proposition 4.3.2 is that if E is a set with finite measure and $X_n \xrightarrow{a.e.} 0$, then $X_n \xrightarrow{m} 0$ on E, i.e., $\mu(\{\omega \in E \mid |X_n(\omega)| > \epsilon\} \to 0$ for all $\epsilon > 0$. It amounts to restricting the measure to E and using the resulting bounded measure where $\Omega = E, \mathfrak{G} = \{A \in \mathfrak{F} \mid A \subset E\}$, and the measure being μ restricted to $\mathfrak{G}$. Without the requirement that $\mu(E) < \infty$, Proposition 4.3.2 is false. Lebesgue measure on the real line gives an example: let $X_n = 1_{[n, +\infty)}$.

While convergence of $(X_n)_{n \geq 1}$ in measure does not imply convergence a.e., it does imply that a subsequence converges a.e., as stated in the next exercise.

3. POINTWISE CONVERGENCE

Exercise 4.3.5. Let $(X_n)_{n\geq 1}$ be a sequence of random variables for which $X_n \xrightarrow{pr} 0$. Prove that there is a subsequence $(X_{n_k})_{k\geq 1}$ such that $X_{n_k} \xrightarrow{a.s.} 0$ by showing the following:
  (1) there is a sequence $(n_k)_{k\geq 1}$ of integers with $\mathbf{P}\big[|X_{n_k}| \geq \frac{1}{2^k}\big] < \frac{1}{2^k}$;
  (2) if $\epsilon > \frac{1}{2^\ell}$, $\{\sup_{k\geq \ell}|X_{n_k}| \geq \epsilon\} \subset \cup_{k=\ell}^\infty\{|X_{n_k}| \geq \frac{1}{2^k}\}$;
  (3) if $\epsilon > \frac{1}{2^\ell}$, $\mathbf{P}\big[\sup_{k\geq \ell}|X_{n_k}| \geq \epsilon\big] \leq \sum_{k=\ell}^\infty \frac{1}{2^k} = \frac{1}{2^{\ell-1}}$.

Conclude that $\lim_{k\to\infty} X_{n_k} \to 0$ a.s.
[Hint: if $X_{n_k}(\omega) \not\to 0$, for some $m \geq 1$, $\omega \in \{\sup_{k\geq k_0}|X_{n_k}| \geq \frac{1}{m}\}$ for all large k_0; alternatively, use the Borel–Cantelli Lemma.]

Proposition 4.3.6. (**Chebychev's inequality**) Let $X \in L^p, 1 \leq p < +\infty$. Then

$$\mathbf{P}\big[|X| \geq \epsilon\big] \leq \frac{\|X\|_p^p}{\epsilon^p}.$$

Proof. $\|X\|_p^p = \int |X|^p d\mathbf{P} \geq \int_{\{|X|\geq \epsilon\}} |X|^p d\mathbf{P} \geq \epsilon^p \mathbf{P}\big[|X| \geq \epsilon\big]$. □

Corollary 4.3.7. If $(X_n)_{n\geq 1} \subset L^p$, and $X_n \xrightarrow{L^p} 0$, then $X_n \xrightarrow{pr} 0$ and, in the case of a measure space, $X_n \xrightarrow{m} 0$.

Proof. If $1 \leq p < \infty$, then for any $\epsilon > 0$, $\mathbf{P}[|X_n| \geq \epsilon] \leq \epsilon^{-p} \|X_n\|_p^p$, which tends to zero as $n \to \infty$.
  If $p = +\infty$, then $\|X_n\|_\infty < \epsilon$ if $n \geq n(\epsilon)$, and so $\mathbf{P}[|X_n| \geq \epsilon] = 0$ if $n \geq n(\epsilon)$. □

Exercise 4.3.8. Modify Example 4.3.3 to show that convergence in probability does not imply convergence in any $L^p, 1 \leq p \leq +\infty$. Use Example 4.3.3 to show that convergence in $L^p, 1 \leq p \leq \infty$, does not imply convergence a.s. On the other hand, show that convergence in L^∞ implies convergence a.s.

Remark. In §5 a closer connection between convergence in probability and in L^p is established using the concept of **uniform integrability**.

Exercise 4.3.9. Use Exercise 4.3.5 to show that if a sequence $X_n \xrightarrow{L^p} 0$ on a probability space $(\Omega, \mathfrak{F}, \mathbf{P})$, then a subsequence converges a.s. Show that Exercise 4.3.5 holds for arbitrary measure spaces and hence that if a sequence converges in L^p, a subsequence converges a.e.

In Chapter III independent random variables were defined, and the result on infinite products, Theorem 3.4.14, shows that, given a fixed probability $\mathbf{Q}$ on $(\mathbb{R}, \mathfrak{B}(\mathbb{R}))$, there are a probability space $(\Omega, \mathfrak{F}, \mathbf{P})$ and a sequence $(X_n)_{n\geq 1}$ of random variables defined on Ω such that they all have the common distribution $\mathbf{Q}$ and are independent (see Exercise 3.4.16). Such a sequence is called an **i.i.d. sequence**, i.e., an **independent and identically distributed sequence**. For such a sequence,

$E[(X_k - m)(X_\ell - m)] = 0, k \neq \ell$, where m is their common expectation if they are integrable. For example, if it is conceivable to toss a fair coin an infinite number of times, this "experiment" may be modeled by a sequence of i.i.d. random variables with common distribution $\mathbf{Q} = \frac{1}{2}\{\varepsilon_0 + \varepsilon_1\}$, where 1 corresponds to a head and 0 to a tail. The result of running the "experiment" is an infinite sequence of 0's and 1's, i.e., a point in an infinite product space.

One expects the average number of 1's in the sequence to tend to the common mean $m = \frac{1}{2}$ as $n \to \infty$. In practice, this average number is determined by computing the average of n samples of the population consisting of $\{0, 1\}$, i.e., computing the average number of heads after n tosses of the coin. The following proposition shows that this sample average converges to the population mean m in L^2.

Proposition 4.3.10. *Let $(X_n)_{n \geq 1}$ be a sequence of i.i.d. random variables. Assume that they have second moments, i.e., $(X_n)_{n \geq 1} \subset L^2$. Let m be their common expectation. Then the sequence $(\frac{S_n}{n})_{n \geq 1}$ converges to m in L^2, where $S_n = \sum_{i=1}^n X_i$.*

Proof. Let $U_n = S_n - nm = \sum_{k=1}^n (X_k - m) = \sum_{k=1}^n Y_k$, where $Y_k = X_k - m$. The random variables Y_k are in L^2, are independent (by Exercise 3.4.5) and have mean zero. Thus, $E[Y_k Y_\ell] = 0, k \neq \ell$. Now $U_n^2 = \sum_{k=1}^n Y_k^2 + 2 \sum_{k < \ell} Y_k Y_\ell$, and so $E[U_n^2] = \sum_{k=1}^n E[Y_k^2] = n\sigma^2$, where σ^2 is the common variance of the variables X_n.

Therefore, $\| U_n \|_2 = \sqrt{n}\sigma$ and so $\| \frac{1}{n} U_n \|_2 = \frac{\sigma}{\sqrt{n}} \to 0$ as $n \to \infty$. □

Corollary 4.3.11. (**A weak law of large numbers**) *Let $(X_n)_{n \geq 1}$ be a sequence of i.i.d. random variables in L^2 (i.e., they have finite second moments). Then, if m denotes the common mean,*

$$\frac{S_n}{n} \xrightarrow{pr} m, \quad \text{where } S_n = \sum_{k=1}^n X_k,$$

i.e., for all $\epsilon > 0$,

$$\mathbf{P}\left[\left|\frac{S_n}{n} - m\right| \geq \epsilon\right] \to 0 \quad \text{as } m \to \infty.$$

Proof. It is a consequence of Proposition 4.3.10 and Corollary 4.3.7. □

Remarks. (1) A **weak law of large numbers** says that under suitable hypotheses on $(X_n)_{n \geq 1}$, $\frac{S_n}{n} \xrightarrow{pr} m$.

(2) A **strong law of large numbers** is a stronger statement. It says that under suitable hypotheses, $\frac{S_n}{n} \xrightarrow{a.s.} m$ (this is a stronger statement because of Proposition 4.3.2.) A simple strong law for L^2-random variables is given in Proposition 4.4.1. An even simpler one due to Borel, stated in

Exercise 4.7.19, has to do with normal numbers. Consider an arbitrary number x in $[0, 1]$. It has a decimal expansion $x = 0.a_1 a_2 \cdots a_n \cdots$. Fix a digit from 0 to 9, say 1, and count the number of occurrences of 1 among the first n digits a_i in the decimal expansion. Divide this number by n. One expects that this ratio will tend to $\frac{1}{10}$ as n tends to infinity. The number x is said to be a **normal number** if this is the case. Borel showed that a.s. every number in $[0, 1]$ is normal. It is a consequence of Borel's strong law (Exercise 4.7.19) and the fact that in a measure-theoretic sense, $([0, 1], \mathfrak{B}(\mathbb{R}), dx)$ is essentially isomorphic to an infinite product of finite probability spaces (Exercise 4.7.20).

These laws of large numbers correspond to what one thinks of intuitively when "sampling" many times and then taking the average of the sample. One expects that in some sense it should be close to the actual mean of the "population". For example, the students at a certain university have an average height. Select a student at random and then measure that person's height. Repeat this procedure a large number of times, say 100 times. The sample average is the average of the observed heights. One way to say that the observed average is close to the true average is to say that this is so in probability if n is large.

Of course, it is desirable to know something about the error that is being made, i.e., how well does $\frac{S_n}{n}$ approximate the mean m? The usual way in statistics makes use of confidence intervals: given an acceptable level of error $\alpha > 0$, determine $a < b$ so that with probability $1 - \alpha$ the random variable $\frac{\sqrt{n}}{\sigma}(\frac{S_n}{n} - m)$ lies in $[a, b]$. The reason for the scaling by $\frac{\sqrt{n}}{\sigma}$ is that, in general, one does not know the underlying distribution and hence the distribution of $\frac{S_n}{n} - m$. As a result, one is forced to rely upon the unit normal distribution to determine a and b (usually with $a = -b$). The reason for the use of the unit normal is that, by the central limit theorem (Theorem 6.7.4), its distribution is the limit (in the weak sense) of the distribution of $\frac{\sqrt{n}}{\sigma}(\frac{S_n}{n} - m)$ as $n \to \infty$ since the scaling by $\frac{\sqrt{n}}{\sigma}$ ensures that $\frac{\sqrt{n}}{\sigma}(\frac{S_n}{n} - m)$ has mean zero, variance one, and is in fact $\frac{1}{\sqrt{n}}$ times the sum of n i.i.d. random variables $(Y_k)_{k \geq 1}$ with mean zero and variance one: namely, $Y_k = \frac{1}{\sigma}(X_k - m)$, where σ is the common variance of the original random variables X_k (whose existence is an additional hypothesis). Having determined a and b to correspond to α by using the unit normal, it follows that $a \leq \frac{\sqrt{n}}{\sigma}(\frac{S_n}{n} - m) \leq b$ with an approximate probability of $1 - \alpha$. Hence, $\frac{\sigma a}{\sqrt{n}} \leq \frac{S_n}{n} - m \leq \frac{\sigma b}{\sqrt{n}}$ with approximate probability $1 - \alpha$. Another, but cruder, way to obtain estimates of error is by using Chebychev's inequality (Proposition 4.3.6) for the random variable $\frac{S_n}{n} - m$, as it gives an upper bound on the probability that the error is at most α.

(3) As Chung [C] points out, the hypothesis that $(X_n)_{n \geq 1}$ is i.i.d. can easily be weakened: provided the random variables are all in L^2, it suffices that they have mean zero, are orthogonal, and are bounded in L^2, i.e., their

L^2-norms are uniformly bounded. According to Chung, this result is due to Chebychev. In addition with these hypotheses, Rajchman proved that the strong law holds (see Proposition 4.4.1).

(4) The natural moment condition to impose on the X_n is that they are all in L^1, as then they have means. Khintchine's weak law (see Theorem 4.3.12) is the standard result in this case. While the random variables are required to be identically distributed, for independence one only requires it pairwise, i.e., for any two of the random variables.

Theorem 4.3.12. (Khintchine's weak law of large numbers). *Let $(X_n)_{n\geq 1}$ be a sequence of pairwise independent, identically distributed random variables on a probability space $(\Omega, \mathfrak{F}, \mathbf{P})$. If the random variables are integrable (i.e., $a = \int |x|\mathbf{Q}(dx) < \infty$ where $\mathbf{Q}$ is the common distribution), then $\frac{1}{n}\sum_{i=1}^{n} X_i \to m$ in probability, where $m = \int x\mathbf{Q}(dx)$ is the common mean. (Note that random variables $X_n, n \geq 1$, are said to be **pairwise independent** if any two are independent.)*

Proof I. (see Feller [F1], pp. 246–248) One can assume $m = 0$ (replace X_n by $X_n - m$ if $m \neq 0$). Let $\delta > 0$ be chosen (its value will be fixed later).

For each n truncate $X_1, \ldots, X_n$ by δn, i.e., for $1 \leq k \leq n$, define

$$U_{k,n}(\omega) = \begin{cases} X_k(\omega) & \text{if } |X_k(\omega)| \leq \delta n, \\ 0 & \text{if } |X_k(\omega)| > \delta n. \end{cases}$$

Let $V_{k,n} = X_k - U_{k,n}, 1 \leq k \leq n$.

Exercise 4.3.13. Let Y and Z be two random variables and let $X = Y + Z$. Show that

(1) $\mathbf{P}[|X| \geq 2\epsilon] \leq \mathbf{P}[|Y| \geq \epsilon] + \mathbf{P}[|Z| \geq \epsilon]$. [Hint: the triangle inequality (Exercise 1.1.16) implies that one of $|a|, |b| \geq \epsilon$ if $|a + b| \geq 2\epsilon$.]

If (Y_n) and (Z_n) are two sequences of random variables, show that

(2) $Y_n + Z_n \xrightarrow{pr} 0$ in probability if both $Y_n \xrightarrow{pr} 0$ and $Z_n \xrightarrow{pr} 0$.

In view of Exercise 4.3.13, it suffices, given $\eta > 0$, to show that for large n,

(1) $\mathbf{P}[\frac{1}{n}\sum_{k=1}^{n} U_{k,n} \geq \epsilon] < \eta$, and
(2) $\mathbf{P}[\frac{1}{n}\sum_{k=1}^{n} V_{k,n} \geq \epsilon] < \eta$.

First consider (1). Let $S'_n = \sum_{k=1}^{n} U_{k,n}$. Then

$$E[(S'_n)^2] = \sum_{k=1}^{n} E[U_{k,n}^2] + 2\sum_{i<j} E[U_{i,n}U_{j,n}]$$

$$= \sum_{k=1}^{n} E[U_{k,n}^2] + 2\sum_{i<j} E[U_{i,n}]E[U_{j,n}],$$

3. POINTWISE CONVERGENCE

as the $U_{k,n}$, $1 \leq k \leq n$, are pairwise independent by Exercise 3.4.5 since $U_{k,n} = \phi_n \circ X_k$, where

$$\phi_n(x) = \begin{cases} x & \text{if } |x| \leq \delta n, \\ 0 & \text{if } |x| > \delta n. \end{cases}$$

Now $E[U_{k,n}^2] = \int_{\{|x| \leq \delta n\}} x^2 \mathbf{Q}(dx)$ by Lemma 4.1.3, as $U_{k,n}^2 = \phi_n^2 \circ X_k$. Hence,

$$E[U_{k,n}^2] \leq \delta n \int_{\{|x| \leq \delta n\}} |x| \mathbf{Q}(dx) \leq a \delta n,$$

where $a = \int |x| \mathbf{Q}(dx)$. Also,

$$E[U_{i,n}] = \int_{\{|x| \leq \delta n\}} x \mathbf{Q}(dx) = \int \phi_n(x) \mathbf{Q}(dx),$$

and $|\phi_n(x)| \leq |x| = f(x)$, where $f \in L^1(\mathbf{Q})$ and $\phi_n(x) \to x$ as $n \to \infty$. Dominated convergence (Theorem 2.1.38) implies that

$$\int \phi_n(x) \mathbf{Q}(dx) \to \int x \mathbf{Q}(dx) = 0 \quad \text{(by assumption) as } n \to \infty.$$

Hence, for large $n \geq n(\delta)$, $|E[U_{i,n}]| \leq a\delta$. Consequently,

$$E[(S'_n)^2] \leq n^2 a \delta + (n^2 - n) a^2 \delta^2 \leq 2 n^2 a \delta,$$

as one may assume that $a\delta < 1$. Hence, by Chebychev's inequality (Proposition 4.3.6),

$$\epsilon^2 \mathbf{P} \left[\frac{|S'_n|}{n} \geq \epsilon \right] \leq \frac{1}{n^2} \int_{\{|S'_n| \geq n\epsilon\}} (S'_n)^2 d\mathbf{P} \leq \frac{E[(S'_n)^2]}{n^2} \leq 2 a \delta.$$

Choose a positive number η and set $\delta = \delta(\epsilon, \eta) = \min\{\frac{\epsilon^2 \eta}{2a}, \frac{1}{a}\}$. Then, for $n \geq n(\delta)$,

(†) $$P\left[\frac{|S'_n|}{n} \geq \epsilon \right] = P\left[\frac{|S'_n(\delta)|}{n} \geq \epsilon \right] \leq \frac{2 a \delta}{\epsilon^2} < \eta,$$

as long as the truncation parameter $\delta < \delta(\epsilon, \eta)$.

It remains to prove (2). Let $S''_n = \sum_{k=1}^n V_{k,n}$. Now

$$\left\{ \omega \mid \frac{|S''_n(\omega)|}{n} \geq \epsilon \right\} \subset \left\{ \frac{S''_n}{n} \neq 0 \right\} \subset \cup_{k=1}^n \{V_{k,n} \neq 0\},$$

and so

$$\mathbf{P}\left[\frac{|S''_n|}{n} > \epsilon \right] \leq \sum_{k=1}^n \mathbf{P}\{V_{k,n} \neq 0\} = n \int_{\{|x| > \delta n\}} \mathbf{Q}(dx)$$

$$\leq \frac{1}{\delta} \int_{\{|x| > \delta n\}} |x| \mathbf{Q}(dx).$$

As $n \to \infty$, this term tends to zero either by dominated convergence or by Exercise 2.4.2, as $E \to \int_E |x| \mathbf{Q}(dx)$ is a measure on $\mathcal{B}(\mathbb{R})$ and $\bigcap_n \{|x| > \delta n\} = \emptyset$. □

Remark. This proof does not show that either $\frac{1}{n}\sum_{k=1}^{n} U_{k,n} \xrightarrow{pr} 0$ or $\frac{1}{n}\sum_{k=1}^{n} V_{k,n} \xrightarrow{pr} 0$ as $n \to \infty$, since the truncation procedure changes with n and so affects these two expressions.

Proof II. (see Chung [C], pp. 109–110). The subtlety of proof I lies in the idea of repeated truncation, i.e., for each n, one truncates $X_1, \ldots, X_n$ at height δn.

As in Proof I, one can assume $E[X_n] = 0$ for all n. In this proof, each variable is truncated once. Let

$$Y_n(\omega) = \begin{cases} X_n(\omega) & \text{if } |X_n(\omega)| \leq n, \\ 0 & \text{otherwise.} \end{cases}$$

Step (1). Let $A_n = \{X_n \neq Y_n\}$. If $\sum_{n=1}^{\infty} \mathbf{P}(A_n) < +\infty$, then by the Borel–Cantelli lemma (Proposition 4.1.29 (1)), a.s. $Y_n(\omega) = X_n(\omega)$ for sufficiently large n.

Now $\mathbf{P}(A_n) = \mathbf{P}(\{|X_n| > n\}) = \int_{\{|x|>n\}} \mathbf{Q}(dx)$. Consider $\mathbf{Q} \times$ (counting measure) on $\mathbb{R} \times \mathbb{N}$. Let $A = \cup_{n=1}^{\infty} \{(-\infty, -n] \cup [n, +\infty)\} \times \{n\}$. By Fubini's theorem (Theorem 3.3.16), $\sum_{n=1}^{\infty} \mathbf{P}(A_n) = \sum_{n=1}^{\infty} \left(\int_{\{|x|>n\}} \mathbf{Q}(dx) \right) < +\infty$ since $\int 1_A d\mathbf{Q} \times dn = \sum_{n=1}^{\infty} \left(\int_{\{|x|>n\}} \mathbf{Q}(dx) \right) = \int n 1_{\{n<|x|\leq n+1\}} \mathbf{Q}(dx) \leq \int |x| \mathbf{Q}(dx) < +\infty$. Alternatively, one may use Exercise 4.7.1 (6) to show that $\sum_{n=1}^{\infty} \mathbf{P}(A_n) < \infty$. Hence, a.s. $\frac{1}{n}\sum_{k=1}^{n}\{X_k(\omega) - Y_k(\omega)\} \to 0$, since a.s. there are only a finite number of non-zero terms $X_k(\omega) - Y_k(\omega)$, where the number of terms depends on ω. Therefore, if $\frac{1}{n}\sum_{k=1}^{n} Y_k \xrightarrow{pr} 0$, then $\frac{1}{n}\sum_{k=1}^{n} X_k \xrightarrow{pr} 0$ by Exercise 4.3.13 (2), i.e., the weak law of large numbers holds.

Step (2). Here one shows that $\frac{1}{n}\sum_{k=1}^{n} Y_k \xrightarrow{pr} 0$. Let $m_k = E[Y_k] = \int_{\{|x|\leq k\}} x \mathbf{Q}(dx)$. Then, by the theorem of dominated convergence (Theorem 2.1.38), $m_k \to m = 0$ as $k \to \infty$.

Exercise 4.3.14. Let $m_n \to L$ as $n \to \infty$. Then $\frac{m_1 + m_2 + \cdots + m_n}{n} \to L$ as $n \to \infty$. In other words, if m_n converges to L, then it converges to L in the sense of **Cesaro**. [*Hint:* choose N so large that $|m_n - L| < \varepsilon$ if $n \geq N$. Observe that $\frac{m_1 + m_2 + \cdots + m_n}{n} = \frac{m_1 + m_2 + \cdots + m_N}{n} + \frac{m_{N+1} + \cdots + m_n}{n}$.]

Exercise 4.3.14 shows that $\frac{1}{n}\sum_{k=1}^{n} E[Y_k] \to 0$. Thus, to show that $\frac{1}{n}\sum_{k=1}^{n} Y_k \xrightarrow{pr} 0$, it suffices to show that $T_n = \frac{1}{n}\sum_{k=1}^{n}\{Y_k - E[Y_k]\} \xrightarrow{pr} 0$ as $n \to \infty$. To do this, one estimates the variance $E[T_n^2]$ of the expression T_n, making use of the independence of the centered random variables $Y_k - E[Y_k]$. One has

$$E[T_n^2] = \frac{1}{n^2}\left[\sum_{k=1}^{n} E[\{Y_k - E[Y_k]\}^2] + 2\sum_{i<j} E[(Y_i - E[Y_i])(Y_j - E[Y_j])]\right]$$

$$= \frac{1}{n^2}\sum_{k=1}^{n} \sigma^2(Y_k) \leq \frac{1}{n^2}\sum_{k=1}^{n} E[Y_k^2] = \frac{1}{n^2}\sum_{k=1}^{n} \int_{\{|x|\leq k\}} x^2 \mathbf{Q}(dx).$$

3. POINTWISE CONVERGENCE

The argument now continues as in Chung [C].

Choose $(a_n)_{n\geq 1}$ a sequence of integers $\to +\infty$ such that $\frac{a_n}{n} \to 0$; e.g., $a_n = [\log n] + 1$, where $[x]$ is the greatest integer $\leq x$. Then,

$$\sum_{k=1}^{n} \int_{\{|x|\leq k\}} x^2 \mathbf{Q}(dx) = \sum_{k=1}^{a_n} \int_{\{|x|\leq k\}} x^2 \mathbf{Q}(dx) + \sum_{k=a_n+1}^{n} \int_{\{|x|\leq k\}} x^2 \mathbf{Q}(dx).$$

The second term

$$\sum_{k=a_n+1}^{n} \int_{\{|x|\leq k\}} x^2 \mathbf{Q}(dx)$$

$$(*) \qquad \leq \sum_{k=a_n+1}^{n} \left[a_n \int_{\{|x|\leq a_n\}} |x| \mathbf{Q}(dx) + \int_{\{a_n<|x|\leq k\}} x^2 \mathbf{Q}(dx) \right],$$

and the first term

$$\sum_{k=1}^{a_n} \int_{\{|x|\leq k\}} x^2 \mathbf{Q}(dx) \leq \sum_{k=1}^{a_n} \int_{\{|x|\leq a_n\}} a_n |x| \mathbf{Q}(dx).$$

Combining this with the first expression in $(*)$, it follows that

$$\sum_{k=1}^{n} \int_{\{|x|\leq k\}} x^2 \mathbf{Q}(dx) \leq na_n \int |x| \mathbf{Q}(dx) + n^2 \int_{\{a_n<|x|\}} |x| \mathbf{Q}(dx).$$

Hence,

$$\frac{1}{n^2} \sum_{k=1}^{n} E[Y_k^2] \leq \frac{a_n}{n} \left[\int |x| \mathbf{Q}(dx) \right] + \int_{\{a_n<|x|\}} |x| \mathbf{Q}(dx).$$

The first term goes to zero as $n \to \infty$ by the choice of the sequence a_n. The second goes to zero as $\bigcap_{n=1}^{\infty} \{a_n < |x|\} = \emptyset$, and $E \to \int_E |x| \mathbf{Q}(dx)$ is a measure on $B(\mathbb{R})$ by Exercise 2.4.2, or by dominated convergence. $\square$

A very famous application of the weak law of large numbers is S. Bernstein's proof of the **Weierstrass approximation theorem** on $[0, 1]$. One form of the Weierstrass approximation theorem states that every continuous function on $[0, 1]$ can be uniformly approximated by a polynomial. Bernstein's proof gives an explicit sequence of polynomials (constructed from the function f) that converge uniformly (Definition 3.6.15) to f.

Theorem 4.3.15. (S. Bernstein) Let f be a continuous function on $[0, 1]$. Then the sequence of polynomials $(P_n)_{n\geq 1}$ converges uniformly to f, where

$$P_n(x) = \sum_{k=0}^{n} f\left(\frac{k}{n}\right) \binom{n}{k} x^k (1-x)^{n-k} \quad \text{(the nth Bernstein polynomial)}.$$

Proof. Let $x \in [0,1]$ and let $(X_n)_{n \geq 1}$ be a sequence of i.i.d. random variables on a probability space $(\Omega, \mathfrak{F}, \mathbf{P})$ with $\mathbf{P}[X_n = 1] = x$ and $\mathbf{P}[X_n = 0] = 1 - x$. The weak law of large numbers in Corollary 4.3.11 applies (as does the far simpler strong law of Borel (Exercise 4.7.19)) and so $\frac{S_n}{n} = \frac{1}{n} \sum_{i=1}^n X_i \to x$ in probability.

Since by Exercise 3.3.21, $\mathbf{P}[S_n = k] = \binom{n}{k} x^k (1-x)^{n-k}$, it follows from the hints in Exercise 4.1.11 that $E[f(\frac{S_n}{n})] = P_n(x)$. Now

$$|P_n(x) - f(x)| \leq E[|f\left(\frac{S_n}{n}\right) - f(x)|] = \int_{\{|\frac{S_n}{n} - x| \geq \epsilon\}} |f\left(\frac{S_n}{n}\right) - f(x)| d\mathbf{P}$$
$$+ \int_{\{|\frac{S_n}{n} - x| < \epsilon\}} |f\left(\frac{S_n}{n}\right) - f(x)| d\mathbf{P}.$$

The first integral $\leq 2 \| f \|_\infty \mathbf{P}[|\frac{S_n}{n} - x| \geq \epsilon]$, which tends to zero as n tends to ∞. It does so uniformly in x since by Chebychev's inequality (Proposition 4.3.6) and independence,

$$\mathbf{P}\left[\left|\frac{S_n}{n} - x\right| \geq \epsilon\right] \leq \frac{1}{\epsilon^2} E\left[\left\{\frac{S_n}{n} - x\right\}^2\right]$$
$$= \frac{1}{\epsilon^2 n^2} \sum_{i=1}^n E[(X_i - x)^2] = \frac{1}{n\epsilon^2} x(1-x) \leq \frac{1}{n\epsilon^2}$$

(see Exercise 4.1.11 for the last equality).

Since f is uniformly continuous (see Property 2.3.8 (3)), for any $\delta \geq 0$ there exists $\epsilon > 0$ such that $|f(x) - f(y)| < \delta$ if $|x - y| < \epsilon$, $0 \leq x, y \leq 1$. Hence, the second integral is at most δ. □

4. KOLMOGOROV'S INEQUALITY AND THE STRONG LAW OF LARGE NUMBERS

If a sequence of random variables converges a.s. to zero (say), then it converges in probability to zero (Proposition 4.3.2). Hence, convergence a.s. is "stronger" than convergence in probability. The following is a simple strong law. It is a modification of an even simpler argument that may be used to prove Borel's strong law for i.i.d. Bernoulli random variables (see Exercise 4.7.19).

Proposition 4.4.1. *Let $(X_n)_{n \geq 1}$ be a sequence of random variables in $L^2(\Omega, \mathfrak{F}, \mathbf{P})$. Assume that they are orthogonal, with mean zero, and are bounded in L^2, i.e., $E[X_n] = 0$ for all $n \geq 1$, $E[X_m X_n] = 0$ if $m \neq n$, and for some $M > 0$ one has $E[X_n^2] \leq M$ for all $n \geq 1$. Then $\frac{S_n}{n} \xrightarrow{a.s.} 0$.*

Proof. Orthogonality of the sequence implies that the square of the L^2-norm of S_n is dominated by Mn as $\| \sum_{k=1}^n X_k \|_2^2 = \sum_{k=1}^n \| X_k \|_2^2$.

Chebychev's inequality implies that if one looks at the sequence $(S_{k^2})_{k\geq 1}$ then, for any $\epsilon > 0$, one has

$$\sum_{k=1}^{\infty} \mathbf{P}[|S_{k^2}| \geq k^2\epsilon] \leq \sum_{k=1}^{\infty} \frac{\|S_{k^2}\|_2^2}{k^4\epsilon^2} \leq \sum_{k=1}^{\infty} \frac{M}{k^2\epsilon^2} < \infty.$$

By the Borel–Cantelli lemma (Proposition 4.1.29), $\frac{S_{k^2}}{k^2} \xrightarrow{a.s.} 0$. Given that this subsequence converges, one needs only show that the maximum, for $k^2 \leq n < (k+1)^2$, of the difference $k^{-2}|S_n - S_{k^2}| \xrightarrow{a.s.} 0$ as

$$\frac{|S_n|}{n} \leq \frac{|S_n|}{k^2} \leq \frac{|S_{k^2}|}{k^2} + \frac{|S_n - S_{k^2}|}{k^2} \quad \text{if } k^2 \leq n < (k+1)^2.$$

For each k, let $M_k \stackrel{\text{def}}{=} \max_{k^2 \leq n < (k+1)^2} |S_n - S_{k^2}|$. Then

$$M_k^2 \leq \sum_{n=k^2+1}^{k^2+2k} |S_n - S_{k^2}|^2 \quad \text{and so}$$

$$\|M_k\|_2^2 \leq \sum_{n=k^2+1}^{k^2+2k} \|S_n - S_{k^2}\|_2^2 \leq 4k^2 M.$$

One now uses Chebychev's inequality again in the same way to get

$$\sum_{k=1}^{\infty} \mathbf{P}[M_k \geq k^2\epsilon] \leq \sum_{k=1}^{\infty} \frac{\|M_k\|_2^2}{k^4\epsilon^2} \leq M \sum_{k=1}^{\infty} \frac{4}{k^2\epsilon^2} < \infty.$$

Hence, by the Borel–Cantelli lemma (Proposition 4.1.29), it follows that $\frac{M_k}{k^2} \xrightarrow{a.s.} 0$. □

Remark. The technique used in this proof of looking at subsequences is used in Exercise 4.7.19 to prove Borel's strong law, which itself is a special case of this result of Rajchman (see also Loève [L2], p. 19). The argument there is simpler because the random variables are uniformly bounded, which of course implies that they are bounded in L^2.

The classic strong law, which only requires first absolute moments, is due to Kolmogorov. The independence condition is more restrictive than in Khintchine's weak law as it requires the random variables to be i.i.d. and not merely identically distributed and pairwise independent.

Theorem 4.4.2. (Kolmogorov's strong law) *Let $(X_n)_{n\geq 1}$ be a sequence of i.i.d. random variables. Let $\mathbf{Q}$ be their common distribution. Assume $\int |x|\mathbf{Q}(dx) < \infty$, i.e., the X_n are all integrable.*
Then $\frac{S_n}{n} \to m$ a.s., where $m = \int x\mathbf{Q}(dx)$ and $S_n = \sum_{k=1}^n X_k$.

As in the earlier discussion of the weak law, by centering the random variables, one may assume $m = 0$.

In order to control the behavior of the sequence $\frac{S_n(\omega)}{n}$ for all large values of n, as ω varies, one uses the following generalization of Chebychev's inequality.

Theorem 4.4.3. (Kolmogorov's inequality) *Let $(X_k)_{1 \leq k \leq n}$ be a finite collection of independent (not necessarily identically distributed) random variables in L^2 with $E[X_k] = 0$ for all n. Then,*

$$\mathbf{P}\big[\max_{1 \leq k \leq n} |S_k| > \epsilon\big] \leq \frac{\sigma^2(S_n)}{\epsilon^2} = \frac{E[S_n^2]}{\epsilon^2}.$$

Proof. Fix ω, and let $\nu(\omega)$ denote the first time ℓ that $|S_\ell(\omega)| > \epsilon$. This defines a function ν of ω on $\Lambda \stackrel{\text{def}}{=} \{\omega \mid \max_{1 \leq k \leq n} |S_k(\omega)| > \epsilon\} = \{\omega \mid \nu(\omega) \leq n\}$. Further, $\{\omega \mid \nu(\omega) \leq k\} \in \mathfrak{F}_k = \sigma(\{X_i \mid 1 \leq i \leq k\})$. Consequently, $\Lambda_k \stackrel{\text{def}}{=} \{\nu = k\} \in \mathfrak{F}_k$.

As a result, using independence, one has

$$\int_{\Lambda_k} S_k(S_n - S_k)d\mathbf{P} = E[S_k 1_{\Lambda_k}(S_n - S_k)] = E[S_k 1_{\Lambda_k}]E[S_n - S_k] = 0$$

since $S_k 1_{\Lambda_k} \in \mathfrak{F}_k$ and $S_n - S_k \in \sigma(\{X_j \mid k+1 \leq j \leq n\})$, with $E[S_n - S_k] = 0$.

Now $\Lambda = \bigcup_{1 \leq k \leq n} \Lambda_k$ and

$$\sigma^2(S_n) = E[S_n^2] \geq \int_\Lambda S_n^2 d\mathbf{P} = \sum_{k=1}^n \int_{\Lambda_k} [S_k + (S_n - S_k)]^2 d\mathbf{P}$$
$$= \sum_{k=1}^n \int_{\Lambda_k} S_k^2 d\mathbf{P} + \int_{\Lambda_k} (S_n - S_k)^2 d\mathbf{P}$$

in view of what has been proved. Hence,

$$\sigma^2(S_n) \geq \sum_{k=1}^n \int_{\Lambda_k} S_k^2 d\mathbf{P} \geq \epsilon^2 \sum_{k=1}^n \mathbf{P}(\Lambda_k) = \epsilon^2 \mathbf{P}(\Lambda). \quad \square$$

Remarks. (1) For random variables in L^2, Chebychev's inequality states that

$$\mathbf{P}\big[|S_n| > \epsilon\big] \leq \frac{\sigma^2(S_n)}{\epsilon^2} = \frac{E[S_n^2]}{\epsilon^2}.$$

As $\{|S_n| > \epsilon\} \subset \{\max_{1 \leq k \leq n} |S_k| > \epsilon\}$, the inequality of Kolmogorov is stronger than that of Chebychev (Proposition 4.3.6).

(2) Kolmogorov's inequality is a special case of a more general one for martingales due to Doob (see Theorem 5.4.20) that involves stopping times.

4. KOLMOGOROV'S INEQUALITY

Proposition 4.4.4. (Kolmogorov's criterion) *(See Feller [F1].)* Let $(X_n)_{n\geq 1}$ be a sequence of independent (but not necessarily identically distributed) random variables with finite second moments.
If $\sum_{n=1}^{\infty} \frac{\sigma^2(X_n)}{n^2} < \infty$, then $\frac{S_n - E[S_n]}{n} \to 0$ a.s.

Proof. To begin with, it can be assumed as usual that $E[X_n] = 0$ for all n. Let $\epsilon > 0$ and let $A_\nu = \{\omega \mid \text{there exists } n \in (2^{\nu-1}, 2^\nu] \text{ with } \frac{|S_n(\omega)|}{n} > \epsilon\}$. Assume $\sum_{\nu=1}^{\infty} \mathbf{P}(A_\nu) < +\infty$. Then $\mathbf{P}(A_\nu \text{ i.o.}) = 0$ by the Borel–Cantelli lemma (Proposition 4.1.29) (1). Hence, for a given $\epsilon > 0$, a.s. there exists $\nu = \nu(\omega)$ such that $n \geq 2^{\nu(\omega)-1}$ implies that $\frac{|S_n(\omega)|}{n} \leq \epsilon$. Since $\epsilon > 0$ is arbitrary, this shows that a.s. for each $\omega \in \Omega$ the sequence $\frac{S_n(\omega)}{n} \to 0$: consider a sequence $\epsilon_k = \frac{1}{k}$, say, that tends to zero as $k \to \infty$.

Now Kolmogorov's inequality (Theorem 4.4.3) implies that $\sum_{\nu=1}^{\infty} \mathbf{P}(A_\nu) < +\infty$. To see this, first observe that Theorem 4.4.3 applied to $Y_1 = S_{2^{\nu-1}+1}, Y_k = X_{2^{\nu-1}+k}, 2 \leq k \leq 2^\nu - 2^{\nu-1}$, implies that

$$\mathbf{P}(A_\nu) = \mathbf{P}\left[\left\{\max_{2^{\nu-1}+1 \leq k \leq 2^\nu} \left|\frac{S_k}{k}\right| > \epsilon\right\}\right]$$

$$\leq \mathbf{P}\left[\left\{\max_{2^{\nu-1}+1 \leq k \leq 2^\nu} |S_k| > \epsilon 2^{\nu-1}\right\}\right]$$

$$\leq \frac{1}{\epsilon^2 2^{2\nu-2}} \sigma^2[S_{2^\nu}].$$

Therefore, since $\sigma^2(S_n) = \sum_{k=1}^{n} \sigma^2(X_k)$,

$$\sum_{\nu=1}^{\infty} \mathbf{P}(A_\nu) \leq \frac{4}{\epsilon^2} \sum_{\nu=1}^{\infty} \frac{1}{2^{2\nu}} \left[\sum_{k=1}^{2^\nu} \sigma^2(X_k)\right]$$

$$\leq \frac{4}{\epsilon^2} \sum_{k=1}^{\infty} \sigma^2(X_k) \left[\sum_{2^\nu \geq k} \frac{1}{2^{2\nu}}\right]$$

$$\leq \frac{6}{\epsilon^2} \sum_{k=1}^{\infty} \frac{\sigma^2(X_k)}{k^2} < \infty, \text{ as}$$

$$\sum_{2^\nu \geq k} \frac{1}{2^{2\nu}} = \frac{1}{2^{2\nu_0}} \frac{1}{1 - \frac{1}{4}} \leq \frac{4}{3k^2} \quad \text{if } 2^{\nu_0} \geq k \geq 2^{\nu_0-1}. \quad \square$$

Remark. Kolmogorov's criterion is trivially satisfied if the sequence is bounded in L^2. It therefore extends the simpler result (Proposition 4.4.1) of Rajchman when applied to an i.i.d. sequence $(X_n)_{n\geq 1}$ whose common mean is zero. Note that the criterion is a variation of the fact that $\sum_{k=1}^{\infty} \frac{1}{k^2} < +\infty$.

Proof of Theorem 4.4.2. (see Feller [F1], p. 260) As usual, one can assume the mean $m = 0$. Let

$$Y_n(\omega) = \begin{cases} X_n(\omega) & \text{if } |X_n(\omega)| \leq n, \\ 0 & \text{otherwise.} \end{cases}$$

Since the $(X_n)_{n\geq 1}$ are identically distributed, $\sum_{n=1}^\infty \mathbf{P}(\{X_n \neq Y_n\}) < \infty$. Hence $\frac{S_n}{n} \to 0$ a.s. if $\frac{1}{n}\sum_{i=1}^n Y_i \to 0$ a.s. (see Step (1) in Proof II of Khintchine's weak law (Theorem 4.3.12)). Assume the sequence $(Y_n)_{n\geq 1}$ satisfies Kolmogorov's criterion (Proposition 4.4.4). Then, $\frac{1}{n}\sum_{i=1}^n Y_i - \frac{1}{n}\sum_{i=1}^n E[Y_i] \to 0$ a.s. It follows from Exercise 4.3.14 that $\frac{1}{n}\sum_{i=1}^n E[Y_i] \to 0$ as $E[Y_n] \to 0$ when $n \to \infty$. Hence, $\frac{1}{n}\sum_{i=1}^n Y_i \to 0$ a.s.

It therefore remains to verify the hypothesis of Proposition 4.4.4 for the sequence $(Y_n)_{n\geq 1}$, in other words, that $\sum_{n=1}^\infty \frac{\sigma^2(Y_n)}{n^2} < \infty$. Now

$$\sigma^2(Y_n) \leq E[Y_n^2] = \int_{\{|x|\leq n\}} x^2 \mathbf{Q}(dx) = \sum_{k=1}^n \int_{\{k-1<|x|\leq k\}} x^2 \mathbf{Q}(dx)$$

$$\leq \sum_{k=1}^n k \int_{\{k-1<|x|\leq k\}} |x| \mathbf{Q}(dx).$$

Let $a_k = \int_{\{k-1<|x|\leq k\}} |x|\mathbf{Q}(dx)$. Then,

$$\sum_{n=1}^\infty \frac{\sigma^2(Y_n)}{n^2} \leq \sum_{n=1}^\infty \frac{1}{n^2} \sum_{k=1}^n k a_k = \sum_{k=1}^\infty k a_k \left[\sum_{n=k}^\infty \frac{1}{n^2}\right]$$

$$< 2\sum_{k=1}^\infty a_k = 2\int |x|\mathbf{Q}(dx) < \infty,$$

where the last inequality holds since $\sum_{n=k}^\infty \frac{1}{n^2} < \int_{k-1}^\infty \frac{dx}{x^2} = \frac{1}{k-1} \leq \frac{2}{k}$ if $k > 1$. □

5. Uniform Integrability and Truncation*

In the proofs of the laws of large numbers, truncation was used to reduce the study of sums of independent random variables to the case of bounded random variables. Truncation cuts the "tail" off the random variable, i.e., the "tail" of its distribution. Given a random variable X and a positive constant c, define its **truncation** X_c **at height** c by

$$X_c(\omega) = \begin{cases} X(\omega) & \text{if } |X(\omega)| \leq c, \\ 0 & \text{if } |X(\omega)| > c. \end{cases}$$

5. UNIFORM INTEGRABILITY AND TRUNCATION

Lemma 4.5.1. *Let $X \in L^1$ and $c > 0$. Then*
(1) $\| X - X_c \|_1 = \int_{\{|X|>c\}} |X| d\mathbf{P}$, *and*
(2) $\int_{\{|X|>c\}} |X| d\mathbf{P} \to 0$ *as* $c \to \infty$.

Proof. (1) is obvious and (2) follows from the dominated convergence theorem (Theorem 2.1.38) since $1_{\{|X|>c\}}|X| \to 0$ as $c \to \infty$. □

Remarks. (1) There is a corresponding L^p result, for all $p, 1 \leq p < \infty$. In what follows, things will be stated for L^1, and the reader is left to formulate the corresponding L^p statement (see Chung [C], p. 97 and Exercises 4.7.14, 4.7.15, and 4.7.16).

(2) It is also useful to use a **Lipschitz truncation** at height $c > 0$, namely $\phi_c \circ X$, where

$$\phi_c(x) = \begin{cases} -c & \text{if } x < -c, \\ x & \text{if } -c \leq x \leq c, \\ c & \text{if } c < x. \end{cases}$$

(3) The function ϕ_c is the "natural" extension of the function $f(x) = x, -c \leq x \leq c$ in the sense of Proposition 2.8.13: one merely extends with the constant value given at the endpoint.

For large $c > 0$, this Lipschitz truncation differs very little from the previous truncation X_c as the following exercise shows.

Exercise 4.5.2. Define

$$\tau_c(x) = \begin{cases} -c & \text{if } x < -c, \\ 0 & \text{if } -c \leq x \leq c, \\ c & \text{if } c < x. \end{cases}$$

Show that
(1) $|\phi_c(x) - \phi_c(y)| \leq |x - y|$ for all $x, y \in \mathbb{R}$,
(2) $X \in L^1$ implies $\phi_c \circ X \in L^1$ [Hint: $\phi_c(0) = 0$ and so $|\phi_c(x)| \leq |x|$.],
(3) $X_c + \tau_c \circ X = \phi_c \circ X$, and
(4) $\| X_c - \phi_c \circ X \|_1 \leq c\mathbf{P}[|X| > c] \leq \int_{\{|X|>c\}} |X| d\mathbf{P}$.

Given a sequence $(X_n)_{n \geq 1}$ of integrable random variables, truncating all the variables at a height $c > 0$ gives an approximation in L^1 to each term X_n of the sequence. When is this approximation uniform? One may also ask this question for the Lipschitz truncation at height c. The answer to these questions involves the concept of uniform integrability.

Definition 4.5.3. *A sequence $(X_n)_{n \geq 1}$ of random variables in $L^1(\Omega, \mathfrak{F}, P)$ is said to be **uniformly integrable** if, for $\epsilon > 0$, there is a constant $c = c(\epsilon) > 0$ such that*

$$\int_{\{|X_n|>c\}} |X_n| dP \leq \epsilon \text{ for all } n \geq 1.$$

It follows from Lemma 4.5.1 (1) and Exercise 4.5.2 that uniform integrability and uniform approximation by truncation at a large level are equivalent:

Lemma 4.5.4. *Let $(X_n)_{n\geq 1}$ be a sequence in L^1. The following are equivalent:*

(1) *for large c, the truncation of the sequence $(X_n)_{n\geq 1}$ at height c gives a bounded sequence that approximates the original sequence uniformly in L^1;*

(2) *for large c, the Lipschitz truncation of the sequence $(X_n)_{n\geq 1}$ at height c gives a bounded sequence that approximates the original sequence uniformly in L^1; and*

(3) *the sequence $(X_n)_{n\geq 1}$ is uniformly integrable.*

Remark. If $(X_n)_{n\geq 1} \subset L^p$, the truncations at height $c > 0$ approximate uniformly in L^p if and only if $(|X_n|^p)_{n\geq 1}$ is uniformly integrable. One could refer to this property by saying $(X_n)_{n\geq 1}$ is uniformly integrable in L^p, but this is not standard terminology.

The point of introducing uniformly integrable sequences is that it enables one to reduce convergence questions to questions about bounded sequences. In particular, it gives insight (Remark 4.5.10) into exactly which sequences converge in L^1 and thereby extends the theorem of dominated convergence (see Exercise 4.5.11).

In order to characterize uniformly integrable sequences, it is important to prove the following continuity lemma.

Lemma 4.5.5. *Let $X \in L^1$ and $A \in \mathfrak{F}$.*
Then
$$\int_A |X| d\mathbf{P} \to 0 \text{ as } \mathbf{P}(A) \to 0.$$

Proof. If $c > 0$,
$$\int_A |X| d\mathbf{P} = \int_{A \cap \{|X| \leq c\}} |X| d\mathbf{P} + \int_{A \cap \{|X| > c\}} |X| d\mathbf{P}$$
$$\leq c\mathbf{P}(A) + \int_{\{|X| > c\}} |X| d\mathbf{P}.$$

The result follows from Lemma 4.5.1. □

This lemma states that the measure $\nu(A) \stackrel{\text{def}}{=} \int |X| d\mathbf{P}$ is continuous with respect to $\mathbf{P}$ in the sense that if $\epsilon > 0$ then there is a $\delta > 0$ such that $\nu(A) < \epsilon$ if $\mathbf{P}(A) < \delta$.

Hence, given the Radon–Nikodym theorem (Theorem 2.7.19), it follows that if ν is absolutely continuous with respect to $\mathbf{P}$ (Definition 2.7.16), then ν is continuous with respect to $\mathbf{P}$ in the above sense. However, this observation does not depend upon the Radon–Nikodym theorem, as the following version of Lemma 4.5.5 shows

5. UNIFORM INTEGRABILITY AND TRUNCATION

Lemma 4.5.5*. (see Halmos [H1], p. 125) Let ν be a finite measure on $(\Omega, \mathfrak{F})$ that is absolutely continuous with respect to a probability $\mathbf{P}$. If $\epsilon > 0$, then there exists a $\delta > 0$ such that $\nu(A) < \epsilon$ if $\mathbf{P}(A) < \delta$.

Proof. Assume the statement is false. Then there is a positive number $\epsilon > 0$ and a sequence of sets $A_n \in \mathfrak{F}$ such that, for all $n \geq 1$, (i) $\nu(A_n) \geq \epsilon$ and (ii) $\mathbf{P}(A_n) < \frac{1}{2^n}$.

Let $B_m = \cup_{n \geq m} A_n$. Then, for all $m \geq 1$, one has $\nu(B_m) \geq \epsilon$ and $\mathbf{P}(B_m) \leq \sum_{n \geq m}^{\infty} \mathbf{P}(A_n) < (\frac{1}{2})^{m-1}$. Clearly, $B_m \supset B_{m+1}$ and so if $B = \cap_{m=1}^{\infty} B_m$, then (i) $\nu(B) \geq \epsilon$ and (ii) $\mathbf{P}(B) = 0$. This contradicts the hypothesis that ν is absolutely continuous with respect to $\mathbf{P}$. □

Combining this lemma with the definition of uniform integrability gives the following characterization of uniform integrability.

Proposition 4.5.6. $(X_n)_{n \geq 1}$ is uniformly integrable if and only if

(1) there is a constant M such that $E[|X_n|] \leq M$ for all $n \geq 1$, and
(2) $\int_A |X_n| d\mathbf{P} \to 0$ uniformly in n as $\mathbf{P}(A) \to 0$.

Proof. Assume that the sequence is uniformly integrable. Then $E[|X_n|] = \int_{\{|X_n| \leq c\}} |X_n| d\mathbf{P} + \int_{\{|X_n| > c\}} |X_n| d\mathbf{P} \leq c + 1$ if $c = c(1)$ in Definition 4.5.3. This establishes (1).

To show (2), observe that

$$\int_A |X_n| d\mathbf{P} = \int_{A \cap \{|X_n| \leq c\}} |X_n| d\mathbf{P} + \int_{A \cap \{|X_n| > c\}} |X_n| d\mathbf{P}$$

$$\leq c\mathbf{P}(A) + \int_{\{|X_n| > c\}} |X_n| d\mathbf{P}.$$

Hence, given $\epsilon > 0$, if $\mathbf{P}(A) < \epsilon/2c$ and $\int_{\{|X_n| > c\}} |X_n| d\mathbf{P} < \epsilon/2$, then one has $\int_A |X_n| d\mathbf{P} < \epsilon$ for all n.

Conversely, by (1), $c\mathbf{P}[|X_n| > c] \leq M$ and so $\mathbf{P}[|X_n| > c] \to 0$ uniformly in n as $c \to \infty$. Property (2) implies that $(X_n)_{n \geq 1}$ is uniformly integrable. More explicitly, if $\epsilon > 0$, condition (2) states that there is a $\delta > 0$ such that, for all $n \geq 1$, $\int_A |X_n| d\mathbf{P} < \epsilon$ if $\mathbf{P}(A) < \delta$. Since, for all n, $\mathbf{P}[|X_n| \geq c] < \delta$ if $c > \frac{M}{\delta}$, the result follows. □

Proposition 4.5.7. If $X_n \xrightarrow{L^1} X$, then

(1) $X_n \xrightarrow{pr} X$, and
(2) $(X_n)_{n \geq 1}$ is uniformly integrable.

Proof. (1) is a repetition of Corollary 4.3.7.
To prove (2), note that

$$\left| \int_A |X_n| d\mathbf{P} - \int_A |X| d\mathbf{P} \right| \leq \int_A |X_n - X| d\mathbf{P} \leq \| X_n - X \|_1 .$$

Let $\varepsilon > 0$, and choose δ_0 such that $\int_A |X| d\mathbf{P} < \varepsilon/2$ if $\mathbf{P}(A) < \delta_0$. Then $\int_A |X_n| d\mathbf{P} < \varepsilon, n \geq n(\varepsilon)$, provided $\| X_n - X \|_1 < \varepsilon/2$, for $n \geq n(\varepsilon)$ and $\mathbf{P}(A) < \delta_0$. Choose $\delta \leq \delta_0$ such that $\int_A |X_n| d\mathbf{P} < \varepsilon, 1 \leq n \leq n(\epsilon)$, if $\mathbf{P}(A) < \delta$. Then $\mathbf{P}(A) < \delta$ implies $\int_A |X_n| d\mathbf{P} < \epsilon$, for all $n \geq 1$. Since $\| X_n \|_1 \to \| X \|_1$, the result follows from Proposition 4.5.6. □

As stated in Exercise 4.3.8, convergence in probability does not necessarily imply convergence in L^1. However, as shown in Theorem 4.5.8, for uniformly bounded random variables, it does. Consequently, for a uniformly bounded sequence $(X_n)_{n\geq 1}$ of random variables X_n, the sequence converges in L^1 if and only if it converges in probability.

Theorem 4.5.8. *Let $(X_n)_{n\geq 1}$ be a sequence of uniformly bounded random variables (i.e., there is a constant M with $|X_n| \leq M$, for all $n \geq 1$). Let $1 \leq p < \infty$. If $X_n \xrightarrow{pr} X$, then $X_n \xrightarrow{L^p} X$.*

Proof. Since $\{|X| \geq M + 1/m\} \subset \{|X_n - X| \geq 1/m\}$, convergence in probability implies that $\mathbf{P}[|X| > M] = 0$. Hence, when integrating, one may assume $|X| \leq M$, and so

$$\int |X_n - X|^p d\mathbf{P} = \int_{\{|X_n - X| \leq \varepsilon\}} |X_n - X|^p d\mathbf{P} + \int_{\{|X_n - X| > \varepsilon\}} |X_n - X|^p d\mathbf{P}$$
$$\leq \varepsilon^p + 2^p M^p \mathbf{P}[|X_n - X| > \varepsilon]. \quad \square$$

Remark. It suffices for the above result that $X_n \in L^\infty$ with $\| X_n \|_\infty \leq M$ for all n.

This theorem has a corollary, which, as pointed out in Exercise 4.5.11, is a generalisation of the theorem of dominated convergence (Theorem 2.1.38).

Corollary 4.5.9. *Assume that X and X_n, $n \geq 1$, are in L^1. If*

(1) $X_n \xrightarrow{pr} X$, and

(2) $(X_n)_{n\geq 1}$ is uniformly integrable,

then

$$X_n \xrightarrow{L^1} X.$$

Proof. By Lemma 4.5.4, the Lipschitz truncation at height c given by $\phi_c \circ X_n$ gives a uniform approximation in L^1 to the X_n. Let $\varepsilon > 0$. Choose $c > 0$ so that $\| X_n - \phi_c \circ X_n \|_1 < \frac{\varepsilon}{3}, n \geq 1$, and $\| X - \phi_c \circ X \|_1 < \frac{\varepsilon}{3}$. Since $|\phi_c(x) - \phi_c(y)| \leq |x - y|$, for all $x, y \in \mathbb{R}$, it follows that $\{|X_n - X| > \eta\} \supset \{|\phi_c \circ X_n - \phi_c \circ X| > \eta\}$. Therefore, the sequence $(\phi_c \circ X_n)_{n\geq 1}$ converges in probability to $\phi_c \circ X$. It follows from Theorem 4.5.8 that $\phi_c \circ X_n \xrightarrow{L^1} \phi_c \circ X$.

Consequently, $\| X_n - X \|_1 \leq \| X_n - \phi_c \circ X_n \|_1 + \| \phi_c \circ X_n - \phi_c \circ X \|_1 + \| \phi_c \circ X - X \|_1 \leq \frac{2\varepsilon}{3} + \| \phi_c \circ X_n - \phi_c \circ X \|_1 \leq \varepsilon$ if $n \geq n(\varepsilon, c)$. □

5. UNIFORM INTEGRABILITY AND TRUNCATION 185

Remark 4.5.10. Conditions (1) and (2) in Corollary 4.5.9 imply that $X \in L^1$ (see Exercise 4.7.13). Hence, in view of Proposition 4.5.7, a sequence $(X_n)_{n \geq 1}$ of random variables in L^1 converges in L^1 if and only if (i) it converges in probability and (ii) it is uniformly integrable (Exercise 4.7.13).

Exercise 4.5.11. Let $(X_n)_{n \geq 1}$ be a sequence of random variables uniformly bounded by $Y \in L^1$, i.e., $|X_n| \leq Y$ for all $n \geq 1$. Show that

(1) the sequence $(X_n)_{n \geq 1}$ is uniformly integrable, and

(2) if, in addition, $X_n \xrightarrow{pr} X$, then $X_n \xrightarrow{L^1} X$.

Conclude that Corollary 4.5.9 is a generalization of the theorem of dominated convergence. In addition, show that

(3) the sequence $(X_n)_{n \geq 1}$ is uniformly integrable if there is a constant $M < \infty$ and $p > 1$ with $E[|X_n|^p] \leq M$ for all $n \geq 1$. [Hint: make use of Hölder's inequality when establishing Proposition 4.5.6 (2)].

Exercise 4.5.12. Show that if X and X_n, $n \geq 1$, are in L^p, then $X_n \xrightarrow{L^p} X$ if $X_n \xrightarrow{pr} X$ and $(X_n)_{n \geq 1}$ is uniformly integrable in L^p, i.e., $(|X_n|^p)_{n \geq 1}$ is uniformly integrable.

This discussion of uniform integrability concludes with a result stating that given convergence in probability, convergence occurs in L^1 when the L^1-norms converge. Thinking in Euclidean terms, this is obvious for a sequence of vectors in $\mathbb{R}^n$: they converge if and only if their components converge and their lengths converge. In fact the second condition is superfluous. However, in an infinite-dimensional space like L^1, where such things are not so evident, it is intuitively appealing that a sequence converges if it converges pointwise and the norms $\| X_n \|_1$, which determine the sphere in L^1 on which X_n lies, also converge. As pointwise convergence implies convergence in probability, the following result says something stronger: namely, convergence in probability suffices in order that convergence of the L^1-norm implies convergence in L^1.

Theorem 4.5.13. Assume that X and X_n, $n \geq 1$, are in L^1 and that $X_n \xrightarrow{pr} X$.

The following are equivalent:

(1) $\| X_n \|_1 \to \| X \|_1$; and

(2) $X_n \xrightarrow{L^1} X$.

Proof. It suffices to show that (1) implies (2) since the converse is clear. If (2) is false, then there is a subsequence with $\| X_{n_k} - X \|_1 \geq \delta$, $k \geq 1$, for some $\delta > 0$. Since every sequence converging in probability contains a subsequence that converges a.s. (Exercise 4.3.5), it follows that if (2) is false, there is a subsequence that converges a.s. for which $\| X_{n_k} - X \|_1 \geq \delta$, $k \geq 1$. This is impossible in view of the following result, Scheffé's lemma.

Proposition 4.5.14. (Scheffé's lemma) *Let $(X_n)_{n\geq 1}$ be a sequence in L^1 that converges a.s. to $X \in L^1$. Then $X_n \xrightarrow{L^1} X$ if $\| X_n \|_1 \to \| X \|_1$.*

Proof. First, assume that all the random variables are non-negative. As a result, on $\{X_n - X \leq 0\}$, $X - X_n \leq X$. Since $X_n - X \xrightarrow{a.s.} 0$ it follows from dominated convergence (Theorem 2.1.38) that $\int_{\{X_n - X \leq 0\}} (X_n - X) d\mathbf{P} \to 0$ as $n \to \infty$. The assumption that the random variables are non-negative, the identity

$$\int (X_n - X) d\mathbf{P} = \int_{\{X_n - X > 0\}} (X_n - X) d\mathbf{P} + \int_{\{X_n - X \leq 0\}} (X_n - X) d\mathbf{P},$$

and the hypothesis that $\| X_n \|_1 - \| X \|_1 = \int (X_n - X) d\mathbf{P} \to 0$, imply that $\int_{\{X_n - X > 0\}} (X_n - X) d\mathbf{P} \to 0$ as $n \to \infty$. Hence,

$$\| X_n - X \|_1 = \int_{\{X_n - X > 0\}} (X_n - X) d\mathbf{P} - \int_{\{X_n - X \leq 0\}} (X_n - X) d\mathbf{P} \to 0$$

as $n \to \infty$.

The general case follows by applying the result for non-negative random variables to the random variables X_n^+, X_n^- and X^+, X^-. This presupposes that if $\| X_n \|_1 = \int |X_n| d\mathbf{P} \to \int |X| d\mathbf{P} = \| X \|_1$, then $\int X_n^+ d\mathbf{P} \to \int X^+ d\mathbf{P}$ and $\int X_n^- d\mathbf{P} \to \int X^- d\mathbf{P}$ (i.e., $\| X_n^\pm \|_1 \to \| X^\pm \|_1$). Using Fatou's lemma (Proposition 2.1.25), one sees that, in any case,

$$\int X^\pm d\mathbf{P} \leq \liminf_n \int X_n^\pm d\mathbf{P}.$$

Since $\int (X_n^+ + X_n^-) d\mathbf{P} = \| X_n \|_1 \to \| X \|_1 = \int (X^+ + X^-) d\mathbf{P}$, it follows from the next exercise that $\int X_n^\pm d\mathbf{P} \to \int X^\pm d\mathbf{P}$. □

Exercise 4.5.15. *Let $(a_n)_{n\geq 1}$ and $(b_n)_{n\geq 1}$ be two real sequences. Let $a \leq \liminf_n a_n$ and $b \leq \liminf_n b_n$. Assume that $a_n + b_n \to a + b$. Show that $a_n \to a$ and $b_n \to b$. [Hint: make use of Exercise 2.1.10.]*

6. Differentiation: the Hardy–Littlewood maximal function*

The Hardy–Littlewood maximal function. If f is a continuous, real-valued function defined on $\mathbb{R}$, then it is clear that, for any symmetric interval $(x - h, x + h)$ about x with $h > 0$, the mean-value $\frac{1}{2h} \int_{x-h}^{x+h} f(u) du$ of f over this interval converges to $f(x)$ as $h \to 0$. Furthermore, the fundamental theorem of calculus states that

$$\frac{d}{dx} \int_a^x f(u) du = \lim_{h \downarrow 0} \frac{1}{h} \int_x^{x+h} f(u) du = \lim_{h \downarrow 0} \frac{1}{h} \int_{x-h}^x f(u) du = f(x).$$

6. DIFFERENTIATION: THE HARDY–LITTLEWOOD MAXIMAL FUNCTION

Since the continuous functions with compact support are dense in $L^1(\mathbb{R})$ (Theorem 4.2.5), it is natural to ask to what extent these results hold for an arbitrary function $f \in L^1$. Since an L^1-function can be modified on a set of measure zero without changing its integral, it is clear that the best one can hope for is a result a.e.

Suppose that $\varphi_n \xrightarrow{L^1} f$ with the functions $\varphi_n \in C_c(\mathbb{R})$. Then

$$\left| \frac{1}{2h} \int_{x-h}^{x+h} f(u)du - f(x) \right| \leq \left| \frac{1}{2h} \int_{x-h}^{x+h} \{f(u) - \varphi_n(u)\}du \right| +$$

$$\left| \frac{1}{2h} \int_{x-h}^{x+h} \varphi_n(u)du - \varphi_n(x) \right| + |\varphi_n(x) - f(x)|.$$

Let $\epsilon > 0$ and $E_\epsilon = \{x \mid \limsup_{h \downarrow 0} |\frac{1}{2h} \int_{x-h}^{x+h} f(u)du - f(x)| > \epsilon\}$. Since the second term goes to zero as φ_n is continuous, it is clear that $E_\epsilon \subset E_{\epsilon,1} \cup E_{\epsilon,2}$, where

$$E_{\epsilon,1} = \left\{ x \mid |\varphi_n(x) - f(x)| > \frac{\epsilon}{2} \right\} \text{ and }$$

$$E_{\epsilon,2} = \left\{ x \mid \limsup_{h \downarrow 0} \left| \frac{1}{2h} \int_{x-h}^{x+h} \{f(u) - \varphi_n(u)\}du \right| > \frac{\epsilon}{2} \right\}.$$

The first set $E_{\epsilon,1}$ has small measure for large n by Chebychev's inequality (Proposition 4.3.6) as $|E_{\epsilon,1}| \leq \frac{2}{\epsilon} \| \varphi_n - f \|_1$. To control the measure of E_ϵ, it suffices to show that the measure of $E_{\epsilon,2}$ can be made small. It is a remarkable fact that the measure of this set can be shown to be small by showing that a much larger set has small measure if $\| \varphi_n - f \|_1$ is small. This larger set is a sort of worst-case scenario: it is $\{x \mid \sup_{h > 0} \frac{1}{2h} \int_{x-h}^{x+h} |f(u) - \varphi_n(u)| du > \frac{\epsilon}{2}\}$, which certainly contains $E_{\epsilon,2}$ since $\int_{x-h}^{x+h} |f(u) - \varphi_n(u)| du \geq |\int_{x-h}^{x+h} f(u) - \varphi_n(u) du|$.

Definition 4.6.1. *Let $\psi \in L^1(\mathbb{R})$. The **Hardy–Littlewood maximal function** ψ^* of ψ is defined by setting*

$$\psi^*(x) = \sup_{h > 0} \frac{1}{2h} \int_{x-h}^{x+h} |\psi(u)| du.$$

If $\psi = f - \varphi_n$, then $\{x \mid \sup_{h>0} \frac{1}{2h} \int_{x-h}^{x+h} |f(u) - \varphi_n(u)| du > \frac{\epsilon}{2}\} = \{x \mid \psi^*(x) > \frac{\epsilon}{2}\}$. As will be shown later, the maximal function is not in L^1 and, so, one cannot use Chebychev's inequality to estimate the measure $\{x \mid \psi^*(x) > \frac{\epsilon}{2}\}$. However, ψ^* belongs to what is called weak $L^1(\mathbb{R})$.

Definition 4.6.2. *A measurable function is said to be in **weak $L^1(\mathbb{R})$** or to be of **weak type** (1,1) if there is a constant $c > 0$ such that, for all $\epsilon > 0$, one has*

$$|\{|f(x)| > \epsilon\}| \leq \frac{c}{\epsilon}.$$

Remark. The function defined by $f(x) = \frac{1}{x}, x \neq 0$, and $f(0) = 0$ is of weak type (1,1) but not integrable.

Not only does ψ^* belong to weak $L^1(\mathbb{R})$, but in fact, as proved by Hardy and Littlewood, one has the following inequality.

Proposition 4.6.3. *If $\psi \in L^1(\mathbb{R})$, there is a constant $c > 0$ independent of ψ such that for any $\epsilon > 0$*

$$|\{\psi^*(x) > \epsilon\}| \leq \frac{c}{\epsilon} \|\psi\|_1.$$

Combining all of this information gives a proof of the following famous result.

Theorem 4.6.4. (Lebesgue's differentiation theorem) *If $f \in L^1(\mathbb{R})$, then*

$$\lim_{h \downarrow 0} \frac{1}{2h} \int_{x-h}^{x+h} f(u) du = f(x) \text{ a.e.}$$

Proof. Fix $\epsilon > 0$, and let $\delta > 0$. If n is sufficiently large, $\|f - \varphi_n\|_1 < \min\{\frac{1}{2}\epsilon\delta, \frac{1}{c}\epsilon\delta\}$, and so $|E_{\epsilon,1}| < \delta$ and $|E_{\epsilon,2}| < \delta$. Hence, $|E_\epsilon| = 0$ for any $\epsilon > 0$. □

Remark. This theorem holds in $\mathbb{R}^n$ using the same proof with cubes centered at x replacing symmetric intervals (see Wheeden and Zygmund [W1], p. 100). It also holds with balls replacing cubes (see Stein and Weiss [S1], p. 60, Theorem 31.2 for a very general theorem of this type).

To prove the analogue of the fundamental theorem of calculus, a further refinement of Lebesgue's differentiation theorem is needed.

Definition 4.6.5. *If $f \in L^1(\mathbb{R})$, a point x is called a **Lebesgue point** of f if*

$$\lim_{h \downarrow 0} \frac{1}{2h} \int_{x-h}^{x+h} |f(u) - f(x)| du = 0.$$

*The collection of Lebesgue points of f is called the **Lebesgue set** of f.*

It is clear that at a Lebesgue point $\lim_{h \downarrow 0} \frac{1}{2h} \int_{x-h}^{x+h} f(u) du = f(x)$. It is not hard to prove using, the differentiation theorem that almost every point is a Lebesgue point of f (see Proposition 4.6.9). Using this fact one can prove the following theorem.

Theorem 4.6.6. *If $f \in L^1(\mathbb{R})$ and x is a Lebesgue point of f, then*

$$f(x) = \lim_{h \downarrow 0} \frac{1}{h} \int_x^{x+h} f(u) du \text{ and}$$

$$f(x) = \lim_{h \downarrow 0} \frac{1}{h} \int_{x-h}^x f(u) du.$$

Hence, for any $a \in \mathbb{R}$,
$$\frac{d}{dx}\int_a^x f(u)du = f(x) \quad \text{a.e.}$$

Proof. Since x is a Lebesgue point of f,
$$\left|\frac{1}{h}\int_x^{x+h} f(u)du - f(x)\right| \le \frac{1}{h}\int_x^{x+h} |f(u) - f(x)|du$$
$$\le \frac{1}{h}\int_{x-h}^{x+h} |f(u) - f(x)|du$$
$$= \frac{2}{2h}\int_{x-h}^{x+h} |f(u) - f(x)|du \to 0 \quad \text{as } h \downarrow 0.$$

As the proof for the other limit is essentially the same, the result follows from the fact, established in Proposition 4.6.9, that almost every point is a Lebesgue point of f. □

Corollary 4.6.7. *Let ν be a signed measure on $\mathfrak{B}(\mathbb{R})$ that is absolutely continuous with respect to Lebesgue measure dx. Let G be any function of bounded variation such that $\nu((a,b]) = G(b) - G(a)$ if $a < b$. Then, if $f \in L^1(\mathbb{R})$ is such that $f(x)dx = \nu(dx)$, it follows that $f(x) = G'(x)$ a.e.*

In particular, if ν is an absolutely continuous probability on $\mathfrak{B}(\mathbb{R})$, then $f(x) = F'(x)$ a.e., where F is the distribution function of ν. In other words, its distribution function F is a.e. differentiable and the derivative F' is its probability density function.

Proof. Up to a constant,
$$G(x) = \begin{cases} \int_0^x f(u)du & \text{if } 0 < x, \\ 0 & \text{if } x = 0, \\ \int_x^0 f(u)du) & \text{if } x < 0, \end{cases}$$

where $f \in L^1(\mathbb{R})$ is the Radon–Nikodym derivative of ν with respect to dx. It follows from Theorem 4.6.6 that $G'(x)$ exists and equals $f(x)$ a.e.

The case of a probability is now immediate. □

Remark 4.6.8. Since these differentiation results make use of the integrability of the function only over a closed, bounded interval, it follows that they all carry over to measurable functions that for each $a \in \mathbb{R}$ are integrable on $(a - \delta, a + \delta)$ for some $\delta = \delta(a) > 0$. These are the so-called **locally integrable** functions. The Heine–Borel theorem implies that a function is locally integrable if and only if it is integrable on any compact set (equivalently, any bounded set). Any non-zero constant function is locally integrable although not in L_1. If $f \in L^1$, then $|f - r|$ is locally integrable for any constant $r \ne 0$ and not integrable as $|f| + |f - r| \ge |r|$.

Proposition 4.6.9. *If $f \in L^1(\mathbb{R})$ or is even locally integrable, then almost every point is a Lebesgue point of f.*

Proof. Let $r \in \mathbb{Q}$. Then $|f - r|$ is locally integrable, and so by the differentation theorem (Theorem 4.6.4),

$$\lim_{h \downarrow 0} \frac{1}{2h} \int_{x-h}^{x+h} |f(u) - r| du = |f(x) - r| \text{ a.e.}$$

Since the rationals are countable, there is a set E with $|E| = 0$ such that

$$\lim_{h \downarrow 0} \frac{1}{2h} \int_{x-h}^{x+h} |f(u) - r| du = |f(x) - r| \text{ for all } x \notin E \text{ and all } r \in \mathbb{Q}.$$

Every point in the complement of E is a Lebesgue point of f since if $x \notin E$ and r is rational,

$$\limsup_{h \downarrow 0} \frac{1}{2h} \int_{x-h}^{x+h} |f(u) - f(x)| du$$

$$\leq \limsup_{h \downarrow 0} \frac{1}{2h} \int_{x-h}^{x+h} \{|f(u) - r| + |r - f(x)|\} du$$

$$= 2|f(x) - r|.$$

Since for $x \notin E$ the rational r may be arbitrary, it follows that r can be chosen with $|f(x) - r|$ arbitrarily small. □

It remains to do the hard work: establish the properties of maximal functions and the Hardy–Littlewood result (Proposition 4.6.3).

Maximal functions. One has the following more or less obvious properties of maximal functions.

Exercise 4.6.10. *Let $f, g, \ldots \in L^1$. Show that*
  (1) $f^* = |f|^*$,
  (2) *if $|f| \leq |g|$, then $f^* \leq g^*$, and*
  (3) *if $0 \leq f_n \uparrow f$, then $f_n^* \uparrow f^*$.*

To show that the maximal function is measurable, first observe that for $h > 0$, fixed the mean value $\frac{1}{2h} \int_{x-h}^{x+h} f(u) du$ is a continuous function of x. This follows immediately from dominated convergence: if $x_n \to x$, then $1_{[x_n - h, x_n + h]}(u) \to 1_{[x-h, x+h]}(u)$ as long as $u \neq x \pm h$ and, hence, a.e.

Exercise 4.6.11. *Let μ_h denote the uniform distribution on $[-h, h]$, i.e. $\mu_h(dx) = \frac{1}{2h} 1_{[-h,h]}(x) dx$. Show that*

$$\frac{1}{2h} \int_{x-h}^{x+h} f(u) du = (f * \mu_h)(x).$$

The maximal function $f^* = \sup_{h>0} (\check{f} * \mu_h)$ is the supremum of a family of continuous functions. As a result, it is lower semicontinuous (see Definition 4.6.12) and, hence, measurable, as the next exercise shows.

6. DIFFERENTIATION: THE HARDY–LITTLEWOOD MAXIMAL FUNCTION

Definition 4.6.12. *A function $\ell : \mathbb{R} \to \mathbb{R} \cup \{\pm\infty\}$ is defined to be* **lower semicontinuous** *at x_0 if for any $\epsilon > 0$ there is a $\delta > 0$ such that $\ell(x) > \ell(x_0) - \epsilon$ when $|x - x_0| < \delta$. It is said to be* **lower semi continuous** *if it is lower semicontinuous at every point.*

Exercise 4.6.13. Let ℓ be a lower semicontinuous function. Show that
(1) $\{x \mid \ell(x) > \lambda\}$ is open for any $\lambda \in \mathbb{R}$,
(2) any function satisfying (1) is lower semicontinuous.

Let $(\varphi_\alpha)_{\alpha \in I}$ be a family of continuous functions and let $\ell \stackrel{\text{def}}{=} \sup_\alpha \varphi_\alpha$. Show that

(3) ℓ satisfies (1) and so is lower semicontinuous,
(4) a lower semicontinuous function is Borel measurable.

Remark. This suggests that maximal functions can be profitably associated to any family $(\mu_t)_{t>0}$ of probabilities for which $\varphi * \mu_t \to \varphi$ pointwise, where φ is any continuous function of compact support, e.g., $\mu_t(dx) = n_t(x)dx$, the Gaussian semigroup. Such a family is called **an approximate identity** or **an approximation to the identity**. See Wheeden and Zygmund [W1] and Stein and Weiss [S1] for further details.

For any measurable set E, the mean value

$$\frac{1}{2h} \int_{x-h}^{x+h} 1_E(u) du = \frac{1}{2h} |E \cap [x-h, x+h]|.$$

Exercise 4.6.14. Let E be measurable and a subset of $[-N, N]$. Show that
(1) if $|x| > N$, then $1_E^*(x) \geq \frac{|E|}{2(|x|+N)} \geq \frac{1}{4} \frac{|E|}{|x|}$.
[*Hint*: for $x > N$, consider $[-N, 2x+N]$ the smallest symmetric interval about x containing $[-N, N]$.]

If $f \in L^1(\mathbb{R})$ and $f(x) = 0$ whenever $|x| > N$ (i.e., the support of f is a subset of $[-N, N]$), show that

(2) if $|x| > N$, then $f^*(x) \geq \frac{1}{4} \frac{\|f\|_1}{|x|}$.

Conclude that, if $f \in L^1(\mathbb{R})$ is not the zero function, then

(3) f^* is not integrable. [*Hint*: use Exercise 4.6.10 (2).]

To get an upper estimate of the maximal function of a bounded, measurable set, consider the case of a bounded interval $[a, b]$: if $x > b$, one has

$$|[a,b] \cap [x-h, x+h]| = \begin{cases} 0 & \text{if } h < x - b, \\ b - x + h & \text{if } x - b \leq h \leq x - a, \\ b - a & \text{if } x - a < h; \end{cases}$$

and if $x < a$, one has

$$|[a,b] \cap [x-h, x+h]| = \begin{cases} 0 & \text{if } h < a-x, \\ x+h-a & \text{if } a-x \leq h \leq b-x, \\ b-a & \text{if } b-x < h. \end{cases}$$

Exercise 4.6.15. Show that
(1) $1^*_{[a,b]}(x) \sim \frac{1}{|x|}$ for large x, i.e, there is a constant $c > 0$ such that $\frac{1}{c|x|} \leq 1^*_{[a,b]}(x) \leq \frac{c}{|x|}$ for large x,
(2) if E is bounded and measurable, then $1^*_E(x) \sim \frac{1}{|x|}$ for large x,
(3) if f has compact support (i.e., for some $N > 0$ its support is contained in $[-N, N]$), then $f^*(x) \sim \frac{1}{|x|}$ for large x, and
(4) if f has compact support, then $|\{f^* > \epsilon\}| < +\infty$ for any $\epsilon > 0$.

Proof of (4.6.3) (Hardy–Littlewood) that f^* is of weak type (1,1). Assume $f \in L^1$ has compact support and that $\epsilon > 0$. If $f^*(x) > \epsilon$, there is a closed interval $[x-h, x+h]$ with $\frac{1}{2h}\int_{x-h}^{x+h} |f(u)|du > \epsilon$. Hence, $\{f^* > \epsilon\}$ is contained in the union of such intervals, i.e., it is covered by these intervals.

Let $([x_i-h_i, x_i+h_i])_{1 \leq i \leq m}$ be a finite disjoint collection of these intervals. If $A = \cup_{i=1}^m [x_i - h_i, x_i + h_i]$, then

$$\| f \|_1 \geq \int_A |f(u)|du = \sum_{i=1}^m \int_{x_i-h_i}^{x_i+h_i} |f(u)|du \geq \sum_{i=1}^m \epsilon \, 2h_i = \epsilon|A|.$$

The key step is a part of the Vitali covering lemma (see Proposition 4.6.16): it implies that, since $|\{f^* > \epsilon\}| < +\infty$, there is a fixed number $\beta > 0$, independent of f and ϵ, such that the collection of disjoint intervals $[x_i - h_i, x_i + h_i]$ can be chosen so that $|A| \geq \beta|\{f^* > \epsilon\}|$. Taking $c = \beta^{-1}$ proves Proposition 4.6.3 in case the support of f is compact.

Now for any $f \in L^1$ there is a sequence $(g_n)_{n \geq 1}$ of non-negative functions with compact support such that $g_n \uparrow |f|$. Since $g_n^* \uparrow f^*$ by Exercise 4.6.10 (3), it follows that $\{g_n^* > \epsilon\} \subset \{g_{n+1}^* > \epsilon\}$ and $\{f^* > \epsilon\} = \cup_{n=1}^\infty \{g_n^* > \epsilon\}$. Since $\| g_n \|_1 \uparrow \| f \|_1$, the result follows as $\beta^{-1} = c$ is independent of the g_n and

$$|\{f^* > \epsilon\}| = \lim_{n \to \infty} |\{g_n^* > \epsilon\}| \leq \frac{c}{\epsilon} \lim_{n \to \infty} \| g_n \|_1 = \frac{c}{\epsilon} \| f \|_1. \quad \square$$

Remark. For a shorter proof of Proposition 4.6.3 see Folland [F3] (p. 91).

It remains to prove the covering lemma. Following Wheeden and Zygmund [W1], the Vitali covering lemma will be proved by fine-tuning the proof of the following simpler covering lemma, which suffices to complete Proposition 4.6.3.

Proposition 4.6.16. *Let E be a measurable set with $|E| < +\infty$. Assume that $\mathcal{I}$ is a collection of closed finite intervals I whose union contains E. Then, if $0 < \beta < \frac{1}{5}$, there is a finite disjoint $I_1, I_2, \ldots, I_m$ collection of these intervals such that*

$$\sum_{i=1}^m |I_i| \geq \beta |E|.$$

Proof. Consider the lengths of the intervals in $\mathcal{I}$. If they are unbounded, there is even one interval $I \in \mathcal{I}$ with $|I| \geq |E| \geq \beta|E|$.

If I is a closed finite interval, let $5I$ denote the closed interval with the same midpoint but 5 times its length. Then, if J is any closed interval that intersects I and has $|J| < 2|I|$, it follows that $J \subset 5I$.

Assume that the supremum ℓ_1 of the lengths of the intervals in $\mathcal{I}$ is finite. Choose $I_1 \in \mathcal{I}$ with $|I_1| > \frac{1}{2}\ell_1$. Then the union of all the intervals that intersect I_1 is a subset of $5I_1$. The remaining intervals are all disjoint from I_1. Let ℓ_2 be the supremum of their lengths, and choose $I_2 \in \mathcal{I}$ disjoint from I_1 with $|I_2| > \frac{1}{2}\ell_2$. Then $5I_1 \cup 5I_2$ contains all the intervals in $\mathcal{I}$ that intersect $I_1 \cup I_2$.

In this way, for each $m \geq 1$, one obtains disjoint intervals $I_1, I_2, \ldots, I_m$ such that all the intervals in $\mathcal{I}$ that intersect $I_1 \cup I_2 \cup \cdots \cup I_m$ are contained in $5I_1 \cup 5I_2 \cup \cdots \cup 5I_m$. Let ℓ_{m+1} be the supremum of the lengths of the intervals in $\mathcal{I}$ that are disjoint from $I_1, I_2, \ldots, I_m$.

It could happen that at some stage the process terminates. This means that, for some m, every interval intersects $I_1 \cup I_2 \cup \cdots \cup I_m$. As a result $E \subset \cup_{i=1}^m 5I_i$, and so $|E| \leq |\cup_{i=1}^m 5I_i| \leq 5\sum_{i=1}^m |I_i|$, which completes the proof in this case.

If the process never terminates, then either $\sum_{i=1}^\infty |I_i| = +\infty$, in which case some partial sum of this series will be larger than $|E|$ (which gives the result), or $\sum_{i=1}^\infty |I_i| < +\infty$.

If $\sum_{i=1}^\infty |I_i| < +\infty$, then $E \subset \cup_{i=1}^\infty 5I_i$: this is so if no $I \in \mathcal{I}$ is disjoint from all the I_i; if $I \in \mathcal{I}$ is disjoint from all the I_i, then $\ell_i \geq |I| > 0$ for all $i \geq 1$. But this is impossible as $\ell_i < 2|I_i|$ and $|I_i| \to 0$ as the series converges.

Since $E \subset \cup_{i=1}^\infty 5I_i$, it follows that $|E| \leq 5\sum_{i=1}^\infty |I_i|$. Hence, if $\beta^{-1} > 5$ is fixed, $|E| \leq \beta^{-1} \sum_{i=1}^m |I_i|$ for some $m \geq 1$, which proves the result. $\square$

Remark. It is not necessary that E be measurable in the above covering lemma. It suffices that it be a subset of a measurable set of finite measure, or equivalently, it have finite outer measure $\lambda^*(E)$. The statement of the result then involves $\lambda^*(E)$ rather than $|E|$.

The Vitali covering lemma: Differentiation of monotone functions.

In Corollary 4.6.7, it was shown that, if F is the distribution function of an absolutely continuous probability on $\mathbb{R}$, then it is differentiable a.e.

and $F(x) = \int_{-\infty}^{x} f(u)du$, where $f = F'$. It turns out that any distribution function is differentiable a.e., regardless of whether or not it is absolutely continuous. This is proved by using the Vitali covering lemma, which is a refinement of the covering lemma (Proposition 4.6.16).

Given any function F on (say) $[a, b]$ and a point $x \in (a, b)$, the difference quotient $\frac{F(x+h)-F(x)}{h} \stackrel{\text{def}}{=} DF(x; h)$ is defined for sufficiently small h. The function is differentiable at x if and only if the following four numbers agree and are finite:

$$DF^+(x) \stackrel{\text{def}}{=} \limsup_{h \downarrow 0} DF(x; h);$$

$$DF_+(x) \stackrel{\text{def}}{=} \liminf_{h \downarrow 0} DF(x; h);$$

$$DF^-(x) \stackrel{\text{def}}{=} \limsup_{h \uparrow 0} DF(x; h); \text{ and}$$

$$DF_-(x) \stackrel{\text{def}}{=} \liminf_{h \uparrow 0} DF(x; h).$$

These are the four so-called **Dini derivatives** of F at x. Note that

$DF^+(x) > q$ implies $DF(u; h) > q$ for arbitrarily small positive h;

$DF_+(x) < p$ implies $DF(u; h) < p$ for arbitrarily small positive h;

$DF^-(x) > q$ implies $DF(u; k) > q$ for arbitrarily small negative k; and

$DF_-(x) < p$ implies $DF(u; k) < p$ for arbitrarily small negative k,

where, for example, $DF(u; h) > q$ for arbitrarily small positive h means that, for any $\delta > 0$, there is an h with $0 < h < \delta$ such that $DF(u; h) > q$.

Since $DF_+(x) \leq DF^+(x)$ and $DF_-(x) \leq DF^-(x)$, the four Dini derivatives agree if and only if $DF^+(x) \leq DF_-(x)$ and $DF^-(x) \leq DF_+(x)$.

Now assume F is non-decreasing. Then the four Dini derivatives are all non-negative. Assume that $DF^+(x) > DF_-(x)$. Then there are two rational numbers $p < q$ with $DF^+(x) > q > p > DF_-(x)$, and the point x belongs to

$$E_{p,q} \stackrel{\text{def}}{=} \{u \in (a, b) \mid DF(u; h) > q \text{ for arbitrarily small positive } h\}$$
$$\cap \{u \in (a, b) \mid DF(u; k) < p \text{ for arbitrarily small negative } k\}.$$

Since there are a countable number of sets $E_{p,q}$ as $p < q$ run over the pairs of positive rational numbers, it follows that $DF^+(x) \leq DF_-(x)$ a.e. on (a, b) if $|E_{p,q}| = 0$ for any pair of positive rationals $E_{p,q}$.

In view of the following exercise, which essentially reverses the order on $\mathbb{R}$, if $DF^+(x) \leq DF_-(x)$ a.e. on (a, b) for any non-decreasing function F, then it also follows that $DF^-(x) \leq DF_+(x)$ a.e. on (a, b) and, hence, that the four Dini derivatives agree a.e. on (a, b).

Exercise 4.6.17. Let F be a non-decreasing, real-valued function on $[a, b]$. Define $G(u) = -F(-u)$. Show that
  (1) G is a non-decreasing function on $[-b, -a]$,
  (2) $DG(-x; -h) = DF(x; h)$ if $h \neq 0$,
  (3) $DG^-(-x) = DF^+(x)$,
  (4) $DG_-(-x) = DF_+(x)$,
  (5) $DG^+(-x) = DF^-(x)$, and
  (6) $DG_+(-x) = DF_-(x)$.

In order to prove that $|E_{p,q}| = 0$, one makes use of the coverings that are given by the intervals $[u, u+h]$, with $DF(u;h) > q$, and the intervals $[u+k, u]$, with $DF(u;k) < p$.

Let $\mathcal{V}_+$ be the collection of closed intervals $[u, u+h]$ where $u \in E_{p,q}$ and $h > 0$ is such that $DF(u;h) > q$. Let $\mathcal{V}_-$ be the collection of closed intervals $[u+k, u]$, where $u \in E_{p,q}$ and $k < 0$ is such that $DF(u;k) < p$. Then both collections of intervals cover $E_{p,q}$, and any point u of $E_{p,q}$ lies in an interval from $\mathcal{V}_\pm$ of arbitrarily small length. All of these intervals are non-trivial, i.e., they all contain more than one point, (equivalently, they all have positive length). In other words, the collections $\mathcal{V}_\pm$ are Vitali covers of $E_{p,q}$.

Definition 4.6.18. *A collection of (non-trivial) closed intervals is said to be a **Vitali cover** of a set E if their union contains E and for any $x \in E$ there is an interval in $\mathcal{V}$ of arbitrarily small length that contains x (the Vitali property).*

Remark 4.6.19. If $\mathcal{V}$ is a Vitali cover of a set E, and if O is an open set, then the Vitali property ensures that the intervals in $\mathcal{V}$ contained in O form a Vitali cover of $E \cap O$. This allows one to localize in a certain sense.

The **Vitali covering lemma** is the following theorem.

Theorem 4.6.20. *(Vitali) Let $\mathcal{V}$ be a Vitali cover of a subset E of $\mathbb{R}$. Then there is a pairwise disjoint countable family of intervals $(I_i)_{1 \leq i \leq N}$, $1 \leq N \leq \infty$, from $\mathcal{V}$ such that*

$$|E \backslash \{\cup_{i=1}^N I_i\}| = 0.$$

Further, if $0 < \lambda^*(E) < +\infty$ then, for each $\epsilon > 0$, there is a finite number $m \geq 1$ such that the outer measure

$$\lambda^*(E \backslash \{\cup_{i=1}^m I_i\}) < \epsilon \quad \text{and} \quad |\cup_{i=1}^N I_i| < (1+\epsilon)\lambda^*(E).$$

Proof. Let $E_n \stackrel{\text{def}}{=} E \cap (n, n+1)$, where $n \in \mathbb{Z}$. Then E differs from $\cup_n E_n$ by at most a countable set. In view of Remark 4.6.19, the intervals of $\mathcal{V}$ that lie in $(n, n+1)$ constitute a Vitali cover $\mathcal{V}_n$ of E_n. If the theorem holds for each E_n, then, by using $\frac{\epsilon}{2^{n+2}}$ for each E_n, the theorem follows for E: one merely puts together the collections of intervals obtained for each $n \in \mathbb{Z}$.

Since Lebesgue measure is translation-invariant, it suffices to prove the theorem when $E \subset (0,1)$.

Consider the proof of Proposition 4.6.16 for $E \subset (0,1)$, where by the Vitali property, one may assume that all the intervals in $\mathcal{V}$ are subsets of $(0,1)$. To begin with, $\ell_1 \leq 1$, so that the selection process needs to be used. Here this process is modified by cutting down the family of intervals used at each stage: after having selected $I_1, I_2, \ldots, I_m$, let $A_m \stackrel{\text{def}}{=} I_1 \cup I_2 \cup \cdots \cup I_m$; the next interval is taken from the Vitali cover of $E \backslash A_m$ given by the intervals of $\mathcal{V}$ that are subsets of the open set $(0,1) \backslash A_m$; and ℓ_{m+1} is taken to be the supremum of the lengths of these intervals. Now if the process of selection of the intervals I_i terminates at stage m, it means that there are no intervals in $\mathcal{V}$ disjoint from the closed set A_m. As a result, $E \subset A_m$ since the intervals in $\mathcal{V}$ contained in the open set $(0,1) \backslash A_m$ form a Vitali cover of $E \backslash A_m$ by Remark 4.6.19.

If the selection process never terminates, then $\sum_{i=1}^{\infty} |I_i| \leq 1 < +\infty$ and $E \subset \cup_{i=1}^{\infty} 5I_i$. Let $A = \cup_{i=1}^{\infty} I_i$. If $x \in E \backslash A$, then $x \in 5I_i$ infinitely often (i.e., $x \in \cup_{i=m+1}^{\infty} 5I_i$ for each $m \geq 1$): since $x \in E \backslash A_m$, by the Vitali property, there is an interval I in $\mathcal{V}$ containing x that is disjoint from A_m; this interval intersects some $A_k, k > m$, as $|I| > 0$ and $\ell_i \to 0$ as $i \to \infty$; hence, $x \in 5I_j$ if $j > m$ is the first time that $I \cap A_j \neq \emptyset$.

Now $|\cup_{i=m+1}^{\infty} 5I_i| \leq 5 \sum_{i=m+1}^{\infty} |I_i| \to 0$ as $m \to \infty$. Hence, $|E \backslash A| = 0$. This completes the proof of the first statement.

Since $E \backslash \{\cup_{i=1}^{m} I_i\} \subset \{\cup_{i=m+1}^{N} I_i\} \cup E \backslash A$, this also proves that, for any $\epsilon > 0$, there is an integer m with $\lambda^*(E \backslash \{\cup_{i=1}^{m} I_i\}) < \epsilon$.

Finally, if $0 < \lambda^*(E)$ let $\delta = \epsilon \lambda^*(E) > 0$. Since $E \subset (0,1)$, there is an open set O with $E \subset O \subset (0,1)$ and $|O| \leq \lambda^*(E) + \delta = (1+\epsilon)\lambda^*(E)$. Now replace the original Vitali cover by the Vitali cover of intervals in $\mathcal{V}$ that are subsets of O. The sequence $(I_i)_{1 \leq i \leq N}$ can then be taken from this new Vitali cover. Hence, $|A| = \sum_{i=1}^{N} |I_i| \leq |O| \leq (1+\epsilon)\lambda^*(E)$.

The last part of Vitali's theorem follows from what has been established for $E \subset (0,1)$: it suffices to observe that, in general, $\lambda^*(E) = \sum_n \lambda^*(E_n)$ is greater than zero if and only if at least one of the $\lambda^*(E_n) > 0$. Then one uses what has been proved to find for each $n \in \mathbb{Z}$ with $\lambda^*(E_n) > 0$ (i) an open set O_n with $E_n \subset O_n \subset (n, n+1)$ and $|O_n| \leq (1+\epsilon)\lambda^*(E_n)$ and (ii) a finite disjoint family $(I_{1 \leq i \leq N}^n)$ of intervals from the Vitali cover of E_n by intervals in $\mathcal{V}$ that are subsets of O_n such that

$$\lambda^*(E_n \backslash \{\cup_{i=1}^{m} I_i^n\}) < \frac{\epsilon}{2^{n+2}} \quad \text{and} \quad |\cup_{i=1}^{N} I_i^n| < (1+\epsilon)\lambda^*(E_n).$$

Then, putting all these disjoint collections together, one obtains the desired result. The fact that $\lambda^*(E) < +\infty$ implies that one can ignore all

6. DIFFERENTIATION: THE HARDY–LITTLEWOOD MAXIMAL FUNCTION   197

but a finite number of the sets E_n in putting together a finite subcollection $(I_j)_{1 \leq j \leq m}$ such that $\lambda^*(E \setminus \{\cup_{j=1}^m I_j\}) < \epsilon$. □

Returning to the question of the differentiability of a monotone function F, assume that $\lambda^*(E_{p,q}) > 0$ for some pair of positive rational numbers $p < q$. Let $\epsilon > 0$ and let O be an open subset of (a,b) containing $E_{p,q}$ such that $|O| < \lambda^*(E_{p,q}) + \epsilon$. The sets in the Vitali cover $\mathcal{V}_-$ of $E_{p,q}$ that are subsets of O are again a Vitali cover of $E_{p,q}$. Hence, there is a finite set of disjoint intervals $[x_i + k_i, x_i]$ in $\mathcal{V}_-$, $1 \leq i \leq m$, that are subsets of O with $\lambda^*(E_{p,q} \setminus \{\cup_{i=1}^m [x_i + k_i, x_i]\}) < \epsilon$. Note that this implies

$$\text{(1)} \qquad \sum_{i=1}^m |k_i| \leq |O| \leq \lambda^*(E_{p,q}) + \epsilon.$$

Let $U = \cup_{i=1}^m (x_i + k_i, x_i) \subset O$. The intervals in $\mathcal{V}_+$ that are subsets of U constitute a Vitali cover of $E_{p,q} \cap U$, and so there is a finite set of disjoint intervals $[y_j, y_j + h_j]$ in $\mathcal{V}_+$, $1 \leq j \leq n$, that are subsets of U with $\lambda^*(E_{p,q} \cap U \setminus \{\cup_{i=1}^n [y_j, y_j + h_j]\}) < \epsilon$. This implies that

$$\lambda^*(E_{p,q} \cap U) \leq \sum_{j=1}^n h_j + \epsilon,$$

and so

$$\text{(2)} \qquad \lambda^*(E_{p,q}) \leq \lambda^*(E_{p,q} \cap U) + \lambda^*(E_{p,q} \setminus U) \leq \sum_{j=1}^n h_j + 2\epsilon,$$

since $\lambda^*(E_{p,q} \setminus U) = \lambda^*(E_{p,q} \setminus \{\cup_{i=1}^m [x_i + k_i, x_i]\}) < \epsilon$.
From this, it follows that

$$\sum_{j=1}^n h_j = |\cup_{j=1}^n [y_j, y_j + h_j]| \leq |U| = |\cup_{i=1}^m [x_i + k_i, x_i]| = \sum_{i=1}^m |k_i|.$$

In addition, because $DF(x_j; h_j) > q$, $DF(y_i; k_i) < p$, and $\cup_{i=1}^m [x_i + k_i, x_i] \subset \cup_{j=1}^n (y_j, y_j + h_j)$, plus the assumption that F is non-decreasing, one has

$$q \sum_{j=1}^n h_j < \sum_{j=1}^n \{F(x_j + h_j) - F(x_j)\}$$

$$\text{(3)} \qquad \leq \sum_{i=1}^m \{F(y_i) - F(y_i + k_i)\} < p \sum_{i=1}^m |k_i|.$$

Let $h = \sum_{j=1}^{n} h_j$, $|k| = \sum_{i=1}^{m} |k_i|$, and $\alpha = \lambda^*(E_{p,q})$. Then by (1), (2), and (3), one has that

$$|k| \leq \alpha + \epsilon,$$
$$0 < \alpha \leq h + 2\epsilon, \text{ and}$$
$$qh < p|k|.$$

Assume that $\epsilon < \frac{\alpha}{2}$. Then

$$\frac{q}{p}(\alpha - 2\epsilon) < \alpha + \epsilon \quad \text{and so} \quad \frac{(q-p)}{(p+2q)}\alpha < \epsilon < \frac{1}{2}\alpha.$$

Therefore, the assumption that $\alpha = \lambda^*(E_{p,q}) > 0$ leads to a contradiction, since, in the above argument, ϵ may be arbitrarily small.

This completes most of the proof of the next result.

Theorem 4.6.21. (Lebesgue) *Let F be any real-valued, non-decreasing function on $[a,b]$. Then F is differentiable a.e. Further, if $f = F'$, then f is Lebesgue measurable and $\int_a^b f(u)du \leq F(b-) - F(a+)$.*

Proof. It remains to prove that a.e. the common value of the four Dini derivatives is finite, and that the resulting function f, which is defined a.e., is Lebesgue measurable with $\int_a^b f(u)du$ as specified.

Let

$$D_n F(x) = \begin{cases} n\{F(x + \frac{1}{n}) - F(x)\} & \text{if } a \leq x \leq b - \frac{1}{n}, \\ 0 & \text{if } b - \frac{1}{n} < x \leq b. \end{cases}$$

Then $D_n F(x) \geq 0$, and it converges to f a.e. by what has been proved. Fatou's lemma implies that

$$\int_a^b f(u)du \leq \liminf_n \int_a^b D_n F(u)du$$
$$= \liminf_n \left[n \int_{b-\frac{1}{n}}^b F(u)du - n \int_a^{a+\frac{1}{n}} F(u)du \right]$$
$$= F(b-) - F(a+).$$

This implies that the non-negative function f is a.e. finite and so F has a derivative a.e. on (a,b). □

Corollary 4.6.22. *Let F be a distribution function on $\mathbb{R}$. The derivative f of F is the density of the absolutely continuous measure in the Lebesgue decomposition of the probability $\mathbf{P}$ determined by F.*

Proof. Let $G(x) = F(x) - \int_{-\infty}^x f(u)du$. Then G is a non-decreasing function: $\{F(b) - F(a)\} - \int_a^b f(u)du \geq \{F(b) - F(a)\} - \{F(b-) - F(a+)\} =$

$F(b) - F(b-)$ as F is right continuous. Hence, the measure μ given by $\mu(A) = \int_A f(u)du$ is dominated by $\mathbf{P}$, i.e., for all A, $\mu(A) \leq \mathbf{P}(A)$.

On the other hand, by Theorem 2.7.22, the probability $\mathbf{P}(dx) = \phi(x)dx + \eta(dx)$ with η a unique singular finite measure. Let $H(x) = \eta((-\infty, x])$. Then, $F(x) = \int_{-\infty}^{x} \phi(u)du + H(x)$ and, so, a.e. $f(x) = \phi(x) + H'(x)$. Since H is non-decreasing, it follows that a.e. $f \geq \phi$. Thus, the difference $f - \phi$ defines a measure ν: set $\nu(A) = \int_A \{f(u) - \phi(u)\}du$. Since

$$\mathbf{P}(A) = \int_A \phi(u)du + \eta(A) \geq \int_A f(u)du = \int_A \phi(u)du + \nu(A),$$

it follows that ν is dominated by η and so ν is singular. This implies that $f = \phi$ a.e. since $\nu = 0$. □

7. ADDITIONAL EXERCISES*

Exercise 4.7.1. Let X be a random variable defined on a probability space $(\Omega, \mathfrak{F}, \mathbf{P})$. Show that

(1) $\sum_{n=0}^{\infty}(n+1)\mathbf{P}[n \leq |X| < (n+1)] = \sum_{n=0}^{\infty} \mathbf{P}[|X| \geq n]$.
 [Hints: observe that $\mathbf{P}[0 \leq |X|] = \mathbf{P}[0 \leq |X| < 1] + \mathbf{P}[1 \leq |X|]$, or use Fubini's theorem on the product of $(\mathbb{N} \cup \{0\}, \mathfrak{P}(\mathbb{N} \cup \{0\}), dn)$ and $(\Omega, \mathfrak{F}, \mathbf{P})$, where dn is counting measure.]

Use (1) to show that

(2) $\sum_{n=1}^{\infty} n\mathbf{P}[n \leq |X| < (n+1)] = \sum_{n=1}^{\infty} \mathbf{P}[|X| \geq n]$, and
(3) $\sum_{n=1}^{\infty} \mathbf{P}[|X| \geq n] \leq E[|X|] \leq 1 + \sum_{n=1}^{\infty} \mathbf{P}[|X| \geq n]$.

Modify (1) and show that

(4) $\sum_{n=1}^{\infty} n\mathbf{P}[n < |X| \leq (n+1)] = \sum_{n=1}^{\infty} \mathbf{P}[|X| > n]$, and
(5) $\sum_{n=1}^{\infty} \mathbf{P}[|X| > n] \leq E[|X|] \leq 1 + \sum_{n=1}^{\infty} \mathbf{P}[|X| > n]$.

Conclude that

(6) $X \in L^1(\Omega, \mathfrak{F}, \mathbf{P})$ if and only if $\sum_{n=1}^{\infty} \mathbf{P}[|X| \geq n] < \infty$ and if and only if $\sum_{n=1}^{\infty} \mathbf{P}[|X| > n] < \infty$.

Use (5) to show that, if $m \geq 1$, then

(7) $\frac{1}{m}\sum_{k=1}^{\infty} \mathbf{P}[|X| > \frac{k}{m}] \leq E[|X|] \leq \frac{1}{m} + \frac{1}{m}\sum_{k=1}^{\infty} \mathbf{P}[|X| > \frac{k}{m}]$.

Conclude that

(8) $\int \mathbf{P}[|X| > x]dx = E[|X|]$.

Assume that X is non-negative, i.e., $F(0-) = 0$. Show that if $F(0-) = 0$ then

(9) $\int \{1 - F(x)\}dx = \int x dF(x) = E[X]$ (see Exercise 3.7.21 for a proof based on integration by parts), and
(10) $E[X^p] = \int x^p dF(x) = \int \{1 - F(x^{\frac{1}{p}})\}dx = \int \{1 - F(u)\}pu^{p-1}du$.
 [Hint: see Remark (3) following Lemma 4.1.3 for the last equality.]

Exercise 4.7.2. (Alternate methods to do Exercise 4.7.1 (10)). Let Y be a non-negative random variable on $(\Omega, \mathfrak{F}, \mathbf{P})$. Show that for any $p, 1 \leq p < \infty$,

$$E[Y^p] = \int_0^\infty p\lambda^{p-1} \mathbf{P}[Y > \lambda] d\lambda.$$

Do this two ways:

(1) first verify it for a simple function $s = \sum_{i=1}^n a_i 1_{A_i}$, where $a_1 < a_2 < \cdots < a_n$, and then pass to the limit; and
(2) use the identity

$$(Y(\omega) \wedge n)^p = \int_0^{Y(\omega) \wedge n} p\lambda^{p-1} d\lambda = \int_0^n p\lambda^{p-1} 1_{\{Y(\omega) > \lambda\}} d\lambda$$

and Fubini's theorem (Theorem 3.3.5) to compute $E[(Y \wedge n)^p]$ in two ways.

Show also that

$$E[Y^p] = \int_0^\infty p\lambda^{p-1} \mathbf{P}[Y \geq \lambda] d\lambda.$$

Remark. $\mathbf{P}[Y > \lambda] = \mathbf{P}[Y \geq \lambda]$ except for at most a countable number of values of λ. This is because for a non-decreasing or non-increasing real-valued function ϕ defined on an interval $[a, b]$, the left limit $\phi(t-)$ equals the right limit $\phi(t+)$ at all but a countable number of values of t: consider for how many points t the magnitude of the "jump" $\phi(t+) - \phi(t-) \geq 1/n$.

Exercise 4.7.3. Let f be a non-negative Lebesgue integrable function on $\mathbb{R}$. Then there is a unique non-negative measure η on $\mathbb{R}$ such that $\eta(B) = |f^{-1}(B)|$ for all Borel subsets B of $\mathbb{R}$. [This measure is the analogue of the distribution $\mathbf{Q}$ of a random variable $X : \mathbf{Q}(B) = \mathbf{P}(X^{-1}(B))$; see the proof of Proposition 2.1.9.] Show that

(1) $B \to |f^{-1}(B)|$ is a non-negative measure η on $\mathfrak{B}(\mathbb{R})$ that is the image of Lebesgue measure under f (see the remark following Proposition 2.1.19 and Lemma 4.1.3),
(2) $\eta((-\infty, 0)) = 0$ and $\eta((a, b]) < \infty$ if $0 < a < b < +\infty$,
(3) $\eta(\{0\}) = +\infty$ if $|f^{-1}((0, +\infty))| < \infty$ and that no conclusion can be drawn about $\eta(\{0\})$ if $|f^{-1}((0, +\infty))| = \infty$.

By (2) η is a measure on $[0, +\infty)$. Let ν be its restriction to $(0, +\infty)$, i.e., to the Borel subsets of $(0, +\infty)$. Then ν is a σ-finite measure on $((0, +\infty), \mathfrak{B}((0, +\infty)))$. The measure ν is determined by what analysts call the distribution function of f (see [W1], p. 77 and [S1], p. 57) which will be referred to here as the **analyst's distribution function**. This is the function ω_f defined by $\omega_f(\lambda) = |\{f > \lambda\}|$. Notice that the integrability of f implies that $|\{f > \lambda\}| < +\infty$ but that $|\{f \leq \lambda\}|$ may very well be infinite. Show that

(4) the analyst's distribution function ω_f is decreasing and right continuous,
(5) ν is the unique σ-finite measure μ on $((0, +\infty), \mathfrak{B}((0, +\infty)))$ such that $\mu((a, b]) = \omega_f(a) - \omega_f(b) = |\{a < f \leq b\}|$ if $0 < a < b < +\infty$,
(6) $\int \phi(y)\nu(dy) = \int \phi(f(x))dx$ for all non-negative Borel functions $\phi : (0, +\infty) \to \mathbb{R}$, and
(7) $\int f(x)^p dx = \int_0^{+\infty} p\lambda^{p-1}\omega_f(\lambda)d\lambda$. [*Hint*: the method of Exercise 4.7.2 (1) can be used.]

Remark. Let f be a Lebesgue measurable function on $\mathbb{R}$. The preceding exercise shows that if $f \in L^1$ and is non-negative, then the analyst's distribution function ω_f determines the measure on $(0, +\infty)$ that is the image of Lebesgue measure under f.

In addition, if $E \subset \mathbb{R}$ has finite Lebesgue measure $|E|$ and $g = f_{|E}$, then the image η_E of Lebesgue measure under g is determined by the analyst's distribution function ω_g of G. In fact, $\omega_g(\lambda) = |\{f > \lambda\} \cap E|$ and if $M = |E|$, then $G(\lambda) = M - \omega_g(\lambda)$ is non-negative, bounded, and right continuous. The measure η_E is the finite measure determined by G (see Theorem 2.2.2).

Exercise 4.7.4. Prove Proposition 4.1.14 by reducing to the case $a+b = 1$ and then using a Lagrange multiplier to locate the minimum of $G(a, b)$.

Exercise 4.7.5. (Young's inequality) Let $y = \varphi(x)$ be a continuous, strictly increasing function on $\mathbb{R}^+ = [0, \infty)$ with $\varphi(0) = 0$ and $\lim_{x \to \infty} \varphi(x) = +\infty$. Let $\psi(y) = x$ denote the inverse function (i.e., $\psi(y) = x$ if and only if $y = \varphi(x)$). Set $\Phi(x) = \int_0^x \varphi(u)du$ and $\Psi(y) = \int_0^y \psi(u)du$.

Show that, for $x \geq 0$, $x\varphi(x) = \Phi(x) + \Psi(\varphi(x))$. [*Hint*: interpret $x\varphi(x)$ as the area of the rectangle determined by $(0,0)$ and $(x, \varphi(x))$.] Compare the area $\int_0^c \varphi(u)du$ under the curve $y = \varphi(x), 0 \leq x \leq c$, with the area of the rectangle determined by $(0, 0)$ and (c, d) when $d \leq \varphi(c)$. Conclude that

(1) if $c, d \geq 0$, then $cd \leq \Phi(c) + \Psi(d)$, and hence that
(2) $cd \leq \frac{c^p}{p} + \frac{d^q}{q}$ if $c, d \geq 0$ and $\frac{1}{p} + \frac{1}{q} = 1$.

Exercise 4.7.6. Assume that Jensen's Inequality in Proposition 4.1.19 holds for any convex function φ and $X \in L^1(\Omega, \mathfrak{F}, \mu)$, where μ is a positive measure. Show that μ is a probability. [*Hint*: let φ be an affine function, i.e., $\varphi(x) = ax + b$.]

Exercise 4.7.7. Show that

(1) if $f \in L^1(\mathbb{R}), g \in L^p(\mathbb{R}), 1 < p \leq \infty$, then $f \star g \in L^p(\mathbb{R})$.

Also show that

(2) $\|f \star g\|_p \leq \|f\|_1 \|g\|_p$. [Hint: for $1 < p < \infty$, use Jensen's inequality (Proposition 4.1.19) to estimate $\left(\int |g| d\nu\right)^p$, where $\nu(dy) = \|f\|_1^{-1} |f(x-y)| dy$.]

Exercise 4.7.8. (Egorov's theorem) Let $(f_n)_{n \geq 1}$ be a sequence of measurable functions f_n on a finite measure space $(\Omega, \mathfrak{F}, \mu)$. Assume that $f_n \to 0$ μ a.e. Let $\epsilon > 0$. Show that there is a measurable set Λ such that

(1) $\mu(\Lambda^c) < \epsilon$, and
(2) on Λ, the sequence $(f_n)_{n \geq 1}$ converges to zero uniformly.

[Hints: one may assume that μ is a probability; modify the argument of Proposition 4.3.2 to show that for each k there is an integer $N(k, \epsilon)$ with $\mu(\Gamma(\frac{\epsilon}{2^k}, N)) > 1 - \frac{\epsilon}{2^k}$ if $N \geq N(\epsilon, k)$; set $\Lambda = \cap_{k=1}^\infty \Gamma(\frac{\epsilon}{2^k}, N(\epsilon, k))$.]

Exercise 4.7.9. (Lusin's theorem)(see Exercise 4.2.1)Let $E \subset \mathbb{R}$ be a Lebesgue measurable set, and let f denote a function $f : E \to \mathbb{R}$. Then f is Lebesgue measurable if and only if, for $\epsilon > 0$, there is a closed set A with $A \subset E$ such that (i) the restriction of f to A is continuous on A and (ii) $|E \backslash A| < \epsilon$.

The proof of this result is a fairly long exercise and will be done in two parts. Recall that the necessity of the condition is fairly easy to prove and has already been done as Exercise 2.3.7. It remains to prove the hard part, i.e., the sufficiency of the condition. To begin, one reduces the result to the case where $|E| < \infty$.

Part A. Let $E_n = E \cap \{x \mid n-1 < |x| \leq n\}, n \geq 1$. Then $E\backslash\{0\} = \cup_{n=1}^\infty E_n$. Assume that Lusin's theorem holds for E bounded. Let $\epsilon > 0$, and for each $n \geq 1$, let $A_n \subset E_n$ be a closed (and hence compact) set such that (i) the restriction of f to A_n is continuous and (ii) $|E_n \backslash A_n| < \frac{\epsilon}{2^n}$. Show that

(1) $A \stackrel{\text{def}}{=} \cup_{n=1}^\infty A_n$ is a closed set [Hint: if a sequence in A converges in $\mathbb{R}$, show that it is bounded and conclude that, for some N, it is contained in the closed set $\cup_{n=1}^N A_n$.],
(2) $|E\backslash A| < \epsilon$, and
(3) the restriction of f to A is continuous on A. [Hint: use the suggestion for (1).]

Conclude that Lusin's theorem holds if it holds for bounded measurable sets.

To prove Lusin's theorem when $|E| < \infty$, one uses Egorov's theorem to pass from the simple case discussed in Exercise 4.2.1 to the final result. The idea of the proof is simple enough. First, it is enough to prove it for a non-negative function f. Such a function is a limit of a sequence $(s_n)_{n \geq 1}$ of simple functions s_n, and Egorov's theorem implies that, except on a set of small measure, these functions converge uniformly to f. Lusin's theorem is valid for each simple function s_n (Exercise 4.2.1), and since the uniform limit of continuous functions is continuous (see the comment following Exercise 4.2.11), one hopes to be able to control the exceptional sets corresponding to each simple function s_n and thereby prove the result.

Part B. Assume that $|E| < \infty$. Let $f \geq 0$ and $(s_n)_{n \geq 1}$ be a sequence of simple functions that converges to f. Let $\epsilon > 0$. By Egorov's theorem, there is a subset F of E such that (1) $|E \backslash F| < \epsilon$ and (2) $s_n \xrightarrow{u} f$ on F (see Definition 3.6.15). For each n, by using Lusin's theorem for simple functions (Exercise 4.2.1), choose a closed set A_n with $A_n \subset F$ such that (i) the restriction of s_n to A_n is continuous on A_n and (ii) $|F \backslash A_n| < \frac{\epsilon}{2^n}$. Show that

(1) $A \stackrel{\text{def}}{=} \cap_{n=1}^{\infty} A_n$ is a closed set,
(2) $|F \backslash A| < \epsilon$; and
(3) the restriction of f to A is continuous.

Conclude that Lusin's theorem holds if $f \geq 0$, and thus deduce its validity for any finite, measurable function f.

Exercise 4.7.10. Let $|X|$ be a random variable. By integrating over $\{|X| \geq \varepsilon\}$ and $\{|X| < \varepsilon\}$, determine upper and lower bounds for $E\left[\frac{|X|}{1+|X|}\right]$ that involve ε and $\mathbf{P}[|X| \geq \varepsilon]$. Conclude that for a sequence of random variables $(X_n)_{n \geq 1}$,

$$X_n \xrightarrow{pr} X \quad \text{if and only if} \quad E\left[\frac{|X_n - X|}{1 + |X_n - X|}\right] \longrightarrow 0.$$

Show that $d(X, Y) = E\left[\frac{|X-Y|}{1+|X-Y|}\right]$ is a metric on the vector space L of finite random variables, with the usual proviso about a random variable being zero if it is equal to zero a.s. Note that this shows that on L convergence in probability is given by a metric.

Exercise 4.7.11. Let $1 \leq p_1 < p_2 < \infty$ and $u \in L^{p_2}(\Omega, \mathfrak{F}, \mu)$. Let $A \in \mathfrak{F}$ with $\mu(A) < \infty$. Assume that $u = u 1_A$. Show that

(1) $\left[\frac{1}{\mu(A)}\right]^{\frac{1}{p_1}} \| u \|_{p_1} \leq \left[\frac{1}{\mu(A)}\right]^{\frac{1}{p_2}} \| u \|_{p_2}$.

If $f \in L^1(\Omega, \mathfrak{F}, \mu) \cap L^{\infty}(\Omega, \mathfrak{F}, \mu)$, show that

(2) $f \in L^p(\Omega, \mathfrak{F}, \mu)$ for all $p \in (0, +\infty)$, and
(3) $\| f \|_p \to \| f \|_{\infty}$ as $p \to +\infty$. [Hints: prove this first when $\mu(\mathbb{R}) < \infty$, and then, by using $\nu(d\omega) = |f(\omega)| \mu(d\omega)$, reduce to the case of a finite measure.]

Finally, show that

(4) the conclusion (3) holds for any function $f \in \mathfrak{F}$ if $\| f \|_{\infty} = +\infty$.

Remark. This exercise plays a key role in Cramér's theory of large deviations (see Stroock [S2]).

Exercise 4.7.12. Let $(a_n)_{n\geq 1}$ be a Cauchy sequence of real numbers relative to the usual distance $d(a, b) = |a - b|$. Show that
  (1) $\limsup_n a_n < +\infty$ [Hint: show that there is an integer N such that $a_n \leq a_N + 1, n \geq N$.],
  (2) $\liminf_n a_n > -\infty$,
  (3) if $\varepsilon > 0$, then $\limsup_n a_n - \liminf_n a_n < \varepsilon$.
Conclude from Exercise 2.1.10 that
  (4) $(a_n)_{n\geq 1}$ converges, i.e., $\mathbb{R}$ is complete (relative to the usual distance).

If $\mathbf{x} \in \mathbb{R}^n$, let $x(i)$ denote the i-th coordinate of $\mathbf{x}$. Let $(\mathbf{x}_n)_{n\geq 1}$ be a sequence in $\mathbb{R}^n$. Show that
  (5) it converges (Definition 4.1.25) relative to the metric $d(\mathbf{x}, \mathbf{y}) = \| \mathbf{x} - \mathbf{y} \|$, where $\| \mathbf{x} \|^2 = \sum_{i=1}^n x(i)^2$, if and only if, for each $i, 1 \leq i \leq n$, the sequence $(x_n(i))_{n\geq 1}$ converges relative to the usual metric of $\mathbb{R}$,
  (6) it is Cauchy relative to the Euclidean metric (the metric in (5)) if and only if for each $i, 1 \leq i \leq n$, the sequence $(x_n(i))_{n\geq 1}$ is Cauchy relative to the usual metric of $\mathbb{R}$.

Conclude that
  (7) $\mathbb{R}^n$ is complete relative to the Euclidean metric.

Exercise 4.7.13. Let $(X_n)_{n\geq 1}$ be a sequence of random variables such that
  (1) $X_n \xrightarrow{pr} X$, and
  (2) $(X_n)_{n\geq 1}$ is uniformly integrable.
Show that $X \in L^1$. [Hint: use the proof of Corollary 4.5.9 and get an upper estimate for $E[|X_c|]$ in terms of $E[|X_n|]$]. Conclude that a sequence $(X_n)_{n\geq 1}$ of random variables in L^1 converges in L^1 if and only if (i) it converges in probability and (ii) it is uniformly integrable.

Exercise 4.7.14. Verify Corollary 4.5.9 for random variables in L^p, i.e., show that $X_n \xrightarrow{L^p} X$ if the sequence is uniformly integrable in L^p and converges in probability to X.

Exercise 4.7.15. State and prove the L^p-analogue of Proposition 4.5.7. [Hint: make use of Theorem 4.5.8 and a Lipschitz contraction to uniformly approximate in L^p.] Conclude that a sequence $(X_n)_{n\geq 1}$ of random variables in L^p converges in L^p if and only if (i) it converges in probability and (ii) it is uniformly integrable in L^p.

Exercise 4.7.16. State and prove the analogue of Exercise 4.7.13 for a sequence $(X_n)_{n\geq 1}$ of random variables in L^p.

7. ADDITIONAL EXERCISES

Exercise 4.7.17. Let A be a closed subset of $\mathbb{R}$ and define $\text{dist}(x, A) = \inf\{|x - y| \mid y \in A\}$. Show that

(1) $|\text{dist}(x_1, A) - \text{dist}(x_2, A)| \leq |x_1 - x_2|$ [*Hint*: use the triangle inequality (Exercise 1.1.16).] (Note that this proves the continuity of $x \to \text{dist}(x, A)$.),

(2) $x \in A$ if and only if $\text{dist}(x, A) = 0$.

Let C be a compact set (i.e., closed and bounded as stated in part A of Exercise 1.5.6) and assume $C \subset O$, where O is an open set. Then, for any $x \in C$, there is an $r_x > 0$ such that $B(x, r_x) = (x - r_x, x + r_x) \subset O$. Show that

(3) there is an $R > 0$ such that $\text{dist}(x, C) < R$ implies $x \in O$. [*Hint*: use the Heine–Borel theorem (Theorem 1.4.5).]

Conclude that

(4) if $f(x) = \min\{\text{dist}(x, C), R\}$, then f is a continuous function with $0 \leq f(x) \leq R$ such that $f(x) = 0$ if and only $x \in C$ and $f(x) = R$ if $x \notin O$.

Exercise 4.7.18. Let E be a bounded, Lebesgue measurable set with $0 < |E| < \infty$, and let $0 < \alpha < 1$. Use Exercise 4.2.2 to show that

(1) there is an open set $O \supset E$ and $\alpha|O| \leq |E|$.

Use the fact that the open set O is a disjoint union of at most countably many open intervals (Exercise 1.3.11 (d)) to show that

(2) there is a bounded open interval (a, b) with $\alpha(b - a) = \alpha|(a, b)| \leq |E \cap (a, b)|$.

Let $E \cap (a, b) \stackrel{\text{def}}{=} E_1$ and $\delta_1 = \delta(b - a), \delta > 0$. Show that if $(-\delta_1 < x < \delta_1)$, then

(3) $E_1 + x = \{y + x \mid y \in E_1\} \subset (a - \delta_1, b + \delta_1)$,

(4) $x = u - v, u, v \in E_1$ if and only if $E_1 \cap (E_1 + x) \neq \emptyset$,

(5) $E_1 \cap (E_1 + x) = \emptyset$ implies $1 + 2\delta \geq 2\alpha$.

Conclude that if $\frac{3}{4} \leq \alpha < 1$ and $0 < \delta < \frac{1}{4}$, the open interval $(-\delta_1, \delta_1) \subset E_1$. Deduce that for any Lebesgue measurable set E, if $|E| > 0$, then $D = \{y_1 - y_2 \mid y_i \in E\}$ contains a non-void open interval.

Exercise 4.7.19. (Borel's strong law of large numbers)

Let $(X_n)_{n \geq 1}$ be an i.i.d. sequence of Bernoulli random variables X_n with $\mathbf{P}[X_n = 1] = p$ and $\mathbf{P}[X_n = 0] = q$. It follows from Kolmogorov's strong law (Theorem 4.4.2) that $\frac{S_n}{n} \xrightarrow{a.s.} p$. This exercise, following Loève [L2], outlines a simple proof of the result exploiting the fact that the common distribution is Bernoulli, (i.e., $\mathbf{Q} = q\varepsilon_0 + p\varepsilon_1$). Let $Z_n = \frac{S_n}{n}$. Show that

(1) $\sigma^2(Z_n) = \frac{pq}{n}$,

(2) $\sum_{k=1}^{\infty} \sigma^2(Z_{k^2}) < +\infty$, and

(3) if $k^2 \leq n < (k+1)^2$, then $|Z_n - Z_{k^2}| \leq \frac{4}{k}$. [Hint: Observe that $Z_n = \frac{1}{n}\sum_{i=1}^{k^2} X_i + \frac{1}{n}\sum_{i=k^2+1}^{n} X_i$, and estimate, for example, $\sum_{i=k^2+1}^{n} X_i$ by $n - k^2$.]

Use (2) to show that $Z_{k^2} \xrightarrow{a.s.} p$. [Hints: note that a sequence $z_n \to 0$ if and only if, for all $m \geq 1$, one has $|z_n| < \frac{1}{m}$ for large n. Chebychev's inequality implies that $\mathbf{P}[|Z_n - p| \geq \frac{1}{m}] \leq m^2 \sigma^2(Z_n)$. (2) implies that $\sum_{k=1}^{\infty} \mathbf{P}[|Z_{k^2} - p| \geq \frac{1}{m}] < \infty$.] Finally, use (3) to prove that $Z_n \xrightarrow{a.s.} p$ as $n \to \infty$.

Now assume that the i.i.d. random variables are all bounded with (say) $|X_n| \leq M$ for all $n \geq 1$. Show that

(4) $\sigma^2(Z_n) \leq \frac{(M+|m|)^2}{n}$, where $m = E[X_n]$,
(5) $\sum_{k=1}^{\infty} \sigma^2(Z_{k^2}) < +\infty$, and
(6) $|Z_n - Z_{k^2}| < \frac{4M}{k}$ if $k^2 \leq n < (k+1)^2$.

Conclude, as in the Bernoulli case, that $Z_n \xrightarrow{a.s.} m$.

Exercise 4.7.20. This exercise explains the relation between normal numbers and Borel's strong law by showing that $([0,1], \mathfrak{B}([0,1]), dx)$ is **almost isomorphic** as a probability space to an infinite product space. This terminology is now explained.

Definition 4.7.21. Let $(\Omega_1, \mathfrak{F}_1, \mu_1)$ and $(\Omega_2, \mathfrak{F}_2, \mu_2)$ be measure spaces. They are said to be **isomorphic** if there is a bijection (i.e., a 1:1 onto map) $\Phi : \Omega_1 \to \Omega_2$ such that

(1) Φ and Φ^{-1} are both measurable (i.e., $A_2 \in \mathfrak{F}_2$ if and only if $\Phi^{-1}(A_2) \in \mathfrak{F}_1$), and
(2) $\mu_1(\Phi^{-1}(A_2)) = \mu_2(A_2)$ for all $A_2 \in \mathfrak{F}_2$.

Let $(\Omega, \mathfrak{F}, \mu)$ be a measure space and Λ be a measurable subset with $\mu(\Omega \backslash \Lambda) = 0$. Set $\mathfrak{G} \stackrel{\text{def}}{=} \{A \cap \Lambda \mid A \in \mathfrak{F}\}$, and let $\nu(A \cap \Lambda) \stackrel{\text{def}}{=} \mu(A)$ for all $A \in \mathfrak{F}$. Then $(\Lambda, \mathfrak{G}, \nu)$ is a measure space that is said to be almost isomorphic to $(\Omega, \mathfrak{F}, \mu)$.

Finally, two measure spaces $(\Omega_1, \mathfrak{F}_1, \mu_1)$ and $(\Omega_2, \mathfrak{F}_2, \mu_2)$ are said to be almost isomorphic if there are measurable subsets $\Lambda_1 \subset \Omega_1$ and $\Lambda_2 \subset \Omega_2$ with $\mu_1(\Omega_1 \backslash \Lambda_1) = 0$ and $\mu_2(\Omega_2 \backslash \Lambda_2) = 0$ such that the measure spaces $(\Lambda_1, \mathfrak{G}_1, \nu_1)$ and $(\Lambda_2, \mathfrak{G}_2, \nu_2)$ are isomorphic.

The exercise has several parts.

Part A. For each $n \geq 1$, consider the dyadic rational numbers of the form $\frac{k}{2^n}$, $0 \leq k \leq 2^n$. They partition $(0,1]$ into 2^n intervals $(\frac{k}{2^n}, \frac{k+1}{2^n}]$, each of length $\frac{1}{2^n}$. Define the random variable X_n on $((0,1], \mathfrak{B}((0,1]), dx)$ by setting

$$X_n(x) = \begin{cases} 0 \text{ if } \frac{2\ell}{2^n} < x \leq \frac{2\ell+1}{2^n}, \\ 1 \text{ if } \frac{2\ell+1}{2^n} < x \leq \frac{2\ell+2+1}{2^n}. \end{cases}$$

Show that

(1) the random variables $X_n, n \geq 1$ are i.i.d. Bernoulli random variables with $|\{X_n = 0\}| = |\{X_n = 1\}| = \frac{1}{2}$,
(2) $X_n(x_1) = X_n(x_2)$ for all $n \geq 1$ if and only if $x_1 = x_2$,
(3) $\sum_{n=1}^{\infty} X_n(x) 2^{-n} = x$.

If $a_n \in \{0, 1\}$ for $n \geq 1$, let $0.a_1 a_2 \cdots a_n \cdots \stackrel{\text{def}}{=} \sum_{n=1}^{\infty} a_n 2^{-n} = a \in [0, 1]$. Then $0.a_1 a_2 \cdots a_n \cdots$ is called a **binary expansion** of a. Show that

(4) the binary expansion of x given by $x = 0.X_1(x) X_2(x) \cdots X_n(x) \cdots$ is the unique binary expansion of x that does not terminate in an unbroken string of zeros, and
(5) if $t = \sum_{n=1}^{N} X_n(x) 2^{-n}$, then $t = \frac{k}{2^N} < x \leq \frac{k+1}{2^N}$.

Part B. Let $\Omega = (\Omega, \mathfrak{F}, \mathbf{P})$ be the countable product of an infinite number of copies of the probability spaces $(\Omega_0, \mathfrak{F}_0, \mathbf{P}_0)$, where $\Omega_0 = \{0, 1\}$ and the probability $\mathbf{P}_0$ defined on all the subsets of Ω_0 gives each point weight $\frac{1}{2}$. Let $\Lambda \subset \Omega$ be the set of functions $\omega : \mathbb{N} \to \{0, 1\}$ such that $\omega(n) = 1$ infinitely often. Define the random variables $Y_n : \Omega \to \{0, 1\}$ for $n \geq 1$ by setting $Y_n(\omega) = \omega(n)$. Show that

(1) $\Omega \backslash \Lambda$ is countable and $\mathbf{P}(\Lambda) = 1$.

Define $X : (0, 1] \to \Omega$ by setting $X(x)(n) = X_n(x)$ for all $n \geq 1$. Show that

(2) X is a measurable map (or Ω-valued random variable). [Hint: verify that the map X into Ω is measurable if and only if $Y_n \circ X$ is measurable for all $n \geq 1$.]

The random variable X maps $(0, 1]$ to Λ in view of part A (4). It has an inverse $S : \Lambda \to (0, 1]$ given by $S(\omega) \stackrel{\text{def}}{=} \sum_{n=1}^{\infty} \omega(n) 2^{-n}$. Show that

(3) $S(X(x)) = x$ for all $x \in (0, 1]$, and
(4) $X(S(\omega)) = \omega$ for all $\omega \in \Lambda$.

Hence, $X : (0, 1] \to \Lambda$ is a bijection.

The goal now is to show that S is measurable. Let $\mathfrak{G}$ denote the σ-algebra on Λ consisting of sets of the form $A \cap \Lambda$, $A \in \mathfrak{F}$. It is shown in Exercise 5.4.6 that $\mathfrak{B}((0, 1])$ is the smallest σ-field that contains all the dyadic intervals. It follows from Exercise 2.1.18 that S is measurable if $S^{-1}((\frac{k}{2^n}, \frac{k+1}{2^n}])$ is the intersection of Λ with a set in the product σ-algebra $\mathfrak{F}$. Show that

(5) $S^{-1}((\frac{k}{2^n}, \frac{k+1}{2^n}]) = \{\omega \mid \omega(i) = a_i, 1 \leq i \leq n\} \cap \Lambda$, where the coefficients $a_i \in \{0, 1\}$ are uniquely determined by the property that $k 2^{-n} = \sum_{i=1}^{n} a_i 2^{-i}$. [Hint: use part A (5)].

This proves that the measurable spaces $(\Lambda, \mathfrak{G})$ and $((0, 1], \mathfrak{B}((0, 1])$ are isomorphic.

Part C. It remains to show that the distribution of S is Lebesgue measure on $(0,1]$ and that the distribution of X is the measure $\mathbf{P}$ restricted to $\mathfrak{G}$ (the product σ-algebra $\mathfrak{F}$ restricted to Λ). At the end of part B, it was shown that a dyadic interval $(\frac{k}{2^n}, \frac{k+1}{2^n}]$ corresponds under X and S to the intersection with Λ of the so-called cylinder set $\{\omega \mid \omega(i) = a_i, \ 1 \le i \le n\}$, where the a_i are the coefficients of the finite binary expansion of $\frac{k}{2^n}$. These sets in $(0,1]$ and Ω, respectively, generate the corresponding σ-algebras. Show that

(1) $|(\frac{k}{2^n}, \frac{k+1}{2^n}]| = \mathbf{P}(\{\omega \mid \omega(i) = a_i, \ 1 \le i \le n\})$.

Conclude that

  (1) the distribution of S is Lebesgue measure, and
  (2) $\mathbf{P}$ is the distribution of X.

This shows that $(\Lambda, \mathfrak{G}, \mathbf{P})$ and $((0,1], \mathfrak{B}((0,1], dx)$ are isomorphic measure spaces. Since $\mathbf{P}(\Lambda) = 1$, the measure space $(\Lambda, \mathfrak{G}, \mathbf{P})$ is almost isomorphic to $(\Omega, \mathfrak{F}, \mathbf{P})$. It follows that $([0,1], \mathfrak{B}([0,1], dx)$ and the infinite product space $(\Omega, \mathfrak{F}, \mathbf{P})$ are almost isomorphic measure spaces since $((0,1], \mathfrak{B}((0,1], dx)$ is clearly almost isomorphic to $([0,1], \mathfrak{B}([0,1], dx)$.

Exercise 4.7.22. Let $(X_n)_{n \ge 1}$ be a sequence of independent random variables in $L^1(\Omega, \mathfrak{F}, \mathbf{P})$ with $E[X_n] = 0$ for all $n \ge 1$. Assume that this sequence satisfies the strong law of large numbers, i.e. $\frac{1}{n}\sum_{i=1}^n X_i \xrightarrow{a.s.} 0$. Show that $\sum_{n=1}^\infty \mathbf{P}[\frac{|X_n|}{n} \ge \epsilon] < \infty$ for any $\epsilon > 0$. [*Hint:* make use of the Borel–Cantelli lemma (Proposition 4.1.29)]

Exercise 4.7.23. Let $(X_i)_{1 \le i \le n}$ be a finite set of i.i.d. random variables in $L^2(\Omega, \mathfrak{F}, \mathbf{P})$. Let σ^2 denote the common variance. If $\overline{X} \stackrel{\text{def}}{=} \sum_{i=1}^n$, show that $\frac{1}{n-1}E[\sum_{i=1}^n (X_i - \overline{X})^2] = \sigma^2$. [**Comment:** this shows that $\frac{1}{n-1}\sum_{i=1}^n (X_i - \overline{X})^2$ (the sample variance) is a so-called **unbiased estimator** of σ^2].

Exercise 4.7.24. Let $(X_n)_{n \ge 1}$ be a sequence of independent random variables on $(\Omega, \mathfrak{F}, \mathbf{P})$. Assume that $\mathbf{P}[X_n = 1] = \mathbf{P}[X_n = -1] = \frac{1}{2}(1 - \frac{1}{2^n})$ and that $P[X_n = 2^n] = P[X_n = -2^n] = \frac{1}{2^{n+1}}$.

(1) Show that $\sum_{n=1}^\infty \frac{\sigma^2(X_n)}{n^2}$ diverges, and that
(2) the sequence $(X_n)_{n \ge 1}$ satisfies the strong law of large numbers. [*Hint:* use a suitable truncation.]

Exercise 4.7.25. (**Convex functions**) A function $\varphi : (a,b) \to \mathbb{R}$ is **convex** if, for any $a < x < y < b$ and $t \in [0,1]$, $\varphi(tx + (1-t)y) \le t\varphi(x) + (1-t)\varphi(y)$. This means that the line segment joining $(x, \varphi(x))$ to $(y, \varphi(y))$ lies above the graph of φ (it may coincide in part). Let $L_n(x) = a_n x + b_n$ for $n \ge 1$. Then, if $\varphi(x) = \sup_n L_n(x)$, it is a convex function on $\{x \mid \varphi(x) < +\infty\}$. The aim of this exercise is to prove the converse: namely, if $\varphi : (a,b) \to \mathbb{R}$ is convex, there is a sequence $(L_n)_{n \ge 1}$ of affine functions $L_n(x) = a_n x + b_n$ such that $\varphi = \sup_n L_n$ on (a,b). This is a long exercise and is presented in several parts.

7. ADDITIONAL EXERCISES

Part A. Assume that φ is a convex function on (a,b). Show that, for any $x \in (a,b)$,

(1) the function $h \to \frac{\varphi(x+h)-\varphi(x)}{h}$ is a non-decreasing function of h for $h > 0$ and for $h < 0$;

(2) $\frac{\varphi(x+k)-\varphi(x)}{k} \leq \frac{\varphi(x+h)-\varphi(x)}{h}$ if $a < x+k < x < x+h < b$.

Let $\varphi_+^{(1)}(x) \stackrel{\text{def}}{=} \lim_{h \downarrow 0} \frac{\varphi(x+h)-\varphi(x)}{h}$ and $\varphi_-^{(1)}(x) \stackrel{\text{def}}{=} \lim_{k \uparrow 0} \frac{\varphi(x+k)-\varphi(x)}{k}$.

Part B. Show that

(1) $\varphi_+^{(1)}(x)$ and $\varphi_-^{(1)}(x)$ both exist, are finite non-decreasing functions for $a < x < b$, and $\varphi_+^{(1)}(x) \geq \varphi_-^{(1)}(x)$ if $x \in (a,b)$,

(2) φ is continuous at each $x \in (a,b)$ [Hint: use (1).],

(3) if $a < x_0 < b$, then the graph of φ lies above the graph of each of the two affine functions $L^+(x) = \varphi_+^{(1)}(x_0)(x-x_0) + \varphi(x_0)$ and $L^-(x) = \varphi_-^{(1)}(x_0)(x-x_0) + \varphi(x_0)$. [Hint: sketch the situation for a typical convex function, e.g., $\varphi(x) = x^2$.]

Part C. Let $(r_n)_{n \geq 1}$ be an enumeration (i.e., a counting) of $\mathbf{Q} \cap (a,b)$, the rationals in (a,b). Define $L_n(x) = a_n(x - r_n) + b_n$, where $a_n = \varphi_-^{(1)}(r_n)$ and $b_n = \varphi(r_n)$. Show that

(1) if for some $\delta > 0$, $\varphi(x_0) \geq L_n(x_0) + \delta$ for any n, then $\varphi_-^{(1)}(x_0) = +\infty$. [Hint: compute $\varphi_-^{(1)}(x_0)$ using a sequence of rationals $(r_{n_k})_k$ that increases to x_0, and use part B (3).]

Conclude that

(2) $\varphi(x) = \sup_{n \geq 1} L_n(x)$ if $a < x < b$.

Exercise 4.7.26. If $\phi, \psi \in L^1(m)$, let $(\phi \star \psi)(g) \stackrel{\text{def}}{=} \int \phi(gh^{-1})\psi(h)m(dh)$. Deonte by f and k the 2π-periodic functions on $\mathbb{R}$ such that $f(t) = \phi(g)$ if $g = e^{it}$ and $k(t) = \psi(h)$ if $h = e^{iy}$. Show that $(\phi \star \psi)(g) = \int_0^{2\pi} f(x-y)k(y)dy$.

CHAPTER V

CONDITIONAL EXPECTATION AND AN INTRODUCTION TO MARTINGALES

1. Conditional Expectation and Hilbert space

In this chapter, the conditional expectation operator will be defined and then used in the study of martingales. To begin, one considers the simplest cases of conditional expectation, which are closely related to conditional probability. Then, one proves the Riesz representation theorem for continuous linear functionals on Hilbert space as a tool for defining conditional expectation for square integrable random variables. Given this, it is easy to then define the conditional expectation of integrable random variables.

Note that this way of explaining conditional expectation can be completely avoided if one wants to use the Radon–Nikodym theorem (Theorem 2.7.19). However, in the author's opinion, the use of Hilbert space ideas to define the conditional expectation operator is simpler and more direct. In addition, it emphasizes the point that conditional expectation is a projection. The basic idea in the Riesz representation theorem is a simple geometric one: given any closed convex set, it contains a unique point closest to the origin. This is intuitively clear in the plane once one sketches some closed convex sets. The proof for the plane carries over without change to any real Hilbert space.

Conditional expectation: introduction and definition.

Let $(\Omega, \mathfrak{F}, \mathbf{P})$ be a probability space and $B \in \mathfrak{F}$ with $0 < \mathbf{P}(B) < 1$. Then, for any $A \in \mathfrak{F}$, the conditional probability of A given B is defined to be $\frac{\mathbf{P}(A \cap B)}{\mathbf{P}(B)}$. If one writes $\mathbf{P}(A \cap B)$ as $E[1_A 1_B]$, then this formula extends from random variables of the form 1_A to any positive random variable X. What results is the conditional expectation of X given B: namely, $E[X|B] \stackrel{\text{def}}{=} \frac{1}{\mathbf{P}(B)} \int_B X \, d\mathbf{P}$. Similarly, one may define the conditional expectation of X given B^c: $E[X|B^c] \stackrel{\text{def}}{=} \frac{1}{\mathbf{P}(B^c)} \int_{B^c} X \, d\mathbf{P}$. In this way, to each non-negative random variable X, one associates two values: $E[X|B]$ and $E[X|B^c]$. With these two values, one may define a random variable Y on the probability space $(\Omega, \{\phi, B, B^c, \Omega\}, \mathbf{P})$ by the formula

$$Y(\omega) = \begin{cases} E[X|B] & \text{if } \omega \in B, \\ E[X|B^c] & \text{if } \omega \in B^c. \end{cases}$$

Note that if $\mathfrak{G} = \{\phi, B, B^c, \Omega\}$, the random variable Y satisfies the condition

(CE) $\int_C Y \, d\mathbf{P} = \int_C X \, d\mathbf{P}$ for all $C \in \mathfrak{G}$.

210

If $\mathfrak{G}$ denotes the σ-field $\{\phi, B, B^c, \Omega\}$, the random variable Y will be denoted by $E[X|\mathfrak{G}]$. It will be called a **conditional expectation of** X **given** $\mathfrak{G}$. This operation of passing from a non-negative random variable X to its conditional expectation Y given $\mathfrak{G}$ has the following properties, which will be shown to follow from the defining property (CE):

(CE_1) $E[\lambda X|\mathfrak{G}] = \lambda E[X|\mathfrak{G}]$ for any $\lambda \geq 0$;
(CE_2) $E[X_1|\mathfrak{G}] + E[X_2|\mathfrak{G}] = E[X_1 + X_2|\mathfrak{G}]$ if the X_i are non-negative;
(CE_3) $E[X_1|\mathfrak{G}] \leq E[X_2|\mathfrak{G}]$ **P**-a.s. if $0 \leq X_1 \leq X_2$;
(CE_4) if $X \in \mathfrak{G}$, $E[X|\mathfrak{G}] = X$, and hence $E[E[X|\mathfrak{G}]|\mathfrak{G}] = E[X|\mathfrak{G}]$;
(CE_5) $E[X] = E[E[X|\mathfrak{G}]]$.

In effect, it is a linear projection of the convex cone $\mathfrak{F}^+$ of non-negative random variables on $(\Omega, \mathfrak{F}, \mathbf{P})$ onto the convex cone $\mathfrak{G}^+$ of non-negative random variables on $(\Omega, \mathfrak{G}, \mathbf{P})$, where property (CE_4) guarantees that it is a projection: $E[E[X|\mathfrak{G}]] = E[X|\mathfrak{G}]$ for all $X \in \mathfrak{F}^+$. Furthermore, this projection depends upon the probability $\mathbf{P}$.

If $B_1, B_2, \ldots, B_n$ are n disjoint sets in Ω with $P(B_i) > 0$ for all i and $\cup_{i=1}^n B_i = \Omega$, the smallest σ-field $\mathfrak{G}$ that contains them is the collection of finite unions of these sets. If $A \in \mathfrak{F}$, the conditional probabilities $\mathbf{P}[A|B_i] = a_i$ define a random variable Y on $(\Omega, \mathfrak{G}, \mathbf{P})$ whose value on B_i is a_i. Similarly, one may define, for each non-negative random variable X, a random variable Y on $(\Omega, \mathfrak{G}, \mathbf{P})$ by setting it equal to $E[X|B_i]$ on B_i. If one denotes Y by $E[X|\mathfrak{G}]$, it is easy to see that this again defines a projection of $\mathfrak{F}^+$ onto $\mathfrak{G}^+$ that satisfies properties (CE_1) to (CE_5) since (CE) holds with $\mathfrak{G}$ the smallest σ-field containing all the sets B_i.

Example 5.1.1. Let $(\Omega, \mathfrak{F}, \mathbf{P}) = ([0,1], \mathfrak{B}([0,1]), dx)$ and, for a fixed $n \geq 1$, let $B_0 = [0, \frac{1}{2^n}]$ and $B_k = (\frac{k}{2^n}, \frac{k+1}{2^n}]$ for $1 \leq k < 2^n$. Let $X(x) = x$. Then $E[X|B_k] = 2^{-n-1}(2k+1)$. Hence, $Y(x) = 2^{-n-1}, 0 \leq x \leq \frac{1}{2^n}$, and $Y(x) = 2^{-n-1}(2k+1), \frac{k}{2^n} < x \leq \frac{k+1}{2^n}$. Here the σ-algebra $\mathfrak{G} \stackrel{\text{def}}{=} \mathfrak{F}_n$ depends upon n and, as one will see later, each X defines in this way a martingale since $E[E[X|\mathfrak{F}_{n+1}]|\mathfrak{F}_n] = E[X|\mathfrak{F}_n]$. Replacing the uniform distribution on $[0,1]$ by the uniform distribution $1_{[0,\frac{1}{2}]}(x) 2 dx$ on $[0, \frac{1}{2}]$ changes the conditional expectation of X. Among other things, it can take any values on the interval $(\frac{1}{2}, 1]$, as pointed out later.

It is not necessary that all the $\mathbf{P}(B_i)$ be strictly positive. If, for instance, $\mathbf{P}(B_1) = 0$ then one may define the value of $E[X|\mathfrak{G}]$ on B_1, to be any value (say zero). The really important and defining property (CE) is unaffected, as integration over B_1 has no effect. As a consequence, the conditional expectation is defined up to a null function.

This defining property for non-negative, integrable X is equivalent to stating that the conditional expectation Y of X given $\mathfrak{G}$ is a density for the measure on $\mathfrak{G}$ defined by X: namely, $C \to \int_C X d\mathbf{P}$. In fact, one motivation for studying conditional expectation is it's relation to the Radon–Nikodym theorem, as mentioned earlier. Given any measure μ on $\mathfrak{F}$ and a finite subfield $\mathfrak{G}$, then, just as in the case of the measure given by X, there is a random variable Y on $\mathfrak{G}$ for which $\mu(C) = \int_C Y d\mathbf{P}$ for all $C \in \mathfrak{G}$ provided that $\mu(C) = 0$ if $\mathbf{P}(C) = 0$. The random variable Y is some kind of approximate derivative of μ with respect to $\mathbf{P}$. By taking larger and larger finite subfields, it is shown at the end of this chapter that these approximate derivatives converge to what is called the Radon–Nikodym derivative of μ with respect to $\mathbf{P}$.

This sets the stage for seeking to extend the conditional expectation operator when $\mathfrak{G}$ is an arbitrary sub σ-field of $\mathfrak{F}$, not necessarily determined by a finite number of sets. For example, if $(X_i)_{i \in I}$ is a stochastic process on $(\Omega, \mathfrak{F}, \mathbf{P})$, then $\mathfrak{G} = \sigma(\{X_i \mid i \in I\})$ is a sub σ-field not necessarily determined by a finite number of sets, even if I consists of one point.

The aim is to show that for any sub-σ-field $\mathfrak{G}$ of $\mathfrak{F}$, to each probability $\mathbf{P}$ on $\mathfrak{F}$ there corresponds a linear projection of $\mathfrak{F}^+$ onto $\mathfrak{G}^+$ that satisfies the defining property (CE) and hence the properties (CE_1) to (CE_5).

Because conditional expectation is to be linear on $\mathfrak{F}^+$, as is the integral, it has an obvious extension to $L^1(\Omega, \mathfrak{F}, \mathbf{P})$ that imitates the extension of the integral from non-negative random variables to integrable ones.

Definition 5.1.2. *Let X be a random variable on $(\Omega, \mathfrak{F}, \mathbf{P})$ that is either non-negative or integrable, and let $\mathfrak{G} \subset \mathfrak{F}$ be a sub σ-field.* **A conditional expectation of X given $\mathfrak{G}$** *is a random variable Y on $(\Omega, \mathfrak{G}, \mathbf{P})$ such that*

$$(CE) \qquad \int_C X d\mathbf{P} = \int_C Y d\mathbf{P} \quad \text{for all } C \in \mathfrak{G}.$$

If Y is a random variable satisfying (CE), this will be denoted by writing $Y = E[X|\mathfrak{G}]$.

A conditional expectation, if it exists, is essentially unique. Furthermore, a conditional expectation of a non-negative or of an integrable random variable is, respectively, non-negative or integrable. These facts are direct consequences of the definition as shown next.

Remark. It is implicit in the definition of a conditional expectation Y that $\int_C Y d\mathbf{P}$ be defined for all $C \in \mathfrak{G}$: in other words, if Y takes both positive and negative values, then either Y^+ or Y^- is integrable, as then $\int_C Y d\mathbf{P} \stackrel{\text{def}}{=} \int_C Y^+ d\mathbf{P} - \int_C Y^- d\mathbf{P}$ is well defined. This point is clarified in Exercise 2.9.19 and in the following lemma.

Lemma 5.1.3. *Let X be a random variable on $(\Omega, \mathfrak{F}, \mathbf{P})$ that is either non-negative or integrable. If Y_1 and Y_2 are two conditional expectations of X given $\mathfrak{G}$, then $Y_1 = Y_2$ $\mathbf{P}$-a.s.*

If X is non-negative or integrable, then a conditional expectation of X is, respectively, non-negative or integrable.

Proof. First, consider the case of a non-negative random variable X. Let Y be a conditional expectation of X given $\mathfrak{G}$. If $C_n = \{Y \leq -\frac{1}{n}\}$, then $C_n \in \mathfrak{G}$ and

$$\left(-\frac{1}{n}\right)\mathbf{P}(C_n) \geq \int_{C_n} Y\,d\mathbf{P} = \int_{C_n} X\,d\mathbf{P} \geq 0.$$

Hence, $\mathbf{P}(C_n) = 0$ for all $n \geq 1$ and so $Y \geq 0$ **P**-a.s. Also, if X is a non-negative, integrable random variable, then a conditional expectation Y of X given $\mathfrak{G}$ is integrable since $\int X\,d\mathbf{P} = \int Y\,d\mathbf{P}$.

Now assume that Y_1 and Y_2 are two conditional expectations of a non-negative random variable X. Let $C_{mk} = \{m \geq Y_1 \geq Y_2 + \frac{1}{k}\}$. Then $C_{mk} \in \mathfrak{G}$, and so $\int_{C_{mk}} Y_1\,d\mathbf{P} = \int_{C_{mk}} Y_2\,d\mathbf{P}$. Also, this common value is finite. Furthermore, since by the definition of C_{mk},

$$\int_{C_{mk}} Y_1\,d\mathbf{P} \geq \int_{C_{mk}} Y_2\,d\mathbf{P} + \left(\frac{1}{k}\right)\mathbf{P}(C_{mk}),$$

it follows that $\mathbf{P}(C_{mk}) = 0$. Hence, $Y_2 \geq Y_1$ **P**-a.s., which implies, by symmetry, that $Y_1 = Y_2$ **P**-a.s.

Now assume that X is integrable and that Y is a conditional expectation of X given $\mathfrak{G}$. Then, if $C_+ = \{Y \geq 0\}$ and $C_- = \{Y \leq 0\}$, it follows that $\int Y^+ d\mathbf{P} = \int_{C_+} Y\,d\mathbf{P} = \int_{C_+} X\,d\mathbf{P}$ and $\int Y^- d\mathbf{P} = \int_{C_-} Y\,d\mathbf{P} = \int_{C_-} X\,d\mathbf{P}$ since $C_\pm \in \mathfrak{G}$. From this, it follows that Y^+ and Y^- are integrable and so Y is integrable.

Finally, if Y_1 and Y_2 are two conditional expectations of an integrable random variable X, it follows (as in the case of non-negative X) that $Y_1 = Y_2$ **P**-a.s. $\square$

The defining property of a conditional expectation implies that properties (CE_1) to (CE_5) are automatically satisfied when they make sense.

Proposition 5.1.4. (Elementary properties of conditional expectation) *Let $X, X_1,$ and X_2 denote non-negative random variables for which the conditional expectations $E[X|\mathfrak{G}], E[X_1|\mathfrak{G}]$, and $E[X_2|\mathfrak{G}]$, respectively, are defined. Then*

(CE_1) $\lambda E[X|\mathfrak{G}] = E[\lambda X|\mathfrak{G}]$ if $\lambda \geq 0$,
(CE_2) $E[X_1|\mathfrak{G}] + E[X_2|\mathfrak{G}] = E[X_1 + X_2|\mathfrak{G}]$,
(CE_3) $E[X_1|\mathfrak{G}] \leq E[X_2|\mathfrak{G}]$ **P**-a.s. if $X_1 \leq X_2$,
(CE_4) if $X \in \mathfrak{G}$, $E[X|\mathfrak{G}] = X$, and hence $E[E[X|\mathfrak{G}]|\mathfrak{G}] = E[X|\mathfrak{G}]$,
(CE_5) $E[X] = E[E[X|\mathfrak{G}]]$.

Proof. All of these properties except (3) are immediate consequences of the definition of a conditional expectation. The third property follows from (2) and the fact (Lemma 5.1.3) that a conditional expectation of a non-negative random variable is itself non-negative. $\square$

Comment. Identities like (2) have to be read modulo null random variables in the same way that they are for functions in, for example, L^1: to say $\|X\|_1 = 0$ means that X is a null function — not that it is identically zero. Similarly, for example, (2) says that the sum of conditional expectations of X_1 and X_2 is a conditional expectation for the sum $X_1 + X_2$.

The issue now is to show that a conditional expectation exists for any random variable X that is either non-negative or integrable.

The easiest way to do this for non-negative integrable random variables is to use the Radon–Nikodym theorem (Theorem 2.7.19) from Chapter II (although that theorem must then first be proved). Assume that X is a non-negative integrable random variable. Then $\nu(C) = \int_C X d\mathbf{P}$, for $C \in \mathfrak{G}$, is a finite measure on $\mathfrak{G}$ by Exercise 2.4.2. Since $\nu(C) = 0$ if $\mathbf{P}(C) = 0$, the Radon–Nikodym theorem implies that there is an integrable, non-negative random variable Y on $(\Omega, \mathfrak{G}, \mathbf{P})$ such that, for all $C \in \mathfrak{G}$, $\int_C Y d\mathbf{P} = \nu(C) = \int_C X d\mathbf{P}$.

In place of the Radon–Nikodym theorem, one may also use the Riesz representation theorem to define the conditional expectation for random variables in the Hilbert space $L^2(\Omega, \mathfrak{F}, \mathbf{P})$ and then extend it to all integrable random variables. The advantage of this approach to conditional expectation is that it avoids having to prove the Radon–Nikodym theorem and shows the connection between conditional expectation and projection.

The Riesz representation theorem.

Let V be a real vector space. A set $C \subset V$ is said to be **convex** if for $x, y \in C$ and all $t \in [0, 1], tx + (1-t)y \in C$. In other words, the line segment joining x to y lies in C.

For example, if $V = \mathbb{R}^2$, then $\{(x,y) | x^2 + y^2 \leq 1\}$, $\{(x,y) | y \geq x^2\}$, and $\{(x,y) | x \geq 0, 0 \leq y \leq mx, m > 0\}$ are examples of convex sets C as are their translates $C + u = \{v + u \mid v \in C\}$ by any vector u. The next two results should appear obvious in view of the familiar geometry of $\mathbb{R}^2$ (draw some diagrams).

Proposition 5.1.5. *Let $C \subset \mathbb{R}^2$ be a closed convex set. Then there is a unique point $c_0 \in C$ closest to the origin. In other words, there is exactly one point $c_0 \in C$ such that $\| c_0 \| \leq \| c \|$ for all $c \in C$.*

Proof. Suppose that c_1, c_2 are two points in C that minimize the distance from 0. The midpoint $\frac{1}{2}(c_1 + c_2)$ of the line segment joining c_1 to c_2 is in C. Hence, if $\delta = \| c_i \|$, $\delta^2 \leq \frac{1}{4} \| c_1 + c_2 \|^2 = \frac{1}{4}\{2\delta^2 + 2c_1 \cdot c_2\}$, and so $\delta^2 \leq c_1 \cdot c_2 \leq \| c_1 \| \| c_2 \| = \delta^2$. Consequently, c_1 and c_2 are collinear, i.e., one is a scalar multiple of the other. Therefore, $c_1 = c_2$ since c_1 (say) $= \lambda c_2$ implies $\lambda = \pm 1$: if $\lambda = -1$, then $c_1 = -c_2$ and so $0 \in C$; this implies $c_1 = c_2 = 0$ as $\delta = 0$. Consequently, there is at most one point in C that is closest to the origin.

1. CONDITIONAL EXPECTATION AND HILBERT SPACE

Now let $(c_n)_{n\geq 1} \subset C$ be such that $\| c_n \| \geq \| c_{n+1} \|$ and $\lim_n \| c_n \| = \delta = \inf_{c \in C} \| c \|$. If this sequence converges to a point c_0, then (i) $c_0 \in C$ and (ii) $\| c_0 \| = \lim_n \| c_n \| = \delta$ as $\big| \| c_0 \| - \| c_n \| \big| \leq \| c_0 - c_n \|$ by the triangle inequality.

To prove convergence, it suffices to verify that the Cauchy criterion is verified (see Definition 4.1.25 and Exercise 4.7.12): $\| c_n - c_m \| < \epsilon$ if $n, m \geq N = N(\epsilon)$ for all $\epsilon > 0$.

Consider two vectors u, v and the parallelogram they determine. Then $\| u \|^2 + \| v \|^2 = \frac{1}{2}\{\| u+v \|^2 + \| u-v \|^2\}$. Hence, $\frac{1}{2} \| c_n - c_m \|^2 = (\| c_n \|^2 + \| c_m \|^2) - \frac{1}{2} \| c_n + c_m \|^2 \leq \| c_n \|^2 + \| c_m \|^2 - 2\delta^2$ as $\frac{1}{2}(c_n + c_m) \in C$ and so $\| c_n + c_m \| \geq 2\delta$. Since $\| c_n \|^2 + \| c_m \|^2 < 2\delta^2 + \epsilon$ for any $\epsilon > 0$, if $n, m > N$, it follows that $\| c_n - c_m \| < \epsilon$ for sufficiently large n and m. □

Corollary 5.1.6. *Let $L \subset \mathbb{R}^2$ be a (necessarily closed) linear subspace. Then for each $u \in \mathbb{R}^2$ there is a unique vector $u_1 \in L + u$ of minimum length. This vector u_1 is orthogonal to L. Further, if $u_1 = \ell + u$, then $-\ell$ is the point in L that is closest to u.*

Proof. $C = L + u = \{v + u \mid v \in L\}$ is a closed convex subset of $\mathbb{R}^2$. Let u_1 be the unique vector in C of minimum length. It will be shown to be perpendicular to L.

If u_1 is not perpendicular to L, then for some $a \in L$, $u_1 \cdot a \neq 0$. One may write $u_1 = \lambda a + w$ with $w \cdot a = 0$ and $\lambda \neq 0$ ($\lambda = (u_1 \cdot a)/ \| a \|^2$). As $u_1 = v_1 + u$, with $v_1 \in L$, it follows that $w \in L + u$. Now $\| u_1 \|^2 = \lambda^2 \| a \|^2 + \| w \|^2 > \| w \|^2$. This contradicts the defining property of u_1.

If $y \in L$, then $d(u, y) = \| u - y \|$. Since $u - y \in L + u$, it follows that $d(u, y) \geq \| u_1 \|$, with equality if and only if $u - y = u_1$, i.e. $y = -\ell$. □

The proofs of these two results make no use of the dimension of $\mathbb{R}^2$ and so are valid in a much larger context, i.e., namely for Hilbert spaces (see Definition 5.1.9).

A (real) Hilbert space is a real vector space equipped with an inner product (Definition 2.5.6) which satisfies a completeness condition.

Example 5.1.7. The following are examples of vector spaces equipped with an inner product (Definition 2.5.6). Recall that $\langle \, , \, \rangle$ is also denoted by $(\, , \,)$.

(1) $V = \mathbb{R}^n$, $\langle u, v \rangle = (u, v) = u \cdot v = \sum_{i=1}^n u_i v_i$.
(2) $V = \ell_2$, $\langle u, v \rangle = (u, v) = \sum_{i=1}^\infty u_i v_i$ (see Exercise 2.5.7).
(3) $V = L^2(\Omega, \mathfrak{F}, \mathbf{P})$, $(X, Y) = E[XY]$.

Remark. In this last example, $(X, X) = 0$ implies $X = 0$, **P**-a.s., and so X is a null function (see Exercise 2.1.28 and Remark 4.1.2).

Let V be a real vector space with inner product, and let $\| u \|$ denote $\sqrt{(u,u)}$, $u \in V$. Then, as shown in the following proposition, $u \to \| u \|$ is a norm on V (Definition 2.5.3).

Proposition 5.1.8. *Let V be a real vector space equipped with an inner product. The following statements are valid:*

(1) **(Parallelogram law)**
   for all $u, v \in V$, $\| u \|^2 + \| v \|^2 = \frac{1}{2}\{\| u+v \|^2 + \| u-v \|^2\}$;
(2) **(Cauchy–Schwarz inequality)**
   for all $u, v \in V$, $|(u,v)| \leq \| u \| \| v \|$ with equality if and only if u and v are collinear;
(3) **(The triangle inequality)**
   for all $u, v \in V$, $\| u+v \| \leq \| u \| + \| v \|$.

Hence, $\| \cdot \|$ is a norm on V and defines a metric d, where $d(u,v) \stackrel{\text{def}}{=} \| u-v \|$ on V.

Proof. (1) $\| u \pm v \|^2 = \| u \|^2 \pm 2(u,v) + \| v \|^2$. (2) Let $Q(\lambda) = \| u + \lambda v \|^2$. Then $Q(\lambda)$ is a quadratic polynomial in λ. Since $Q(\lambda) \geq 0$ for all $\lambda \in \mathbb{R}$, $4(u,v)^2 - 4\| u \|^2 \| v \|^2 \leq 0$. Further (see Exercise 2.5.8), this discriminant is zero if and only if Q vanishes at $\lambda_0 = \frac{-(u,v)}{\|v\|^2}$. Since $Q(\lambda_0) = 0$, $u = -\lambda_0 v$ (note that v may be assumed $\neq 0$ here, as the inequality is trivial if u or $v = 0$).

Observe that (3) and (2) are equivalent since $\| u+v \|^2 = \| u \|^2 + 2(u,v) + \| v \|^2$. $\square$

Definition 5.1.9. *A real vector space V equipped with an inner product (i.e., an inner product space) is said to be a (real) **Hilbert space** if it is complete with respect to the metric given by the norm $\| u \| = \sqrt{(u,u)}$.*

Remarks 5.1.10. (1) For a complex vector space, for example, $\mathbb{C}^n$, Definition 2.5.6 needs to be modified: the inner product $\langle u, v \rangle \in \mathbb{C}$; in (3), $\lambda \in \mathbb{C}$; and (1) is replaced by $\langle u, v \rangle = \overline{\langle v, u \rangle}$ for all $u, v \in V$. For example, in the case of $\mathbb{C}^n$, one defines $\langle u, v \rangle$ to be $\sum_{i=1}^{n} u_i \bar{v}_i$.

(2) The Riesz–Fischer theorem (Proposition 4.1.28) and Example 5.1.7 (3) imply that $L^2(\Omega, \mathfrak{F}, \mathbf{P})$ is a Hilbert space. So too is any $L^2(\Omega, \mathfrak{F}, \mu)$.

(3) By using complex measurable functions, one also obtains examples of (complex) Hilbert spaces: $L^2_{\mathbb{C}}(\Omega, \mathfrak{F}, \mu)$, the space of complex-valued $\mathfrak{F}$-measurable functions $f : \Omega \to \mathbb{C}$ with $\int |f|^2 d\mu < \infty$: note that if $f = u + iv$ then f is $\mathfrak{F}$-measurable if and only if u and v are (see Proposition 3.1.1); $|f|^2 = u^2 + v^2$ implies that $f \in L^2_{\mathbb{C}}(\Omega, \mathfrak{F}, \mu)$ if and only if u and v belong to $L^2(\Omega, \mathfrak{F}, \mu)$; and finally, $\langle f, g \rangle = \int f \bar{g} d\mu$.

Example 5.1.11. Let $V = \mathcal{C}([0,1])$, the vector space of continuous real-valued functions on $[0,1]$. Define (f,g) to be $\int_0^1 f(x)g(x)dx$. This defines an inner product on V, but V is not a Hilbert space: for example, let

$$f_n(x) = \begin{cases} 0 & \text{for } 0 \leq x \leq \frac{1}{2} - \frac{1}{n}, \\ nx - \left(\frac{n}{2} - 1\right) & \text{for } \frac{1}{2} - \frac{1}{n} < x \leq \frac{1}{2}, \\ 1 & \text{for } \frac{1}{2} < x \leq 1. \end{cases}$$

The sequence $(f_n)_{n\geq 1}$ is Cauchy but fails to converge to a continuous function. Note that by Theorem 4.2.8, $V \subset L^2([0,1])$ is a dense subspace of $L^2([0,1])$, which is a Hilbert space by Proposition 4.1.28 (see Remarks 5.1.10).

The following two results have been established by the proofs of Proposition 5.1.5 and Corollary 5.1.6.

Proposition 5.1.12. *Let C be a closed convex set in a Hilbert space V. Then there is a unique point $c_0 \in C$ closest to the origin.*

Corollary 5.1.13. *Let L be a closed linear subspace of a Hilbert space V. Then for each vector $u \in V$ there is a unique vector $u_1 \in L+u$ of minimum length. This vector is orthogonal to L. Further, if $u_1 = \ell + u$, then $-\ell$ is the point in L that is closest to u.*

Remark. If $L = L_N$ is the closed linear subspace determined by the first N vectors of an orthonormal system $(\phi_n)_{n\geq 1}^\infty$ in V, the vector u_1 equals $u - \sum_{n=1}^N \langle u, \phi_n \rangle \phi_n$ (see Lemma 4.2.15).

Theorem 5.1.14. **(Riesz representation theorem)** *Let V be a Hilbert space and let $\ell : V \to \mathbb{R}$ be a non-trivial, continuous linear functional. Then there is a unique $v_0 \in V$ such that $\ell(v) = (v, v_0)$ for all $v \in V$.*

Proof. Let $\mathrm{Ker}(\ell) \overset{\mathrm{def}}{=} \{v \in V \mid \ell(v) = 0\}$. This is a closed linear subspace L. Since ℓ is non-trivial, there is a vector v_1 with $\ell(v_1) = 1$. Let $E = L + v_1$, and let u_0 be the vector in E of minimum length. Then, since $\ell = 1$ on E, $\| u_0 \| \neq 0$.

By Corollary 5.1.13, $(v, u_0) = 0$ for all $v \in L$. In other words, $\ell(v) = 0$ implies $(v, u_0) = 0$. Now let $\ell(v) = \alpha \neq 0$. Since $\ell(u_0) = 1$, $\ell(v - \alpha u_0) = 0$. Hence, $(v - \alpha u_0, u_0) = 0$, i.e., $(v, u_0) = \alpha(u_0, u_0) = \alpha \| u_0 \|^2$.

Consequently, if $v_0 = \frac{1}{\|u_0\|^2} u_0$, then $\ell(v) = (v, v_0)$ for all $v \in V$.

The vector v_0 is unique because $\ell(v) = (v, v_0')$ for all $v \in V$ implies $(v, v_0 - v_0') = 0$ for all $v \in V$; in particular, for $v = v_0 - v_0'$. Hence, $v_0 - v_0' = 0$. □

2. Conditional Expectation

It follows immediately from the Riesz representation theorem that every square integrable random variable has a conditional expectation.

Proposition 5.2.1. *Let $X \in L^2(\Omega, \mathfrak{F}, \mathbf{P})$ and $\mathfrak{G} \subset \mathfrak{F}$. Then the element Y of $L^2(\Omega, \mathfrak{G}, \mathbf{P})$ that represents the continuous linear functional $Z \to E[XZ]$ on the Hilbert space $L^2(\Omega, \mathfrak{G}, \mathbf{P})$ is a conditional expectation Y of X given $\mathfrak{G}$.*

Proof. Let $Z \in L^2(\Omega, \mathfrak{G}, \mathbf{P})$. The function $Z \to E[XZ]$ is a continuous linear functional on $L^2(\Omega, \mathfrak{G}, \mathbf{P})$ since by the Cauchy–Schwarz inequality

$|E[XZ]| \leq \|X\|_2 \|Z\|_2$. Hence, by the Riesz representation theorem (Theorem 5.1.14) there is a unique (up to null function) $Y \in L^2(\Omega, \mathfrak{G}, \mathbf{P})$ such that $E[XZ] = E[YZ]$ for all $Z \in L^2(\Omega, \mathfrak{G}, \mathbf{P})$.

Since $1_C \in L^2(\Omega, \mathfrak{G}, \mathbf{P})$ for all $C \in \mathfrak{G}$, it follows that $E[1_C X] = E[1_C Y]$ (i.e., $\int_C X d\mathbf{P} = \int_C Y d\mathbf{P}$) for all $C \in \mathfrak{G}$. In other words, Y is a conditional expectation of X given $\mathfrak{G}$. □

Remark 5.2.2. Proposition 5.2.1 is a well-known fact from Hilbert space theory. Given a closed (linear) subspace V of a Hilbert space H, there is a unique linear operator $P : H \to V$ with $\| P \| \leq 1$ and $P^2 = P$. In the case of $H = L^2(\Omega, \mathfrak{F}, \mathbf{P})$ and $V = L^2(\Omega, \mathfrak{G}, \mathbf{P})$, this operator is the conditional expectation operator $E[\cdot|\mathfrak{G}]$. Such an operator is called a **projection operator**. If $P(x) = y$, then y is the closest point in V to x (see Corollary 5.1.13). In other words, y solves a **least squares** problem. This terminology comes from the observation that if $H = \mathbb{R}^n$, then y is the point $v \in V$ that minimizes $\sum_{i=1}^n (v_i - x_i)^2$.

By Lemma 5.1.3, $E[X|\mathfrak{G}]$ is non-negative if $X \in L^2(\Omega, \mathfrak{F}, \mathbf{P})$ is non-negative. This allows one to extend $E[\cdot|\mathfrak{G}]$ to arbitrary positive random variables and so to integrable random variables.

Proposition 5.2.3. *If X is a non-negative random variable on $(\Omega, \mathfrak{F}, \mathbf{P})$ and $\mathfrak{G}$ is a sub σ-algebra let $Y \stackrel{\text{def}}{=} \lim_n E[X \wedge n|\mathfrak{G}]$. Then Y is a conditional expectation of X given $\mathfrak{G}$.*

If $X \in L^1(\Omega, \mathfrak{F}, \mathbf{P})$, then $E[X^\pm|\mathfrak{G}] \in L^1(\Omega, \mathfrak{G}, \mathbf{P})$. Let $Y \stackrel{\text{def}}{=} E[X^+|\mathfrak{G}] - E[X^-|\mathfrak{G}]$. Then $Y \in L^1(\Omega, \mathfrak{G}, \mathbf{P})$ is a conditional expectation of X given $\mathfrak{G}$.

Proof. Assume that X is non-negative. Then, or each $n \geq 1$, $X \wedge n \in L^2(\Omega, \mathfrak{F}, \mathbf{P})$. Let $Y_n = E[X \wedge n|\mathfrak{G}]$. Then $Y_{n+1} \geq Y_n$ $\mathbf{P}$-a.s. Hence, by modifying the variables Y_n on sets in $\mathfrak{G}$ of probability zero (which does not change the fact that they are conditional expectations of the $X \wedge n$), one may assume $0 \leq Y_n \leq Y_{n+1}$ for all $n \geq 1$.

Define $Y = \lim_n Y_n$, and let $C \in \mathfrak{G}$. Then,

$$\int_C X d\mathbf{P} = \lim_{n \to \infty} \int_C (X \wedge n) d\mathbf{P} = \lim_{n \to \infty} \int_C Y_n d\mathbf{P} = \int_C Y d\mathbf{P},$$

i.e., Y is a conditional expectation of X given $\mathfrak{G}$.

If $X \in L^1(\Omega, \mathfrak{F}, \mathbf{P})$, then the conditional expectations $E[X^\pm|\mathfrak{G}]$ are integrable by Lemma 5.1.3, and so $Y = E[X^+|\mathfrak{G}] - E[X^-|\mathfrak{G}] \in L^1(\Omega, \mathfrak{G}, \mathbf{P})$. It is a conditional expectation of X given $\mathfrak{G}$ since

$$\int_C X d\mathbf{P} = \int_C X^+ d\mathbf{P} - \int_C X^- d\mathbf{P} = \int_C Y d\mathbf{P} \quad \text{for all } C \in \mathfrak{G}. \quad \square$$

2. CONDITIONAL EXPECTATION

Remark. The conditional expectation can also be defined for any random variable that is the sum of a non-negative random variable Z and an integrable one X. Let $E[Z|\mathfrak{G}] = W$ and $E[X|\mathfrak{G}] = Y$. Then for any set $C \in \mathfrak{G}$, it follows from Exercise 2.9.19 that $\int_C (Z+X)d\mathbf{P} = \int_C Z d\mathbf{P} + \int_C X d\mathbf{P} = \int_C W d\mathbf{P} + \int_C Y d\mathbf{P} = \int_C (W+y)d\mathbf{P}$. As a result, $E[Z+X|\mathfrak{G}] = E[Z|\mathfrak{G}] + E[X|\mathfrak{G}]$.

Proposition 5.2.4. (Additional properties of conditional expectation) *Assume that each random variable is either non-negative or in L^1. Then the following properties hold:*

(CE_6) *if $(X_n)_{n \geq 1}$ is such that $X_n \leq X_{n+1}$ and $\lim_n X_n = X$, then $E[X|\mathfrak{G}] = \lim_n E[X_n|\mathfrak{G}]$;*

(CE_7) *if $\mathfrak{H}$ is a sub σ-field of $\mathfrak{G}$, then $E[E[X|\mathfrak{G}]|\mathfrak{H}] = E[X|\mathfrak{H}]$;*

(CE_8) **(Jensen's inequality)** *Let $c: \mathbb{R} \to \mathbb{R}$ be convex and let $X \in L^1$. Then $E[c \circ X]$ is defined and*

$$c \circ E[X|\mathfrak{G}] \leq E[c \circ X|\mathfrak{G}];$$

(CE_9) *if $X \in \mathfrak{G}$, then $E[XY|\mathfrak{G}] = X E[Y|\mathfrak{G}]$ (here $X, Y \in L^2$ or both are non-negative);*

(CE_{10}) *if X is independent of $\mathfrak{G}$ (i.e., $\sigma(X)$ and $\mathfrak{G}$ are independent σ-fields), then $E[X|\mathfrak{G}] = E[X]$; and*

(CE_{11}) *if $\mathfrak{H}$ and $\sigma(\sigma(X) \cup \mathfrak{G})$ are independent, then $E[X|\sigma(\mathfrak{G} \cup \mathfrak{H})] = E[X|\mathfrak{G}]$, where $\sigma(\mathfrak{G} \cup \mathfrak{H})$ is the smallest σ-algebra containing $\mathfrak{G} \cup \mathfrak{H}$.*

Proof. (CE_6) and (CE_7) are left to the reader to verify. To prove (CE_8), note that by Exercise 4.7.25, there exists a sequence $(L_n)_{n \geq 1}$ of linear functions $L_n(x) = a_n x + b_n$ with $c(x) = \sup_n L_n(x), x \in \mathbb{R}$. Now $c \circ X \geq L_n \circ X$, and so $c \circ X = Z_n + L_n \circ X, Z_n \geq 0$. It follows from the remark preceding this proposition that $E[c \circ X|\mathfrak{G}]$ is defined (or one may impose the extra condition that $c \circ X \in L^1$) and that $E[c \circ X|\mathfrak{G}] = E[Z_n] + E[L_n \circ X|\mathfrak{G}] \geq E[L_n \circ X|\mathfrak{G}] = a_n E[X|\mathfrak{G}] + b_n$ for all n. Hence, $E[c \circ X|\mathfrak{G}] \geq c(E[X|\mathfrak{G}])$.

If $A \in \mathfrak{G}$, then (CE_9) holds for $X = 1_A$. Hence, it holds for all simple functions in $\mathfrak{G}$ and consequently for all non-negative $X \in \mathfrak{G}$. This proves (CE_9).

(CE_{10}) follows immediately from Proposition 3.2.3 since $\int_C X d\mathbf{P} = \mathbf{P}(C)E[X]$ if $C \in \mathfrak{G}$.

To prove (CE_{11}), first note that $Y = E[X|\mathfrak{G}] \in \mathfrak{G}$ is independent of $\mathfrak{H}$. Thus, by Proposition 3.2.3, for any $B \in \mathfrak{H}$ and $A \in \mathfrak{G}$,

$$\int_{A \cap B} Y d\mathbf{P} = \mathbf{P}[B] \int_A Y d\mathbf{P}$$
$$= \mathbf{P}[B] \int_A X d\mathbf{P} = \int 1_B (1_A X) d\mathbf{P} = \int_{A \cap B} X d\mathbf{P},$$

where the last but one equality uses the hypothesis that $\mathfrak{H}$ and $\sigma(X)\cup\mathfrak{G}$ are independent. Since the class $\mathfrak{C}$ of sets $C \in \sigma(\mathfrak{G} \cup \mathfrak{H})$ for which $\int_C Y d\mathbf{P} = \int_C X d\mathbf{P}$ contains $C_2\setminus C_1$ whenever it contains each C_i and $C_1 \subset C_2$, the result follows from Proposition 3.2.6. □

Corollary 5.2.5. Let $X \in L^p(\Omega, \mathfrak{F}, \mathbf{P})$, $1 \le p \le +\infty$. Then $E[X|\mathfrak{G}] \in L^p(\Omega, \mathfrak{G}, \mathbf{P})$ and
$$\| E[X|\mathfrak{G}] \|_p \le \| X \|_p.$$

Proof. If $1 \le p < \infty$, $t \to |t|^p$ is a convex function, and so by Jensen's inequality, $\bigl(|E[X|\mathfrak{G}]|\bigr)^p \le E[|X|^p|\mathfrak{G}]$. Hence,

$$\| E[X|\mathfrak{G}] \|_p \le \bigl(E[|X|^p|\mathfrak{G}]\bigr)^{\frac{1}{p}} = \| X \|_p.$$

If $p = +\infty$ then $-\| X \|_\infty \le X \le \| X \|_\infty$ **P**-a.e. Hence, by (CE_3), $-\| X \|_\infty \le E[X|\mathfrak{G}] \le \| X \|_\infty$ as $E[1\,|\mathfrak{G}] = 1$. □

Remark. In particular, if $X \in L^2$, $E[X|\mathfrak{G}] \in L^2$ and so as remarked before, it is (up to a null function) the random variable Y in L^2 for which $E[XZ] = E[YZ]$ for all $Z \in L^2(\Omega, \mathfrak{G}, \mathbf{P})$.

When the σ-algebra $\mathfrak{G}$ is generated by a random variable Y (i.e., $\mathfrak{G} = \sigma(Y)$), then one often sees expressions of the type $E[X|Y = y]$ in connection with $E[X|\sigma(Y)] \stackrel{\text{def}}{=} E[X|Y]$. To explain why the expression $E[X|Y = y]$ is meaningful in this context, one needs to see the relation between Borel functions and random variables that are measurable with respect to $\sigma(Y)$.

Exercise 5.2.6. Let Y denote a random variable on a probability space $(\Omega, \mathfrak{F}, \mathbf{P})$. Let $\sigma(Y)$ denote the smallest σ-algebra with respect to which Y is measurable. Show that
  (1) $A \in \sigma(Y)$ if and only if $A = Y^{-1}(B)$; equivalently, $1_A = 1_B \circ Y$ for some Borel subset B of $\mathbb{R}$ (see Exercise 2.9.10),
  (2) if $A_1, A_2 \in \sigma(Y)$ and are disjoint, then $A_i = Y^{-1}(B_i)$, $i = 1, 2$, implies that $A_2 = Y^{-1}(B_2\setminus B_1)$; equivalently, $1_{A_2} = 1_{B_2\setminus B_1} \circ Y$,
  (3) if s is $\sigma(Y)$-simple, then there is a Borel simple function t with $s = t \circ Y$;
  (4) if U is non-negative and $\sigma(Y)$-measurable, then there is a non-negative Borel function ψ with $U = \psi \circ Y$, [Hint: if $s_2 \ge s_1$ and $s_i = t_i \circ Y$ then $s_2 = (t_1 \vee t_2) \circ Y$], and finally,
  (5) if U is finite and $\sigma(Y)$-measurable, then there is a finite Borel function ψ with $U = \psi \circ Y$.

Using the result of this exercise, one defines $E[X|Y = y]$ to be $f(y)$ if $E[X|Y] = f \circ Y$, $f \in \mathfrak{B}(\mathbb{R})$. This kind of notation is used in Karlin and Taylor [K1], p. 5, where a formula is given for $\mathbf{P}[X \le x|Y = y]$ (the **conditional distribution function** of X given $Y = y$) when X and Y

2. CONDITIONAL EXPECTATION

are random variables for which the distribution of the random vector (X,Y) has a density $p(x,y)$. This formula will now be explained as a special case of the formula for the conditional expectation of $g(X) = g \circ X$ given $Y = y$, (i.e., for $E[g(X)|Y = y]$), (see [K1], p. 7 (1.5)), by using the concept of conditional expectation as introduced earlier (here g is a Borel function for which $g \circ X \in L^1(\Omega, \mathfrak{F}, \mathbf{P})$). Note that the density p, which is in principle only Lebesgue measurable on $\mathbb{R}^2$, can be taken to be Borel measurable.

Proposition 5.2.7. *Let X, Y be random variables on a probability space $(\Omega, \mathfrak{F}, \mathbf{P})$, and let $g \in \mathfrak{B}(\mathbb{R})$ be such that $g \circ X \in L^1(\Omega, \mathfrak{F}, \mathbf{P})$. Assume that the joint distribution of (X, Y) has a density $p(x, y)$ (which may be assumed to be a Borel function of x and y). Then $p_2(y) \stackrel{\text{def}}{=} \int p(x,y)dx$ is the density of the (marginal) distribution of Y, and*

$$E[g \circ X | Y = y] = \begin{cases} \frac{1}{p_2(y)} \int g(x) p(x,y) dx & \text{if } p_2(y) \neq 0, \\ 0 & \text{if } p_2(y) = 0. \end{cases}$$

Proof. If $\mathbf{Q}(dx, dy) = p(x,y)dxdy$ is the joint distribution of (X,Y), its second marginal is the distribution of Y, i.e., $\mathbf{P}[Y \leq b] = \mathbf{Q}[\mathbb{R} \times (-\infty, b]] = \int_{-\infty}^{b} [\int_{-\infty}^{+\infty} p(x,y)dx] dy$, which has as density the Borel function $p_2(y) = \int p(x,y)dx$ (see Fubini's theorem (Theorem 3.3.5)).

If $E[g \circ X | \sigma(Y)] = U$, then for all $A = Y^{-1}(B) \in \sigma(Y)$, it follows that

$$\int_A (g \circ X) d\mathbf{P} = \int_A U d\mathbf{P},$$

that is,

$$\int (1_B \circ Y)(g \circ X) d\mathbf{P} = \int (1_B \circ Y) U d\mathbf{P}$$

for all $B \in \mathfrak{B}(\mathbb{R})$.

Let $\Phi(x,y) = 1_B(y)g(x)$. Then

$$\int (1_B \circ Y)(g \circ X) d\mathbf{P} = \int \Phi(X, Y) d\mathbf{P} = \int \int 1_B(y) g(x) p(x,y) dx dy.$$

Also, by Exercise 5.2.6, one has $U = \phi \circ Y$, where $\phi(y)$ is a Borel function and $\int |\phi(y)| \mathbf{Q}(dx, dy) < +\infty$ since U is integrable.

Hence,

$$\int (1_B \circ Y) U d\mathbf{P} = \int (1_B \circ Y)(\phi \circ Y) d\mathbf{P} = \int 1_B(y) \phi(y) p_2(y) dy,$$

and so

$$\int \int 1_B(y) g(x) p(x,y) dx dy = \int 1_B(y) \phi(y) p_2(y) dy \quad \text{for all } B \in \mathfrak{B}(\mathbb{R}).$$

The left-hand side is the integral of $\Phi(x,y)p(x,y)$ with respect to Lebesgue measure on $\mathfrak{B}(\mathbb{R}^2)$. Obviously, one wants to compute it by Fubini's theorem and thus to get

$$\int 1_B(y)\left[\int g(x)p(x,y)dx\right]dy = \int 1_B(y)\phi(y)p_2(y)dy \quad \text{for all } B \in \mathfrak{B}(\mathbb{R}).$$

Assuming this is possible (which indeed is the case: see Exercise 5.2.8 (2)) if one sets

$$\psi(y) \stackrel{\text{def}}{=} \frac{\int_{-\infty}^{+\infty} g(x)p(x,y)dx}{\int_{-\infty}^{+\infty} p(x,y)dx} = \frac{\int_{-\infty}^{+\infty} g(x)p(x,y)dx}{p_2(y)} \quad \text{when } p_2(y) \neq 0,$$

and equal to zero otherwise, then $\psi(y)p_2(y) = \int_{-\infty}^{+\infty} g(x)p(x,y)dx$ for all $y \in \mathbb{R}$ and so

$$\int 1_B(y)\psi(y)p_2(y)dy = \int 1_B(y)\phi(y)p_2(y)dy \quad \text{for all } B \in \mathfrak{B}(\mathbb{R}).$$

Provided ψ is a Borel function (see Exercise 5.2.8 (1)), it follows that

$$\int_{Y^{-1}(B)} (g \circ X)d\mathbf{P} = \int_{Y^{-1}(B)} Ud\mathbf{P} = \int_{Y^{-1}(B)} (\phi \circ Y)d\mathbf{P}$$
$$= \int_{Y^{-1}(B)} (\psi \circ Y)d\mathbf{P} \quad \text{for all } B \in \mathfrak{B}(\mathbb{R}).$$

This implies that $\psi \circ Y = E[g \circ X | \sigma(Y)]$, and so $E[g \circ X | Y = y] = \psi(y)$. □

When $g = 1_A$, $g \circ X = 1_{\{X \in A\}}$, which implies that

$$\mathbf{P}[X \in A | Y = y] = \frac{\int_{-\infty}^{+\infty} 1_A(u)p(u,y)du}{p_2(y)} \quad \text{if } p_2(y) \neq 0$$

and is equal to zero otherwise. In particular, the conditional probability distribution function of X given $Y = y$ is

$$\mathbf{P}[X \leq x | Y = y] = \frac{\int_{-\infty}^{x} p(u,y)du}{p_2(y)} \quad \text{if } p_2(y) \neq 0,$$

and equals zero otherwise. Notice that the normalisation by $p_2(y)$ is exactly what is needed to give a distribution function. Finally, when $X \in L^1$,

$$E[X|Y=y] = \frac{\int_{-\infty}^{+\infty} up(u,y)du}{p_2(y)} \quad \text{if } p_2(y) \neq 0,$$

and as usual equals zero when this is not the case. To complete this discussion, it remains to settle the technical issues concerning the use of the Fubini theorem and the measurability of ψ. They are given as an exercise.

Exercise 5.2.8. (Technical details for Proposition 5.2.7) If V is a non-negative measurable function on $(\Omega, \mathfrak{F})$, show that

(1) Z is also a measurable function on $(\Omega, \mathfrak{F})$, where

$$Z(\omega) = \begin{cases} \frac{1}{V(\omega)} & \text{if } V(\omega) > 0, \\ 0 & \text{if } V(\omega) = 0. \end{cases}$$

[Hint: first show that $\frac{1}{V}$ is a measurable function if one defines $\frac{1}{0}$ to be $+\infty$.]

Using the notations of Proposition 5.2.7, show that

(2) $|g(x)|p(x,y) \in L^1(\mathbb{R}^2)$, i.e., relative to Lebesgue measure on $\mathfrak{B}(\mathbb{R}^2)$ since $g \circ X \in L^1(\Omega, \mathfrak{F}, \mathbf{P})$,

(3) there is a Borel set $\Gamma \subset \mathbb{R}$ with $|\Gamma| = 0$ such that the function γ defined by setting

$$\gamma(y) = \begin{cases} \int g(x)p(x,y)dx & \text{if } y \notin \Gamma, \\ 0 & \text{if } y \in \Gamma, \end{cases}$$

is a Borel function and $\int 1_B(y)\gamma(y)dy = \int \int 1_B(y)g(x)p(x,y)dxdy$.

If one lets this function $\gamma(y)$ stand for $\int g(x)p(x,y)dx$, the computation in Proposition 5.2.7 goes through. The function $\psi(y)$ is then the product of $\gamma(y)$ and a Borel function by (1).

This illustrates the technical difficulties of using Fubini's theorem for L^1-functions. The formalism is clear, but the fine print needs inspection.

The initial examples of $E[\,\cdot\,|\mathfrak{G}]$ used a procedure for obtaining $E[X|\mathfrak{G}]$ from X that involved integration. In the case of n sets B_i each of positive probability, define $\mathbf{N}(\omega, d\eta)$ by $\mathbf{N}(\omega, A) = \mathbf{P}(A|B_i)$ if $\omega \in B_i$. Then $\mathbf{N}$ is a Markov kernel on $(\Omega, \mathfrak{F})$ to $(\Omega, \mathfrak{G})$ and $E[X|\mathfrak{G}](\omega) = \int \mathbf{N}(\omega, d\eta)X(\eta)$. This last equality holds because if $\mathbf{P}(d\omega|B)$ denotes the conditional probability $\mathbf{P}(A|B)$, for $A \in \mathfrak{F}$, where $\mathbf{P}(B) > 0$, then $\int X(\omega)\mathbf{P}(d\omega|B) = \frac{1}{\mathbf{P}(B)}\int XdP$. To see this, note that it suffices to consider $X = 1_A$, $A \in F$. Then $\int 1_A(\omega)\mathbf{P}(d\omega|B) = \mathbf{P}(A|B) = \frac{\mathbf{P}(A \cap B)}{\mathbf{P}(B)} = \frac{1}{\mathbf{P}(B)}\int_B 1_A d\mathbf{P}$.

It is important (and useful) to know when the conditional expectation operator may be realized in this way by a Markov kernel. In particular, this Markov kernel automatically selects a conditional expectation of a random variable, as indicated in the following definition.

Definition 5.2.9. Let $(\Omega, \mathfrak{F}, \mathbf{P})$ be a probability space, and let $\mathfrak{G} \subset \mathfrak{F}$. A Markov kernel $\mathbf{N}$ from $(\Omega, \mathfrak{F})$ to $(\Omega, \mathfrak{G})$ is called a **regular conditional probability** if, for all non-negative X, the random variable Y defined by

$$Y(\omega) = \int \mathbf{N}(\omega, d\omega')X(\omega')$$

is a conditional probability $E[X|\mathfrak{G}]$.

From the previous remarks, it is clear that if $\mathfrak{G}$ is a finite σ-field, $E[\,\cdot\,|\mathfrak{G}]$ is realized by a regular conditional probability. This is because for a finite σ-algebra $\mathfrak{G}$, the whole space Ω decomposes into a finite number of disjoint sets $B_i \in \mathfrak{G}$ each of which has no proper subset in $\mathfrak{G}$ (they are called the **atoms** of $\mathfrak{G}$). If $\mathbf{P}(B_i) > 0$ for each atom, then a regular conditional probability is defined as above. If some of the $\mathbf{P}(B_i) = 0$, then define $\mathbf{N}(\omega, A) = \mathbf{P}(A)$ for $\omega \in B_i$ when $\mathbf{P}(B_i) = 0$ and otherwise as above. A regular conditional probability $\mathbf{N}$ is then defined.

Another case when a regular conditional probability may be defined is for $\mathfrak{G} = \sigma(\{Y\})$, where Y is a random variable that assumes at most a countable number of values a_i, $i \geq 1$. The sets $B_i = \{Y = a_i\}$ are the atoms of $\mathfrak{G}$, and one may define (as before)

$$\mathbf{N}(\omega, A) = \begin{cases} \mathbf{P}(A\,|B_i) & \text{if } \omega \in B_i, \mathbf{P}(B_i) > 0, \\ \mathbf{P}(A) & \text{if } \omega \in B_i, \mathbf{P}(B_i) = 0. \end{cases}$$

Remark 5.2.10. While it is not always possible to realize a conditional expectation operator by means of a regular conditional probability, for a large class of measurable spaces this is the case (see [I], p. 13, Theorem 3.1) These are the so-called **standard measurable spaces** — they are measure isomorphic to one of the following spaces — see Exercise 4.7.21. Ikeda and Watanabe, [I], p. 13 or Neveu [N1]):

(i) $(\{1, 2, \ldots, n\}, \mathfrak{P}(\{1, 2, \ldots, n\}))$;
(ii) $(\mathbb{N}, \mathfrak{P}(\mathbb{N}))$;
(iii) $(\mathbb{R}, \mathfrak{B}(\mathbb{R}))$.

If Ω is a complete separable metric space (a so-called **Polish space**), then $(\Omega, \mathfrak{B}(\Omega))$ is a standard measure space ("**separable**" means that there is a countable subset S such that every non-void open set contains a point of S: e.g., $\mathbb{R}^n$ is a separable metric space relative to the Euclidean metric; the set S of points all of whose coordinates is rational has this property). In addition, every measurable subset C of a standard measurable space $(E, \mathfrak{E})$ is itself a standard measurable space when equipped with the σ-algebra of sets $C \cap A, A \in \mathfrak{E}$ (see Parthasarathy [P1], p. 135, Theorem 2.3).

Remark. The formula (1.50) in [K1] for $E[g(X)|Y = y]$ that was established in Proposition 5.2.7 is not given by a regular conditional probability, as it depends upon the density for the pair (X, Y) and so varies with X.

Remark 5.2.11. In Proposition 3.5.8, a probability $\mathbf{Q}$ is constructed on $(\mathbb{R} \times \mathbb{R}^n, \mathfrak{B}(\mathbb{R}^{n+1}))$ from a Markov kernel $\mathbf{N}$ on $(\mathbb{R}, \mathfrak{B}(\mathbb{R}))$ and an initial probability $\mathbf{P}_0$ on $\mathbb{R}$. In Remark 3.5.9 (2), the following formula is given, where ψ is a non-negative function of x_{k+1}, $\tilde{E}_0$ is a Borel subset of $\mathbb{R} \times \mathbb{R}^k$, and $\mathbf{N}\psi(u) = \int \mathbf{N}(u, dv)\psi(v)$:

$$E[1_{\tilde{E}_0}(x)\psi(x_{k+1})] = E[1_{\tilde{E}_0}(x)\mathbf{N}\psi(x_k)].$$

Define the random variables X_k on $\mathbb{R} \times \mathbb{R}^n$, for $0 \leq k \leq n$ by setting $X_k(x_0, x_1, \ldots, x_n) = x_k$. The above formula can be restated as follows:

$$\int_{\tilde{E}_0} \psi(X_{k+1})d\mathbf{Q} = \int_{\tilde{E}_0} (\mathbf{N}\psi)(X_k)d\mathbf{Q} \text{ if } \tilde{E}_0 \in \mathfrak{B}(\mathbb{R}^{k+1}).$$

In other words, if $\mathfrak{F}_k = \sigma\{X_i \mid 0 \leq i \leq k\}$, then

$$E[\psi \circ X_{k+1} | \mathfrak{F}_k] = \mathbf{N}\psi \circ X_k$$

since $\Lambda \in \mathfrak{F}_k$ if and only if $\Lambda = \tilde{E}_0 \times \mathbb{R}^{n-k}$ with $\tilde{E}_0 \in \mathfrak{B}(\mathbb{R}^{k+1})$ (see Proposition 3.4.9).

Exercise 5.2.12. (**Markov processes**) Let $(X_n)_{n \geq 0}$ be a (real-valued) stochastic process on $(\Omega, \mathfrak{F}, \mathbf{P})$, and let $\mathbf{N}$ be a transition kernel from $(\mathbb{R}, \mathfrak{B}(\mathbb{R}))$ to $(\mathbb{R}, \mathfrak{B}(\mathbb{R}))$. Denote by $\mathfrak{F}_n$ the σ-field $\sigma(\{X_i \mid 0 \leq i \leq n\})$. The stochastic process is called a **Markov process with transition kernel N** if

(*) $$E[\psi \circ X_{n+1} | \mathfrak{F}_n] = \mathbf{N}\psi \circ X_n$$

for any bounded Borel function ψ.

The σ-fields $\mathfrak{F}_n$ increase with n and can be replaced by any increasing sequence (referred to as a **filtration** in §4), provided that $X_n \in \mathfrak{F}_n$ for each n (the process is then said to be **adapted** (Definition 5.4.1)).

Let $\mathfrak{T}_n \stackrel{\text{def}}{=} \sigma(\{X_j \mid j \geq n\})$. This σ-field represents the future relative to time n and $\mathfrak{F}_n$ corresponds to the past. For a Markov process, the conditional expectation of any future event given the past depends only upon the present. The exercise explains this fact.

First, show that

(1) if $(\psi_\ell)_{1 \leq \ell \leq m}$ is any finite collection of bounded Borel functions, then $E[\prod_{\ell=1}^m (\psi_\ell \circ X_{n+\ell}) | \mathfrak{F}_n] = \Phi \circ X_n$ for some Borel function Φ on $\mathbb{R}$. [Hint: use induction on m and the fact that $E[U | \mathfrak{F}_{n+j}] = E[E[U | \mathfrak{F}_{n+j+1}] | \mathfrak{F}_{n+j}]$ to find a formula for Φ.]

In particular, this implies that if F is a finite subset of $\{j \geq n+1\}$ and $A_i \in \mathfrak{B}(\mathbb{R})$ for each $i \in F$, then $E[\prod_{i \in F}(1_{A_i} \circ X_i) | \mathfrak{F}_n] = \Psi \circ X_n$ for some Borel function Ψ. Use this to show that

(2) if $\Lambda \in \mathfrak{T}_{n+1}$, then $E[1_\Lambda | \mathfrak{F}_n] = \Theta \circ X_n$ for some Borel function Θ. [Hint: use Proposition 3.2.6.]

This shows that for every future event, its conditional expectation given the past is a Borel function of the present. Make use of (2) to show that

(3) if U is a non-negative random variable measurable with respect to $\mathfrak{T}_n$, then $E[U | \mathfrak{F}_n]$ is a Borel function of X_n.

In addition, show the following:
(4) if $X \in \mathfrak{F}_n^+$ and $Y \in \mathfrak{T}_n^+$, then

$$E[XY|X_n] = E[X|X_n]E[Y|X_n].$$

[Hint: first condition the product on $\mathfrak{F}_n$ and then use Proposition 5.2.4 (CE_7).]

Whenever this last identity holds, one says that the past and the future are **conditionally independent** given the present (compare (4) with the formula in Proposition 3.2.3). A process for which this holds is said to have the **Markov property**. What this exercise shows is that the presence of the transition kernel implies that the process is Markov.

3. SUFFICIENT STATISTICS*

An important statistical concept that involves conditional expectation is the notion of a sufficient statistic.

Consider a probability space $(\Omega, \mathfrak{F}, \mathbf{P})$, and a sub σ-field $\mathfrak{G}$ of $\mathfrak{F}$. If X is a non-negative random variable (i.e., $X \in \mathfrak{F}$ is non-negative), the probability $\mathbf{P}$ determines the conditional expectation $E[X|\mathfrak{G}] = Y$. In general, for a given X, there is no reason for Y not to depend upon the particular probability $\mathbf{P}$.

Example 5.3.1. Let Ω denote $\{0,1\}$, $\mathfrak{F}$ denote the collection of all subsets of Ω, and $\mathfrak{G} = \{\Omega, \emptyset\}$. Let $X(0) = 0$, $X(1) = a$. Then $E[X|\mathfrak{G}] = E[X] = a\mathbf{P}(\{1\})$ if $\mathbf{P}$ is a probability on $(\Omega, \mathfrak{F})$ and so depends on $\mathbf{P}$.

However, when several variables are involved, it is easy to construct examples of families $\mathcal{P}$ of probabilities $\mathbf{P}$ for which the conditional expectation is independent of $\mathbf{P}$.

Example 5.3.2. Let $\Omega = \mathbb{R}^2$, $\mathfrak{F} = \mathfrak{B}(\mathbb{R}^2)$, and $\mathcal{P} = \{\mathbf{P} = n \times \mu \mid \mu \text{ a probability on } \mathbb{R}\}$, where $n(dx) = \frac{1}{\sqrt{2\pi}}e^{-x^2/2}dx$ is the unit normal. Let $\mathfrak{G} = \{\mathbb{R} \times B \mid B \in \mathfrak{B}(\mathbb{R})\} = X_2^{-1}(\mathfrak{B}(\mathbb{R})) = \sigma(X_2)$, where $X_2(x_1, x_2) = x_2$. Then, for any non-negative random variable X, it follows from Fubini's theorem (Theorem 3.3.5) that $E[X|\mathfrak{G}](x_1, x_2) = \int X(x_1, x_2)n(dx_1)$, which is independent of the particular probability $\mathbf{P} \in \mathcal{P}$.

In the situation of Example 5.3.2, the random variable X_2 is called a **sufficient statistic**. To formulate the definition, it is convenient to introduce the concept of a sufficient σ-field.

Definition 5.3.3. *Let $\mathcal{P}$ be a collection of probabilities $\mathbf{P}$ on $(\Omega, \mathfrak{F})$, and let $\mathfrak{G} \subset \mathfrak{F}$ be a sub σ-field. The σ-field $\mathfrak{G}$ is said to be* **sufficient for the collection** $\mathcal{P}$ *if, for all non-negative $\mathfrak{F}$-measurable functions X on Ω, $E^{\mathbf{P}}[X|\mathfrak{G}] = Y$ is independent of $\mathbf{P} \in \mathcal{P}$, where $E^{\mathbf{P}}[X|\mathfrak{G}]$ denotes the conditional expectation of X relative to $\mathbf{P}$.*

If $T : (\Omega, \mathfrak{F}) \to (E, \mathfrak{E})$, where $\mathfrak{E}$ is a σ-algebra on E, is a measurable function, that is, if $T^{-1}(\mathfrak{E}) \subset \mathfrak{F}$ and if $\mathfrak{G} = T^{-1}(\mathfrak{E})$ is sufficient for $\mathcal{P}$, then T is said to be a **sufficient statistic for** $\mathcal{P}$.

Intuitively, a σ-field $\mathfrak{G}$ is sufficient for a collection $\mathcal{P}$ of probabilities $\mathbf{P}$ if enough information to distinguish between any two of them is to be found in $\mathfrak{G}$.

For example, one has the following result.

Exercise 5.3.4. Let $\mathfrak{G}$ be sufficient for a collection $\mathcal{P}$ of probabilities on $(\Omega, \mathfrak{F})$. Show that if $\mathbf{P}_1, \mathbf{P}_2 \in \mathcal{P}$ and are distinct (that is, for some $A \in \mathfrak{F}$, $\mathbf{P}_1(A) \neq \mathbf{P}_2(A)$), then when viewed as probabilities on $(\Omega, \mathfrak{G})$ they are distinct. [Hint: $\mathbf{P}_1(A) = \int E_1[1_A|\mathfrak{G}]d\mathbf{P}_1$, where E_1 indicates conditional expectation relative to $\mathbf{P}_1$.]

However, the fact that a sub σ-field $\mathfrak{G}$ is capable of distinguishing between probabilities is not enough to guarantee that it is sufficient. This is not so surprising if one recalls that conditional expectation has to do with projection.

Exercise 5.3.5. Construct an example on the measurable space $(\Omega, \mathfrak{F}) = (\{0, 1, 2\}, \mathcal{P}(\{0, 1, 2\}))$ to show that a σ-field $\mathfrak{G} \subset \mathfrak{F}$ is not sufficient for a class $\mathcal{P}$ of probabilities on $\mathfrak{F}$ if it merely distinguishes between probabilities in $\mathcal{P}$.

According to Zacks ([Z], p. 29), "the main formulation of the concept 'sufficient statistic' is based on the assertion that if the conditional distribution of X, given T, is independent of (i.e., does not depend upon) the actual (non-conditional) distribution of X, then T is sufficiently informative for $\mathcal{P}$."

Example 5.3.6. (see Zacks [Z], p. 30), Let $\mathbf{P}^\lambda$ be the probability on $(\mathbb{R}^n, \mathfrak{B}(\mathbb{R}^n))$ that is the product of n copies of the Poisson distribution on $\mathbb{R}$ with mean λ. Then $\mathbf{P}^\lambda$ is concentrated on the points $(k_1, k_2, \ldots, k_n)$ with k_i non-negative integers, and $\mathbf{P}^\lambda(\{(k_1, k_2, \ldots, k_n)\}) = e^{-n\lambda} \frac{\lambda^{k_1+k_2+\ldots+k_n}}{k_1! k_2! k_3! \ldots k_n!}$. Let $T(x_1, x_2, \ldots, x_n) = \sum_{i=1}^n x_i$. Then T is a sufficient statistic for the collection $\mathcal{P}$ of probabilities $\mathbf{P}^\lambda, \lambda \in \mathbb{R}$.

To see this, let $B_\ell = \{T = \ell\}, \ell \in \mathbb{N} \cup \{0\}$ and let $B = \mathbb{R}^n \setminus \cup_{\ell=0}^\infty B_\ell$. Then $\mathbf{P}^\lambda(B) = 0$ and $\mathbf{P}^\lambda(B_\ell) = e^{-n\lambda} \frac{\lambda^\ell n^\ell}{\ell!}$, as T has a Poisson distribution with parameter $n\lambda$ (see Exercise 3.3.22). Hence, as far as $\mathbf{P}^\lambda$ is concerned, T has only a countable number of values, and so a regular conditional probability N may be computed.

For $\omega \in B$ it does not matter what probability one associates with ω since $\mathbf{P}^\lambda(B) = 0$. Define $N(\omega, A) = \varepsilon_\mathbf{0}(A)$ if $\omega \in B$, where $\varepsilon_\mathbf{0}$ is the unit point mass at $\mathbf{0} \in \mathbb{R}^n$.

228 V. CONDITIONAL EXPECTATION AND MARTINGALES

For $\omega \in B_\ell$, define

$$N(\omega, A) = \mathbf{P}^\lambda(A \cap B_\ell)/\mathbf{P}^\lambda(B_\ell)$$

$$= \frac{1}{\mathbf{P}^\lambda(B_\ell)} \times \sum_{\substack{(k_1,k_2,\ldots,k_n)\\ \in A \cap B_\ell}} \mathbf{P}^\lambda(\{k_1, k_2, \ldots, k_n\})$$

$$= \frac{\ell!}{e^{-n\lambda}\lambda^\ell n^\ell} \times \sum_{\substack{(k_1,k_2,\ldots,k_n)\\ \in A \cap B_\ell}} \frac{e^{-n\lambda}\lambda^{k_1+k_2+\cdots+k_n}}{k_1! k_2! \cdots k_n!}$$

$$= \frac{\ell!}{n^\ell} \times \sum_{\substack{(k_1,k_2,\ldots,k_n)\\ \in A \cap B_\ell}} \frac{1}{k_1! k_2! \cdots k_n!}, \quad \text{since } \sum_{i=1}^n k_i = \ell.$$

This regular conditional probability does not depend upon λ, and so $E^\lambda[X|T^{-1}(B(\mathbb{R}))] = \int N(\omega, d\sigma)X(\sigma)$ also does not depend upon λ and so T is sufficient.

Another example, also taken from Zacks [Z], p. 30, is the following:

Example 5.3.7. Let P^θ be the probability on $(\mathbb{R}^n, \mathfrak{B}(\mathbb{R}^n))$ which is the product of n copies of the normal law $n^\theta(dx) = \frac{1}{\sqrt{2\pi}}e^{-\frac{(x-\theta)^2}{2}}dx$. Then

$$\mathbf{P}^\theta(d\mathbf{x}) = \frac{1}{(2\pi)^{\frac{n}{2}}}e^{-\frac{\|\mathbf{x}-\theta\mathbf{e}\|^2}{2}}d\mathbf{x},$$

where $\mathbf{e} = (1, 1, \ldots, 1)$. Define $\overline{X} = \frac{1}{n}\sum_{i=1}^n X_i$. Then $\overline{X}$ is a sufficient statistic for the collection of probabilities $\mathbf{P}^\theta$, $\theta \in \mathbb{R}$.

To see this, first note that $\mathbf{P}^\theta$ is invariant under rotations about the point $\theta\mathbf{e}$. Choose an orthonormal basis $\mathbf{f}_1, \mathbf{f}_2, \ldots, \mathbf{f}_n$ for $\mathbb{R}^n$, where $\mathbf{f}_1 = \frac{1}{\sqrt{n}}\mathbf{e}$ and $\mathbf{f}_2, \mathbf{f}_3, \ldots, \mathbf{f}_n$ is an orthonormal basis for the subspace $x_1 + x_2 + \cdots + x_n = 0$. Then, if $\mathbf{x} = \sum_{i=1}^n y_i \mathbf{f}_i$, it follows that

$$\mathbf{P}^\theta(d\mathbf{x}) = \frac{1}{(2\pi)^{\frac{n-1}{2}}}e^{-\frac{\|\tilde{y}\|^2}{2}}d\tilde{y} \times \frac{1}{\sqrt{2\pi}}e^{-\frac{\|y_1-\theta\sqrt{n}\|^2}{2}}dy_1,$$

where $\tilde{y} = (y_2, \ldots, y_n)$ (translate to the point $\theta\mathbf{e} = \theta\sqrt{n}\mathbf{f}_1$ and then rotate). In this system of coordinates, the probabilities $\mathbf{P}^\theta$ are products $n(d\tilde{y})n_\theta(dy_1)$ and the random variable $\overline{X}$ maps $y = (y_1, \ldots, y_n)$ to y_1 since (i) $\sum_{i=2}^n y_i = 0$ and (ii) $\sum_{i=1}^n x_i = ny_1$. Consequently, as in Example 5.3.2, $\overline{X}$ is a sufficient statistic. So too is $n\overline{X}$.

In Zacks [Z], a statistic $T : (\Omega, \mathfrak{F}) \to (E, \mathfrak{E})$ is said to be sufficient for a class $\mathcal{P}$ of probabilities $\mathbf{P}$ if the regular conditional probability $N^\mathbf{P}$ for which $E^\mathbf{P}[X|T^{-1}(\mathfrak{E})](\omega) = \int N^\mathbf{P}(\omega, d\sigma)X(\sigma)$ does not depend upon $\mathbf{P} \in \mathcal{P}$. This implies that T is sufficient in the sense of Definition 5.3.3. Note that this definition assumes a situation where all these regular conditional probabilities exist.

Exercise 5.3.8. Let $\mathcal{P}$ be any family of probabilities on $(\mathbb{R}, \mathfrak{B}(\mathbb{R}))$. Show that the statistic $T(x) = x$ is sufficient. Is it important that the measurable space is $(\mathbb{R}, \mathfrak{B}(\mathbb{R}))$?

For further information on this topic, one may consult Zacks [Z] or an article by Halmos and Savage.[6]

4. MARTINGALES

The most important concept that makes use of conditional expectation is that of a martingale.

Let $(\Omega, \mathfrak{F}, \mathbf{P})$ be a probability space and $(\mathfrak{F}_t)_{t \in I}$ be an increasing family of sub σ-fields of $\mathfrak{F}$, where $I \subset [0, +\infty)$ (that is, $s < t$ in I implies that $\mathfrak{F}_s \subset \mathfrak{F}_t \subset \mathfrak{F}$). Such a family will be referred to as a **filtration**.

Definition 5.4.1. *A process* $(X_t)_{t \in I}$ *on* $(\Omega, \mathfrak{F}, \mathbf{P})$ *will be called a* **martingale** *(respectively,* **supermartingale***), with respect to the filtration* $(\mathfrak{F}_t)_{t \in I}$, *if*

(M1) *each* $X_t \in \mathfrak{F}_t$ *(such a process is said to be* **adapted** *to the filtration) and is integrable,*

(M2) *for all* $s < t$ *in* I, $E[X_t|\mathfrak{F}_s] = X_s$ *(respectively,* $\leq X_s$*).*

A process X is called a **submartingale** *if* $-X$ *is a supermartingale.*

Remarks. (1) When $I = \mathbb{N} \cup \{0\}$ and $(X_n)_{n \in I}$ is a martingale, the process may be thought of as the fortune of a gambler in a fair game: the expected value of his fortune does not change with the number of times the game is played. (For additional information on the relations between martingales and gambling, see Billingsley [B1], p. 407, an entertaining article by Snell[7], and also the book of Williams [W2].)

(2) If the filtration is trivial with each $\mathfrak{F}_t = \mathfrak{F}$, for all $t \in I$, then a process $X = (X_t)_{t \in I}$ is a martingale if and only if it is constant: each $X_t = Y$, a fixed random variable. Similarly, it is a supermartingale if and only if it is a non-increasing family of random variables, i.e., $X_s \geq X_t$ if $s < t$ and $s, t \in I$.

Example 5.4.2M. Let $(X_n)_{n \geq 1}$ be a sequence of independent integrable random variables on a probability space $(\Omega, \mathfrak{F}, \mathbf{P})$ with $E[X_n] = 0$ for all $n \geq 1$. Define $\mathfrak{F}_n = \sigma(\{X_i \mid 1 \leq i \leq n\})$ and $S_n = \sum_{i=1}^n X_i$. Then $(S_n)_{n \geq 1}$ is a martingale, with respect to the given filtration: $E[S_{n+1}|\mathfrak{F}_n] = S_n + E[X_{n+1}|\mathfrak{F}_n]$; the hypothesis of independence implies by Proposition 5.2.4 (CE_{10}) that $E[X_{n+1}|\mathfrak{F}_n] = E[X_{n+1}] = 0$ and so $E[S_{n+1}|\mathfrak{F}_n] = S_n$ (this martingale appears in the study of the laws of large numbers in Chapter IV).

---

[6]*Application of the Radon–Nikodym Theorem to the theory of sufficient statistics* Ann. Math. Stat. **20** (1972), 225–251.

[7]*Gambling, probability and martingales* The Math. Intelligencer **4** (1982), 118–124.

Exercise 5.4.2S. Let $(X_n)_{n\geq 1}$ be a sequence of independent integrable random variables on a probability space $(\Omega, \mathfrak{F}, \mathbf{P})$ with $E[X_n] \leq 0$ for all $n \geq 1$. Define $\mathfrak{F}_n = \sigma(\{X_i \mid 1 \leq i \leq n\})$ and $S_n = \sum_{i=1}^n X_i$. Show that $(S_n)_{n\geq 1}$ is a supermartingale with respect to the given filtration. Modify this exercise so that $(S_n)_{n\geq 1}$ is a submartingale.

Exercise 5.4.3. Let $X \in L^1(\Omega, \mathfrak{F}, \mathbf{P})$, and let $(\mathfrak{F}_n)_{n\geq 1}$ be a filtration. Show that

(1) $(X_n)_{n\geq 1}$ is a martingale if $X_n \stackrel{\text{def}}{=} E[X|\mathfrak{F}_n]$ for all $n \geq 1$ (see Example 5.1.1).

Let $(\mathfrak{G}_n)_{n\geq 1}$ be any sequence of σ-fields (not necessarily a filtration). Define the sequence of random variables $(Y_n)_{n\geq 1}$ by setting $Y_n \stackrel{\text{def}}{=} E[X|\mathfrak{G}_n]$. Show that

(2) the sequence $(Y_n)_{n\geq 1}$ is uniformly integrable. [*Hints*: use the fact that $|E[X|\mathfrak{G}_n]| \leq E[|X||\mathfrak{G}_n]$ and Chebychev's inequality to show that $\mathbf{P}[|Y_n| \geq c]$ is uniformly small for large c; make use of Lemma 4.5.5.]

Conclude that

(3) the martingale $(X_n)_{n\geq 1}$ in (1) is uniformly integrable.

Example 5.4.4. Let $(X_n)_{n\geq 1}$ be a martingale with respect to $(\mathfrak{F}_n)_{n\geq 1}$ and assume $X_n \in L^2$ for each $n \geq 1$. Then $(X_n^2)_{n\geq 1}$ is a submartingale: by Jensen's inequality, $X_n^2 = (E[X_{n+1}|\mathfrak{F}_n])^2 \leq E[X_{n+1}^2|\mathfrak{F}_n]$. In particular, if S_n is as in Example 5.4.2M with $X_n \in L^2$ for all n, then $(S_n^2)_{n\geq 1}$ is a submartingale.

Example 5.4.5. Let μ be a probability on $(\Omega, \mathfrak{F}, \mathbf{P})$ such that $\mu(A) = 0$ if $\mathbf{P}(A) = 0$ (i.e., by Definition 2.7.16, μ is absolutely continuous with respect to $\mathbf{P}$), and assume that $\mathfrak{F}$ is the smallest σ-algebra containing $\mathfrak{A} = \cup_{n\geq 1}\mathfrak{F}_n$, where $\mathfrak{F}_n \subset \mathfrak{F}_{n+1}$ is an increasing sequence of finite σ-fields. For each n, the set $W_n(\omega) = \cap\{A \in \mathfrak{F}_n \mid \omega \in A\}$ is in $\mathfrak{F}_n$, and the sets $W_n(\omega)$ are identical or pairwise disjoint as ω varies through Ω (they are the **atoms** of $\mathfrak{F}_n$: sets in $\mathfrak{F}_n$ which are non-void and are minimal relative to inclusion).

Define X_n to be constant on the atoms A of $\mathfrak{F}_n$ with

$$X_n(\omega) = \begin{cases} \frac{\mu(A)}{\mathbf{P}(A)} & \text{if } \omega \in A \text{ and } \mathbf{P}(A) \neq 0, \\ 0 & \text{if } \omega \in A \text{ and } \mathbf{P}(A) = 0. \end{cases}$$

Then $\int_C X_n d\mathbf{P} = \mu(C)$ for any set $C \in \mathfrak{F}_n$. To see this, observe that it is enough to verify it when C is an atom.

The sequence $(X_n)_{n\geq 1}$ is a martingale: if $C \in \mathfrak{F}_n$, then $\int_C X_n d\mathbf{P} = \mu(C) = \int_C X_{n+1} d\mathbf{P}$. This martingale will be used later in Exercise 5.7.6 to show that there is an L^1-density f for μ, that is, a function such that $\mu(C) = \int_C f d\mathbf{P}$ for any $C \in \mathfrak{F}$. In fact, $f = \lim_n X_n$. This is another way to prove the Radon–Nikodym theorem (Theorem 2.7.19) when the σ-algebra $\mathfrak{F}$ is generated by a Boolean algebra $\mathfrak{A}$ that is the union of a sequence of finite σ-algebras. Note that as n increases, the martingale approximates some kind of "derivative" of the measure μ with respect to $\mathbf{P}$.

The hypothesis that μ is absolutely continuous plays no role in the definition of this martingale. For a general measure μ, the martingale converges to the L^1-density of the absolutely continuous measure in the Lebesgue decomposition of μ (see Theorem 2.7.22).

Remark. As an illustration of the above example, consider Example 5.1.1, where $\Omega = [0,1]$, $\mathfrak{F} = \mathfrak{B}([0,1])$, and $d\mathbf{P} = dx$. Let $\mathfrak{F}_n$ be the smallest σ-field containing the dyadic intervals $[0, \frac{1}{2^n}]$, $(\frac{k}{2^n}, \frac{k+1}{2^n}]$, $1 \le k \le 2^n - 1$: these intervals are in fact the atoms of $\mathfrak{F}_n$. The following exercise states that $\mathfrak{B}([0,1])$ is the smallest σ-field that contains all the dyadic intervals.

Exercise 5.4.6. Let $\Omega = [0,1]$, and let $\mathfrak{F}_n$ be the finite σ-field whose atoms are the dyadic intervals $[0, \frac{1}{2^n}]$ and $(\frac{k}{2^n}, \frac{k+1}{2^n}]$, $1 \le k \le 2^n - 1$. Show that $\mathfrak{B}([0,1])$ is the smallest σ-algebra containing $\mathfrak{A} = \cup_{n \ge 1} \mathfrak{F}_n$, or, equivalently, containing all the dyadic intervals. [*Hint*: if $0 < a < b \le 1$, show that (a,b) is a countable union of dyadic intervals.]

Exercise 5.4.7. Let $(X_n)_{n \ge 0}$ be a Markov chain with transition kernel $\mathbf{N}$ (Exercise 5.2.12). Let $\mathfrak{F}_n = \sigma(\{X_k \mid 0 \le k \le n\})$. Show that
   (1) if ψ is a non-negative Borel function such that $\mathbf{N}\psi \le \psi$ (a so-called **excessive function**), then $(\psi \circ X_n)_{n \ge 0}$ is a supermartingale with respect to the filtration $(\mathfrak{F}_n)_{n \ge 0}$,
   (2) if ψ is a bounded Borel function such that $\mathbf{N}\psi = \psi$ (a so-called bounded **harmonic function**), then $(\psi \circ X_n)_{n \ge 0}$ is a martingale with respect to the same filtration.

Remarks. (1) The function $\psi(n) = \frac{p^n}{q^n}$ is a harmonic function on $\mathbb{Z}$ relative to the convolution kernel $\mathbf{N}$ given by $\mathbf{N}(0, dx) = p\varepsilon_1(dx) + q\varepsilon_{-1}(dx)$, where $0 < p, q < 1$, and $p + q = 1$. Consequently, if $(X_n)_{n \ge 0}$ is a Markov chain with this transition kernel and $X_0 = n_0$ for some fixed $n_0 \in \mathbb{Z}$, then $\psi(X_n) = \frac{p^{X_n}}{q^{X_n}}$ is a martingale. Since the function ψ is unbounded, this is not an immediate consequence of Exercise 5.4.7 (2) even though $\mathbf{N}\psi = \psi$. However, since X_n a.s. assumes only a finite number of values for each n, it follows that $\psi \circ X_n$ is integrable and, as a result, is a martingale.

(2) This connection between supermartingales and martingales on the one hand and excessive and harmonic functions on the other is at the basis of the close connection between classical potential theory and Brownian motion (see Dynkin and Yushekevič [D4]).

A martingale $(X_n)_{n \ge 0}$ with respect to a filtration $(\mathfrak{F}_n)_{n \ge 0}$ can always be viewed as the sequence of partial sums of an infinite series $\sum_{n=0}^{\infty} D_n$ whose terms $D_n \stackrel{\text{def}}{=} X_n - X_{n-1}$, $n \ge 1$, $D_0 = 0$ are such that $E[D_n | \mathfrak{F}_{n-1}] = 0$

for all $n \geq 1$. It is of interest to ask how one can "weight" the martingale differences D_n so as to still have a sequence of martingale differences. This will be useful for optional stopping and is a precursor of the kind of sums needed to define a stochastic integral. This transformation of the martingale by weights is called a **martingale transform**.

Proposition 5.4.8. (**Martingale transform**) *Let $(X_n)_{n\geq 0}$ be a martingale with respect to $(\mathfrak{F}_n)_{n\geq 0}$ and $f = (f_n)_{n\geq 1}$ be a sequence of random variables with $f_n \in \mathfrak{F}_{n-1}$ for all $n \geq 1$ (f is usually called a **predictable process**). Assume the f_n are bounded. Define $(Y_n)_{n\geq 0}$ by the formulas*
$$Y_0 = X_0,$$
$$Y_{n+1} = Y_n + f_{n+1}(X_{n+1} - X_n),\ n \geq 0,$$
that is,
$$Y_{n+1} = X_0 + \sum_{k=0}^{n} f_{k+1}(X_{k+1} - X_k).$$
Then $Y = (Y_n)_{n\geq 0}$ is a martingale and will be denoted by $f \cdot X$.

In addition, if X is a supermartingale (respectively, a submartingale) and f is non-negative, then $f \cdot X$ is also a supermartingale (respectively, a submartingale).

Proof. Consider first the case of a martingale X. Then $f \cdot X$ is a martingale since
$$E[Y_{n+1}|\mathfrak{F}_n] = Y_n + E[f_{n+1}(X_{n+1} - X_n)|\mathfrak{F}_n]$$
$$= Y_n + f_{n+1}(E[X_{n+1}|\mathfrak{F}_n] - X_n) = Y_n.$$
This also proves that $f \cdot X$ is a supermartingale if X is a supermartingale and f is non-negative since then $f_{n+1}(E[X_{n+1}|\mathfrak{F}_n] - X_n)$ is non-positive. The submartingale case follows by the same argument with the proviso that $f_{n+1}(E[X_{n+1}|\mathfrak{F}_n] - X_n)$ is then non-negative. $\square$

Remarks. (1) Note that if $X_0 = 0$, then $f \to f \cdot X$ is linear, and replacing X_n by $U_n = X_n - X_0$, one has $f.X = f.U + X_0$ since $U_{k+1} - U_k = X_{k+1} - X_k$, $k \geq 0$.

(2) In Billingsley [B1], p. 412, $(f_n)_{n\geq 1}$ is called a **betting system**. Intuitively, the original process represents the gambler's fortune, and the betting system allows scaling of the increment $X_{n+1} - X_n$ of the gambler's fortune before knowing the outcome of the $(n+1)$st game. See also Williams [W2], p. 96.

(3) As mentioned, the martingale transform is the discrete analogue of the **stochastic integral** in the sense of Itô, which is defined as a limit of Riemann like sums, each of which is a martingale transform (see Ikeda and Watanabe [I], p. 48, and Protter [P2], p. 44) relative to a sequence of stopping times.

The classes of martingales and supermartingales have some obvious elementary properties, which are now stated as two propositions.

4. MARTINGALES

Proposition 5.4.9M. *Let $X = (X_t)_{t \in I}$ and $Y = (Y_t)_{t \in I}$ be martingales with respect to the same filtration $(\mathfrak{F}_t)_{t \in I}$. Then*

(1) *$Z = aX + bY$ is a martingale, where $Z_t = aX_t + bY_t$ for all real a, b,*
(2) *if $c : \mathbb{R} \to \mathbb{R}$ is convex and each $c \circ X_t \in L^1$ for all $t \in I$, then $C = c \circ X$ is a submartingale, where $C_t = c \circ X_t$.*

Proof. (1) is obvious; (2) follows from Jensen's inequality (Proposition 5.2.4 (CE_8)): if $s < t$, then $c \circ X_s = c \circ E[X_t|\mathfrak{F}_s] \leq E[c \circ X_t|\mathfrak{F}_s]$. □

Proposition 5.4.9S. *Let $X = (X_t)_{t \in I}$ and $Y = (Y_t)_{t \in I}$ be supermartingales with respect to the same filtration $(\mathfrak{F}_t)_{t \in I}$. Then*

(1) *$Z = aX + bY$ is a supermartingale, $Z_t \stackrel{\text{def}}{=} aX_t + bY_t$ if $a, b \geq 0$,*
(2) *$Z = X \wedge Y$ is a supermartingale, $Z_t \stackrel{\text{def}}{=} X_t \wedge Y_t, t \in I$, and*
(3) *if $c : \mathbb{R} \to \mathbb{R}$ is concave and increasing and $c \circ X_t \in L^1$ for all $t \in I$, then $C = c \circ X$ is a supermartingale, where $C_t = c \circ X_t$.*

Proof. (1) is clear since $a \geq 0$ implies $aX_s \geq aE[X_t|\mathfrak{F}_s]$ if $s < t$. (2) is obvious once one checks that each Z_t is integrable as $s < t$ implies that $E[Z_t|\mathfrak{F}_s] \leq E[X_t|\mathfrak{F}_s] \leq X_s$ and similarly $E[Z_t|\mathfrak{F}_s] \leq Y_s$. Property (3) corresponds to (2) in Proposition 5.4.9M, where c is replaced by $(-c)$. Jensen's inequality implies that $E[c \circ X_t|\mathfrak{F}_s] \leq c \circ E[X_t|\mathfrak{F}_s]$ and that, as c is increasing, $c \circ E[X_t|\mathfrak{F}_s] \leq c \circ X_s$, since $E[X_t|\mathfrak{F}_s] \leq X_s$ if $s < t$. □

Consider now $(X_n)_{n \geq 1}$, a sequence of independent identically distributed (i.i.d.) random variables on $(\Omega, \mathfrak{F}, \mathbf{P})$ with $\mathbf{P}[X_n = +1] = \mathbf{P}[X_n = -1] = \frac{1}{2}$. Let $X_0(\omega) = 0$ for all $\omega \in \Omega$. Then $S_n = \sum_{i=0}^{n} X_i$ is a martingale with respect to the natural filtration $\mathfrak{F}_n = \sigma(\{X_i | 0 \leq i \leq n\})$. The random variable represents the position of a particle moving on $\mathbb{Z}$ under the influence of the symmetric random walk started from 0 at time 0. As time passes, the value $S_n(\omega)$ varies over all of $\mathbb{Z}$, and so if one fixes an integer $N \geq 0$ there is a first time $T = T(\omega)$ at which $|S_n(\omega)|$ attains the value $N+1$ (that is, at which the particle visits $N+1$ or $-(N+1)$). This time T is random (in other words, it depends upon ω). Furthermore, since $T(\omega) = \inf\{n \mid |S_n(\omega)| > N\}$, it follows that $\{T = k\} = \{\omega \mid |S_\ell(\omega)| \leq N, 0 \leq \ell \leq k-1, |S_k(\omega)| = N+1\}$. This set is in $\mathfrak{F}_k$ and so $\{\omega \mid T(\omega) \leq k\} \in \mathfrak{F}_k$ (in other words, this set is determined by the information available up to and including time k). Random times with this property are called **stopping times** and are very common in martingale theory. A random time of this type was used in the proof of Kolmogorov's inequality (Theorem 4.4.3).

Definition 5.4.10. *Let $(\mathfrak{F}_t)_{t \in I}$ be a filtration on $(\Omega, \mathfrak{F}, \mathbf{P})$. A random variable $T : \Omega \to I \cup \{+\infty\} \subset [0, +\infty]$ such that, for all $t \in I$, $\{T \leq t\} \in \mathfrak{F}_t$ is called a **stopping time** for the filtration $(\mathfrak{F}_t)_{t \in I}$.*

Remark. The event associated with a stopping time T need not occur. When this happens, it is often convenient to define T to be $+\infty$. It is sometimes useful to define $\mathfrak{F}_\infty$ to be the smallest σ-algebra containing all the $\mathfrak{F}_t, t \in I$, in which case the set I can be viewed as a subset of $[0, +\infty]$.

Examples 5.4.11.

(1) If $T(\omega) = s \in I$ for all $\omega \in \Omega$, then T is a stopping time.
(2) Let $(X_n)_{n\geq 1}$ be an adapted process and $c > 0$. Then $T = \inf\{n \mid |X_n| > c\}$ is a stopping time. For example, the first time that a symmetric random walk on $\mathbb{Z}$, started at 0, leaves $[-N, N]$ is a stopping time —see the discussion preceding Definition 5.4.10.

Assume that $I = [0, +\infty)$ and that $X = (X_t)_{t\geq 0}$ is an adapted process such that for almost every ω, the function $t \to X_t(\omega)$ is continuous. Assume that $(\Omega, \mathfrak{F}, \mathbf{P})$ is complete and that each $\mathfrak{F}_t$ contains all the sets in $\mathfrak{F}$ of probability zero. Then

(3) $T(\omega) = \inf\{t \mid |X_t| \geq 1\}$, with $\inf \emptyset = +\infty$, is a stopping time.

Assume that $t \to X_t(\omega)$ is continuous. Then $T(\omega) \leq t$ if and only if $\max_{0\leq s\leq t} |X_s(\omega)| \geq 1$. Hence, $T(\omega) > t$ if and only if for some n, $\max_{0\leq s\leq t} |X_s(\omega)| \leq 1 - \frac{1}{n}$, equivalently, by continuity, $\sup_{0\leq r\leq t} |X_r(\omega)| \leq 1 - \frac{1}{n}, r \in \mathbb{Q}$. Since $\mathbb{Q}$ is countable, it follows that $\{T > t\} \cap \Omega_1 \in \mathfrak{F}_t$, where $\Omega_1 = \{\omega \mid t \to X_t(\omega) \text{ is continuous}\}$. Since by hypothesis $\mathbf{P}(\Omega \backslash \Omega_1) = 0$, $\{T > t\} \in \mathfrak{F}_t$. Consequently, $\{T \leq t\} \in \mathfrak{F}_t$.

Under the same hypotheses as in (3), one can prove that $\{T < t\} \in \mathfrak{F}_t$, for all $t \geq 0$ if $T = \inf\{t \mid |X_t| > 1\}$. This implies that

(4) $T = \inf\{t \mid |X_t| > 1\}$ is a stopping time with respect to the filtration $(\mathfrak{F}_t^+)_{t\geq 0}$, where $\mathfrak{F}_t^+ = \cap_{s>t} \mathfrak{F}_s$.

Remark. These last two examples indicate something of the technical complications that arise when looking at processes indexed by $[0, \infty)$ rather than a finite or a countable time set I, the classical example being Brownian motion in $\mathbb{R}^n$. From now on, I will be finite or countable.

Exercise 5.4.12. Show that T in Example 5.4.11 (4) is a stopping time.

Definition 5.4.13. *A set $A \in \mathfrak{F}$ is said to* **occur prior to a stopping time** *T if, for all $t \in I$, $A \cap \{T \leq t\} \in \mathfrak{F}_t$. Let $\mathfrak{F}_T$ denote the sets in $\mathfrak{F}$ that occur prior to T.*

Exercise 5.4.14. Show that $\mathfrak{F}_T$ is a σ-algebra. [Hint: $A^c \cup \{T > t\} \cap \{T \leq t\} = A^c \cap \{T \leq t\}$.]

Exercise 5.4.15. If $S \leq T$ are two stopping times, show that $\mathfrak{F}_S \subset \mathfrak{F}_T$. [Hint: $S \leq T$ implies that $\{S \leq t\} \cap \{T \leq t\} = \{T \leq t\}$.]

Example 5.4.16. In Example 5.4.11 (2), the following sets A occur prior to T:

(1) $A = \{\omega \mid |S_n(\omega)| \leq N \text{ for all } n\}$ as $A \cap \{T = k\} = \emptyset$ for all k;

(2) $A_{n_1} = \{\omega \mid |S_n(\omega)| \leq N \text{ for } 0 \leq n \leq n_1\}$ as $A_{n_1} \cap \{T = k\} = \emptyset$ if $k \leq n_1$ and $\{T = k\}$ if $k > n_1$, and

(3) the set $A = \{\omega \mid |S_k(\omega)| \leq N + 1,\ 0 \leq k \leq 2N+2, S_{2N+2}(\omega) = 0\}$ does not occur prior to T, as $A \cap \{T = N+1\} = \{\omega \mid X_n(\omega) = +1,\ 1 \leq n \leq N+1,\ X_n(\omega) = -1,\ N+2 \leq n \leq 2N+2\}$, which is not in $\mathfrak{F}_{N+1}$.

Exercise 5.4.17. Let $I \subset [0, +\infty)$ be finite and $T : \Omega \to I$ be a stopping time for a filtration $(\mathfrak{F}_t)_{t \in I}$. Let $(X_t)_{t \in I}$ be a real-valued process that is adapted to the filtration. Define $X_T(\omega) = X_{T(\omega)}(\omega)$. Show that X_T is $\mathfrak{F}_T$-measurable. [Hint: find an alternate expression for $\{X_T < \lambda\} \cap \{T = k\}$].

Theorem 5.4.18. (Doob's optional stopping theorem: finite case) (See [M2], [I].) Let $I \subset [0, +\infty)$ be finite and let S, T be two stopping times for a filtration $(\mathfrak{F}_t)_{t \in I}$ on $(\Omega, \mathfrak{F}, \mathbf{P})$. Let $(X_t)_{t \in I}$ be an $(\mathfrak{F}_t)_{t \in I}$-supermartingale (respectively, submartingale). If $S \leq T$ then

$$E[X_T | \mathfrak{F}_S] \leq X_S \quad (\text{respectively}, \geq X_S).$$

Proof I. (See [M2]). One may as well assume that $I = \{0, 1, \ldots, n\}$. Let $A \in \mathfrak{F}_S$. Then

$$\int_A X_T d\mathbf{P} = \sum_{j=0}^n \int_{\{T=j\} \cap A} X_j d\mathbf{P} = \sum_{j=0}^n \sum_{i=0}^n \int_{\{T=j\} \cap \{S=i\} \cap A} X_j d\mathbf{P}$$

$$= \sum_{i=0}^n \sum_{j=i}^n \int_{\{T=j\} \cap \{S=i\} \cap A} X_j d\mathbf{P} \text{ as } S \leq T$$

$$= \sum_{i=0}^n \int_{A_i \cap \{T \geq i\}} X_T d\mathbf{P},$$

where $A_i = \{S = i\} \cap A \in \mathfrak{F}_i$, since S is a stopping time. Now

$$\int_{A_i \cap \{T \geq i\}} X_T d\mathbf{P} = \int_{A_i \cap \{T = i\}} X_i d\mathbf{P} + \int_{A_i \cap \{T > i\}} X_T d\mathbf{P}.$$

Note that $A_i \cap \{T > i\} \in \mathfrak{F}_i$. Assume that $T \leq S+1$. Then,

$$\int_{A_i \cap \{T > i\}} X_T d\mathbf{P} = \int_{A_i \cap \{T > i\}} X_{i+1} d\mathbf{P} \leq \int_{A_i \cap \{T > i\}} X_i d\mathbf{P}$$

(respectively, $\geq$) as $(X_t)_{t \in I}$ is a supermartingale (respectively, submartingale).

Therefore, if $S \leq T \leq S+1$,

$$\int_A X_T d\mathbf{P} \leq \sum_{i=0}^n \int_{A_i \cap \{T \geq i\}} X_i d\mathbf{P} = \int_A X_S d\mathbf{P} \ (\text{respectively}, \geq).$$

236   V. CONDITIONAL EXPECTATION AND MARTINGALES

For the general case of $S \leq T$, define a sequence of stopping times $T^{(k)}, 1 \leq k \leq n$, as follows: $T^{(k)} = T \wedge (S+k)$. Then $T^{(k)} \leq T^{(k+1)} \leq T^{(k)} + 1$ and $\mathfrak{F}_{T^{(k)}} \subset \mathfrak{F}_{T^{(k+1)}}$. Hence, for $A \in \mathfrak{F}_S$, and $1 \leq k \leq n$, $\int_A X_{T^{(k)}} d\mathbf{P} \leq \int_A X_{T^{(k+1)}} d\mathbf{P}$ (respectively, $\geq$). Since $T^{(n)} = T$, the result follows. □

Proof II. (See [I]). If T is a stopping time for $(\mathfrak{F}_t)_{t \in I}$, $I = \{0, 1, \ldots, n\}$, and if $h_k = 1_{\{k \leq T\}}$, then the process $h = (h_k)_{1 \leq k \leq n}$ is predictable as $\{k \leq T\} = \{T \leq k-1\}^c \in \mathfrak{F}_{k-1}$. Let $f_k = 1_{\{S < k \leq T\}} = 1_{\{k \leq T\}} - 1_{\{k \leq S\}}$. Then $(f_k)_{1 \leq k \leq n}$ is also predictable and non-negative.

Let $(X_t)_{t \in I}$ be a supermartingale. First assume $X_0 = 0$. Then $(h \cdot X)_0 = 0$ by definition (see Proposition 5.4.8) and for $k \geq 1$, $(h \cdot X)_k = X_{T \wedge k}$ by induction since

$$(h \cdot X)_{k+1} = (h.X)_k + h_{k+1}(X_{k+1} - X_k)$$
$$= X_{T \wedge k} + h_{k+1}(X_{k+1} - X_k)$$
$$= X_{T \wedge k} + \begin{cases} X_{k+1} - X_k & \text{if } T \geq k+1, \\ 0 & \text{if } T < k+1 \end{cases}$$
$$= X_{T \wedge (k+1)}.$$

Consequently, by linearity (see Proposition 5.4.8),

$$(f \cdot X)_k = X_{T \wedge k} - X_{S \wedge k}, \ 0 \leq k \leq n,$$

is a supermartingale. Therefore,

$$0 \geq E[(f \cdot X)_n] = E[X_T] - E[X_S].$$

If $X_0 \neq 0$, let $U_t = X_t - X_0$. Then $(U_t)_{t \in I}$ is a supermartingale with $U_0 = 0$ and $U_T = X_T - X_0, U_S = X_S - X_0$. Hence, $X_T - X_S = U_T - U_S$ and so $E[X_T] \leq E[X_S]$.

Exercise 5.4.19. For any $A \in \mathfrak{F}_S$, define

$$S_A(\omega) = \begin{cases} S(\omega) & \text{if } \omega \in A, \\ n & \text{if } \omega \notin A, \end{cases}$$

and

$$T_A(\omega) = \begin{cases} T(\omega) & \text{if } \omega \in A, \\ n & \text{if } \omega \notin A. \end{cases}$$

Show that S_A and T_A are stopping times.

Hence, by what has been proved,

$$\int_A X_T d\mathbf{P} + \int_{A^c} X_n d\mathbf{P} = E[X_{T_A}] \leq E[X_{S_A}] = \int_A X_S d\mathbf{P} + \int_{A^c} X_n d\mathbf{P},$$

and so

$$\int_A X_T d\mathbf{P} \leq \int_A X_S d\mathbf{P}. \quad \Box$$

4. MARTINGALES

Theorem 5.4.20. (Doob's maximal inequality: finite case) *Let $I \subset [0, +\infty)$ be finite and $(X_t)_{t \in I}$ be a submartingale with respect to a filtration $(\mathfrak{F}_t)_{t \in I}$ on $(\Omega, \mathfrak{F}, \mathbf{P})$. Let $\lambda > 0$. Then*

$$\lambda \mathbf{P}[\max_{t \in I} X_t \geq \lambda] \leq E[1_{\{\max_{t \in I} X_t \geq \lambda\}} X_\tau^+] \leq E[X_\tau^+],$$

where $X_\tau^+ = X_\tau \vee 0$, and τ is the largest element of I.

Proof. One may as well assume $I = \{0, 1, \ldots, N\}$. Let $T(\omega)$ be the first time in I that $X_t(\omega) \geq \lambda$ (if it ever occurs) and N otherwise.

Then T is a stopping time and, by Theorem 5.4.19, $X_N^+ \geq E[X_T^+ | \mathfrak{F}_N]$, as $(X_t^+)_{t \in I}$ is a supermartingale. Hence, if $B = \{\max_{t \in I} X_t \geq \lambda\}$,

$$E[X_N^+] \geq E[X_N^+ 1_B] \geq E[X_T^+ 1_B]$$
$$\geq \sum_{n=0}^{N} E[X_n^+ 1_{B \cap \{T=n\}}]$$
$$\geq \sum_{n=0}^{N} \lambda \mathbf{P}[B \cap \{T = n\}] = \lambda \mathbf{P}[\max_{t \in I} X_t \geq \lambda]. \quad \square$$

Remark. This inequality is called a "maximal" inequality as it involves $\sup_n |X_n| \stackrel{\text{def}}{=} X^*$, the so-called **maximal function** of the submartingale. This maximal function is a direct analogue of the Hardy–Littlewood maximal function (Definition 4.6.1).

Corollary 5.4.21. (Kolmogorov's inequality). *Let $(S_n)_{n \geq 1}$ be a martingale with respect to a filtration $(\mathfrak{F}_n)_{n \geq 1}$ on $(\Omega, \mathfrak{F}, \mathbf{P})$. Assume $S_n \in L^2$ for all n. Then, for any $\lambda > 0$,*

$$\lambda^2 \mathbf{P}[\max_{1 \leq n \leq N} |S_n| \geq \lambda] \leq E[S_N^2].$$

Proof. $(S_n^2)_{n \geq 1}$ is a positive submartingale and

$$\left\{ \max_{1 \leq n \leq N} S_n^2 \geq \lambda^2 \right\} = \left\{ \max_{1 \leq n \leq N} |S_n| \geq \lambda \right\}. \quad \square$$

Remark. Doob's maximal inequality is an improvement of Chebychev's inequality (Proposition 4.3.6):

$$\lambda^2 \mathbf{P}[|S_n| \geq \lambda] \leq E[S_N^2].$$

It also generalizes Kolmogorov's inequality for sums of independent random variables (Theorem 4.4.3).

5. An introduction to martingale convergence

If $(X_n)_{n\geq 1}$ is a sequence of i.i.d. integrable random variables with expectation zero, the process $(U_n)_{n\geq 1}$, where $U_n = \sum_{k=1}^{n} \frac{X_k}{k}$, is a martingale. Kronecker's lemma (Proposition 5.5.5) states that $\frac{1}{n}\sum_{k=1}^{n} X_k(\omega) \to 0$ if $\sum_{k=1}^{n} \frac{X_k(\omega)}{k}$ converges. Consequently, to prove the strong law of large numbers for $(X_n)_{n\geq 1}$, it would suffice to show that the martingale $(U_n)_{n\geq 1}$ converges a.s. as $n \to \infty$.

In case the X_n are in L^2, the martingale $(U_n)_{n\geq 1}$ converges as a consequence of the next result since $E[U_n^2] = C \sum_{k=1}^{n} \frac{1}{k^2}$, where $C = E[X_n^2]$. (This proves the i.i.d. version of Proposition 4.4.1.) This result is a special convergence theorem and is by no means the most general result of this type (see Corollary 5.7.3 for a general result).

Proposition 5.5.1. (L^2-*martingale convergence theorem*). *Let $(X_n)_{n\geq 0}$ be a martingale, and assume that for some constant M, $E[X_n^2] \leq M$ (i.e., the martingale is L^2-**bounded**). Then there is a random variable $Y \in L^2$ such that $X_n \xrightarrow{a.s.} Y$. In addition, $X_n \xrightarrow{L^2} Y$.*

Proof. (see Feller [F2], p. 236). Since $(X_n^2)_{n\geq 0}$ is a submartingale, $E[X_n^2]$ increases to a limit $\mu \leq M$. Let $\lambda > 0$. Kolmogorov's inequality implies that

$$\lambda^2 \mathbf{P}[\max_{1\leq r\leq R} |X_{N+r} - X_N| \geq \lambda] \leq E[(X_{N+R} - X_N)^2], \tag{1}$$

as $(X_{N+r})_{r\geq 1}$ is a martingale with respect to $(\mathfrak{G}_r)_{r\geq 1}$, $\mathfrak{G}_r = \mathfrak{F}_{N+r}$.

For any martingale in L^2, $E[(X_{n+r} - X_n)^2 | \mathfrak{F}_n] = E[X_{n+r}^2 - 2X_{n+r}X_n + X_n^2 | \mathfrak{F}_n] = E[X_{n+r}^2 - X_n^2 | \mathfrak{F}_n]$ as $E[X_{n+r}X_n | \mathfrak{F}_n] = X_n^2$ by Proposition 5.2.4 (CE_9). Hence,

$$E[(X_{N+R} - X_N)^2] = E[X_{N+R}^2 - X_N^2]. \tag{2}$$

Since $E[X_n^2] \to \mu$ as $n \to \infty$, (2) implies (i) that the martingale is Cauchy in L^2 and (ii), in view of (1), that for any $\varepsilon > 0, \lambda > 0$, there exists $N_0 = N_0(\varepsilon)$ such that for all R if $N \geq N_o$, then

$$\lambda^2 \mathbf{P}[\max_{1\leq r\leq R} |X_{N+r} - X_N| \geq \lambda] \leq \varepsilon. \tag{3}$$

As shown by the following exercise, (3) implies that a.s. every sequence $(X_n(\omega))_{n\geq 1}$ is Cauchy. $\square$

Exercise 5.5.2. Let $k \geq 1$ and $E_N \stackrel{\text{def}}{=} \{\omega \mid \text{for some } r \geq 1, |X_{N+r}(\omega) - X_N(\omega)| \geq \frac{1}{k}\}$. Use (3) to show that

(1) there is an integer N_k with $\mathbf{P}(E_{N_k}) \leq \frac{1}{2^k}$.

Use the Borel-Cantelli lemma to show that

(2) a.s. $|X_{N+r}(\omega) - X_N(\omega)| < \frac{1}{k}$ for all $r \geq 1$ if N is sufficiently large.

Conclude that

(3) a.s. every sequence $(X_n(\omega))_{n \geq 1}$ is Cauchy.

Remarks. (1) As observed at the beginning of this section, the simple strong law for sums of independent L^2-random variables (Proposition 4.4.1) follows from this convergence theorem.

(2) In Doob [D1] there is a related result (see [D1], p. 108, Theorem 2.3).

(3) The L^p-analogue, for $1 < p < \infty$, of this theorem is also true. It follows from the martingale convergence theorem (5.7.3) that an L^p-bounded martingale converges a.s. However, it does not follow from the proof (as in the case $p = 2$) that this gives convergence in L^p. To get L^p-convergence, it suffices to determine a dominating function in L^p. There is a natural candidate for this function, the **maximal function** $M^* \stackrel{\text{def}}{=} \sup_n |M_n|$. As shown in the next exercise, this function is in L^p (see Williams [W2], p. 143, and Ikeda and Watanabe, [I], p. 28).

Exercise 5.5.3. (Doob's L^p-inequality: countable case) (see Revuz and Yor [R2], p. 51 and Ikeda and Watanabe [I], p. 28). Let $(X_n)_{0 \leq n \leq N}$ be a submartingale and let $X^* = \sup |X_n|$. Use Doob's inequality (Theorem 5.4.20), Exercise 4.7.2, and Fubini's theorem (Theorem 3.3.5) to show that, for any integer k and $1 < p < +\infty$,

(1)
$$E[(X^* \wedge k)^p] \leq \int_0^k p\lambda^{p-2} E[|X_N|1_{\{X^* \geq \lambda\}}]d\lambda$$
$$= \frac{p}{p-1} E[|X_N|(X^* \wedge k)^{p-1}].$$

(2) Apply Hölder's inequality (Proposition 4.1.15) to the last term (recall that $Z \in L^p$ if and only if $Z^{p-1} \in L^q$) and conclude that

$$E[(X^* \wedge k)^p] \leq \left(\frac{p}{p-1}\right) \left(E[|X_N|^p]\right)^{\frac{1}{p}} \left(E[(X^* \wedge k)^p]\right)^{\frac{p-1}{p}}.$$

(3) Hence show that
$$\| X^* \|_p \leq q \| X_N \|_p,$$

where q is the conjugate index to p.

Use this to deduce **Doob's L^p-inequality**: if $X = (X_n)_{n \geq 0}$ is a martingale that is bounded in L^p, $1 < p < +\infty$, and $X^* = \sup_{n \geq 0} X_n$, then

(5) $X^* \in L^p$ and

$$\sup_n \| X_n \|_p \leq \| X^* \|_p \leq q\left(\sup_n \| X_n \|_p\right).$$

Remark. If $f \in L^p(\mathbf{R}), 1 < p < \infty$ then the Hardy–Littlewood maximal function f^* is in L^p (see [W1], p. 155, Theorem 9.16). The proof of this is closely related to the above proof of the analogous result for martingales.

For the finite set $I = \{0, 1, \ldots, N\}$, it is obvious that $\sup_{0 \leq n \leq N} |X_n| \in L^p$ as it is dominated by $\sum_{n=0}^{N} |X_n|$; what is not obvious but crucial is that its norm in L^p satisfies (3) and so the maximal function $X^* = \sup_{n \geq 0} |X_n| \in L^p$ with a norm equivalent to the bound in L^p of the martingale.

By using truncation to get an L^2-bounded martingale, Kolmogorov's strong law of large numbers will now be proved for $(X_n)_{n \geq 1}$ in L^1 by making use of the martingale convergence theorem for L^2-bounded martingales.

Theorem 5.5.4. (Kolmogorov's strong law of large numbers) *Let $(X_n)_{n \geq 1}$ be a sequence of i.i.d. integrable random variables. Then $\frac{1}{n} S_n \to m$ a.s., where $m = \int x Q(dx)$, $\mathbf{Q}$ the common distribution, and $S_n = \sum_{k=1}^{n} X_k$.*

Proof. (Compare this martingale argument with the proof of Theorem 4.4.2.) To begin, as usual, one may assume $m = 0$ by replacing X_n with $X_n - m$.

Let Y_n be the truncation of X_n at height n, (i.e., $Y_n(\omega) = X_n(\omega)$ if $X_n(\omega) \leq n$ and equals 0 otherwise). Let $A_n = \{Y_n \neq X_n\}$. Then, as was shown in Proof II of Theorem 4.3.12, $\sum_{n=1}^{\infty} \mathbf{P}(A_n) < +\infty$ and so, with probability 1, for each ω there exists $n_0(\omega) = n_0$ such that $Y_n(\omega) = X_n(\omega)$ for $n \geq n_0$. Hence, it suffices to show that $\frac{1}{n} \sum_{k=1}^{n} Y_k \to 0$ a.s.

Let $m_n = E[Y_n] = \int_{\{|x| \leq n\}} x Q(dx)$. Define $Z_n = \frac{1}{n}(Y_n - m_n)$. Then, if $T_n = \sum_{k=1}^{n} Z_k$, the process $(T_n)_{n \geq 1}$ is a martingale with respect to $(\mathfrak{F}_n)_{n \geq 1}$, $\mathfrak{F}_n = \sigma(\{X_i | 1 \leq i \leq n\})$ (see Exercise 5.4.2M). Since each Z_n is bounded, $T_n \in L^2(\Omega, \mathfrak{F}, \mathbf{P})$ for all $n \geq 1$. In fact, $T = (T_n)_{n \geq 1}$ is an L^2-bounded martingale: using independence, one has $E[T_n^2] = \sum_{k=1}^{n} E[Z_k^2] = \sum_{k=1}^{n} \frac{\sigma_k^2}{k^2}$, where σ_k^2 is the variance of Y_k. In the earlier proof of this strong law it was shown that $\sum_{n=1}^{\infty} \frac{\sigma_n^2}{n^2} < +\infty$ (see the end of the proof of Theorem 4.4.2), which proves the L^2-boundedness of T.

By Proposition 5.5.1, there is a random variable Z such that $T_n = \sum_{k=1}^{n} Z_k \to Z$ a.s., that is, $\sum_{k=1}^{n} \frac{(Y_k - m_k)}{k} \to Z$ a.s. Kronecker's lemma (see Proposition 5.5.5) implies that $\frac{1}{n} \sum_{k=1}^{n} (Y_k - m_k) \to 0$ a.s.

Hence, it suffices to show that $\frac{1}{n} \sum_{k=1}^{n} m_k \to 0$. This is immediate since m_k tends to $\int x \mathbf{Q}(dx) = 0$ as $|x| \in L^1(\mathbf{Q})$, and so it follows that m_k tends to zero in the sense of Cesaro, i.e., $\frac{1}{n} \sum_{k=1}^{n} m_k \to 0$ (see Exercise 4.3.14). □

Proposition 5.5.5. (Kronecker's lemma). *Let $(a_n)_{n \geq 1}$ be a sequence of real numbers, and assume that $\sum_{n=1}^{\infty} \frac{a_n}{n}$ converges. Then, $\frac{1}{n} \sum_{i=1}^{n} a_i \to 0$.*

Proof. The first problem is how to relate the two series. Let $b_n = \sum_{k=1}^{n} \frac{a_k}{k}$ and $b_0 = 0$. Then $a_n = \{b_n - b_{n-1}\}n$ for all $n \geq 1$, and so the partial sum $\sum_{k=1}^{n} a_k = \sum_{k=1}^{n} \{b_k - b_{k-1}\}k$. Formally, this looks like $\int_1^n b'(x)k(x)dx$, which suggests that one should try to use integration by parts to evaluate it.

The following lemma gives the formula for **integration by parts** of a series.

Lemma 5.5.6. *Let* $(C_n)_{n\geq 1}$ *and* $(D_n)_{n\geq 1}$ *be two sequences of real numbers. Then, for all* $n \geq 2$,

$$C_n D_n - C_1 D_1 = \sum_{k=1}^{n-1} \{C_{k+1} - C_k\} D_{k+1} + \sum_{k=1}^{n-1} \{D_{k+1} - D_k\} C_k.$$

Proof of the lemma.

$$C_{k+1} D_{k+1} - C_k D_k = \{C_{k+1} - C_k\} D_{k+1} + \{D_{k+1} - D_k\} C_k. \quad \square$$

Applying Lemma 5.5.6 to integrate $\frac{1}{n}\sum_{k=1}^{n} a_k = \frac{1}{n}\sum_{k=1}^{n}\{b_k - b_{k-1}\}k$ by parts, one obtains

$$\frac{1}{n}\sum_{k=1}^{n} a_k = \frac{1}{n}\left[\sum_{k=1}^{n}\{b_k - b_{k-1}\}k\right] = \frac{1}{n}\left[(b_1 - b_0) + \sum_{k=1}^{n-1}\{b_{k+1} - b_k\}(k+1)\right]$$

$$= \frac{1}{n}\left[b_1 + (nb_n - b_1) - \sum_{k=1}^{n-1} b_k\right], \text{ letting } C_k = b_k \text{ and } D_k = k,$$

$$= b_n - \frac{(n-1)}{n}\left[\frac{1}{(n-1)}\sum_{k=1}^{n-1} b_k\right].$$

This last expression tends to zero as b_n converges by assumption and so $(a_n)_{n\geq 1}$ converges to zero in the sense of Cesaro (Exercise 4.3.14). $\quad \square$

Remark. There is a martingale proof of this theorem, due to Doob, that makes no use of truncation. In place of truncation, it uses the concept of a reversed, or backward, martingale. It differs from a martingale in that the σ-algebras decrease rather than increase. This proof is given later as Theorem 5.7.9 (see [D1], p. 341).

6. THE THREE-SERIES THEOREM AND THE DOOB DECOMPOSITION

A) The three-series theorem.

Consider a series $\sum_n X_n$ of independent random variables. If the X_n are all integrable with mean zero, then the partial sums of this series form a martingale and the series converges if and only if the martingale converges (which by the martingale convergence theorem (Theorem 5.7.3) is the case if the martingale is bounded in L^1). The question of convergence involves an event in the σ-field $\mathfrak{T}_\infty \stackrel{\text{def}}{=} \cap_n \sigma(\{X_k \mid k \geq n\})$, as indicated by the following exercise.

Exercise 5.6.1. Show that $\{\omega \mid \sum_n X_n(\omega) \text{ converges }\}$ belongs to $\mathcal{T}_\infty$.

Kolmogorov's 0–1 law (Proposition 3.4.4) implies that either the series converges a.s. or the series diverges a.s.

Example 5.6.2. The harmonic series $\sum_n \frac{1}{n}$ diverges. Let $\epsilon_n, n \geq 1$ be i.i.d. Bernoulli random variables with distribution $\frac{1}{2}(\varepsilon_{-1} + \varepsilon_1)$. It turns out that the random series $\sum_n \epsilon_n(\frac{1}{n})$ does converge a.s. Its partial sums $S_n = \sum_{k=1}^n \epsilon_k(\frac{1}{k}), n \geq 1$, define a martingale by Example 5.4.2M, and since $\sigma^2(S_n) = \sum_{k=1}^n \frac{1}{k^2}$ is the partial sum of the convergent series $\sum_n \frac{1}{n^2}$, this martingale is bounded in L^2. As a result, it follows from Proposition 5.5.1 that it converges a.s.

There is a well-known general criterion for the convergence of an infinite series of independent random variables that makes use of truncation. It is stated as the next result.

Theorem 5.6.3. (**The three-series theorem**) *Let $\sum_n X_n$ be a series of independent random variables. The following are equivalent*

(1) $\sum_n X_n$ *converges a.s.;*
(2) *for every truncation height $c > 0$, the three series*
   (i) $\sum_n \mathbf{P}[X_n \neq Y_n]$,
   (ii) $\sum_n E[Y_n]$, and
   (iii) $\sum_n \sigma^2(Y_n)$
   all converge;
(3) *for some truncation height $c > 0$, the three series all converge, where Y_n is the truncation of X_n at height $c > 0$.*

Proof. Assume (3). Then, by the Borel–Cantelli lemma (Proposition 4.1.29 (1)), the convergence of (i) implies that a.s. $X_n(\omega) = Y_n(\omega)$ for $n \geq n(\omega)$. Therefore, the series $\sum_n X_n$ converges if $\sum_n Y_n$ converges. Since (ii) converges, $\sum_n Y_n$ converges if and only if the martingale given by $\sum_n Y_n - E[Y_n]$ converges. By independence, if $S_n = \sum_{k=1}^n Y_k - E[Y_k]$, then $\| S_n \|_2^2 = \sum_{k=1}^n \sigma^2(Y_k)$. Since (iii) converges, this martingale is bounded in L^2 and so by Proposition 5.5.1 it converges. Hence (3) implies (1).

Now assume (1) and let $c > 0$ be any truncation height. Since the series $\sum_n X_n$ converges, a.s. $X_n(\omega) = Y_n(\omega)$ for $n \geq n(\omega)$ since a.s. $X_n(\omega) \to 0$. It follows from independence and the Borel–Cantelli lemma (Proposition 4.1.29 (2)) that the series (i) converges. As a result, the series $\sum_n Y_n$ converges a.s.

From this, it follows that the second series (ii) converges if and only if the series $\sum_n Y_n - E[Y_n]$ converges. By Proposition 5.5.1, this series will converge if the third series (iii) converges. To prove (2), it will suffice to prove the following.

Lemma. *If $|Y_n| \leq c$, $n \geq 1$ are independent and $\sum_n Y_n$ converges a.s., then $\sum_n \sigma^2(Y_n)$ converges.*

6. THE THREE-SERIES THEOREM AND THE DOOB DECOMPOSITION

This will be proved as an application of the Doob decomposition (Proposition 5.6.4) of a submartingale. Since (2) implies (3), this completes the proof. □

B) The Doob decomposition of a submartingale.

One way to produce a submartingale would be to start with a martingale $(M_n)_{n\geq 1}$ and add an increasing adapted process $(A_n)_{n\geq 1}$, where **increasing** means that $A_n \leq A_{n+1}$, for all $n \geq 1$. Then $E[M_n + A_n|\mathfrak{F}_{n-1}] = M_{n-1} + E[A_n|\mathfrak{F}_{n-1}] \geq M_{n-1} + E[A_{n-1}|\mathfrak{F}_{n-1}] = M_{n-1} + A_{n-1}$.

The Doob decomposition shows that every submartingale has this form. It also states that the increasing process $(A_n)_{n\geq 1}$ is unique if it is **predictable**.

Proposition 5.6.4. (Doob decomposition) *Let $X = (X_n)_{n\geq 0}$ be a submartingale with respect to a filtration $(\mathfrak{F}_n)_{n\geq 0}$ on a probability space $(\Omega, \mathfrak{F}, \mathbf{P})$. Then there is a unique decomposition of X into the sum of an $\mathfrak{F}_0$ random variable X_0, a martingale M with $M_0 = 0$, and a predictable increasing process A with $A_0 = 0$:*

$$X_n = X_0 + M_n + A_n, \quad A_n = \sum_{k=1}^{n} E[X_k|\mathfrak{F}_{k-1}] - X_{k-1}.$$

Proof. Assume that such a decomposition exists. Then

$$E[X_n|\mathfrak{F}_{n-1}] = X_0 + M_{n-1} + E[A_n|\mathfrak{F}_{n-1}] = X_0 + M_{n-1} + A_n$$
$$\geq X_{n-1} = X_0 + M_{n-1} + A_{n-1}.$$

Hence,

(*) $$E[X_n|\mathfrak{F}_{n-1}] - X_{n-1} = A_n - A_{n-1}.$$

This implies that $A_n = \sum_{k=1}^{n} E[X_k|\mathfrak{F}_{k-1}] - X_{k-1}$.

Conversely, define $A_0 = 0$ and $A_n = \sum_{k=1}^{n} E[X_k|\mathfrak{F}_{k-1}] - X_{k-1}$. Then A is increasing and predictable, and, as (*) holds,

$$E[X_n - A_n|\mathfrak{F}_{n-1}] = E[X_n|\mathfrak{F}_{n-1}] - A_n = X_{n-1} - A_{n-1}.$$

Hence, if $M_n = X_n - A_n$, then $(M_n)_{n\geq 0}$ is a martingale. □

Exercise 5.6.5. Suppose that X is the sum of a martingale M' and an increasing adapted process A' (not necessarily predictable). Determine the Doob decomposition of the submartingale X.

Exercise 5.6.6. Let $(Y_n)_{n \geq 1}$ be a sequence of independent random variables Y_n in L^2 with mean zero. Let $S_n = \sum_{k=1}^n Y_k$ and $\mathfrak{F}_n = \sigma(\{X_k \mid 1 \leq k \leq n\})$. Show that the increasing process of the Doob decomposition for the submartingale $(S_n^2)_{n \geq 1}$ is the sequence of partial sums of the series $\sum_n \sigma^2(Y_n)$. [Hint: recall that $E[(M_n - M_{n-1})^2 | \mathfrak{F}_{n-1}] = E[M_n^2 - M_{n-1}^2 | \mathfrak{F}_{n-1}]$ if M is an L^2-martingale; see the proof of (2) in Proposition 5.5.1.]

The result of this exercise is a key to proving the following proposition which is very similar to the lemma that is needed to complete the three-series theorem.

Proposition 5.6.7. Let $(X_n)_{n \geq 1}$ be a sequence of independent random variables with mean zero that are uniformly bounded (i.e., $|X_n| \leq C, n \geq 1$). If $\sum_n X_n$ converges a.s., then $\sum_n \sigma^2(X_n)$ converges.

Proof. If $S_n = \sum_{k=1}^n X_k$, then S_n is a martingale and $S_n^2 - \sum_{k=1}^n \sigma^2(X_k)$ is a martingale by Exercise 5.6.6. Therefore, for any stopping time T,

(†) $$E[S_{n \wedge T}^2 - A_{n \wedge T}] = 0, \quad \text{where } A_n = \sum_{k=1}^n \sigma^2(X_k).$$

Let $c > 0$ and let $T = \inf\{n \mid |S_n| > c\}$ with $\inf \emptyset = +\infty$. This is a stopping time by Exercise 5.4.11 (2). The key step is the following lemma.

Lemma. *For some $c > 0$, $\mathbf{P}[T = +\infty] = a > 0$.*

Assume the lemma. Since each increment X_n of the martingale S is bounded by C, it follows that $S_{n \wedge T}^2 \leq (C + c)^2$. Consequently, it follows from (†) that

$$a \sum_{k=1}^n \sigma^2(X_k) = E[A_{n \wedge T} \cdot 1_{\{T=+\infty\}}] \leq E[A_{n \wedge T}] = E[S_{n \wedge T}^2] \leq (C+c)^2. \quad \square$$

Proof of the lemma. Let $S = \sum_{n=1}^\infty X_n$ and $\Lambda_{c_0} = \{|S| \leq c_0\}$. Then for some c_0, one has $\mathbf{P}(\Lambda_{c_0}) > \frac{1}{2}$. If $\omega \in \Lambda_{c_0}$, then for some N, $\omega \in \Lambda_{c_0,N} = \cap_{n \geq N} \{|S_n| \leq c_0 + 1\}$. As a result, there is an N with $\mathbf{P}(\Lambda_{c_0,N}) > \frac{1}{4}$ (say). If $c = c_0 + 1 + NC$, and $\omega \in \Lambda_{c_0,N}$, then $|S_n(\omega)| \leq c$ for all $n \geq 1$. $\quad \square$

While this last result is not enough by itself to prove the missing lemma, by a clever trick, involving independence (see Lamperti [L1], p. 36, and Williams [W2]), one may use Proposition 5.6.7 to prove this lemma. The original sequence of independent bounded random variables $(Y_n)_{n \geq 1}$ is defined on a probability space $(\Omega, \mathfrak{F}, \mathbf{P})$. Take a copy of $(\Omega, \mathfrak{F}, \mathbf{P}), (Y_n)$, denote it by $(\Omega', \mathfrak{F}', \mathbf{P}'), (Y'_n)$, and form the product $(\Omega \times \Omega', \mathfrak{F} \times \mathfrak{F}', \mathbf{P} \times \mathbf{P}')$ of the probability spaces. Define two processes on $\Omega \times \Omega'$: the first one, Y, has $Y_n(\omega, \omega') = Y_n(\omega)$ and the second one, Y', has $Y'_n(\omega, \omega') = Y'_n(\omega')$. These two processes on the product space are independent, i.e., $\sigma(\{Y_n \mid n \geq 1\})$ and $\sigma(\{Y_n' \mid n \geq 1\})$ are independent σ-fields and have the same finite-dimensional joint distributions.

As a result, the sequence of random variables $Z_n = Y_n - Y_n'$ is independent (see Exercise 5.6.9) and each Z_n has mean zero since the distribution of Y_n' equals the distribution of Y_n. Also, $|Z_n| \leq 2c$. Provided the series $\sum_n Z_n$ converges a.s., Proposition 5.6.7 shows that $\sum_n \sigma^2(Z_n) = 2\sum_n \sigma^2(Y_n)$ converges, which proves the lemma that completes the proof of the three -series theorem.

This question of convergence amounts to the a.s. convergence of the series $\sum_n Y_n'$. The distribution $\mathbf{Q}$ of the processes Y and Y' coincides: it is the infinite product $\times_{n=1}^{\infty} \mathbf{Q}_n$ on the product $\times_{n=1}^{\infty} \mathbb{R}$ of a countable number of copies of $\mathbb{R}$, where $\mathbf{Q}_n$ is the distribution of Y_n. Now the original series $\sum_n Y_n$ on $(\Omega, \mathfrak{F}, \mathbf{P})$ converges a.s. if and only if $\mathbf{Q}(\Lambda) = 1$, where $\Lambda = \{(a_n)_{n \geq 1}, a_n \in \mathbb{R} \mid \sum_n a_n \text{ converges}\}$ (which begs a measurability question: see the following exercise). Given this, it follows that the series $\sum_n Y_n$ converges a.s. if and only if the series $\sum_n Y_n'$ converges a.s. and the proof of the three series theorem is complete since a.s. both series converge if this is true of one of them.

Exercise 5.6.8. Show that $\Lambda \in \times_{n=1}^{\infty} \mathfrak{B}(\mathbb{R})$.

Exercise 5.6.9. Let $(\Omega, \mathfrak{F}, \mathbf{P})$ be a probability space, and let $\mathfrak{F}_1, \mathfrak{F}_2, \mathfrak{G}_1$ and $\mathfrak{G}_2$ be four σ algebras contained in $\mathfrak{F}$. Assume that

(1) $\mathfrak{F}_1$ and $\mathfrak{F}_2$ are independent,
(2) $\mathfrak{G}_1$ and $\mathfrak{G}_2$ are independent, and
(3) $\sigma(\mathfrak{F}_1 \cup \mathfrak{F}_2)$ and $\sigma(\mathfrak{G}_1 \cup \mathfrak{G}_2)$ are independent.

Show that $\sigma(\mathfrak{F}_1 \cup \mathfrak{G}_1)$ and $\sigma(\mathfrak{F}_2 \cup \mathfrak{G}_2)$ are independent. [Hint: if $A_i \in \mathfrak{F}_i$ and $B_i \in \mathfrak{G}_i$, then $\mathbf{P}(A_1 \cap A_2 \cap B_1 \cap B_2) = \mathbf{P}(A_1)\mathbf{P}(A_2)\mathbf{P}(B_1)\mathbf{P}(B_2)$, and make use of Proposition 3.2.6 as in the proof of Proposition 3.2.8).] Conclude that if $(Y_n)_{n \geq 1}$ is an independent real-valued process Y and $(Y_n')_{n \geq 1}$ is an independent real-valued process Y', then the process $(Z_n)_{n \geq 1}$, $Z_n = Y_n - Y_n'$, is an independent real-valued process provided that Y and Y' are independent, i.e., provided the σ-algebras $\sigma(\{Y_n \mid n \geq 1\})$ and $\sigma(\{Y_n' \mid n \geq 1\})$ are independent.

7. THE MARTINGALE CONVERGENCE THEOREM

A sequence $(a_n)_{n \geq 1}$ of real numbers fails to converge if and only if $\liminf_n a_n < \limsup_n a_n$ (see Exercise 2.1.10). When this happens, there is a pair of rational numbers $p < q$ such that the sequence is infinitely often $\leq p$ and infinitely often $\geq q$. In other words, the sequence crosses the interval $[p, q]$ from p to q an infinite number of times, where exactly one such "**upcrossing**" takes place during the time interval $[k, \ell]$ if $a_k \leq p, a_\ell \geq q$ and in between times the value never gets to be $\geq q$, i.e., ℓ is the first time n after k that $a_n \geq q$. Consequently, to study the convergence of a process, one examines the random variable that describes the upcrossings of an interval $[a, b]$ by the paths of a process.

Proposition 5.7.1. (see Ikeda and Watanabe, [I], p. 29, Theorem 6.3) Let $X = (X_n)_{0 \leq n \leq N}$ be a submartingale. If $a < b$, let $U([a, b]) = U^X([a, b])$ denote the number of upcrossings of $[a, b]$ by X during $\{0, 1, 2, \ldots, N\}$. Then,
$$E[U([a, b])] \leq \frac{1}{b - a}\Big[E[(X_N - a)^+] - E[(X_0 - a)^+]\Big].$$

Proof. Let $T_0 = 0$ and T_1 be the first time that $X_n \leq a$ with default value N if this never occurs. Let T_2 be the first time after T_1 that $X_n \geq b$ with N as default value. By induction, define T_{2k+1} to be the first time after T_{2k} that $X_n \leq a$ with default value N. Similarly, let $T_{2(k+1)}$ denote the first time after T_{2k+1} that $X_n \geq b$, again with default N. This is an increasing sequence of stopping times (see Exercise 5.7.2) all of which equal the default value of N if $2k \geq N$. Further, $U([a, b]) = \sum_{k=1}^{2N} 1_{A_k}$, where $A_k = \{\omega \mid X_{T_{2k-1}}(\omega) \leq a < b \leq X_{T_{2k}}(\omega)\}$, and hence is a random variable.

Replace the submartingale X by the process Y, where $Y_n = (X_n - a)^+$, which is a submartingale by Proposition 5.4.9S (2) or (3). Note that $U^X([a, b]) = U^Y([0, b-a])$ and that by optional stopping (Theorem 5.4.18), $E[Y_{T_{j+1}} - Y_{T_j}] \geq 0$, for all j. Hence, if $2k \geq N$,

$$E[Y_N - Y_0] = \sum_{j=0}^{2k} E[Y_{T_{j+1}} - Y_{T_j}] \geq \sum_{\ell=1}^{k} E[Y_{T_{2\ell}} - Y_{T_{2\ell-1}}]$$

$$= E[\sum_{\ell=1}^{k}(Y_{T_{2\ell}} - Y_{T_{2\ell-1}})] \geq (b-a)E[U^Y([b-a, 0])]$$

since $\sum_{\ell=1}^{k}(Y_{T_{2\ell}} - Y_{T_{2\ell-1}})] \geq (b-a)U^Y([b-a, 0])$. □

Exercise 5.7.2. Show by induction that the times T_j above are stopping times. [Hint: if $m < N$, then $(T_{2k} = m) = \cup_{\ell < m}[(T_{2k-1} = \ell) \cap \{X_{\ell+1} < b, X_{\ell+2} < b, \ldots, X_{m-1} < b, X_m \geq b\}] \in \mathfrak{F}_m$.]

Corollary 5.7.3. (**Martingale convergence theorem: countable case**) Let $(X_n)_{n \geq 0}$ be a submartingale with $\sup_n E[X_n^+] < +\infty$ or, equivalently, bounded in L^1. Then it converges a.s. to a limit in L^1. In particular, if $(X_n)_{n \geq 0}$ is a martingale that is bounded in L^1, it converges a.s. to a limit in L^1.

Proof. Since $|X_n| = 2X_n^+ - X_n$, $E[|X_n|] = 2E[X_n^+] - E[X_n] \leq 2E[X_n^+] - E[X_0]$, the process X is bounded in L^1 (i.e., $\sup_n E[|X_n|] < +\infty$), if and only if $\sup_n E[X_n^+] < \infty$.

If a sequence of random variables X_n is bounded in L^1 and converges a.s. to X, then $X \in L^1$ since by Fatou's lemma (Proposition 2.1.25), $E[|X|] \leq \liminf_n E[|X_n|] \leq \sup_n E[|X_n|] < \infty$.

7. THE MARTINGALE CONVERGENCE THEOREM

In view of the remarks preceding Proposition 5.7.1, the submartingale converges a.s. if the mean or expected number of upcrossings of any interval $[a,b]$ is a.s. finite. Let $U_N([a,b])$ be the number of upcrossings of $[a,b]$ by the submartingale during $\{0,1,2,\ldots,N\}$. Then, the number $U([a,b])$ of upcrossings of $[a,b]$ by the submartingale during $\{0\} \cup \mathbb{N}$ is the limit of the increasing sequence $U_N([a,b])$. Since $(x-a)^+ \leq x^+ + |a|$, it follows from Proposition 5.7.1 that

$$E[U([a,b])] \leq \frac{1}{b-a}\left(\sup_n E[X_n^+] + |a|\right) < +\infty. \qquad \square$$

Corollary 5.7.4. *Let $1 < p < \infty$ and let $(X_n)_{n \geq 1}$ be a martingale that is bounded in L^p, i.e., for some constant M, $E[|X|_n^p] \leq M$. Then there is a random variable $X \in L^p$ such that $X_n \xrightarrow{a.s.} X$. In addition, $X_n \xrightarrow{L^p} X$.*

Proof. Since $E[X_n^+] \leq \|X_n\|_1 \leq \|X_n\|_p \leq M^{1/p}$, the martingale is bounded in L^1 and hence by Corollary 5.7.3 it converges a.s. Since by Exercise 5.5.3, its maximal function is in L^p, it follows from the theorem of dominated convergence (Theorem 2.1.38) that it converges in L^p. $\square$

Remarks. (1) The case of $p = \infty$ is even simpler, as then the maximal function is bounded.

(2) Note that there is nothing in the proof of the martingale convergence theorem that lets one conclude the L^p-convergence, except that in the special case of $p = 2$ a different proof of convergence also gives the L^2-convergence (see Proposition 5.5.1). For this reason, it is important to know that the maximal function is in L^p. For $p = 1$, uniform integrability of the martingale is necessary and sufficient for convergence in L^1 (see Proposition 4.5.7, Corollary 4.5.9, and Exercise 4.7.13). When $1 < p < \infty$, uniform integrability in L^p for a martingale (i.e., the sequence of pth powers is uniformly integrable) is equivalent to its L^p-boundedness since the maximal function is in L^p (see Exercise 4.5.11 (1)).

Applying this convergence theorem to the martingale $X = (X_n)_{n \geq 0}$ of Example 5.4.5 gives a proof of the Radon–Nikodym theorem for a positive bounded measure μ on say $[0,1]$ (see Doob [D1], p. 343). Since $\mu([0,1]) < +\infty$, the martingale is bounded in L^1. Therefore, it converges a.s. to a limit function $X \in L^1$. To prove the Radon–Nikodym Theorem, it is necessary and sufficient to show that this martingale is uniformly integrable (see Exercise 5.7.5). When this is the case, it also converges in L^1 by Corollary 4.5.9.

The significance of this is that, for all $A \in \mathfrak{A} = \cup_{n=1}^{\infty} \mathfrak{F}_n$, one has, $\int_A X_n d\mathbf{P} \longrightarrow \int_A X d\mathbf{P}$ as $\left|\int_A X_n d\mathbf{P} - \int_A X d\mathbf{P}\right| \leq \int |1_A(X_n - X)| d\mathbf{P} \leq \|X_n - X\|_1$. Now if $A \in \mathfrak{F}_k$, $\int_A X_n d\mathbf{P} = \mu(A)$ for all $n \geq k$. Consequently, $\mu(A) = \int_A X d\mathbf{P}$, for all $A \in \mathfrak{A}$.

Since $\mathfrak{A}$ generates $\mathfrak{B}([0,1])$ by Exercise 5.4.6, it follows from a monotone class argument (Exercise 1.4.16) or Proposition 3.2.6, that

$$\mu(A) = \int_A X\,d\mathbf{P}, \quad \text{for all } A \in \mathfrak{B}([0,1]).$$

In other words, the limit of the martingale is the Radon–Nikodym derivative of μ with respect to $\mathbf{P}$.

Exercise 5.7.5. (This exercise explains why the martingale in Exercise 5.4.5 is uniformly integrable.) Show that

(1) $\mathbf{P}[X_n \geq c] \leq \frac{\mu(\Omega)}{c}$. [Hint: $\int_C X_n\,d\mathbf{P} = \mu(C)$ for all $C \in \mathfrak{F}_n$.]

The next step is to recall that the measure μ is "continuous" relative to $\mathbf{P}$ by Lemma 4.5.5*: given $\epsilon > 0$, there exists $\delta > 0$ such that $\mu(A) < \epsilon$ if $\mathbf{P}(A) < \delta$. Conclude from this that

(2) $\int_{\{X_n \geq c\}} X_n\,d\mathbf{P}$ is uniformly small if c is sufficiently large, and
(3) the martingale is uniformly integrable.

This martingale method can be used to prove the Radon–Nikodym theorem on σ-fields that are countably generated: the generators can be used to get an increasing sequence of finite σ-algebras whose union generates $\mathfrak{F}$.

Exercise 5.7.6. Extend the results of Example 5.4.5 and Exercise 5.7.5 to prove the following version of the **Radon–Nikodym theorem** (Theorem 2.7.19). Let $(\Omega, \mathfrak{F})$ be a measure space, and assume that there is a sequence $(A_n)_{n \geq 1} \subset \mathfrak{F}$ such that $\mathfrak{F} = \sigma\{A_n \mid n \geq 1\}$) (i.e., the σ-algebra $\mathfrak{F}$ is **countably generated**). Let μ and ν be two (non-negative, finite) measures on $\mathfrak{F}$ such that $\mu(E) = 0$ implies $\nu(E) = 0$ (i.e, ν is **absolutely continuous** with respect to μ). Then there is a (unique) function $f \in L^1(\Omega, \mathfrak{F}, \mu)$ such that $\nu(E) = \int_E f\,d\mu$ for all $E \in \mathfrak{F}$.

Backward martingales.

Consider a filtration given by $\mathfrak{F}_n$, where $n \in -\mathbb{N}$, i.e, for each $n \in -\mathbb{N}$ one has $\mathfrak{F}_{n-1} \subset \mathfrak{F}_n$. The filtration "provides" less and less "information" as times increases. Such filtrations arise naturally from sequences $(X_k)_{k \geq 1}$ of random variables: set $\mathfrak{F}_n = \sigma(\{X_k \mid -n \leq k\})$. This type of filtration occurred in the proof of Kolmogorov's 0–1 Law (Proposition 3.4.4). A martingale relative to this type of filtration is called a **backward martingale**.

Proposition 5.7.7. *Let $(X_n)_{n \leq -1}$ be a backward martingale with respect to the filtration $(\mathfrak{F}_n)_{n \in -\mathbb{N}}$. Then $(X_n)_{n \leq -1}$ is uniformly integrable and converges a.s. and in L^1.*

Proof. Since $X_n = E[X_{-1}|\mathfrak{F}_n]$ for all $n \leq -1$, it follows from Exercise 5.4.3 (2) that $(X_n)_{n \leq -1}$ is uniformly integrable. Corollary 4.5.9 implies that it converges in L^1 if it converges a.s.

7. THE MARTINGALE CONVERGENCE THEOREM   249

The mean upcrossing inequality in Proposition 5.7.1 applies to this martingale on the time set $\{-N, -N+1, \ldots, -2, -1\}$ and so

$$E[U([a,b])] \leq \frac{1}{b-a}(E[(X_{-1} - a)^+].$$

Hence, a.s. the number $U([a, b])$ of upcrossings of $[a, b]$ by the martingale during $-\mathbb{N}$ is finite and so it converges a.s. □

Remark 5.7.8. In fact, the argument proves that a backward submartingale converges a.s.

Kolmogorov's strong law of large numbers (Theorem 4.4.2) follows from the convergence of backward martingales. Here is the restatement of the theorem in this context.

Theorem 5.7.9. (**Kolmogorov's strong law**). *Let* $(X_n)_{n \geq 1}$ *be a sequence of i.i.d. integrable random variables with mean zero, and set* $S_n = \sum_{k=1}^{n} X_k$. *Let* $\mathfrak{G}_{-n} = \sigma(\{S_n\} \cup \{X_k \mid k \geq n+1\})$. *Then* $(\frac{1}{n} S_n)_{n \geq 1}$ *is a backward martingale relative to the filtration* $(\mathfrak{G}_m)_{m \in -\mathbb{N}}$, *and its a.s. limit is* 0.

Proof. Since $\mathfrak{T}_{n+1} = \sigma(\{X_k \mid k \geq n+1\})$ and $\sigma(\{X_1, S_n\})$ are independent, it follows from Proposition 5.2.4 (CE_{11}) that $E[X_1 | \mathfrak{G}_{-n}] = E[X_1 | S_n]$.

If $A \in \sigma(S_n)$, by Exercise 2.9.10 there is a Borel set $B \subset \mathbb{R}$ with $1_A = 1_B \circ S_n$. Since X_1 and S_n are $\sigma(\{X_j \mid 1 \leq j \leq n\})$-measurable, and the distribution of $(X_1, X_2, \ldots, X_n)$ is $\mathbf{Q} \times \mathbf{Q} \times \cdots \times \mathbf{Q}$ (n copies), where $\mathbf{Q}$ is the common distribution of the X_k, it follows from Lemma 4.1.3 that $\int_A X_1 d\mathbf{P} = \int x_1 1_B(x_1 + x_2 + \cdots + x_n) \mathbf{Q}(dx_1)\mathbf{Q}(dx_2) \cdots \mathbf{Q}(dx_n)$.

It follows by symmetry that for all j, $1 \leq j \leq n$, $\int_A X_1 d\mathbf{P} = \int_A X_j d\mathbf{P}$ and so $\int_A X_1 d\mathbf{P} = \frac{1}{n} \int_A S_n d\mathbf{P}$ (i.e., $E[X_1 | \mathfrak{G}_n] = \frac{1}{n} S_n$). This proves that $(\frac{1}{n} S_n)_{n \geq 1}$ is a backward martingale.

Proposition 5.7.7 implies that the limit Z of $\frac{1}{n} S_n$ exists and also that $E[Z] = 0$ since the convergence is also in L^1. The limit function $Z \in \mathfrak{T} = \cap_{n+1}^{\infty} \mathfrak{T}_n$ since, for any $m \geq 1$, the limit $Z = \lim_n \frac{1}{n} \sum_{k=m}^{n} X_k \in \mathfrak{T}_m$. Kolmogorov's 0–1 Law (Proposition 3.4.4) implies that Z is constant a.s. and so $Z = 0$ a.s. □

Remark. This proof is due to Doob and may be found in his book [D1], pp. 341–342. Many other things having to do with martingales and convergence of sequences and series of random variables are to be found in Doob's book. In addition, it is also worth consulting for the historical notes and references to the literature on these topics.

CHAPTER VI

AN INTRODUCTION TO WEAK CONVERGENCE

1. Motivation: empirical distributions

Let $(X_n)_{n\geq 1}$ be a sequence of i.i.d. random variables on a probability space $(\Omega, \mathfrak{F}, \mathbf{P})$. Let F be the distribution function of the common law $\mathbf{Q}$ of the random variables.

Pick a real number x, and let $Y_n = 1_{(-\infty, x]} \circ X_n$. Then the Y_n are i.i.d. random variables (see Exercise 3.4.5). They are Bernoulli with $\mathbf{P}[Y_n = 1] = F(x)$ and $\mathbf{P}[Y_n = 0] = 1 - F(x)$. It follows from the strong law of large Numbers, either Kolmogorov's strong law (Theorem 4.4.2) or more simply from Borel's strong law (see Exercise 4.7.19), that $\frac{1}{n}\sum_{k=1}^n Y_k \to F(x)$ a.s.

If the $X_k(\omega)_{1\leq k\leq n}$ are viewed as n independent samples from a population with underlying probability distribution $\mathbf{Q}$, then one defines the **empirical distribution function** $F_n(x, \omega)$ to be the random variable $\frac{1}{n}\sum_{k=1}^n Y_k(\omega)$, which equals $\frac{1}{n}$ times the number of k, $1 \leq k \leq n$, for which $X_k(\omega) \leq x$. For each $\omega \in \Omega$, this is a distribution function that takes the values $0, \frac{1}{n}, \frac{2}{n}, \ldots, \frac{n-1}{n}, 1$. Hence, $F_n(x, \omega)$ is a random distribution function.

By the strong law, there is a set $\Lambda \in \mathfrak{F}$ with $\mathbf{P}(\Lambda) = 1$ such that for all $q \in \mathbb{Q}$ one has $\lim_n F_n(q, \omega) = F(q)$ if $\omega \in \Lambda$.

Lemma 6.1.1. *Let F and F_n, $n \geq 1$, be non-decreasing real-valued functions on $\mathbb{R}$. If $\lim_n F_n(q) = F(q)$ for all $q \in \mathbb{Q}$, and F is continuous at x_0, then $\lim_n F_n(x_0) = F(x_0)$.*

Proof. Let $\epsilon > 0$, and assume $\delta > 0$ is such that $|F(x_0) - F(x)| < \epsilon$ if $|x - x_0| < \delta$ (Definition 2.1.13). Since the rationals are dense in $\mathbb{R}$ (Exercise 2.1.12), there exist rational numbers q_1, q_2 with $|q_i - x_0| < \delta$ and, say, $q_1 \leq x_0 \leq q_2$. Then $F(q_1) \leq F(x_0) \leq F(q_2)$ and $F_n(q_1) \leq F_n(x_0) \leq F_n(q_2)$. By assumption, if n is large enough, one has $|F_n(q_1) - F(q_1)| < \epsilon$ and $|F_n(q_2) - F(q_2)| < \epsilon$. Hence, $F_n(q_1)$ and $F_n(q_2)$ are both in $[F(q_1) - \epsilon, F(q_2) + \epsilon]$ for large n. As a result, $F_n(x_0)$ is also in this interval for large n, and so $F_n(x_0) \in [F(x_0) - 2\epsilon, F(x_0) + 2\epsilon]$ (i.e., $|F_n(x_0) - F(x_0)| \leq 2\epsilon$ for n large enough). $\square$

Remark. While this proof makes no use of the right continuity of a distribution function, essential use is made of the fact that a distribution function is non-decreasing.

2. WEAK CONVERGENCE OF PROBABILITIES

Corollary 6.1.2. *Let Λ be the set of probability 1 such that if $q \in \mathbb{Q}$, then $F_n(q, \omega) \to F(q)$ for all $\omega \in \Lambda$. If F is continuous at x_0, then $F_n(x_0, \omega) \to F(x_0)$.*

This motivates the following definition.

Definition 6.1.3. *A sequence of distribution functions F_n on $\mathbb{R}$ converges weakly to a distribution function F if $F_n(x) \to F(x)$ whenever F is continuous at x. This will be denoted by writing $F_n \Rightarrow F$ or $F_n \xrightarrow{w} F$.*

In other words, it follows that a.s. the empirical distribution function of a random sample of a population converges weakly to the underlying distribution function of the population.

2. WEAK CONVERGENCE OF PROBABILITIES: EQUIVALENT FORMULATIONS

Definition 6.2.1. *A sequence $(\mu_n)_{n \geq 1}$ of probabilities on $(\mathbb{R}, \mathcal{B}(\mathbb{R}))$ is said to* **converge weakly** *to a probability μ if $\int f d\mu_n \to \int f d\mu$ for all bounded continuous functions f on $\mathbb{R}$. This will be denoted by writing $\mu_n \Rightarrow \mu$ or $\mu_n \xrightarrow{w} \mu$.*

Exercise 6.2.2. Let μ and ν be two finite measures on $(\mathbb{R}, \mathcal{B}(\mathbb{R}))$ (recall they are non-negative). Assume that $\int \varphi d\mu = \int \varphi d\nu$ for all continuous functions on $\mathbb{R}$ with compact support (see Definition 4.2.4). Show that

(1) $\mu([a, b]) = \nu([a, b])$ for any closed bounded interval $[a, b]$. [Hint: approximate $1_{[a,b]}$ by a larger continuous function whose support lies in $[a - \frac{1}{k}, b + \frac{1}{k}]$.]

Conclude that

(2) $\mu(\mathbb{R}) = \nu(\mathbb{R})$, and
(3) $\mu = \nu$. [Hint: normalize to get probabilities and show that they have the same distribution function or use Proposition 3.2.6.]

Remark. Definition 6.2.1 makes sense on a metric space. It is in this context that weak convergence is usually discussed (see Parthasarathy [P1] and Billingsley [B1]). It will be shown that this concept of weak convergence for probabilities is the same as the one defined earlier (Definition 6.1.3) using distribution functions.

The first step is to show that weak convergence of probabilities implies weak convergence of their distribution functions.

Proposition 6.2.3. *Let $(\mu_n)_{n \geq 1}$ converge weakly to μ, and let F_n be the distribution function associated with μ_n. Then $F_n \xrightarrow{w} F$, where F is the distribution function for μ.*

Proof. Let x be a point of continuity of F. Define

$$\psi_k(t) = \begin{cases} 1 & \text{if } t \leq x - \frac{1}{k}, \\ -k(t-x) & \text{if } x - \frac{1}{k} < t < x, \\ 0 & \text{if } t \geq x, \end{cases}$$

and let $\varphi_k(t) = \psi(t - \frac{1}{k})$. Then $1_{(-\infty, x - \frac{1}{k}]} \leq \psi_k \leq 1_{(-\infty, x]} \leq \varphi_k \leq 1_{(-\infty, x + \frac{1}{k}]}$.

Since ψ_k and φ_k are bounded continuous functions, one has

$$F\left(x - \frac{1}{k}\right) \leq \int \psi_k d\mu = \lim_n \int \psi_k d\mu_n \leq \liminf_n F_n(x), \text{ and}$$

$$\limsup_n F_n(x) \leq \lim_n \int \varphi_k d\mu_n = \int \varphi_k d\mu \leq F\left(x + \frac{1}{k}\right).$$

The continuity of F at x and these estimates imply that $\liminf_n F_n(x) = \limsup_n F_n(x)$. It follows from Exercise 2.1.10 that $\lim_n F_n(x) = F(x)$. □

Remark. The function ψ_k is bounded and uniformly continuous on all of $\mathbb{R}$, i.e., given $\epsilon > 0$, there is a $\delta > 0$ such that $|x - y| < \delta$ implies $|\psi_k(x) - \psi_k(y)| < \epsilon$. This follows from Property 2.3.8 (3) applied to ψ_k on the interval $[x - \frac{1}{k} - 1, x + 1]$ (say) since outside that interval the problem of finding δ is trivial. As a result, $F_n \xrightarrow{w} F$ as long as $\int f d\mu_n \to \int f d\mu$ only for all bounded uniformly continuous functions on $\mathbb{R}$.

By using integration by parts, it is easy to see that weak convergence of distribution functions implies something very similar to weak convergence of the probabilities.

Proposition 6.2.4. *Assume that $F_n \xrightarrow{w} F$ and that $\varphi \in C_c(\mathbb{R})$, i.e., that φ is a continuous function with compact support (Definition 4.2.4). Then $\int \varphi d\mu_n \to \int \varphi d\mu$.*

Proof. First, assume that $\varphi \in C_c^1(\mathbb{R})$, i.e., that $\varphi \in C_c(\mathbb{R})$ is continuously differentiable. The support of φ lies in some interval $(-m, m]$ for sufficiently large m (i.e., $(-m, m] \supset \overline{\{x \mid \varphi(x) \neq 0\}}$).

Lemma 6.2.5. *Let $\mathbf{Q}$ be a probability on $\mathfrak{B}(\mathbb{R})$, and let F denote its distribution function. If $\varphi \in C_c^1(\mathbb{R})$, then*

$$\int \varphi d\mathbf{Q} \stackrel{\text{def}}{=} \int \varphi(x) dF(x) = -\int \varphi'(x) F(x) dx.$$

(An outline of the proof of this integration by parts formula is given in Exercise 6.8.1.)

Assume Lemma 6.2.5. Now $F_n \to F$ a.e. as F is continuous except for most at a countable number of points (Exercise 2.9.14). Thus, by

2. WEAK CONVERGENCE OF PROBABILITIES

dominated convergence (Theorem 2.1.38), it follows that $\int F_n(x)\varphi'(x)dx \to \int F(x)\varphi'(x)dx$ and hence $\int \varphi d\mu_n \to \int \varphi d\mu$.

It is shown in the appendix to this chapter that if $\varphi \in C_c(\mathbb{R})$ and $\epsilon > 0$, there exists $\psi \in C_c^1(\mathbb{R})$ with $\|\varphi - \psi\|_\infty < \epsilon$ (where $\|\varphi - \psi\|_\infty = \sup_{x \in \mathbb{R}} |\varphi(x) - \psi(x)|$; see Exercise 4.2.11). This implies that $|\int \varphi d\mathbf{Q} - \int \psi d\mathbf{Q}| < \epsilon$ for any probability $\mathbf{Q}$. Hence, if $\int \psi d\mu - \epsilon < \int \psi d\mu_n < \int \psi d\mu + \epsilon$ for large n, it follows that $\int \varphi d\mu - 3\epsilon < \int \varphi d\mu_n < \int \varphi d\mu + 3\epsilon$ for large n. □

It turns out that Proposition 6.2.4, in fact, is enough to show that $\mu_n \xrightarrow{w} \mu$ (see Proposition 6.2.9). However, this is not obvious as there are lots of bounded continuous functions f that do not belong to $C_c(\mathbb{R})$, e.g., the constant function 1. To further complicate matters, the fact that $\lim_n \int \varphi d\mu_n$ exists, for all $\varphi \in C_c(\mathbb{R})$, does not imply that $\lim_n \int f d\mu_n$ exists for all bounded continuous functions f on $C_c(\mathbb{R})$. For example, if $\mu_n = \varepsilon_{n\pi}$, then $\lim_n \int \varphi d\mu_n = 0$ and yet, if $f(x) = \cos x$, then $\lim_n \int f d\mu_n$ does not exist as $\int f d\mu_n = \pm 1$.

In the next exercise, an example is given of a sequence (μ_n) of probabilities for which $\lim_n \int \varphi d\mu_n$ exists, for all $\varphi \in C_c(\mathbb{R})$, but which does not converge weakly to a probability μ.

Exercise 6.2.6. Let ε_a denote the point mass at a (Exercise 1.5.2). Show that

(1) if $x_n \to x$, then $\varepsilon_{x_n} \xrightarrow{w} \varepsilon_x$.

Let $F_0(x) = e^x$ for $x < 0$ and $F_0(x) = 1$ for $x \geq 0$, and let F_n denote the distribution function $F_n(x) = \frac{1}{2}F_0(x) + \frac{1}{2}H(x - n)$, where H is the Heaviside function (see Exercise 1.5.2). Show that

(2) $F_n(x) \to \frac{1}{2}F_0(x)$ for all x as n tends to infinity,

(3) $\int \varphi(x) dF_n(x) \to \frac{1}{2}\int_{-\infty}^{0} \varphi(x)e^x dx = \int \varphi d\nu$ as $n \to \infty$.

In this exercise, in each case, there is a finite measure ν such that $\lim_n \int \varphi d\mu_n \to \int \varphi d\nu$. Proposition 6.2.4 states that if $F_n \xrightarrow{w} F$ then the measure ν is the probability μ whose distribution function is F. This rules out the type of thing that occurs in part (2) of Exercise 6.2.6.

Proposition 6.2.7. Let $(\mu_n)_{n \geq 1}$ be a sequence of probabilities on $\mathbb{R}$. Assume that there is a probability μ such that

$$\lim_n \int \varphi d\mu_n = \int \varphi d\mu, \text{ for all } \varphi \in C_c(\mathbb{R}).$$

Then, given $\epsilon > 0$, there exists $M > 0$ such that $\mu_n([-M, M]) > 1 - \epsilon$ for all $n \geq 1$.

Proof. Since $1 = \lim_N \mu([-N, N])$, there exists $N \geq 1$ with $\mu([-N, N]) > 1 - \frac{\epsilon}{2}$. Let θ_N be the function in $C_c(\mathbb{R})$ defined by

$$\theta_N(x) = \begin{cases} 1 & \text{if } |x| \leq N, \\ N + 1 - |x| & \text{if } N < |x| < N + 1, \\ 0 & \text{if } N + 1 \leq |x|. \end{cases}$$

Since $0 \le 1_{[-N,N]} \le \theta \le 1_{[-N-1,N+1]}$ and $\lim_n \int \theta d\mu_n = \int \theta d\mu$, it follows that $\mu_n([-N-1, N+1]) > 1 - \epsilon$ for $n \ge n(\epsilon)$. As in the case of μ, there exists $M \ge N+1$ with $\mu_n([-M, M]) > 1 - \epsilon$ for $1 \le n \le n(\epsilon)$. □

The property of the sequence (μ_n) that is the conclusion of this proposition is referred to as tightness.

Definition 6.2.8. *A sequence $(\mu_n)_{n \ge 1}$ of probabilities on $(\mathbb{R}, \mathfrak{B}(\mathbb{R}))$ is said to be **tight** if, for any $\epsilon > 0$, there is a compact set K such that $\mu_n(\mathbb{R} \setminus K) < \epsilon$ for all $n \ge 1$.*

Remark. Since every compact set K is a subset of $[-a, a]$ for some $a > 0$, one may assume $K = [-a, a]$.

Proposition 6.2.9. *Let $(\mu_n)_{n \ge 1}$ be a tight sequence of probabilities on $(\mathbb{R}, \mathfrak{B}(\mathbb{R}))$, and assume that, for some measure ν, one has*

$$\int \varphi d\mu_n \to \int \varphi d\nu \text{ for all } \varphi \in \mathcal{C}_c(\mathbb{R}).$$

Then, ν is a probability and $\mu_n \xrightarrow{w} \nu$.

Proof. Let $\epsilon > 0$, and let $[-N, N]$ be such that $\mu_n([-N, N]) > 1 - \epsilon$ for all $n \ge 1$. Let $\theta = \theta_N$ be the function used in the proof of Proposition 6.2.7. Since $\int \theta d\nu = \lim_n \int \theta d\mu_n$ and $\int \theta d\mu_n \ge \mu_n([-N, N])$, it follows that $\nu([-N-1, N+1]) > 1 - \epsilon$. Hence, $\nu(\mathbb{R}) = 1$ as, for any $M > 0$, $\nu([-M, M]) \le \int \theta_M d\nu = \lim_n \int \theta_M d\mu_n \le 1$.
Let f be a bounded continuous function on $\mathbb{R}$. Then, for $\theta = \theta_{N+1}$,

$$\int f d\mu_n = \int f\theta d\mu_n + \int (1-\theta) f d\mu_n, \text{ and so}$$

$$\left| \int f d\mu_n - \int f d\nu \right| \le \left| \int f\theta d\mu_n - \int f\theta d\nu \right| + \int (1-\theta)|f| d\mu_n$$

$$+ \int (1-\theta)|f| d\nu$$

$$\le \left| \int f\theta d\mu_n - \int f\theta d\nu \right| + 2\epsilon \| f \|_\infty,$$

where $\| f \|_\infty \stackrel{\text{def}}{=} \sup_{x \in \mathbb{R}} |f(x)|$ (see Exercise 4.2.11).
Since $f\theta \in \mathcal{C}_c(\mathbb{R})$, it follows from the hypothesis that $|\int f d\mu_n - \int f d\nu| < \epsilon + 2\epsilon \|f\|_\infty$, if n is large enough, and so $\lim_n \int f d\mu_n = \int f d\nu$. □

Exercise 6.2.10. Show that if ν is a measure and $(\mu_n)_{n \ge 1}$ is a sequence of probabilities such that $\lim_n \int \varphi_n d\mu_n = \int \varphi d\nu$ for all $\varphi \in \mathcal{C}_c(\mathbb{R})$, then ν is a probability if and only if $(\mu_n)_{n \ge 1}$ is tight.

These results may be summarized as the following theorem.

Theorem 6.2.11. *Let $(\mu_n)_{n\geq 1}$ be a sequence of probabilities on $(\mathbb{R}, \mathfrak{B}(\mathbb{R}))$ with corresponding distribution functions F_n. If μ is a probability on $(\mathbb{R}, \mathfrak{B}(\mathbb{R}))$ with distribution function F, the following are equivalent:*

(1) $\mu_n \xrightarrow{w} \mu$, *i.e.*, $\mu_n \Rightarrow \mu$;
(2) $F_n \xrightarrow{w} F$, *i.e.*, $F_n \Rightarrow F$.

Remark 6.2.12. Two other ways to show that $F_n \xrightarrow{w} F$ implies $\mu_n \xrightarrow{w} \mu$ are presented later in this chapter. One method, due to Gnedenko and Kolmogorov [G] can be found in Exercise 6.8.6. It avoids integration by parts and proves in an elegant manner the equivalence of Definitions 6.1.3 and 6.2.1, and convergence in a metric due to Lévy (Definition 6.3.6). The other method, which also makes implicit use of the Lévy metric, involves a theorem due to Skorohod (Theorem 6.3.4) which states that weak convergence of distribution functions can be realized as a.s. convergence of a sequence of random variables.

3. Weak Convergence of Random Variables

Since statisticians are mainly interested in the distribution of a random variable rather than the random variable itself, it is natural to look at the behaviour of the distributions of a sequence of random variables.

Definition 6.3.1. *A sequence of random variables $(X_n)_{n\geq 1}$ is said to **converge weakly** or in **distribution** or **law** to a random variable X if the distribution of X_n converges weakly to the distribution of X. This will be denoted by writing $X_n \xrightarrow{w} X$, $X_n \xrightarrow{d} X$, or $X_n \Rightarrow X$.*

This notion is related to convergence in probability. First, note that convergence in probability implies tightness of the resulting distributions.

Proposition 6.3.2. *Let $(X_n)_{n\geq 1}$ be a sequence of finite random variables on a probability space $(\Omega, \mathfrak{F}, \mathbf{P})$. Assume that $X_n \xrightarrow{pr} X$, and let μ_n be the distribution of X_n. Then, the sequence $(\mu_n)_{n\geq 1}$ of probabilities μ_n is tight.*

Proof. Let μ be the distribution of X, and assume that $\mu([-M, M]) > 1-\epsilon$. Let $A_n = \{|X_n| > M+1\}$. Since $A_n = (A_n \cap \{|X_n - X| \geq 1\}) \cup (A_n \cap \{|X_n - X| < 1\})$, it follows that

$$\mu_n([-M-1, M+1]^c) = \mathbf{P}(\{|X_n| > M+1\})$$
$$\leq \mathbf{P}(\{|X_n - X| \geq 1\}) + \mathbf{P}(\{|X| > M\})$$
$$\leq \mathbf{P}(\{|X_n - X| \geq 1\}) + \epsilon.$$

Hence, for sufficiently large n, $\mu_n([-M-1, M+1]) \geq 1 - 2\epsilon$. Since any finite set of probabilities is tight, the result follows. □

Not only does convergence in probability of a sequence of random variables ensure that their distributions form a tight sequence, but in fact they converge weakly.

Proposition 6.3.3. *Let $(X_n)_{n\geq 1}$ be a sequence of random variables on a probability space $(\Omega, \mathfrak{F}, \mathbf{P})$. If $X_n \xrightarrow{pr} X$, then $X_n \xrightarrow{d} X$.*

Proof. Since by Proposition 6.3.2, the sequence of distributions is tight, it follows from Proposition 6.2.9 that it is enough to show that $\int \varphi d\mu_n \to \int \varphi d\mu$ for all $\varphi \in C_c(\mathbb{R})$. By Lemma 4.1.3, therefore, it suffices to show that for all $\varphi \in C_c(\mathbb{R})$, $E[(\varphi(X_n)] \to E[\varphi(X)]$. Now for any $\delta > 0$, one has

$$E[\varphi(X_n)] - E[\varphi(X)] = \int_{\{|X_n - X| \geq \delta\}} \{\varphi \circ X_n - \varphi \circ X\} d\mathbf{P}$$
$$+ \int_{\{|X_n - X| < \delta\}} \{\varphi \circ X_n - \varphi \circ X\} d\mathbf{P}.$$

Let $\epsilon > 0$. By uniform continuity (Property 2.3.8 (3)) there exists a $\delta > 0$ such that $|x - y| < \delta$ implies $|\varphi(x) - \varphi(y)| < \epsilon$.

Hence,

$$|E[\varphi(X_n)] - E[\varphi(X)]| \leq 2\|\varphi\|_\infty \mathbf{P}[\,|X_n - X| \geq \delta\,] + \epsilon.$$

Consequently, if $X_n \xrightarrow{pr} X$, it follows that $|E[\varphi(X_n)] - E[\varphi(X)]| = |\int \varphi d\mu_n - \int \varphi d\mu| < 2\epsilon$ for sufficiently large n. $\square$

Since the statement $X_n \xrightarrow{w} X$ gives no information about the behaviour of the random variables themselves, but rather about their distributions, it is remarkable that weak convergence of probabilities may be realized as a.s. convergence of some sequence of random variables. This is stated as the following theorem due to Skorohod.

Theorem 6.3.4. *Let $(\mu_n)_{n\geq 1}$ be a sequence of probabilities on $(\mathbb{R}, \mathfrak{B}(\mathbb{R}))$ that converges weakly to the probability μ. Then there is a sequence $(Y_n)_{n\geq 1}$ of random variables Y_n on $((0,1), \mathfrak{B}((0,1)), dx)$ and a further random variable Y on the same probability space such that*

(1) μ_n *is the distribution of Y_n for all $n \geq 1$, μ is the distribution of Y, and*

(2) $Y_n \xrightarrow{a.s.} Y$.

Proof. The key to the argument is the observation that any distribution function F can be "inverted" to obtain a random variable Y on $(0,1)$ whose distribution function is F, as stated in the following lemma.

Lemma 6.3.5. *Let F be a distribution function on $\mathbb{R}$. Define $Y(t) = \inf\{x \mid t < F(x)\}$. Then Y is a random variable on $((0,1), \mathfrak{B}((0,1)), dx)$ such that*

(1) *it is right continuous and non-decreasing, and*

(2) *its distribution function is F, i.e., $|\{t \mid Y(t) \leq x\}| = F(x)$ for all $x \in \mathbb{R}$.*

3. WEAK CONVERGENCE OF RANDOM VARIABLES

Continuation of the proof. Assuming this lemma, let F_n be the distribution function associated with μ_n and F be the distribution function of μ. Let Y_n and Y be the corresponding random variables on $(0, 1)$ given by the lemma. They obviously satisfy condition (1) of the theorem. It remains to verify that they converge a.s.

This is done by making use of the Lévy distance between two distribution functions.

Definition 6.3.6. *Let F_1 and F_2 be two distribution functions on $\mathbb{R}$. The **Lévy distance** $d(F_1, F_2)$ between them is defined to be the infimum of all positive ϵ such that, for all $x \in \mathbb{R}$, one has $F_1(x - \epsilon) - \epsilon \leq F_2(x) \leq F_1(x + \epsilon) + \epsilon$.*

Note that $F_1(x - \epsilon) - \epsilon \leq F_2(x) \leq F_1(x + \epsilon) + \epsilon$ for all $x \in \mathbb{R}$ if and only if $F_2(x - \epsilon) - \epsilon \leq F_1(x) \leq F_2(x + \epsilon) + \epsilon$ for all $x \in \mathbb{R}$.

Exercise 6.3.7. Show that

(1) $d(F_1, F_2) \leq 1$ for any two distribution functions F_1 and F_2,
(2) the Lévy distance defined above is a metric (Definition 4.1.25) on the set of all distribution functions on $\mathbb{R}$, and
(3) if $d(F_n, F) \to 0$, then $F_n \xrightarrow{w} F$.

Note that the Lévy distance also has the property that

(4) if $F_n \xrightarrow{w} F$, then $d(F_n, F) \to 0$ as n goes to infinity.

Property (4) is established in two exercises in §8 (see Proposition 6.8.3).

By (4), it follows that if $\epsilon > 0$, then, for large n, one has $F(x - \epsilon) - \epsilon \leq F_n(x) \leq F(x + \epsilon) + \epsilon$ for all $x \in \mathbb{R}$. It follows almost immediately from the definition of the inverse of the distribution function given in the lemma that $Y(t - \epsilon) - \epsilon \leq Y_n(t) \leq Y(t + \epsilon) + \epsilon$ for all $t \in (\epsilon, 1 - \epsilon)$.

More explicitly: if $t < F_n(x) \leq F(x + \epsilon) + \epsilon$, then $t - \epsilon < F(x + \epsilon)$ and so $Y(t - \epsilon) \leq x + \epsilon$, which implies that $Y(t - \epsilon) - \epsilon \leq Y_n(t)$. Similarly, $F_n(x) \leq t$ implies that $F(x - \epsilon) - \epsilon \leq t$ and so $Y(t + \epsilon) \geq x - \epsilon$ (i.e., $Y(t + \epsilon) + \epsilon \geq Y_n(t)$).

The inequality $Y(t - \epsilon) - \epsilon \leq Y_n(t) \leq Y(t + \epsilon) + \epsilon$ for all $t \in (\epsilon, 1 - \epsilon)$ implies that $Y_n(t) \to Y(t)$ at all points of continuity of Y in $(0, 1)$. This completes the proof since by part (1) of the lemma, the function Y is right continuous, non-decreasing, and hence is continuous except at a countable number of points in $(0, 1)$ by Exercise 2.9.14. $\square$

Before giving the proof of Lemma 6.3.5 (see Lemma 6.3.10), here are several applications of Skorohod's theorem.

Proposition 6.3.8. *Let $\mu_n \xrightarrow{w} \mu$, and let h be a Borel function that is μ-a.s. continuous. Then the distribution $\mathbf{Q}_n$ of h relative to μ_n converges weakly to the distribution $\mathbf{Q}$ of h relative to μ (in other words, using a common notation for these distributions, $\mu_n \circ h^{-1} \xrightarrow{w} \mu \circ h^{-1}$.)*

Proof. Let D_h denote the set of discontinuities of h, and let $(Y_n)_{n\geq 1}$ be the sequence of random variables given by Skorohod's theorem. Since $Y_n \xrightarrow{a.s.} Y$ and $|\{Y \in D_h\}| = \mu(D_h) = 0$, where Y is the inverse of the distribution function of μ, it follows that $h \circ Y_n \xrightarrow{a.s.} h \circ Y$: let $\Lambda_1 = \{t \in (0,1) \mid Y_n(t) \to Y(t)\}$ and $\Lambda_2 = \{t \in (0,1) \mid h$ is continuous at $Y(t)\}$; then $|\Lambda_1 \cap \Lambda_2| = 1$, and for $t \in \Lambda_1 \cap \Lambda_2$ one has $Y_n(t) \to Y(t)$ and $h(Y_n(t)) \to h(Y(t))$. Proposition 6.3.3 implies that $\mathbf{Q}_n \xrightarrow{w} \mathbf{Q}$. □

Remark. This result can be proved without using Skorohod's theorem, (see Billingsley [B2], p. 343, Theorem 25.7).

For weak convergence of random variables, one has the following consequences.

Proposition 6.3.9. *Let $X_n \xrightarrow{w} X$ be a weakly convergent sequence of random variables on a probability space $(\Omega, \mathfrak{F}, \mathbf{P})$, and let h be a Borel function on $\mathbb{R}$. Denote by D_h its set of discontinuities. Then*

(1) $h \circ X_n \xrightarrow{w} h \circ X$ *if* $\mathbf{P}[X \in D_h] = 0$,
(2) *if $X = a$ a.s. (i.e., if $X_n \xrightarrow{w} a$) and h is continuous at $x = a$, then $h \circ X_n \xrightarrow{w} h(a)$,*
(3) *if $a_n \to a$ and $b_n \to b$, then $a_n X_n + b_n \xrightarrow{w} aX + b$, and*
(4) $E[|X|] \leq \liminf_n E[|X_n|]$.

Proof. (1) and (2) are corollaries of Proposition 6.3.8. Skorohod's theorem and Fatou's lemma applied to the (Y_n) prove (4): since $|Y_n| \xrightarrow{a.s.} |Y|$, by Fatou, $\int_0^1 |Y(t)| dt \leq \liminf_n \int_0^1 |Y_n(t)| dt$. Also, by Lemma 4.1.3, $\int_0^1 |Y(t)| dt = E[|X|]$ and $\int_0^1 |Y_n(t)| dt = E[|X_n|]$ since, for example, the law of Y is the law $\mathbf{Q}$ of X and so $E[\varphi(X)] = \int \varphi(x) \mathbf{Q}(dx) = \int_0^1 \varphi(Y(t)) dt$ for any bounded Borel function φ.

To verify (3), observe that if $h_n(x) = a_n x + b_n$ and $h(x) = ax + b$, then $Y_n \xrightarrow{a.s.} Y$ implies that $h_n(X_n) \xrightarrow{a.s.} h(Y)$. Hence, (3) follows from Proposition 6.3.3. □

Equivalence of (6.1.3) and (6.2.1). (see Billingsley [B2], pp. 344–345) The most interesting application of Skorohod's theorem, however, is another proof of the equivalence of Definitions 6.1.3 and 6.2.1, which was already alluded to in Remark 6.2.12. While these two ways of describing weak convergence have already been shown to be equivalent, an inspection of the proof of Skorohod's theorem and its consequences, Propositions 6.3.8 and 6.3.9, shows that, in fact, everything can be formulated in terms of the distribution functions and Definition 6.1.3, so that no use is made of Definition 6.2.1. As shown earlier (Proposition 6.2.3), it is straightforward to show that if $\mu_n \xrightarrow{w} \mu$, then $F_n \xrightarrow{w} F$. The converse is a consequence of Proposition 6.3.8. Let f be a bounded continuous function and let Y_n and Y be the random variables in Skorohod's theorem corresponding to the distribution functions F_n and F. Then by Proposition 6.3.8, $f \circ Y_n \xrightarrow{a.s.} f \circ Y$ and these random variables on $(0,1)$ are uniformly bounded by $\|f\|_\infty$. Hence, by dominated convergence (or Theorem 4.5.8 if one prefers) it follows that $\int f d\mu_n = \int_0^1 f(Y_n(t))dt \to \int_0^1 f(Y(t))dt = \int f d\mu$, and so $\mu_n \xrightarrow{w} \mu$. □

Before proving Lemma 6.3.5, it will be restated as follows.

Lemma 6.3.10. *Let F be a distribution function on $\mathbb{R}$. If $0 < t < 1$, let $Y(t) \stackrel{\text{def}}{=} \inf\{x \mid t < F(x)\}$. Then*

(1) $Y(t) = \sup\{y \mid F(y) \leq t\}$,
(2) $F(x) < t$ *implies that* $x < Y(t)$ *and, hence,* $F(Y(t)) \geq t$ *for all* $t \in (0,1)$,
(3) Y *is non-decreasing and right continuous, and*
(4) $F(a) < t < F(b)$ *implies* $a < Y(t) \leq b$ *and* $a < Y(t) \leq b$ *implies* $F(a) \leq t \leq F(b)$.

Hence, as a random variable on $(0,1)$, the distribution of Y is F, i.e., $|\{Y(t) \leq x\}| = F(x)$.

Proof. First, note that $Y(t)$ has a finite value for each $t \in (0,1)$ since a distribution function F has infimum equal to zero and supremum equal to one. For any $t \in (0,1)$, the distribution function F may be used to split $\mathbb{R}$ into the disjoint union of two intervals, namely $I_1 = \{x \mid F(x) \leq t\}$ and $I_2 = \{x \mid t < F(x)\}$. As a result, (1) follows from Exercise 1.1.19 (2), which implies that $Y(t)$ is a boundary point of both intervals.

To prove the first part of (2), observe that by right continuity of F, if $F(x) < t$ there is an x' with $x < x'$ and $F(x') < t$. Hence, $Y(t) \geq x' > x$.

Now, as observed above, each $t \in (0,1)$ splits $\mathbb{R}$ into two disjoint intervals with $Y(t)$ the common boundary point. If $x = Y(t)$ is in the left-hand interval, it follows from what has just been proved that $F(Y(t)) = F(x) = t$. Hence, for all t, one has $F(Y(t)) \geq t$.

Clearly, Y is non-decreasing. If $t_n \downarrow t$ and $t < F(x)$, then, for large n, $t_n < F(x)$ and so $\cup_n \{x \mid t_n < F(x)\} = \{x \mid t < F(x)\}$. The right continuity of Y follows from Exercise 1.1.19 (4) with $I_n = \{x \mid t_n < F(x)\}$.

It is clear that (4) amounts to the following four statements: (i) $a < Y(t)$ implies $F(a) \leq t$; (ii) $Y(t) \leq b$ implies $t \leq F(b)$; (iii) $F(a) < t$ implies $a < Y(t)$; and (iv) $t < F(b)$ implies $Y(t) \leq b$.

The first one is obvious; the second follows from (2) as $t \leq F(Y(t)) \leq F(b)$; the third one is part of (2); and the last one is also obvious.

It follows from (4) that $|\{a < Y(t) \leq b\}| = F(b) - F(a)$ and so the distribution function of Y is F. □

Remark 6.3.11. The operation that constructs the "inverse" Y of a distribution function F is often referred to as the **inverse** or **quantile transformation** (see an informative review by Csörgő[8] of a monograph on em-

---

[8]Bull. (new series) AMS **17** (1987), 189–200.

pirical processes). There is a certain arbitrariness in taking Y to be right continuous and there is a corresponding left continuous version of the quantile transformation: one defines $\tilde{Y}(t) = \inf\{x \mid F(x) \geq t\}$, which is the one discussed in the review by Csörgö. In case the distribution function F is continuous, then $Y(t)$ is the last "time" x that $F(x) = t$, whereas for the left continuous version of the quantile transformation it is the first "time" x that $F(x) = t$.

Note also that the "analyst's distribution function" of an integrable function (4.7.3) has an inverse that is referred to (see [S1], p. 189) as the **non-increasing rearrangement** or **monotone rearrangement** of the original measurable function.

Remark 6.3.12. Let F be a distribution function on $\mathbb{R}$ with associated probability μ. If $A \in \mathfrak{B}(\mathbb{R})$, then by Lemma 4.1.3, $\int 1_A(Y(t))dt = \mu(A)$. Hence, as pointed out by Billingsley ([B2], p. 190), one obtains the measure μ on the Borel subsets of $\mathbb{R}$ from the distribution function F, its inverse Y, and Lebesgue measure on $(0,1)$ without going through the procedure given in Chapter I. However, this is not an entirely free ride, as in any case one needs to get hold of Lebesgue measure on $(0,1)$, which amounts to proving the key Theorem 1.4.13 for the distribution function of the uniform distribution on $(0,1)$ (see Example 1.3.9 (1)).

Exercise 6.3.13. Let X be a random variable on a probability space $(\Omega, \mathfrak{F}, \mathbf{P})$. Let F denote its distribution function. Show that if F is continuous, then $F \circ X = F(X)$ is uniformly distributed on $[0,1]$. [*Hint*: observe that $F(X(\omega)) \leq t$ if and only if $X(\omega) \in \{x \mid F(x) \leq t\} = (\infty, x_0]$; compute $\mathbf{P}[F \circ X \leq t]$ in terms of the distribution of X.]

4. Empirical distributions again: the Glivenko–Cantelli theorem

Exercises 6.8.4 and 6.8.5 show that the Lévy distance of F_n from F goes to zero as $n \to \infty$. When the distribution function F is continuous it is a fact, as indicated in the next exercise, that F_n converges to F uniformly in x (i.e., given $\epsilon > 0$, there is an integer $n(\epsilon)$ such that $|F_n(x) - F(x)| < \epsilon$ for all $x \in \mathbb{R}$ provided $n \geq n(\epsilon)$).

Exercise 6.4.1. Let $F_n \xrightarrow{w} F$, and assume that F is continuous. Let $\epsilon > 0$. Show that

(1) there is a finite set of points $t_0 < t_1 < \cdots < t_{k+1}$ such that $F(t_0) < \epsilon$, $F(t_{i+1}) - F(t_i) < \epsilon$ for $0 \leq i \leq k$ and $F(t_{k+1}) > 1 - \epsilon$ [*Hints*: first determine t_0 and t_{k+1}, then use either uniform continuity (Property 2.3.8 (3)) on $[t_0, t_{k+1}]$ or, having chosen $t_0 < t_1 < \cdots < t_i$, choose t_{i+1} to be the last time that $F(t) \leq F(t_i) + \epsilon$, which amounts to looking at the quantile transformation or inverse Y of F.],

(2) there is an $n(\epsilon)$ such that $n \geq n(\epsilon)$ implies $|F_n(t_i) - F(t_i)| < \epsilon$ for $0 \leq i \leq k+1$.

4. EMPIRICAL DISTRIBUTIONS: THE GLIVENKO–CANTELLI THEOREM

Make use of the non-decreasing property of F and of the F_n to show that

(3) $|F_n(x) - F(x)| < 2\epsilon$ for all $x \in \mathbb{R}$ if $n \geq n(\epsilon)$.

In the case of empirical distributions, where a.s. the random distribution function $F_n(x,\omega) \xrightarrow{w} F$, in fact this can be improved to give uniform convergence a.s. regardless of whether or not the "population" distribution function F is continuous. This is the Glivenko–Cantelli theorem, which is often referred to as the "fundamental theorem of statistics", as it says that the (random) empirical distribution function, constructed from observations, converges uniformly a.s. to the "population" distribution function F.

Recall that the empirical distribution $F_n(x, \cdot)$ is defined to be $\frac{1}{n}\sum_{k=1}^{n} Y_k$, where the $Y_k = 1_{(-\infty,x]} \circ X_k = 1_{\{X_k \leq x\}}$ are Bernoulli random variables with $\mathbf{P}[Y_k = 1] = F(x)$ and $\mathbf{P}[Y_k = 0] = 1 - F(x)$ and, as mentioned earlier, the strong law implies that $\frac{1}{n}\sum_{k=1}^{n} Y_k = F_n(x,\cdot) \xrightarrow{a.s.} F(x)$.

The random variables $Z_n = 1_{(-\infty,x)} \circ X_n = 1_{\{X_n < x\}}$ are also Bernoulli random variables with $\mathbf{P}[Z_n = 1] = F(x-)$, the left limit at x of F, and $\mathbf{P}[Z_n = 0] = 1 - F(x-)$. In the same way as for the random variables Y_n, it follows from the strong law that $\lim_{n\to\infty} \frac{1}{n}\sum_{k=1}^{n} 1_{\{X_k < x\}} \stackrel{\text{def}}{=} F_n(x-,\cdot) \xrightarrow{a.s.} F(x-)$ for any fixed value of x.

Theorem 6.4.2. (Glivenko–Cantelli) Let $(X_n)_{n\geq 1}$ be a sequence of i.i.d. random variables on a probability space $(\Omega, \mathfrak{F}, \mathbf{P})$. Let $F_n(x,\omega) = \frac{1}{n}\sum_{k=1}^{n} 1_{\{X_k \leq x\}}$ denote the corresponding empirical distribution. Given $\epsilon > 0$, there exists an integer $n(\epsilon,\omega)$ such that a.s.

$$\sup_{x\in\mathbb{R}} |F_n(x,\omega) - F(x)| < \epsilon \text{ if } n \geq n(\epsilon,\omega).$$

Proof. Let $k \geq 1$ and let $x_{j,k} = Y(\frac{j}{k})$ for $1 \leq j \leq k-1$, where Y is the quantile transformation of F (i.e., its inverse). The idea of the proof is to use the a.s. convergence of F_n to F at the values $x_{j,k}$ and $x_{j,k}-$ to control the potential discontinuity of F and thus get uniform convergence.

To begin with, notice that the change in $F(x)$ between $x_{j,k}$ and $x_{j+1,k}-$ is at most $\frac{1}{k}$ since $F(x_{j,k}) = \mathbf{P}[X \leq x_{j,k}] \geq \frac{j}{k}$ by Lemma 6.3.10 (2). Also, (i) in the proof of Lemma 6.3.10 (4) implies that $F(x_{j+1,k}-) = \mathbf{P}[X < x_{j+1,k}] \leq \frac{j+1}{k}$. In other words, if $1 \leq j \leq k-2$, and $x_{j,k} < x < x_{j+1,k}$, then

(1) $$F(x_{j,k}) \leq F(x) \leq F(x_{j+1,k}-) \leq F(x_{j,k}) + \frac{1}{k}.$$

Now, for each k and $j, 1 \leq j \leq k-1$, $F_n(x_{j,k},\cdot) \xrightarrow{a.s} F(x_{j,k})$ and $F_n(x_{j,k}-,\cdot) \xrightarrow{a.s} F(x_{j,k}-)$. Denote by Λ the set where this occurs for $1 \leq j \leq k-1$, and let $n_k = n_k(\omega)$ for $\omega \in \Lambda$ be such that, when $n \geq n_k$

and $1 \leq j \leq k-2$, then

(2) $\quad |F_n(x_{j,k}-,\omega) - F(x_{j,k}-)| < \dfrac{1}{k}$ and $|F_n(x_{j,k},\omega) - F(x_{j,k})| < \dfrac{1}{k}$,

(3) $\quad F(x_{j,k}) - \dfrac{1}{k} < F_n(x_{j,k},\omega) \leq F_n(x_{j+1,k}-,\omega) < F(x_{j+1,k}-) + \dfrac{1}{k}.$

Assume that $x_{j,k} < x < x_{j+1,k}$, where $1 \leq j \leq k-2$. It follows from (1) and (3) that if $n \geq n_k(\omega)$, then

(4) $\quad F(x) - \dfrac{2}{k} < F_n(x_{j,k},\omega) \leq F_n(x,\omega) \leq F_n(x_{j+1,k}-,\omega) < F(x) + \dfrac{2}{k}.$

Combining (4) with (2), it follows that $|F_n(x,\omega) - F(x)| < \frac{2}{k}$ if $x_{j,k} \leq x \leq x_{j+1,k}$ for $1 \leq j \leq k-2$.

It remains to consider what happens on $(-\infty, x_{1,k})$ and $(x_{k-1,k}, +\infty)$. On the first interval, $F(x)$ and $F_n(x,\omega)$ both lie in $[0, \frac{2}{k})$ if $n \geq n_k(\omega)$ since this is true at the right-hand endpoint: $F(x) \leq F(x_{1,k}-) \leq \frac{1}{k}$ and $F_n(x,\omega) \leq F_n(x_{1,k}-) < \frac{2}{k}$. Similarly, on $(x_{k-1,k}, +\infty)$ it follows that $F(x)$ and $F_n(x,\omega)$ both lie in $(\frac{k-2}{k}, 1]$ if $n \geq n_k(\omega)$. This completes the argument. □

5. THE CHARACTERISTIC FUNCTION

The limit theorems that are called laws of large numbers are geared to approximating the mean of a distribution.

If a probability $\mathbf{Q}$ on $\mathbb{R}$ admits a second moment, then random variables X with distribution $\mathbf{Q}$ are necessarily in L^2. Hence, they have variances.

Lemma 6.5.1. (See Proposition 4.3.10.) *Let $(X_n)_{n \geq 1}$ be a sequence of i.i.d. random variables, and assume their common distribution admits a second moment. If $S = \sum_{k=1}^n X_k$, then the variance $\sigma^2(S_n)$ of S_n equals $n\sigma^2$, where σ^2 is the common variance of the X_n.*

It is clear that $\sigma^2(\lambda Y) = \lambda^2 \sigma^2(Y)$. As a result, if one wants to scale S_n so that the variance is constant, one has to divide it by $\sqrt{n}$. Dividing it by $\sigma\sqrt{n}$ normalizes it to have variance equal to 1. The scaling used for the laws of large numbers (dividing by n) is appropriate to leaving the mean constant, but $\frac{\sigma^2}{n}$ is the variance of $\frac{1}{n}S_n$ and it goes to zero as n tends to infinity, which is what one expects from the weak law of large numbers.

The random variable $\frac{1}{\sigma\sqrt{n}}S_n$ converges weakly (i.e., in distribution) to the unit normal. This fact is called the **central limit theorem**. The basic tool for obtaining this result is the Fourier transform, or rather the particular form of it used in probability that is called the characteristic function. The key property of this transform is that it converts convolutions into products, which are easier to analyze than convolutions (or sums of independent random variables).

The first central limit theorem that was proved was for Bernoulli random variables. It is called the de Moivre–Laplace limit theorem and can be proved directly without using characteristic functions (see Feller [F1], p. 182).

Definition 6.5.2. Let μ be a probability on $\mathbb{R}$. The **characteristic function** of μ is defined to be $f(t) = \int e^{itx}\mu(dx), t \in \mathbb{R}$, where $e^{iu} \stackrel{\text{def}}{=} \cos u + i \sin u$ (see the Appendix). If X is a random variable with distribution μ, the characteristic function of X is defined to be the characteristic function of μ.

Remark 6.5.3. The Fourier transform of μ, denoted by $\hat{\mu}(t)$, is defined to be $f(-t)$ in Körner [K3] and $f(-2\pi t)$ in Stein and Weiss [S1]. In analysis, the minus sign is omnipresent, but the use of the scaling constant is not fixed. In analysis, it is used more for functions in an L^p-space than for finite measures, which is where it is used in probability theory. In probability, the characteristic function is not spoken of as a transform.

Proposition 6.5.4. (Basic properties of characteristic functions)

(1) If f is the characteristic function of μ, then it is continuous and bounded with $|f(t)| \le 1 = f(0)$.

(2) If f is the characteristic function of μ, then its complex conjugate (see the Appendix) $\overline{f(t)} = f(-t)$ and $\overline{f}$ is the characteristic function of $\check{\mu}$, where $\check{\mu}(A) \stackrel{\text{def}}{=} \mu(-A)$ for all $A \in \mathfrak{B}(\mathbb{R})$.

(3) If f_1 and f_2 are the characteristic functions of μ_1 and μ_2, then $\lambda f_1 + (1-\lambda)f_2$ is the characteristic function of $\lambda\mu_1 + (1-\lambda)\mu_2$ if $0 \le \lambda \le 1$.

(4) If $\sum_{n=1}^{\infty} \lambda_n = 1$, $0 \le \lambda_n \le 1$, and f_n is the characteristic function of μ_n for $n \ge 1$, then $\sum_{n=1}^{\infty} \lambda_n f_n$ is the characteristic function of $\nu = \sum_{n=1}^{\infty} \lambda_n \mu_n$.

(5) If f is the characteristic function of μ and X is a random variable with distribution μ, then $f(t) = E[e^{itX}]$.

(6) If f_1 and f_2 are the characteristic functions of μ_1 and μ_2, then $f_1 f_2$ is the characteristic function of $\mu_1 \star \mu_2$.

(7) If f is the characteristic function of μ, then $|f|^2 = f\overline{f}$ is the characteristic function of $\mu \star \check{\mu}$.

(8) If f is the characteristic function of X, then $e^{ibt}f(t)$ is the characteristic function of $X + b$ and of μ_b, where $\int \psi(x)\mu_b(dx) \stackrel{\text{def}}{=} \int \psi(x+b)\mu(dx)$.

(9) If f is the characteristic function of X and $a \in \mathbb{R}$, then $f(at)$ is the characteristic function of aX.

Proof. (1) $|\int e^{itx}\mu(dx)| \le \int |e^{itx}|\mu(dx) = \int \mu(dx) = 1 = f(0)$. The continuity of f follows by dominated convergence since $e^{it_n x} \to e^{itx}$ if $t_n \to t$

and $|e^{itx}| = 1$ for all x and t.

(2) Clearly, $\overline{f(t)} = \overline{\int e^{itx}\mu(dx)} = \int e^{-itx}\mu(dx) = f(-t)$. As $\int \psi(x)\check{\mu}(dx) = \int \psi(-x)\mu(dx)$, it follows that $f(-t) = \int e^{itx}\check{\mu}(dx)$.

(3) Property (3) follows immediately from the definition.

(4) If the probability ν is given by $\nu(A) = \lim_{n\to\infty}\sum_{k=1}^n \lambda_k\mu_k(A)$, then $\int \psi d\nu = \lim_{n\to\infty}\sum_{k=1}^n \lambda_k \int \psi d\mu_k$. Hence,

$$\int e^{ixt}d\nu = \lim_n \sum_{k=1}^n \lambda_k \int e^{itx}\mu_k(dx) = \lim_n \sum_{k=1}^n \lambda_k f_k(t).$$

(5) is a particular case of Lemma 4.1.3: $E[\varphi \circ X] = \int \varphi(x)\mu(dx)$.

(6) follows from Proposition 3.2.3: let X_1 and X_2 be two independent random variables with distributions μ_1 and μ_2; if f is the characteristic function of $\mu_1 \star \mu_2$, then $f(t) = E[e^{it(X_1+X_2)}] = E[e^{itX_1}]E[e^{itX_2}] = f_1(t)f_2(t)$.

(7) follows immediately from (6) and (2).

(8) $E[e^{it(X+b)}] = e^{itb}E[e^{itX}] = e^{itb}f(t)$. Also, if F is the distribution function of X, then $F(x-b)$ is the distribution function of $X+b$. Hence, the corresponding probability μ_b is related to μ by the formula $\mu_b((c,d]) = F(d-b) - F(c-b) = \mu((c-b, d-b])$. This implies that for any Borel set A one has $\mu_b(A) = \mu(A-b)$. In terms of integrals, this says $\int 1_A(x)\mu_b(dx) = \int 1_{A-b}(x)\mu(dx) = \int 1_A(x+b)\mu(dx)$. Hence, for any positive Borel function ψ, one has $\int \psi d\mu_b = \int \psi(x+b)\mu(dx)$.

(9) $E[e^{it(aX)}] = f(at)$. □

Exercise 6.5.5. Determine the characteristic functions of

(1) ε_a,

(2) $p\varepsilon_0 + q\varepsilon_1$,

(3) the binomial distribution $b(n,p)$ (see Example 4.1.11), and

(4) the Poisson distribution with mean λ.

Exercise 6.5.6. Compute the characteristic function of a normal distribution by calculating

(1) $\frac{1}{\sqrt{2\pi}}\int e^{-tx}e^{-(\frac{x^2}{2})}dx$ (this is the value at t of the Laplace transform of the unit normal distribution: one completes the square of the exponent),

(2) $\frac{1}{\sqrt{2\pi}}\int \cosh(tx)e^{-(\frac{x^2}{2})}dx$ (recall that $\cosh u \stackrel{\text{def}}{=} \frac{e^u + e^{-u}}{2}$).

Recall the power series for e^u, and so

(3) determine the power series for $\cosh u$.

The well-known power series for the cosine is $\cos u = \sum_{k=0}^\infty (-1)^k \frac{u^{2k}}{(2k)!}$. Show that, for any $n \geq 1$,

(4) $|\sum_{k=0}^n (-1)^k \frac{u^{2k}}{(2k)!}| \leq \cosh u$.

5. THE CHARACTERISTIC FUNCTION

Conclude that

(5) $\frac{1}{\sqrt{2\pi}} \int \cos(tx) e^{-(\frac{x^2}{2})} dx = \sum_{k=0}^{\infty} (-1)^k \frac{t^{2k}}{(2k)!} \left(\frac{1}{\sqrt{2\pi}} \int x^{2k} e^{-(\frac{x^2}{2})} dx \right).$

Use integration by parts to show that

(6) $\int x^{2n} e^{-(\frac{x^2}{2})} dx = \frac{2n(2n-1)}{2n} \int x^{2n-2} e^{-(\frac{x^2}{2})} dx$ if $n > 1$.

Compute

(7) the even moments $m_{2n} \stackrel{\text{def}}{=} \frac{1}{\sqrt{2\pi}} \int x^{2n} e^{-(\frac{x^2}{2})} dx$ of the unit normal distribution $n(x)dx$.

Show that

(8) the characteristic function $f(t) = \int \cos tx\, n(x) dx$ of the unit normal $n(x)dx$ is $e^{-(\frac{t^2}{2})}$

and

(9) the characteristic function $f(t)$ of an $N(0, \sigma^2)$ random variable (i.e., of $n_{\sigma^2}(x)dx$) is $e^{-(\frac{\sigma^2 t^2}{2})}$.

Exercise 6.5.7. This is a method for computing the characteristic function of a normal distribution by using complex variable theory. Let $\phi(w) = e^{-\frac{(x-w)^2}{2}}$, where $w = u + iv$. Show that

(1) ϕ is analytic.

By integrating counterclockwise along the contour that is the boundary of the rectangle in $\mathbb{R}^2 = \mathbb{C}$ determined by $(R, 0), (R, t), (-R, t)$, and $(-R, 0)$, show that

(2) $\frac{1}{\sqrt{2\pi}} \int e^{-\frac{(x-it)^2}{2}} dx = \frac{1}{\sqrt{2\pi}} \int e^{-(\frac{x^2}{2})} dx = 1.$

By completing the square in the exponent, show that

(3) $\frac{1}{\sqrt{2\pi}} \int e^{itx} e^{-(\frac{x^2}{2})} dx = e^{-(\frac{t^2}{2})}.$

Use (3) to show that

(4) the characteristic function of $n_{\sigma^2}(dx) = \frac{1}{\sigma\sqrt{2\pi}} e^{-\frac{x^2}{2\sigma^2}} dx$ is $e^{-(\frac{\sigma^2 t^2}{2})}$.

Remark 6.5.8. If you are not familiar with complex variable theory, you may also proceed as follows. First, compute the Laplace transform $\mathcal{L}(n)(t)$ of the unit normal distribution (see Exercise 6.5.6 (1)). Except for the sign of the exponent, it is the same as the characteristic function. One gets the correct formula by substituting it for t. This substitution can be justified by complex variable theory. A word of warning: it is not true that the characteristic function can always be obtained from the Laplace transform in this way; for this to work, one needs the domain of the Laplace transform of the probability to be a neighbourhood of zero.

Exercise 6.5.9. Let τ be the probability density function where
$$\tau(x) = \begin{cases} 0 & \text{if } x < -1, \\ (x+1) & \text{if } -1 \leq x \leq 0, \\ (1-x) & \text{if } 0 \leq x \leq 1, \\ 0 & \text{if } 1 < x. \end{cases}$$

Show that

(1) the characteristic function of $\tau(x)dx$ is
$$f(t) = \frac{2(1-\cos t)}{t^2} = \left(\frac{\sin\frac{t}{2}}{\frac{t}{2}}\right)^2.$$

Let τ_a be the probability density $\tau_a(x) = \frac{1}{a}\tau(\frac{1}{a}x)$, where $a > 0$. Show that

(2) if the distribution of X has probability density function τ, then the distribution function of aX has probability density function τ_a.

As a result,

(3) determine the characteristic function of $\tau_a(x)dx$.

For a basic table of characteristic functions see, for example, Feller [F2], p. 416, or Chung [C], p. 147.

6. Uniqueness and Inversion of the Characteristic Function

There is a simple relation between two probabilities and their characteristic functions that has far-reaching effects when used in conjunction with the normal or Gaussian densities (as shown by Feller [F2] and Lamperti [L1]) that illustrates some general summability results in harmonic analysis (see Stein and Weiss [S1], pp. 8–12). To emphasize this relation, the probability $\mu(dx)$ corresponding to a distribution function will sometimes be denoted by $dF(x)$.

Proposition 6.6.1. (**Parseval's relation**) Let f_1 and f_2 be the characteristic functions of the respective probabilities μ_1 and μ_2. Then
$$\int f_1(x)\mu_2(dx) = \int f_2(x)\mu_1(dx), \quad i.e.,$$
$$\int f_1(x)dF_2(x) = \int f_2(x)dF_1(x),$$
where F_1 and F_2 are the corresponding distribution functions.

Proof. This is a simple application of Fubini's theorem (Corollary 3.3.17). One has
$$\int f_1(x)\mu_2(dx) = \int \left[\int e^{ixy}\mu_1(dy)\right]\mu_2(dx)$$
$$= \int \left[\int e^{ixy}\mu_2(dx)\right]\mu_1(dy) = \int f_2(y)\mu_1(dy). \quad \square$$

6. UNIQUENESS AND INVERSION OF THE CHARACTERISTIC FUNCTION

Corollary 6.6.2. $\int e^{-ity} f_1(y) dF_2(y) = \int f_2(x-t) dF_1(x)$.

Proof. By Proposition 6.5.4 (8), the function $e^{-ity} f_1(y)$ is the characteristic function of $(\mu_1)_{-t}$, the translate of μ_1 by $-t$ (i.e., the probability ν such that $\int \psi d\nu \stackrel{\text{def}}{=} \int \psi(x-t) \mu_1(dx) = \int \psi(x-t) dF_1(x)$ for any bounded function ψ). □

Corollary 6.6.3. Let $n_{\sigma^2}(x) = \frac{1}{\sigma\sqrt{2\pi}} e^{-(\frac{x^2}{2\sigma^2})}$, and let f be the characteristic function of μ. Then

$$\frac{1}{2\pi} \int e^{-ity} f(y) e^{-(\frac{\sigma^2 y^2}{2})} dy = \int n_{\sigma^2}(t-x) \mu(dx).$$

Proof. By Exercise 6.5.6 (9), the characteristic function of $n_{\frac{1}{\sigma^2}}(y) dy$ is $e^{-\frac{t^2}{2\sigma^2}}$. Hence, by Corollary 6.6.2, if $dF_2(y) = \mu_2(dy) = n_{\frac{1}{\sigma^2}}(y) dy$, one has

$$\frac{\sigma}{\sqrt{2\pi}} \int e^{-ity} f(y) e^{-(\frac{\sigma^2 y^2}{2})} dy = \int e^{-(\frac{(x-t)^2}{2\sigma^2})} \mu(dx).$$

Since normal laws are symmetric, the formula follows after multiplying both sides by $\frac{1}{\sigma\sqrt{2\pi}}$. □

The expression $\int n_{\sigma^2}(t-x) \mu(dx)$ is the density of the probability $n_{\sigma^2} \star \mu$ by Exercise 3.3.21 (2), where n_{σ^2} also denotes the normal law with mean zero and variance σ^2 (i.e., $n_{\sigma^2}(dx) = n_{\sigma^2}(x) dx$). In other words, the density for $n_{\sigma^2} \star \mu$ is expressed here in terms of the characteristic function of f as in the Fourier inversion result (Theorem 6.6.10).

Proposition 6.6.4. Let μ be any probability on $\mathbb{R}$. Then $n_{\sigma^2} \star \mu \xrightarrow{w} \mu$ as σ tends to zero, i.e., for any sequence $(\sigma_k)_{k\geq 1}$ of values of σ converging to zero, $n_{\sigma_k^2} \star \mu \xrightarrow{w} \mu$.

Proof. First, observe that, for any $\varphi \in C_c(\mathbb{R})$,

$$\int \varphi d(n_{\sigma^2} \star \mu) = \int \int \varphi(x+y) n_{\sigma^2}(x) dx \mu(dy)$$

(*) $$= \int \left[\int \varphi(y-x) n_{\sigma^2}(x) dx \right] \mu(dy),$$

since $\int \psi(x) n_{\sigma^2}(x) dx = \int \psi(x) \tilde{n}_{\sigma^2}(x) dx = \int \tilde{\psi}(x) n_{\sigma^2}(x) dx$ by Exercise 4.2.21 (4). Furthermore,

$$\left| \int \varphi(y-x) n_{\sigma^2}(x) dx - \varphi(y) \right| \leq \int_{\{|x|<\delta\}} |\varphi(y-x) - \varphi(y)| n_{\sigma^2}(x) dx$$

(†) $$+ \int_{\{|x|\geq\delta\}} |\varphi(y-x) - \varphi(y)| n_{\sigma^2}(x) dx.$$

Since φ is a continuous function with compact support, it is uniformly continuous (see Property 2.3.8 (3)). Given $\epsilon > 0$, let $\delta > 0$ be such that $|u - v| < \delta$ implies that $|\varphi(u) - \varphi(v)| < \epsilon$. It follows that the first integral in (†) is less than ϵ for any $\sigma > 0$. The second integral is less than $2\|\varphi\|_\infty n_{\sigma^2}(\{|x| \geq \delta\})$. This tends to zero as σ tends to zero since

(††)
$$n_{\sigma^2}(\{|x| \geq \delta\}) = \frac{1}{\sigma\sqrt{2\pi}} \int_{\{|x| \geq \delta\}} e^{-(\frac{x^2}{2\sigma^2})} dx = \frac{1}{\sqrt{2\pi}} \int_{\{|u| \geq \frac{\delta}{\sigma}\}} e^{-(\frac{u^2}{2})} du.$$

Hence, $\|\varphi \star n_{\sigma^2} - \varphi\|_\infty \to 0$ as $\sigma \to 0$. It follows from (*) that

$$\int \varphi d(n_{\sigma^2} \star \mu) \to \int \varphi d\mu \text{ as } \sigma \to 0.$$

If $(\sigma_k)_{k \geq 1}$ converges to zero, it follows from Propositions 6.2.7 and 6.2.9 that $n_{\sigma_k^2} \star \mu \xrightarrow{w} \mu$. □

Note that the set of measures $n_{\sigma^2} \star \mu$ is tight when, say, $\sigma \leq 1$. This fact is stated as the following lemma.

Lemma. *The collection $(n_{\sigma^2})_{\sigma \leq c}$ of probabilities n_{σ^2} is tight, and hence the collection $(n_{\sigma^2} \star \mu)_{\sigma \leq c}$ of probabilities $n_{\sigma^2} \star \mu$ is tight.*

Proof. It follows from (††) that there is a constant δ_0 such that $n_{\sigma^2}(\{|x| > \delta_0\}) < \epsilon$ for all $\sigma \leq c$ (i.e., this collection of probabilities is tight).

Let $\epsilon > 0$ and choose $M > 0$ such that $\mu([-M, M]) > 1 - \epsilon$. Hence, $(n_{\sigma^2} \star \mu)([-M - \delta_0, M + \delta_0]) \geq 1 - 2\epsilon$ for all $\sigma \leq c$. This is because, by the triangle inequality, $1_{[-M-\delta_0, M+\delta_0]}(x+y) \geq 1_{\{|x| \leq \delta_0\}}(x) 1_{\{|y| \leq M\}}(y)$ and so

$$\int\int 1_{[-M-\delta_0, M+\delta_0]}(x+y) n_{\sigma^2}(dx) \mu(dy) \geq n_{\sigma^2}(\{|x| \leq \delta_0\}) \mu([-M, M])$$
$$\geq (1-\epsilon)^2 \geq 1 - 2\epsilon. \quad \square$$

An immediate consequence of this result is the fact that a probability is determined by its characteristic function.

Theorem 6.6.5. (Uniqueness theorem) *Let μ_1 and μ_2 be two probabilities with the same characteristic function f. Then $\mu_1 = \mu_2$.*

Proof. It follows from Corollary 6.6.3 that $n_{\sigma^2} \star \mu_1 = n_{\sigma^2} \star \mu_2$. Proposition 6.6.4 implies that $\mu_1 = \mu_2$ since weak limits are unique by Exercise 6.2.2. □

In order to use the transform given by the characteristic function to prove the central limit theorem, it is essential to relate weak convergence to continuity properties of the transform.

6. UNIQUENESS AND INVERSION OF THE CHARACTERISTIC FUNCTION

Theorem 6.6.6. (Continuity theorem) Let $(\mu_n)_{n\geq 1}$ be a sequence of probabilities on $\mathbb{R}$ with corresponding characteristic functions f_n. The following conditions are equivalent:

(1) there is a probability μ such that $\mu_n \xrightarrow{w} \mu$; and
(2) $\lim_{n\to\infty} f_n$ exists and the limit function f is continuous at zero.

The function f is the characteristic function of μ.

Proof. Assume that $\mu_n \xrightarrow{w} \mu$. Since e^{itx} is a bounded continuous function of x for any $t \in \mathbb{R}$, it follows that $\lim_n f_n = f$. Furthermore, f is continuous by Proposition 6.5.4 (1).

The hard part of the proof is to show that (2) implies (1). It makes use of a compactness property due to Helly, which is stated as follows.

Lemma 6.6.7. (Helly's first selection principle) Let $(G_n)_{n\geq 1}$ be any sequence of non-negative, right continuous, non-decreasing functions on $\mathbb{R}$ that are uniformly bounded by 1 (i.e., $G_n(x) \leq 1$ for all $x \in \mathbb{R}$). Then there is a subsequence $(G_{n_k})_{k\geq 1}$ and a non-negative, right continuous, non-decreasing function G such that $G_{n_k}(x) \to G(x)$ for each continuity point x of G.

Assume the lemma. Let F_n be the distribution function of μ_n. Then there is a subsequence $(F_{n_k})_{k\geq 1}$ and a non-negative, right continuous, non-decreasing function G such that $F_{n_k}(x) \to G(x)$ for each continuity point of G, in particular, $F_{n_k} \xrightarrow{a.e.} G$.

At this point, there is no reason to assume that G is a distribution function. However, in any case there is a unique measure ν, with $\nu(\mathbb{R}) \leq 1$, such that $\nu((a,b]) = G(b) - G(a)$ if $a < b$. The aim now is to prove that ν is a probability. Then G will be its distribution function and F_{n_k} will converge weakly to G.

The integration by parts formula Exercise 6.8.2 (2) implies that

(*)
$$\int n_{\sigma^2}(t-x)\mu_{n_k}(dx) = -\int n'_{\sigma^2}(t-x)F_{n_k}(x)dx \text{ and}$$
$$\int n_{\sigma^2}(t-x)\nu(dx) = -\int n'_{\sigma^2}(t-x)G(x)dx.$$

Now by Corollary 6.6.3 and dominated convergence,

$$\int n_{\sigma^2}(t-x)\mu_{n_k}(dx) = \frac{1}{2\pi}\int e^{-ity}f_{n_k}(y)e^{-(\frac{\sigma^2 y^2}{2})}dy$$
$$\to \frac{1}{2\pi}\int e^{-ity}f(y)e^{-(\frac{\sigma^2 y^2}{2})}dy \text{ as } k \to \infty.$$

Since $F_{n_k} \xrightarrow{a.e.} G$, it follows from (*) that

$$\frac{1}{2\pi}\int e^{-ity}f(y)e^{-(\frac{\sigma^2 t^2}{2})}dy = \int n_{\sigma^2}(t-x)\nu(dx).$$

Multiply both sides by $\sigma\sqrt{2\pi}$. Since $(\sigma\sqrt{2\pi})n_{\sigma^2}(x) = e^{-\frac{x^2}{2\sigma^2}} \leq 1$, one obtains a lower bound for the total mass of ν (dependent upon σ) as

$$\frac{\sigma}{\sqrt{2\pi}} \int e^{-ity} f(y) e^{-(\frac{\sigma^2 y^2}{2})} dy = \sigma\sqrt{2\pi} \int n_{\sigma^2}(t-x)\nu(dx) \leq \nu(\mathbb{R}).$$

At this point, the continuity of f at zero comes into play. Let $\sigma \to \infty$. Then

(†) $\quad \dfrac{\sigma}{\sqrt{2\pi}} \displaystyle\int e^{-ity} f(y) e^{-(\frac{\sigma^2 y^2}{2})} dy = \int e^{-ity} f(y) n_{\frac{1}{\sigma^2}}(dy) \to f(0) = 1,$

and so ν is a probability with G as its distribution function.

To prove (†), one uses the argument for (†) in Proposition 6.6.4 (note that one cannot use the weak convergence of $n_{\frac{1}{\sigma^2}}$ to ε_0). Let $\epsilon > 0$. Then,

$$\left| \int e^{-ity} f(y) n_{\frac{1}{\sigma^2}}(y) dy - 1 \right| \leq \int_{\{|y|<\delta\}} |e^{-ity} f(y) - 1| n_{\frac{1}{\sigma^2}}(y) dy$$
$$+ \int_{\{|y|\geq\delta\}} |e^{-ity} f(y) - 1| n_{\frac{1}{\sigma^2}}(y) dy$$
$$\leq \epsilon + 2 n_{\frac{1}{\sigma^2}}(\{|y| \geq \delta\}),$$

for sufficiently small δ as by continuity, $|e^{-ity} f(y) - 1| < \epsilon$ if $|y| < \delta$. As before in the proof of Proposition 6.6.4 (see (††)), one has $n_{\frac{1}{\sigma^2}}(\{|y| \geq \delta\}) < \epsilon$ for all $\sigma \geq 1$ sufficiently large.

It has now been proved that (2) implies that there is a subsequence $(\mu_{n_k})_{k\geq 1}$ that converges weakly to a probability ν and hence (by the first part of the proof) $f = \lim_k f_{n_k}$ is its characteristic function.

The uniqueness theorem (Theorem 6.6.5) implies that ν is the only probability that is the weak limit of a subsequence of the μ_n since f is necessarily the characteristic function of any weak limit. This implies that the whole sequence converges weakly to ν: the set of probabilities is a metric space with respect to the Lévy metric. In a metric space, a sequence $(x_n)_{n\geq 1}$ converges to a point x_0 if every subsequence of that sequence $(x_n)_{n\geq 1}$ has a subsequence that converges to x_0 since, given $\epsilon > 0$, this means it is impossible to find an infinite number of points in the sequence whose distance from x_0 is greater than or equal to ϵ. □

Remark 6.6.8. If $\mu_n \xrightarrow{w} \mu$, the convergence of the characteristic functions f_n to f is better than pointwise. In fact, it occurs uniformly on the compact subsets of $\mathbb{R}$ (see Exercise 6.8.7).

To complete the proof of the continuity theorem, one needs to prove the key lemma, Lemma 6.6.7. Its proof makes use of the diagonal argument used by Cantor to prove that the real numbers are not countable (see Proposition 1.3.14) and is given in the appendix to this chapter.

6. UNIQUENESS AND INVERSION OF THE CHARACTERISTIC FUNCTION

Since the characteristic function determines a unique probability, it is natural to attempt to invert the transform, i.e., to get the probability from its characteristic function. There are many ways to do this (see Feller's comments [F2], p. 484). The one presented here is known as Gauss summability in harmonic analysis (see Stein and Weiss [S1], p. 11, Theorem 1.20).

In the simple case where the characteristic function is in $L^1(\mathbb{R})$, it has a Fourier transform. This gives the following inversion formula.

Theorem 6.6.9. (Fourier inversion) *Let f be the characteristic function of a probability μ, and assume that $f \in L^1(\mathbb{R})$. Then μ has as bounded continuous density the function φ, where*

$$\varphi(x) = \frac{1}{2\pi} \int e^{-itx} f(t) dt.$$

Proof. As remarked earlier, the probability $n_{\sigma^2} \star \mu$ has density

$$\varphi_\sigma(x) = \frac{1}{\sigma\sqrt{2\pi}} \int e^{-(\frac{(x-y)^2}{2\sigma^2})} \mu(dy)$$

$$= \int n_{\sigma^2}(x-y)\mu(dy) = \frac{1}{2\pi} \int e^{-ixt} f(t) e^{-(\frac{\sigma^2 t^2}{2})} dt,$$

where the last equality holds by Corollary 6.6.3.

Since $n_{\sigma^2} \star \mu \xrightarrow{w} \mu$ by Proposition 6.6.4 as $\sigma \to 0$, it looks hopeful that one can get the density φ by setting $\sigma = 0$.

First of all, since $f \in L^1$, it follows from dominated convergence that $\lim_{\sigma \to 0} \varphi_\sigma = \varphi$. This implies that $\varphi \geq 0$.

The formula for φ implies that $|\varphi(x)| \leq \frac{1}{2\pi}\|f\|_1$ for all x and similarly, $|\varphi_\sigma(x)| \leq \frac{1}{2\pi}\|f\|_1$ for all x. Hence, $\int_A \varphi(x) dx = \eta(A)$ defines a σ-finite measure on $\mathfrak{B}(\mathbb{R})$. It follows from Exercise 6.2.2 that $\eta = \mu$ (i.e., φ is the density of μ), if $\int \phi d\eta = \int \phi d\mu$ for all continuous functions ϕ with compact support.

Since $n_{\sigma^2} \star \mu \xrightarrow{w} \mu$, it follows that

$$\int \phi(x) \varphi_\sigma(x) dx = \int \phi d(n_{\sigma^2} \star \mu) \to \int \phi d\mu \text{ as } \sigma \to 0.$$

On the other hand, the densities φ_σ are uniformly bounded, and so by dominated convergence one has

$$\int \phi(x) \varphi_\sigma(x) dx \to \int \phi(x) \varphi(x) dx = \int \phi d\eta \text{ as } \sigma \to 0. \quad \Box$$

In spite of the fact that not every probability has a bounded continuous density, the argument just used shows how to determine an arbitrary probability from its characteristic function f: while f need not be in $L^1(\mathbb{R})$ and, hence, need not be integrable, it is bounded and so $f(t)e^{-\frac{\sigma^2 t^2}{2}}$ is in L^1; its Fourier transform is the density of $n_{\sigma^2} \star \mu$, which converges weakly to μ as $\sigma \to 0$. In this way, one may recuperate μ from its characteristic function.

Theorem 6.6.10. (Inversion theorem) *Let f be the characteristic function of a probability μ. If a and b are continuity points of μ, then*

$$\mu((a,b]) = \lim_{\sigma \to 0} \frac{1}{2\pi} \int f(t) e^{-(\frac{\sigma^2 t^2}{2})} \left[\frac{e^{-ita} - e^{-itb}}{it} \right] dt.$$

Proof. To begin,

$$\mu((a,b]) = \lim_{\sigma \to 0}(n_{\sigma^2} \star \mu)((a,b]) = \lim_{\sigma \to 0} \int_a^b \left[\frac{1}{2\pi} \int e^{-ity} f(t) e^{-(\frac{\sigma^2 t^2}{2})} dt \right] dy$$

in view of Corollary 6.6.3 since $(n_{\sigma^2} \star \mu)(y) = \int n_{\sigma^2}(y-x)\mu(dx)$ is the density of $n_{\sigma^2} \star \mu$ by Exercise 3.3.21 (2).

The result follows provided

$$\int_a^b \left[\frac{1}{2\pi} \int e^{-ity} f(t) e^{-(\frac{\sigma^2 t^2}{2})} dt \right] dy = \frac{1}{2\pi} \int \left[\int_a^b e^{-ity} dy \right] f(t) e^{-(\frac{\sigma^2 t^2}{2})} dt.$$

This follows from Fubini's theorem for Lebesgue measure (Corollary 3.3.17) as $|e^{-iyt} f(t) e^{-(\frac{\sigma^2 t^2}{2})}| \le e^{-(\frac{\sigma^2 t^2}{2})}$, which is in $L^1([a,b] \times \mathbb{R})$. □

Finally, to extend this inversion formula to the case when the endpoints are not continuity points of F, it is necessary to determine the limiting behaviour of the distribution function F_{σ^2} of $n_{\sigma^2} \star \mu$.

Proposition 6.6.11. *Let F_{σ^2} denote the distribution function of $n_{\sigma^2} \star \mu$ and F be the distribution function of μ. Then, for all x,*

$$\lim_{\sigma \to 0} F_{\sigma^2}(x) = \frac{1}{2}\{F(x-) + F(x)\}.$$

Proof. Since $n_{\sigma^2} \star \mu = \mu \star n_{\sigma^2}$ by Exercise 3.3.21, it follows from Exercise 3.3.21 (9) that $F_{\sigma^2}(x) = \int F(x-y) n_{\sigma^2}(y) dy = (F \star n_{\sigma^2})(x)$.

The convoluting density $n_{\sigma^2}(y)$ is an even function of y and so, just as in the proof of Féjer's theorem (Theorem 4.2.27), one may average on each side of x. More specifically, if $a, b \in \mathbb{R}$, then

(1)
$$\left| \int_0^{+\infty} F(x-y) n_{\sigma^2}(y) dy - \frac{1}{2}a \right| = \left| \int_0^{+\infty} \{F(x-y) - a\} n_{\sigma^2}(y) dy \right|$$
$$\le \int_0^{+\infty} |F(x-y) - a| n_{\sigma^2}(y) dy$$
$$= \int_0^{\delta} |F(x-y) - a| n_{\sigma^2}(y) dy$$
$$+ \int_{\delta}^{+\infty} |F(x-y) - a| n_{\sigma^2}(y) dy$$
$$= I_1 + I_2$$

and

$$\left| \int_{-\infty}^0 F(x-y) n_{\sigma^2}(y) dy - \frac{1}{2} b \right| = \left| \int_{-\infty}^0 \{F(x-y) - b\} n_{\sigma^2}(y) dy \right|$$

$$\leq \int_{-\infty}^0 |F(x-y) - b| n_{\sigma^2}(y) dy$$

(2)
$$= \int_{-\delta}^0 |F(x-y) - b| n_{\sigma^2}(y) dy$$

$$+ \int_{-\infty}^{-\delta} |F(x-y) - b| n_{\sigma^2}(y) dy$$

$$= J_1 + J_2.$$

Assume that $F(x-)$ exists, and let $a = F(x-)$ in (1). Given $\epsilon > 0$, for small $\delta > 0$ it follows that $|F(x-y) - F(x-)| < \epsilon$ if $0 < y < \delta$, and so $I_1 < \epsilon$ for small $\delta > 0$. Fix such a $\delta > 0$. The second integral I_2 in (1) is dominated by $\{1 + |a|\} n_{\sigma^2}(\{y \mid y \geq \delta\})$, which goes to zero as $\sigma \to 0$ by (††) in the proof of Proposition 6.6.4.

Let $b = F(x)$ in (2). Then, for the same reasons, for a small $\delta > 0$, $J_1 + J_2 \leq \epsilon + \{1 + |b|\} n_{\sigma^2}(\{y \mid y \leq -\delta\})$, which is dominated by 2ϵ for sufficiently small $\sigma = \sigma(\delta, \epsilon)$.

This shows that $2(F \star n_{\sigma^2})(x) \to F(x-) + F(x)$ as $n \to \infty$. $\square$

Corollary 6.6.12. *Let $a < b$. Then*

$$\mu((a,b)) + \frac{1}{2}\mu(\{a\}) + \frac{1}{2}\mu(\{b\}) = \lim_{\sigma \to 0} \frac{1}{2\pi} \int f(t) e^{-(\frac{\sigma^2 t^2}{2})} \left[\frac{e^{-ita} - e^{-itb}}{it} \right] dt.$$

Proof. Since $F_{\sigma^2}(b) - F_{\sigma^2}(a) = \int_a^b [\frac{1}{2\pi} \int e^{-ity} f(t) e^{-(\frac{\sigma^2 t^2}{2})} dt] dy$, the result follows from Proposition 6.6.11 and the fact that

$$\frac{1}{2}[\{F(b) + F(b-)\} - \{F(a) + F(a-)\}]$$

$$= \{F(b-) - F(a)\} + \frac{1}{2}\{F(a) - F(a-)\} + \frac{1}{2}\{F(b) - F(b-)\}$$

$$= \mu((a,b)) + \frac{1}{2}\mu(\{a\}) + \frac{1}{2}\mu(\{b\}). \quad \square$$

7. THE CENTRAL LIMIT THEOREM

The simple i.i.d. case. The simplest case of the central limit theorem is for sums of i.i.d. random variables $(X_n)_{n\geq 1}$. By centering the random variables (subtracting the mean) and then scaling them by $\frac{1}{\sigma}$, one may assume that they have common mean zero and common variance one. The central limit theorem states that, under these conditions, the distribution of $\frac{1}{\sqrt{n}} S_n$ converges weakly to the unit normal distribution.

Reformulated, without centering and rescaling, this theorem states that $\frac{1}{\sigma\sqrt{n}}(S_n - nm)$ converges in distribution to the unit normal.

To prove this, some basic analytic tools are needed. To begin, it is important to relate the existence of moments to analytic properties of the characteristic function f of the common distribution μ of the X_n.

Proposition 6.7.1. *Let μ be a probability on $\mathfrak{B}(\mathbb{R})$, and assume that it has a second moment (i.e., $\int x^2 \mu(dx) < \infty$). If f is the characteristic function of μ, then*

(1) *f has a continuous second derivative,*
(2) *$f'(t) = \int ix e^{itx} \mu(dx)$, and*
(3) *$f''(t) = -\int x^2 e^{itx} \mu(dx)$.*

In particular, $f'(0) = i \int x \mu(dx)$ and $f''(0) = -\int x^2 \mu(dx)$. In other words, if X is a random variable with distribution μ, then $iE[X] = f'(0)$ and $-E[X^2] = f''(0)$.

Proof. Since μ has a second moment, it also has a first absolute moment (i.e., $\int |x| \mu(dx) < \infty$ (see Proposition 4.1.8)). The difference quotient $\frac{1}{h}\{f(t+h) - f(t)\} = \frac{1}{h} \int \{e^{i(t+h)x} - e^{itx}\} \mu(dx) = \frac{1}{h} \int e^{itx}\{e^{ihx} - 1\} \mu(dx)$. Since $|e^{ihx} - 1| \leq |hx|$ (see Exercise 6.7.2) it follows that $\int |\frac{1}{h}\{e^{i(t+h)x} - e^{itx}\}| \leq |x|$, which is in $L^1(\mu)$. It follows from Theorem 2.6.1 that one can differentiate under the integral sign and so $f'(t) = \int ix e^{itx} \mu(dx)$.

Since $x^2 \in L^2(\mu)$, the same argument shows that one can differentiate $\int ix e^{itx} \mu(dx)$ in t by differentiating the integrand in t and so $f''(t) = -\int x^2 e^{itx} \mu(dx)$. □

Exercise 6.7.2. Show that

(1) $|e^{ix} - 1|^2 = 2(1 - \cos x)$, and
(2) $2(1 - \cos x) \leq x^2$. [*Hint*: the power series for $\cos x$ is an alternating series.]

This result, which obviously extends to higher derivatives in the presence of higher moments, shows that one has a Taylor expansion of $f(t)$ of order 2.

Exercise 6.7.3. Let g be a continuous real-valued function on $\mathbb{R}$ that has a derivative $g'(0)$ at zero. Show that

(1) $g(x) = g(0) + xg'(0) + o(x)$, where $o(x)$ is a continuous function such that $|\frac{o(x)}{x}| \to 0$ as $x \to 0$. [*Hint*: recall the definition of the derivative.]

Let f be a real-valued function such that $f' \in C^1(\mathbb{R})$ and has a second derivative at zero. Show that

(2) $f(x) = f(0) + xf'(0) + \frac{x^2}{2}f''(0) + o(x^2)$, where $o(x^2)$ is a continuous

function such that $|\frac{o(x^2)}{x^2}| \to 0$ as $x \to 0$. [*Hint*: let $g = f'$ in (1) and integrate from 0 to x.]

These results also hold for complex-valued functions $f(x)$ of $x \in \mathbb{R}$. It suffices to consider their real and imaginary parts u and v, where $f(x) = u(x) + iv(x)$.

Let $(X_n)_{n \geq 1}$ be a sequence of i.i.d. random variables with common distribution μ. Assume that the common mean is 0 and that the second moment of μ equals 1 (i.e., the common variance is 1).

Theorem 6.7.4. (Central limit theorem for i.i.d. random variables) Let $S_n = \sum_{k=1}^n X_k$. If μ_n is the distribution of $\frac{1}{\sqrt{n}} S_n$, then $\mu_n \xrightarrow{w} n$, the unit normal distribution $n(dx) = \frac{1}{\sqrt{2\pi}} e^{-(\frac{x^2}{2})} dx$.

Proof. Let f be the common characteristic function. It follows from Proposition 6.5.4 (6) and (9) that the characteristic function of $\frac{1}{\sqrt{n}} S_n$ is $\varphi(t) = [f(\frac{t}{\sqrt{n}})]^n$.

Since μ has a second moment, the mean is 0 and the variance is 1, it follows from Exercise 6.7.3 (2) and Proposition 6.7.1 that

$$f(t) = 1 - \frac{t^2}{2} + o(t^2) \text{ and so } f\left(\frac{t}{\sqrt{n}}\right) = 1 - \frac{1}{n}\left(\frac{t^2}{2}\right) + o\left(\frac{t^2}{n}\right).$$

If one neglects the term $o(\frac{t^2}{n})$, then, from calculus, one has

$$\varphi(t) = \left[f\left(\frac{t}{\sqrt{n}}\right)\right]^n = \left[1 - \frac{1}{n}\left(\frac{t^2}{2}\right)\right]^n \to e^{-(\frac{t^2}{2})} \text{ as } n \to \infty.$$

The theorem follows from the continuity theorem (Theorem 6.6.6) and the fact that $e^{-(\frac{t^2}{2})}$ is the characteristic function of the unit normal. □

The completion of this proof needs some technical work to justify dropping the error term $o(\frac{t^2}{n})$. The natural way to do this is to use the logarithm, although this may be bypassed by using a clever trick (see Billingsley [B2], p. 367, Lemma 1).

Exercise 6.7.5. Recall that if $x > 0$, then $\log x = \int_1^x \frac{du}{u}$. The geometric series gives a power series expansion for $\log(1-x)$ if $|x| < 1$:

(1) $\frac{1}{1-u} = \sum_{n=0}^\infty u^n$ if $|u| < 1$.

Integrate both sides of (1) from 0 to x, and show that

(2) $-\log(1-x) = \sum_{n=0}^\infty \frac{x^{n+1}}{n+1}$ if $|x| < 1$

(observe that if $-1 < x < 1$, then $0 < 1 - x < 2$).

The logarithm in real calculus is the inverse of the exponential function: $y = \log x$ if and only if $x = e^y$. While there is no power series that defines $\log x$ for all $x > 0$, there is a well-known power series for e^y, namely $e^y = \sum_{k=0}^{\infty} \frac{y^k}{k!}$. Hence, if $y = \log(1-x)$, then

$$(1-x) = e^y = \sum_{k=0}^{\infty} \frac{\{\log(1-x)\}^k}{k!} \quad \text{and}$$

$$(1-x)^n = e^{ny} = \sum_{k=0}^{\infty} \frac{\{n\log(1-x)\}^k}{k!}.$$

If now $z, w \in \mathbb{C}$, then one may define $z = e^w$ by the power series $e^w = \sum_{k=0}^{\infty} \frac{w^k}{k!}$ (see the appendix to this chapter) and then attempt to invert this function to determine w as a function of z. It turns out that this is not a simple matter because the complex exponential map is not one-to-one (e.g., $e^{iv} = e^{i(2\pi+v)}$ for any real v). However, if $|z| < 1$, then the power series for $\log(1-x)$ makes sense with x replaced by z and it inverts the exponential function. More explicitly, let

(*) $$w = \log(1-z) \stackrel{\text{def}}{=} -\sum_{n=0}^{\infty} \frac{z^{k+1}}{k+1} \quad \text{if } |z| < 1.$$

Then

$$(1-z) = e^w = \sum_{k=0}^{\infty} \frac{\{\log(1-z)\}^k}{k!} \quad \text{and}$$

$$(1-z)^n = e^{nw} = \sum_{k=0}^{\infty} \frac{\{n\log(1-z)\}^k}{k!}.$$

These identities are proved in complex calculus, for example by analytic continuation from the identities in the real case, or more directly by using the principal branch of the complex logarithm. One direct way to do this is to observe that the algebra involved in substituting the series for $\log(1-x)$ in the exponential series and computing to get $(1-x)$ in the real case is exactly the same as that involved when doing the analogous substitution in the complex case (see the Appendix for a very brief introduction to complex numbers).

The formula (*) for $w = \log(1-z)$ is the technical tool that justifies the statement that

$$\lim_{n\to\infty} \left[1 - \frac{1}{n}\left(\frac{t^2}{2}\right) + o\left(\frac{t^2}{n}\right)\right]^n = \lim_{n\to\infty} \left[1 - \frac{1}{n}\left(\frac{t^2}{2}\right)\right]^n = e^{-\left(\frac{t^2}{2}\right)}.$$

Let $z_n = \frac{1}{n}(\frac{t^2}{2}) - o(\frac{t^2}{n})$. Then $n\log(1-z_n) = -nz_n \left[\sum_{k=0}^{\infty} \frac{z_n^k}{k+1}\right] = -nz_n g(z_n)$, where $g(z) = \sum_{k=0}^{\infty} \frac{z^k}{k+1}$. Note that g is continuous as it is defined by a convergent power series when $|z| < 1$, and $g(0) = 1$.

Now $nz_n = \frac{t^2}{2} - no(\frac{t^2}{n})$, and since $\frac{o(u)}{u} \to 0$ as $u \to 0$, it follows that for $t \neq 0$ fixed $no(\frac{t^2}{n}) \to 0$ as n tends to infinity. Hence, $nz_n \to \frac{t^2}{2}$ and $z_n \to 0$. Since g is continuous in z, it follows that $g(z_n) \to g(0) = 1$ and so $n\log(1 - z_n) \to -\frac{t^2}{2}$. Therefore,

$$(1 - z_n)^n = \left[1 - \frac{1}{n}\left(\frac{t^2}{2}\right) + o\left(\frac{t^2}{n}\right)\right]^n = e^{n\log(1-z_n)} \to e^{-(\frac{t^2}{2})}.$$

This completes the proof of the simple case of the central limit theorem.

The Lindeberg condition. There has been an enormous amount of effort spent in generalizing and extending the above simple case of the central limit theorem. The main classical result in this work is due to Lindeberg. He solved the problem for a sequence $(X_n)_{n \geq 1}$ of independent random variables X_n all with mean zero and finite variance σ_n^2. As before, let $S_n = \sum_{k=1}^n X_k$. Then, if s_n^2 denotes the variance of S_n, it follows from independence that $s_n^2 = \sum_{k=1}^n \sigma_k^2$, where σ_k^2 is the variance of X_k, $k \geq 1$. The analogue of $\frac{1}{\sqrt{n}} S_n$ in the i.i.d. case is $\frac{1}{s_n} S_n$. The characteristic function $\varphi_n(t)$ of S_n is the product of the characteristic functions $f_k(t)$ of the X_k and so the characteristic function of $\frac{1}{s_n} S_n$ is

$$\varphi_n\left(\frac{t}{s_n}\right) = \prod_{k=1}^n f_k\left(\frac{t}{s_n}\right).$$

As before, the question is how to show that $\varphi(\frac{t}{s_n}) \to e^{-(\frac{t^2}{2})}$.

Using Exercise 6.7.3 (2) to expand each characteristic function f_k, one has

(†) $$f_k\left(\frac{t}{s_n}\right) = 1 - \frac{\sigma_k^2}{2}\left(\frac{t^2}{s_n^2}\right) + o_k\left(\frac{t^2}{s_n^2}\right) \stackrel{\text{def}}{=} 1 - z_{nk}.$$

Clearly, one wants to show that $\sum_{k=1}^n \log f_k\left(\frac{t}{s_n}\right) \to -(\frac{t^2}{2})$ as then

$$\varphi_n\left(\frac{t}{s_n}\right) = \prod_{k=1}^n f_k\left(\frac{t}{s_n}\right) = \exp\left\{\sum_{k=1}^n \log f_k\left(\frac{t}{s_n}\right)\right\} \to \exp\left(-\frac{t^2}{2}\right).$$

In order to be able to apply the power series for $\log(1 - z_{nk})$, it is necessary to know that, in fact, $|z_{nk}| < 1$. In the i.i.d. case this was no problem as the variance $s_n = n$ in this case and $\sigma_k^2 = \sigma^2 = 1$ for all k. This condition on the z_{nk} demands that one have a better control of the term $o_k\left(\frac{t^2}{s_n^2}\right)$ in the expansion (†) and that in addition one be able to show that $\frac{\sigma_k^2}{s_n^2} \to 0$ as n tends to infinity.

Proposition 6.7.6. *Let $\mathbf{Q}$ be any probability on $\mathbb{R}$ that has a second moment. Let $f(t)$ be its characteristic function. Then, for any $c > 0$,*

$$\left| f(t) - \left\{ 1 + itf'(0) - \frac{t^2}{2} f''(0) \right\} \right| \le |t|^3 c \int x^2 \mathbf{Q}(dx) + t^2 \int_{\{|x| \ge c\}} x^2 \mathbf{Q}(dx).$$

Proof. Since by Proposition 6.7.1,

$$f(t) - \left\{ 1 + itf'(0) - \frac{t^2}{2} f''(0) \right\} = \int \left[e^{itx} - \left\{ 1 + itx - \frac{t^2 x^2}{2} \right\} \right] \mathbf{Q}(dx),$$

it will suffice to estimate the above integral. The integrand is estimated by using the following lemma.

Lemma 6.7.7. *The following estimates hold:*
  (1) $|e^{it} - 1| \le |t|$;
  (2) $|e^{it} - \{1 + it\}| \le \frac{|t|^2}{2}$; *and*
  (3) $|e^{it} - \{1 + it - \frac{t^2}{2}\}| \le \frac{|t|^3}{6}$.

Proof. The first one follows from the observation that $e^{it} - 1 = i \int_0^t e^{iu} du$ since $|i \int_0^t e^{iu} du| \le |t|$. The second follows in the same way, using (1) since $e^{it} - \{1 + it\} = i \int_0^t [e^{iu} - 1] du$. The third makes use, in the same way, of the identity $e^{it} - \{1 + it - \frac{t^2}{2}\} = i \int_0^t [e^{iu} - \{1 + iu\}] du$. To conclude, note that if $|h(u)| \le |u|^n$, then $|\int_0^t h(u) du| \le \frac{|t|^{n+1}}{n+1}$. □

Continuation of the proof of Proposition 6.7.6.

It follows from the lemma that

$$\left| \int_{\{|x|<c\}} \left[e^{itx} - \left\{ 1 + itx - \frac{t^2 x^2}{2} \right\} \right] \mathbf{Q}(dx) \right| \le \int_{\{|x|<c\}} \frac{|t^3 x^3|}{6} \mathbf{Q}(dx)$$

$$\le |t|^3 c \int x^2 \mathbf{Q}(dx)$$

and

$$\left| \int_{\{|x| \ge c\}} \left[e^{itx} - \left\{ 1 + itx - \frac{t^2 x^2}{2} \right\} \right] \mathbf{Q}(dx) \right|$$

$$\le \int_{\{|x| \ge c\}} \left[|e^{itx} - \{1 + itx\}| + \frac{t^2 x^2}{2} \right] \mathbf{Q}(dx)$$

$$\le \int_{\{|x| \ge c\}} t^2 x^2 \mathbf{Q}(dx).$$

Combining these two estimates gives the result. □

7. THE CENTRAL LIMIT THEOREM

Remark. One cannot immediately use Lemma 6.7.7 (3) in the proof of Proposition 6.7.6 because the probability does not necessarily have a third absolute moment. The estimate is therefore a subtle one, combining third- and second-order terms.

Returning to the problem of controlling the error term o_k, it follows from Proposition 6.7.6 and (†) that for any $c > 0$,

$$\left| o_k \left(\frac{t^2}{s_n^2} \right) \right| = \left| f_k \left(\frac{t}{s_n} \right) - \left\{ 1 - \frac{\sigma_k^2}{2} \left(\frac{t^2}{s_n^2} \right) \right\} \right|$$

$$\leq \frac{|t|^3 c}{s_n^3} \int x^2 \mathbf{Q}_k(dx) + \frac{t^2}{s_n^2} \int_{\{|x| \geq c\}} x^2 \mathbf{Q}_k(dx),$$

where $\mathbf{Q}_k$ is the distribution of X_k. If one takes $c = s_n \epsilon$, then one has

$$\left| o_k \left(\frac{t^2}{s_n^2} \right) \right| \leq \epsilon |t|^3 \frac{\sigma_k^2}{s_n^2} + \frac{t^2}{s_n^2} \int_{\{|x| \geq s_n \epsilon\}} x^2 \mathbf{Q}_k(dx).$$

This implies that

$$(*) \qquad \sum_{k=1}^n \left| o_k \left(\frac{t^2}{s_n^2} \right) \right| \leq \epsilon |t|^3 + \sum_{k=1}^n \frac{t^2}{s_n^2} \int_{\{|x| \geq s_n \epsilon\}} x^2 \mathbf{Q}_k(dx).$$

The Lindeberg condition states that, for any $\epsilon > 0$,

$$(L) \qquad \sum_{k=1}^n \frac{1}{s_n^2} \int_{\{|x| \geq s_n \epsilon\}} x^2 \mathbf{Q}_k(dx) \to 0 \text{ as } n \to \infty.$$

Hence, if t is fixed, it follows that, given $\epsilon > 0$, there is an integer $n_0 = n_0(\epsilon)$ such that for all $n \geq n_0$ one has $\sum_{k=1}^n \left| o_k \left(\frac{t^2}{s_n^2} \right) \right| < 2\epsilon |t|^3$ for all $k \leq n$.

The Lindeberg condition (L) also implies that $\frac{\sigma_k^2}{s_n^2}$ is uniformly small for large n (i.e., given $\epsilon > 0$ one has $\frac{\sigma_k^2}{s_n^2} < 2\epsilon^2$ for all $k \leq n$ as long as n is large).

This is because

$$\sigma_k^2 = \int_{\{|x| < s_n \epsilon\}} x^2 \mathbf{Q}_k(dx) + \int_{\{|x| \geq s_n \epsilon\}} x^2 \mathbf{Q}_k(dx)$$

$$\leq s_n^2 \epsilon^2 + \int_{\{|x| \geq s_n \epsilon\}} x^2 \mathbf{Q}_k(dx).$$

Putting all this together, it follows that $|z_{nk}| \leq \frac{\sigma_k^2}{2} \left(\frac{t^2}{s_n^2} \right) + \left| o_k \left(\frac{t^2}{s_n^2} \right) \right|$ is small for all $k \leq n$ as long as n is large. It follows that

$$\sum_{k=1}^n \log f_k \left(\frac{t}{s_n} \right) = - \sum_{k=1}^n z_{nk} g(z_{nk}) \to - \left(\frac{t^2}{2} \right)$$

as in the simple case of i.i.d. random variables (see Exercise 6.7.8): $g(z_{nk})$ is uniformly close to 1 for all $k \leq n$ for large n and, for fixed t, $\sum_{k=1}^{n} z_{nk} = \frac{t^2}{2} - \sum_{k=1}^{n} o_k(\frac{t^2}{s_n^2}) \to \frac{t^2}{2}$ as $n \to \infty$ since by $(*)$ and the Lindeberg condition (L) it follows that if $n \geq n(\epsilon)$, then $\sum_{k=1}^{n} |o_k(\frac{t^2}{s_n^2})| < 2\epsilon|t|^3$. □

Exercise 6.7.8. Show that

(1) if $|g(z_{nk}) - 1| < \epsilon$ for all $k \leq n$ when $n \geq n(\epsilon)$, then $|\sum_{k=1}^{n} \frac{\sigma_k^2}{s_n^2} g(z_{nk}) - 1| < \epsilon$ if $n \geq n(\epsilon)$ [Hint: $\sum_{k=1}^{n} \sigma_k^2 = s_n^2$.],

(2) $|\sum_{k=1}^{n} o_k(\frac{t^2}{s_n^2}) g(z_{nk})| \leq (1+\epsilon)\alpha$ if $n \geq n(\epsilon)$ and $|\sum_{k=1}^{n} o_k(\frac{t^2}{s_n^2})| < \alpha$,

(3) $\sum_{k=1}^{n} z_{nk} g(z_{nk}) \to \frac{t^2}{2}$.

This completes the proof of the following version of the central limit theorem.

Theorem 6.7.9. (**Central limit theorem: Lindebergh condition**) Let $(X_k)_{k \geq 1}$ be a sequence of independent random variables with mean zero that have second moments or, equivalently, are in L^2. Let σ_k^2 be the variance of X_k, $k \geq 1$, and let $s_n^2 \stackrel{\text{def}}{=} \sum_{k=1}^{n} \sigma_k^2$ and $S_n \stackrel{\text{def}}{=} \sum_{k=1}^{n} X_k$. If μ_n is the distribution of $\frac{1}{s_n} S_n$, then μ_n converges weakly to n, the unit normal distribution, $n(dx) = \frac{1}{\sqrt{2\pi}} e^{-\frac{x^2}{2}} dx$, provided the **Lindebergh condition** (L) is satisfied: for any $\epsilon > 0$,

$$\text{(L)} \qquad \sum_{k=1}^{n} \frac{1}{s_n^2} \int_{\{|x| \geq s_n \epsilon\}} x^2 \mathbf{Q}_k(dx) \to 0 \text{ as } n \to \infty,$$

where $\mathbf{Q}_k$ is the distribution of X_k, $k \geq 1$.

There is another version of this theorem for so-called **triangular arrays** whose proof is essentially the same as the proof for a sequence once the more or less obvious notational changes have been made.

Definition 6.7.10. *A* **triangular array** *of random variables is a sequence of finite collections* $X_{n1}, X_{n2}, \ldots, X_{nk_n}$ *of independent random variables* X_{ni}, $1 \leq i \leq k_n$, *for* $n \geq 1$.

Any sequence $(X_n)_{n \geq 1}$ of independent random variables determines a triangular array in a natural way: let $k_n = n$ and $X_{ni} = X_i$ for $1 \leq i \leq n$.

Given a triangular array where each random variable has mean zero and finite variance, for each $n \geq 1$ let $s_n = \sum_{i=1}^{k_n} \sigma_{ni}^2$ and $S_n = \sum_{i=1}^{k_n} X_{ni}$. If $\mathbf{Q}_{ni}$ is the distribution of X_{ni}, the triangular array will be said to satisfy the Lindebergh condition if

$$\text{(L)} \qquad \sum_{i=1}^{k_n} \frac{1}{s_n^2} \int_{\{|x| \geq s_n \epsilon\}} x^2 \mathbf{Q}_{ni}(dx) \to 0 \text{ as } n \to \infty.$$

By copying the proof of Theorem 6.7.9, with the new definitions of s_n and S_n, one obtains a proof of the central limit theorem for triangular arrays.

Theorem 6.7.11. (Central limit theorem: triangular arrays)
Given a triangular array $X_{n1}, X_{n2}, \ldots, X_{nk_n}, n \geq 1$, of random variables X_{ni}, if $s_n = \sum_{i=1}^{k_n} \sigma_{ni}^2$ and $S_n = \sum_{i=1}^{k_n} X_{ni}$, then the distribution of $\frac{1}{s_n} S_n$ converges weakly to the unit normal distribution n if the following Lindeberg condition (L) is satisfied:

(L) $$\sum_{i=1}^{k_n} \frac{1}{s_n^2} \int_{\{|x| \geq s_n \epsilon\}} x^2 \mathbf{Q}_{ni}(dx) \to 0 \text{ as } n \to \infty,$$

where $\mathbf{Q}_{ni}$ is the distribution of $X_{ni}, 1 \leq i \leq k_n, n \geq 1$.

8. ADDITIONAL EXERCISES*

Exercise 6.8.1. (Integration by parts) Let G be a non-decreasing, right continuous, real-valued function on $\mathbb{R}$ with associated σ-finite measure μ. Let φ be a continuously differentiable function whose support (Definition 4.2.4) is contained in $[a,b]$ and let $a = t_0 < t_1 < t_2 < \cdots < t_n = b$. Show that

(1) $\sum_{i=1}^{n} \varphi(t_i)\{G(t_i) - G(t_{i-1})\} = -\sum_{i=1}^{n}\{\varphi(t_i) - \varphi(t_{i-1})\}G(t_{i-1})$.
[Hint: use integration by parts for series (Lemma 5.5.6).]

Let $\epsilon > 0$. Show that

(2) if $|\varphi(t_{i+1}) - \varphi(x)| < \epsilon$ for $t_i < x \leq t_{i+1}$, then one has $|\int \varphi d\mu - \sum_{i=1}^{n} \varphi(t_i)\{G(t_i) - G(t_{i-1})\}| < \epsilon\mu((a,b]) \leq \epsilon$,
(3) if $G(t_i) \leq G(x) < G(t_i) + \epsilon$ for $t_i \leq x < t_{i+1}$, then $|\int \varphi'(x)G(x)dx - \sum_{i=1}^{n}\{\varphi(t_i) - \varphi(t_{i-1})\}G(t_{i-1})| < \epsilon \int |\varphi'(x)|dx$,
(4) the points t_i can be chosen such that the conditions in (2) and (3) are both satisfied. [Hints: for (2) use uniform continuity and for (3) verify that, for some m, there are m points s_j with $s_0 = a$, $s_m = b$, and $s_{j+1} = \sup\{x \mid G(x) \leq G(s_j) + \epsilon\}$.]

Conclude that

(5) $\int \varphi(x)\mu(dx) = -\int \varphi'(x)G(x)dx$.

Exercise 6.8.2. Let ν be a finite Borel measure and let G be a non-decreasing, right continuous, real-valued function such that $\nu((c,d]) = G(d) - G(c)$ if $c < d$. Let φ be a continuously differentiable function. If $a < b$, show that

(1) $\int_{(a,b]} \varphi d\nu = \varphi(b)G(b) - \varphi(a)G(a) - \int_a^b \varphi'(x)G(x)dx$.

Assume that $\varphi' \in L^1(\mathbb{R})$ and that φ vanishes at infinity. Show that

(2) $\varphi d\nu = -\int \varphi'(x)G(x)dx$.

Remark. These two integration by parts formulas are special cases of well-known results for what are often referred to as Riemann–Stieltjes integrals (see Rudin [R4], p. 123, Theorem 6.30) as any function φ with a continuous derivative is of bounded variation on a bounded interval $[a,b]$ since it satisfies a Lipschitz condition: if $a \leq x < y \leq b$, then $|\varphi(x) - \varphi(y)| \leq M|x-y|$ with $M = \sup_{a \leq t \leq b} |\varphi'(t)|$. These exercises can also be proved by extending to signed measures the integration by parts formula given in Exercise 3.6.21.

Proposition 6.8.3. If $F_n \xrightarrow{w} F$, the Lévy distance of F_n from F goes to zero as n tends to infinity.

The proof consists of two exercises with hints based on an argument in Gnedenko and Kolmogorov ([G], p. 34), which the reader is advised to consult for its clarity and elegance.

Exercise 6.8.4. Let $\epsilon > 0$. Show that
(1) there are two continuity points a and b of F with $F(a) < \frac{\epsilon}{2}$ and $F(b) > 1 - \frac{\epsilon}{2}$ [Hint: recall that the set of continuity points of F is dense as its complement is countable.],
(2) if $|F_n(a) - F(a)| < \frac{\epsilon}{2}$, then $F_n(x) \leq F_n(a) \leq \epsilon$ if $x \leq a$,
(3) if $|F_n(a) - F(a)| < \frac{\epsilon}{2}$, then $F_n(x-\epsilon) - \epsilon \leq F(x)$ and $F(x-\epsilon) - \epsilon \leq F_n(x)$ if $x \leq a$.

Similarly, show that
(4) if $|F_n(b) - F(b)| < \frac{\epsilon}{2}$, then $1 \geq F_n(x) \geq F(x-\epsilon) - \epsilon$ and $1 \geq F(x) \geq F_n(x-\epsilon) - \epsilon$ if $x \geq b$. [Hint: show that $F(x) \geq 1 - \epsilon \geq F_n(x) - \epsilon$ and $F_n(x) \geq 1 - \epsilon \geq F(x) - \epsilon$ if $x \geq b$.]

Exercise 6.8.5. Let $\epsilon > 0$. Show that
(1) there are continuity points t_i of F with $a = t_o < t_1 < \cdots < t_{k+1} = b$ such that $t_{i+1} - t_i < \epsilon$ for $0 \leq i \leq k$, and
(2) if $|F_n(t_i) - F(t_i)| < \epsilon$ for $0 \leq i \leq k+1$, then $t_i \leq x \leq t_{i+1}$ implies that $F(x-\epsilon) - \epsilon \leq F_n(t_i) \leq F_n(x) \leq F_n(t_{i+1}) \leq F(x+\epsilon) + \epsilon$ for $0 \leq i \leq k$.

The above argument of Gnedenko and Kolmogorov is one part of an elegant theorem that shows the equivalence of the two notions of weak convergence (Definition 6.1.3) and (Definition 6.2.1) as well as the equivalence to convergence in the Lévy distance. Their argument that Definition 6.1.3 implies Definition 6.2.1 avoids the use of integration by parts and approximation by smooth functions. It is based on the ideas used above and is outlined in the following exercise.

Exercise 6.8.6. Assume that $F_n \xrightarrow{w} F$, and let μ_n and μ be the corresponding probabilities. Let $\epsilon > 0$ and f be a bounded continuous function. Show that
(1) the continuity points t_i in Exercise 6.8.5 can be chosen (if necessary,

closer together) so that, in addition, for two points x, y in any interval $[t_i, t_{i+1}]$, one has $|f(x) - f(y)| < \epsilon$ [Hint: make use of uniform continuity on $[a, b]$.],

(2) $|\int_{(a,b]} f(x)\mu(dx) - \sum_{i=0}^{k} f(t_{i+1})\mu((t_i, t_{i+1}])| < \epsilon$ and
$|\int_{(a,b]} f(x)\mu_n(dx) - \sum_{i=0}^{k} f(t_{i+1})\mu_n((t_i, t_{i+1}])| < \epsilon$.
[Hint: consider, for example, $\int_{(t_i, t_{i+1}]} f(x)\mu(dx)$.]

Let $n(\epsilon)$ be such that $|F_n(a) - F(a)| < \frac{\epsilon}{2}$, $|F_n(t_i) - F(t_i)| < \frac{\epsilon}{2}$ for $1 \leq i \leq k$, and $|F_n(b) - F(b)| < \frac{\epsilon}{2}$ if $n \geq n(\epsilon)$.

Show that, for $n \geq n(\epsilon)$, if $C = \|f\|_\infty$, one has

(3) $|\sum_{i=0}^{k} f(t_{i+1})\mu_n((t_i, t_{i+1}]) - \sum_{i=0}^{k} f(t_{i+1})\mu((t_i, t_{i+1}])| < \epsilon(k+1)C$
$|\int_{(-\infty, a]} f(x)\mu_n(dx) - \int_{(-\infty, a]} f(x)\mu(dx)| < \epsilon C$, and
$|\int_{(b, +\infty)} f(x)\mu_n(dx) - \int_{(b, +\infty)} f(x)\mu(dx)| < \epsilon C$.

Conclude that

(4) $\int f \, d\mu_n \to \int f \, d\mu$ and hence that $\mu_n \xrightarrow{w} \mu$.

Comment. This exercise is in essence the Helly–Bray lemma (see Loève [L2], p. 182).

Exercise 6.8.7. Let $(\mu_n)_{n \geq 1}$ be a tight sequence of probabilities on $\mathbb{R}$. Show that

(1) the corresponding sequence $(f_n)_{n \geq 1}$ of characteristic functions f_n is **uniformly equicontinuous**, i.e., for each $t \in \mathbb{R}$ and $\epsilon > 0$, there is a $\delta > 0$ that depends only on ϵ such that

$$|f_n(t+h) - f_n(t)| < \epsilon \text{ if } |h| < \delta.$$

[*Comment*: if one δ works for all f_n at one point, the sequence is called **equicontinuous**, and if it also works at all points t, it is said to be uniformly equicontinuous.] [Hint: verify that $|f_n(t+h) - f_n(t)| \leq \int |e^{ihx} - 1| \mu_n(dx)$.]

Now assume that $\mu_n \xrightarrow{w} \mu$. Show that

(2) f_n converges to f uniformly on any finite interval $[a, b]$ (i.e., given $\epsilon > 0$ there exists $n_0 = n(\epsilon)$ such that

$$|f_n(t) - f(t)| < \epsilon \text{ for all } t \in [a, b] \text{ if } n \geq n_0).$$

[Hint: make use of (1), the fact that there are $k+1$ points $a = t_0 < t_1 < \cdots < t_k = b$ with $|t_i - t_{i+1}| < \delta$ for $0 \leq i \leq k-1$, and show that one can find n_0 with $|f_n(t_i) - f(t_i)| < \epsilon$ for $0 \leq i \leq k$ as long as $n \geq n_0$.]

Remark. There is an idea in the argument suggested for the solution of Exercise 6.8.7 (2). The underlying formal result (which has to do with when a pointwise limit of continuous functions is continuous) is called the Ascoli-Arzelá theorem (see Royden [R3]). Notice that, in fact, one does not apparently need to use uniform equicontinuity, as equicontinuity and compactness of $[a, b]$ suffice. However, equicontinuity and compactness of $[a, b]$ imply uniform equicontinuity just as continuity and compactness imply uniform continuity.

9. APPENDIX*

Approximation by smooth functions.

This section of the Appendix is devoted to a proof of the following result, where $C_c^\infty(\mathbb{R})$ denotes the space of infinitely differentiable functions with compact support.

Theorem 6.9.1. Let $\varphi \in C_c(\mathbb{R})$ and $\epsilon > 0$. Then there exists $\psi \in C_c^\infty(\mathbb{R})$ with $\|\varphi - \psi\|_\infty < \epsilon$.

Let $C_o(\mathbb{R})$ denote the continuous functions $\phi(x)$ on $\mathbb{R}$ that converge to zero as $|x| \to \infty$. This is a Banach space with respect to the norm $\|\phi\|_\infty = \sup_{x \in \mathbb{R}} |\phi(x)|$ since it a closed subspace of the Banach space $C_b(\mathbb{R})$ (see Exercise 4.2.11).

Proposition 6.9.2. Let $n_t(x) = \frac{1}{\sqrt{2\pi t}} e^{-\frac{x^2}{2t}}$. If $\phi \in C_0(\mathbb{R})$, then $n_t \star \phi \in C_0^\infty(\mathbb{R})$ and $\|n_t \star \phi - \phi\|_\infty \to 0$ as $t \to 0$.

Proof. Exercise 3.3.24 shows that $n_t \star \phi \in C_0(\mathbb{R})$ for any function $\phi \in C_0(\mathbb{R})$. The fact that $n_t \star \phi \in C_0^\infty(\mathbb{R})$ was established in Exercise 2.9.6 (and also in an easier way using Proposition 4.2.29 and part C of Exercise 2.9.6).

Finally, it follows from the fact that any function $\phi \in C_0(\mathbb{R})$ is uniformly continuous (it suffices to observe that by Property 2.3.8 (3) it is uniformly continuous on $[-N, N]$ for any $N \geq 1$), and the proof of (†) in Proposition 6.6.4 that $\|n_t \star \phi - \phi\|_\infty \to 0$ as $t \to 0$. □

Given this result, to complete the proof of Theorem 6.9.1, it suffices to know how to "cut off" smooth functions in $C_0(\mathbb{R})$ to get smooth functions of compact support.

Proposition 6.9.3. Let $\alpha < \beta < \gamma < \delta$. Then there is a C^∞-function "cut off function" θ with $0 \leq \theta \leq 1$ such that

$$\theta(x) = \begin{cases} 1 & \text{if } \beta \leq x \leq \gamma, \\ 0 & \text{if } x < \alpha \text{ or } x > \delta \end{cases}$$

Proof of Theorem 6.9.1. By Proposition 6.9.2, there is a $t > 0$ such that $\|\varphi - n_t \star \varphi\|_\infty < \frac{\epsilon}{2}$. Since the C^∞-function $n_t \star \varphi$ vanishes at infinity, there is an integer $M > 0$ such that $|n_t \star \varphi(x)| < \frac{\epsilon}{2}$ if $|x| \geq M$. Let θ be a C^∞ cut off function corresponding to $-\beta = \gamma = M$ and $-\alpha = \delta = M+1$. Then $|n_t \star \varphi(x) - \theta(n_t \star \varphi)(x)| \leq |n_t \star \varphi(x)|$ for all x and so $\|n_t \star \varphi - \theta(n_t \star \varphi)\|_\infty < \frac{\epsilon}{2}$ since the difference is non-zero only if $|x| > M$. Hence, $\|\varphi - \psi\|_\infty < \epsilon$ if $\psi = \theta(n_t \star \varphi)$. □

The following technical lemma gives the essential ingredient for constructing the cut off function of Proposition 6.9.3.

Lemma 6.9.4. *The function f defined by*

$$f(x) = \begin{cases} 0 & \text{if } x \leq 0 \\ e^{-\frac{1}{x}} & \text{if } x > 0 \end{cases}$$

is a C^∞-function on $\mathbb{R}$.

Let $\alpha < \beta$, and define $f_l(x) = f(x - \alpha)$ and $f_r(x) = f(\beta - x)$, where f is the function in Lemma 6.9.4. Let $\alpha < \beta$ and $g_{\alpha,\beta} \stackrel{\text{def}}{=} f_l f_r$. Then $g_{\alpha,\beta}$ is in $C_c^\infty(\mathbb{R})$ with compact support (Definition 4.2.4) contained in $[\alpha, \beta]$.

Define $\psi_{\alpha,\beta}(x) = \frac{1}{C(\alpha,\beta)} \int_{-\infty}^x g_{\alpha,\beta}(u) du$, where $C(\alpha, \beta)$ is the normalizing constant equal to $\int_\alpha^\beta g_{\alpha,\beta}(u) du$.

The C^∞-function $\psi_{\alpha,\beta}$ takes values in $[0, 1]$ and

$$\psi_{\alpha,\beta}(x) = \begin{cases} 0 & \text{if } x \leq \alpha, \\ 1 & \text{if } x \geq \beta. \end{cases}$$

Let $\alpha < \beta < \gamma < \delta$. The function $\theta = \psi_{\alpha,\beta}(1 - \psi_{\gamma,\delta})$ is C^∞ with compact support, takes values in $[0, 1]$, and

$$\theta(x) = \begin{cases} 1 & \text{if } \beta \leq x \leq \gamma, \\ 0 & \text{if } x < \alpha \text{ or } x > \delta. \end{cases} \quad \square$$

Helly's selection principle.

Consider the set $\mathcal{S}$ of sequences $a = (a_n)_{n \geq 1}$ of real numbers a_n such that $0 \leq a_n \leq 1$ for all $n \geq 1$. This is a subset of ℓ_∞ (Exercise 4.1.17). It can also be viewed as the product of an infinite number of copies of $[0, 1]$.

Define a metric on $\mathcal{S}$ by setting

$$(*) \qquad d(a, b) = \sum_{n=1}^\infty \frac{1}{2^n} |a_n - b_n|.$$

Exercise 6.9.5. Show that
(1) $(*)$ defines a metric on $\mathcal{S}$,
(2) a sequence $(a^k)_{k \geq 1}$ of points in $\mathcal{S}$ converges to a if and only if $a_n = \lim_k a_n^k$ for all $n \geq 1$,
(3) a subset E of $\mathcal{S}$ is open (relative to this metric) if and only if $a \in E$ implies that there is an integer N and a positive number ϵ such that $\{b \in \mathcal{S} \mid |a_j - b_j| < \epsilon, \ 1 \leq j \leq N\} \subset E$. Recall that a set E in a metric space is **open** (Remarks 4.1.26) if for any $a \in E$ there is an $\epsilon > 0$ such that the **open ball about** a **of radius** ϵ (i.e., $\{x \mid d(x, a) < \epsilon\}$) is a subset of E.

Remarks. Condition (3) shows that the open sets defined by the metric d (i.e., the metric topology), is the so-called **product topology** (see [R3]): the smallest topology such that the functions $a \to a_j$ are continuous for all $j \geq 1$. It is well known that the product of compact spaces is compact (see Tychonov's theorem [R3]). The selection principle of Helly in effect gives a proof of this topological fact for countable products of compact metric spaces. The product topology is then defined by a metric as above. For metric spaces, as is well known (see [R3]), compactness is equivalent to the **Bolzano–Weierstrass property**: every sequence has a convergent subsequence. The fact that the product is countable makes it possible to use Cantor's diagonal procedure (Proposition 1.3.14) to prove this.

Proposition 6.9.6. *Any sequence of points in $\mathcal{S}$ has a convergent subsequence.*

Proof. The idea is as follows: given a sequence $(a^k)_{k\geq 1}$ of points in $\mathcal{S}$, the set $\{a_1^k \mid k \geq 1\}$ of first coordinates of these points is a subset of $[0,1]$ and so, by the compactness of $[0,1]$, it has a convergent subsequence (Exercise 1.5.6). This subsequence is defined by a strictly increasing function $m \to k(1,m)$ on $\mathbb{N}$ (see §1 of Chapter I) with the property that there is a point $a_1 \in [0,1]$ for which $a_1 = \lim_m a_1^{k(1,m)}$. In effect, one now forgets about the points a^k that are not labeled by this function $m \to k(1,m)$ and looks at the second coordinates of the remaining points and applies the same argument, which leads to the dropping of still more points; one continues by applying the argument to what's left, and so on. One finally applies the diagonal procedure to the countable array of subsequences that arises in this way to get a subsequence that converges to the point a whose coordinates were determined by the choice of subsequence at each stage.

More formally, there is a sequence $(k(\ell, \cdot))_{\ell \geq 1}$ of strictly increasing functions $m \to k(\ell, m)$ defined on $\mathbb{N}$ with the following properties valid for each $\ell \geq 1$:

(1) the sequence $(k(\ell+1, m))_{m\geq 1}$ is a subsequence of the previous sequence $(k(\ell, m))_{m\geq 1}$ (i.e., there is a strictly increasing function $m \to n(m)$ with $k(\ell+1, m) = k(\ell, n(m))$ for $m \geq 1$); and

(2) there is a point $a_\ell \in [0,1]$ such that $a_\ell = \lim_m a_\ell^{k(\ell,m)}$.

It follows from property (1) that for any $\ell \geq 1$ one has

(3) $a_j = \lim_m a_j^{k(\ell,m)}$ for all j, $1 \leq j \leq \ell$,

as by (1) the sequence $(k(\ell, m))_{m\geq 1}$ is a subsequence of each of the sequences $(k(j,m))_{m\geq 1}$ for $1 \leq j \leq \ell$.

In other words, one obtains the following array of natural numbers, where each row is strictly increasing and every row is a subsequence of the preceding row:

$$k(1,1) \quad k(1,2) \quad \ldots \quad k(1,m) \quad \ldots \to \infty$$

$$k(2,1) \quad k(2,2) \quad \ldots \quad k(2,m) \quad \ldots \to \infty$$

$$\vdots \qquad \vdots \qquad \vdots \qquad \qquad \vdots$$

$$k(\ell,1) \quad k(\ell,2) \quad \ldots \quad k(\ell,m) \quad \ldots \to \infty$$

$$\vdots \qquad \vdots \qquad \vdots \qquad \qquad \vdots$$

such that for each $\ell \geq 1$, the first ℓ coordinates of the original sequence $(a^k)_{k \geq 1}$ of points $a^k \in \mathcal{S}$ all converge along the subsequence defined by the ℓth row of this array.

By the diagonal procedure of Cantor, consider the strictly increasing function $m \to k(m,m) \overset{\text{def}}{=} k(m)$: the sequence $(k(m))_{m \geq 1}$ is a subsequence of every one of the sequences $(k(\ell,m))_{m \geq 1}$ for $\ell \geq 1$.

The diagonal sequence $(k(m))_{m \geq 1}$ determines a subsequence $(a^{k(m)})_{m \geq 1}$ of the sequence $(a^k)_{k \geq 1}$ of points $a^k \in \mathcal{S}$.

Let a be the sequence for which $a_\ell = \lim_m a_\ell^{k(\ell,m)}$ for each $\ell \geq 1$. Then the sequence $(a^{k(m)})_{m \geq 1}$ converges to a (i.e., for each coordinate ℓ, one has $\lim_m a_\ell^{k(m)} = a_\ell$). This is obvious, as the sequence $(k(m))_{m \geq 1}$ is a subsequence of each of the sequences $(k(\ell,m))_{m \geq 1}$. □

This compactness result is the key to Helly's selection principle.

Proposition 6.9.7. (Helly's first selection principle) *Let $(G_n)_{n \geq 1}$ be a sequence of non-decreasing, right continuous, non-negative functions on $\mathbb{R}$ that are uniformly bounded by 1 (i.e., $0 \leq G_n(x) \leq 1$ for all $x \in \mathbb{R}$). Then there is a non-decreasing, right continuous, function G on $\mathbb{R}$, with $0 \leq G(x) \leq 1$ for all x, and a subsequence $(G_{n_k})_{k \geq 1}$ such that*

$$\lim_k G_{n_k}(x) = G(x)$$

for all continuity points x of G.

Proof. Enumerate the rationals $\mathbb{Q}$ with a 1:1 onto function $m \to r_m$. Each distribution function G_n determines a sequence a^n in $\mathcal{S}$: $a_\ell^n = G_n(r_\ell)$. Let a be a point in $\mathcal{S}$ that is a limit of a subsequence $(a^{n_k})_{k \geq 1}$ of the sequence $(a^n)_{n \geq 1}$. Define $\tilde{G}(r_\ell) = a_\ell$. Then $\lim_k G_{n_k}(r) = \tilde{G}(r)$ for all $r \in \mathbb{Q}$, and the function $\tilde{G}$ defined on the rationals is non-decreasing because each G_{n_k} is non-decreasing.

Define $G(x) = \inf\{\tilde{G}(r) \mid x < r \in \mathbb{Q}\}$. Then G

(1) is non-decreasing,
(2) is right continuous, and
(3) $\lim_k G_{n_k}(x) = G(x)$ at each continuity point x of G.

The first property is obvious. Let $\epsilon > 0$ and consider $G(x_0)$: there is a rational number $r > x_0$ with $G(x_0) \leq \tilde{G}(r) < G(x_0) + \epsilon$. If $x_0 < x < r$, then $G(x_0) \leq G(x) \leq G(r) < G(x_0) + \epsilon$. This proves (2).

The third property is more delicate and is subtle: it would follow immediately from Lemma 6.1.1 if $G(r) = \tilde{G}(r)$ for all $r \in \mathbb{Q}$; this need not be the case; all that is certain is that $\tilde{G}(r) \leq G(r)$ for all $r \in \mathbb{Q}$.

Now it is fairly clear that G is continuous at x_0 (i.e., G is left continuous at x_0), if and only if $G(x_0) = \sup\{\tilde{G}(r) \mid r < x_0\}$: if $x_0 - \delta < x_1 < x_0$ implies $G(x_0) - \epsilon \leq G(x_1) \leq G(x_0)$, then $x_1 < r_1 < x_0$ implies that $G(x_0) - \epsilon \leq G(x_1) \leq \tilde{G}(r_1) \leq G(x_0)$. Conversely, if there is a rational number $r_1 < x_0$ with $G(x_0) - \epsilon < \tilde{G}(r_1) \leq G(x_0)$, then $G(x_0) - \epsilon < G(r_1) \leq G(x) \leq G(x_0)$ if $r_1 < x < x_0$.

Let x_0 be a continuity point of G and let $\epsilon > 0$. Then there are two rational numbers r_1, r_2 with $r_1 < x_0 < r_2$ such that

$$G(x_0) - \epsilon \leq \tilde{G}(r_1) \leq G(x_0) \leq \tilde{G}(r_2) < G(x_0) + \epsilon.$$

Let $k_0 = k(\epsilon, r_1, r_2)$ be such that

$$|G_{n_k}(r_1) - \tilde{G}(r_1)| < \epsilon \text{ and } |G_{n_k}(r_2) - \tilde{G}(r_2)| < \epsilon \text{ if } k \geq k_0.$$

Since $G_{n_k}(r_1) \leq G_{n_k}(x_0) \leq G_{n_k}(r_2)$, it follows that $|G_{n_k}(x_0) - G(x_0)| < 2\epsilon$ if $k \geq k_0$. □

Remark 6.9.8. This result shows that the collection of finite measures ν on $(\mathbb{R}, \mathfrak{B}(\mathbb{R}))$ with total mass at most one (the collection of subprobabilities) is a compact metric space when equipped with the Lévy distance since the proof of the proposition applies without any change to the case when the distribution functions are the "distribution functions" of such measures.

Complex numbers.

To maintain the plan that these notes, be self-contained, it is appropriate to say a word about the field $\mathbb{C}$ of complex numbers. Simply stated, $\mathbb{C}$ is the usual Euclidean plane $\mathbb{R}^2$ equipped with a multiplication that is compatible with the vector space structure of $\mathbb{R}^2$.

More explicitly, any point $(x, y) \in \mathbb{R}^2$ can be written as $x\mathbf{e}_1 + y\mathbf{e}_2$, where $\mathbf{e}_1$ and $\mathbf{e}_2$ are the canonical basis vectors. To "multiply" $x\mathbf{e}_1 + y\mathbf{e}_2$ by $a\mathbf{e}_1 + b\mathbf{e}_2$, it suffices to be able to "multiply" the basis vectors as one expects that

$$\{x\mathbf{e}_1 + y\mathbf{e}_2\}\{a\mathbf{e}_1 + b\mathbf{e}_2\} = xa\mathbf{e}_1\mathbf{e}_1 + xb\mathbf{e}_1\mathbf{e}_2 + ya\mathbf{e}_2\mathbf{e}_1 + yb\mathbf{e}_2\mathbf{e}_2.$$

The convention is made that $\mathbf{e}_1 = (1, 0)$ is denoted by 1, i.e., it is to be the multiplicative unit: multiplication by 1 changes nothing. Further, the "multiplication" is to be commutative. With these simplifications, one has

$$\{x\mathbf{e}_1 + y\mathbf{e}_2\}\{a\mathbf{e}_1 + b\mathbf{e}_2\} = xa + (xb + ya)\mathbf{e}_2 + yb\mathbf{e}_2\mathbf{e}_2.$$

This means that the "multiplication" will be defined by deciding on the value of $\mathbf{e}_2\mathbf{e}_2$. Now looking at things in the plane, since $\mathbf{e}_1\mathbf{e}_2 = 1\mathbf{e}_2 = \mathbf{e}_2$, one observes that multiplication by $\mathbf{e}_2$ rotates $1 = \mathbf{e}_1$ counterclockwise by $\frac{\pi}{2}$. This leads one to define multiplication by $\mathbf{e}_2$ as rotation counterclockwise by $\frac{\pi}{2}$: the upshot of this is that $\mathbf{e}_2\mathbf{e}_2 = (-1, 0) = -1$ since $(-1, 0)$ is obtained from $(0, 1)$ by rotation counterclockwise by $\frac{\pi}{2}$.

The basis vector $\mathbf{e}_2$ is labeled as the complex number i, and one makes the following definition of multiplication for complex numbers.

Definition 6.9.9. *Let $(x, y) = x + iy$ and $(a, b) = a + ib$ be two complex numbers. Their **product** is defined by the formula*

$$(x + iy)(a + ib) = (xa - yb) + i(xb + ya).$$

With this definition, it is easy to verify all the usual rules of arithmetic (the axioms of a field) except possibly those related to division.

If $z = x + iy$ is not zero, its reciprocal $\frac{1}{z}$ is defined by making use of the **complex conjugate** of z. This is by definition the complex number $x - iy$, and it is denoted by $\bar{z}$ (i.e., $\bar{z} \stackrel{\text{def}}{=} x - iy$ if $z = x + iy$). The reason the complex conjugate is so important is that $z\bar{z} = x^2 + y^2 = \|(x, y)\|^2$. The square root of $z\bar{z}$ is denoted by $|z|$ and is called the **modulus** of z (i.e., $|z| = \|(x,y)\|$ if $z = x+iy$). In terms of the modulus and the complex conjugate, one has $\frac{1}{z} = \frac{\bar{z}}{|z|^2}$ if $z \neq 0$.

One of the most important functions of a complex variable is the complex exponential e^z. If t is a real number one sets $e^{it} \stackrel{\text{def}}{=} \cos t + i \sin t$. It satisfies the law of exponents

$$e^{i(t_1 + t_2)} = e^{it_1} e^{it_2}$$

precisely because of the trigonometric formulas

$$\cos(t_1 + t_2) = \cos t_1 \cos t_2 - \sin t_1 \sin t_2, \text{ and}$$
$$\sin(t_1 + t_2) = \sin t_1 \cos t_2 + \cos t_1 \sin t_2.$$

One then sets $e^z \stackrel{\text{def}}{=} e^x e^{iy} = e^x \{\cos y + i \sin y\}$ if $z = x + iy$, where e^x is the real exponential.

The law of exponents

$$e^{z_1 + z_2} = e^{z_1} e^{z_2}$$

is satisfied once again since it holds for real numbers and purely imaginary numbers (those whose real part is zero).

Finally, the convergence of sequences and of series is formulated in exactly the same terms as for the real case. It suffices to "copy" the definitions. In particular, the well-known series for e^x has an exact counterpart for the complex exponential, namely,

$$e^z = \sum_{n=0}^{\infty} \frac{z^n}{n!}.$$

Further information on functions of a complex variable can be found in any of the standard texts on this subject.

BIBLIOGRAPHY

[B1] Billingsley, P., *Convergence of Probability Measures*, John Wiley & Sons, Inc., New York, 1968.
[B2] Billingsley, P., *Probability and Measure*, 2nd edition, John Wiley & Sons, Inc., New York, 1986.
[C] Chung, K. L., *A Course in Probability Theory*, 2nd edition, Academic Press, New York, 1974.
[D1] Doob, J. L., *Stochastic Processes*, John Wiley & Sons, Inc., New York, 1953.
[D2] Dudley, R. M., *Real Analysis and Probability*, Wadsworth & Brooks/Cole, Pacific Grove, Calif., 1989.
[D3] Dynkin, E. B., *Theory of Markov processes*, Prentice–Hall, Inc., Englewood Cliffs, N. J., 1961.
[D4] Dynkin, E. B. and Yushekevič, A. A., *Markov Processes*, Plenum Press, New York, 1969.
[F1] Feller, W., *An Introduction to Probability and Its Applications, Vol. I*, 3rd edition, John Wiley & Sons, Inc., New York, 1968.
[F2] Feller, W., *An Introduction to Probability and Its Applications, Vol. II*, John Wiley & Sons, Inc., New York, 1966.
[G] Gnedenko, B. V. and Kolmogorov, A. N., *Limit Distributions for Sums of Independent Random Variables*, Addison–Wesley, Cambridge, Mass., 1954.
[H1] Halmos, P. R., *Measure Theory*, Van Nostrand, New York, 1950.
[H2] Halmos, P. R., *Naive Set Theory*, Van Nostrand, Princeton, N. J., 1960.
[I] Ikeda, N. and Watanabe, S., *Stochastic Differential Equations and Diffusion Processes*, 2nd edition, North Holland Publ. Co., Amsterdam, 1989.
[K1] Karlin, S. and Taylor, H. M., *A First Course in Stochastic Processes*, 2nd edition, Academic Press, New York, 1975.
[K2] Kolmogorov, A. N., *Foundations of Probability Theory*, Chelsea Publ. Co., New York, 1950.
[K3] Körner, T. W., *Fourier Analysis*, Cambridge University Press, Cambridge, 1988.
[L1] Lamperti, J., *Probability*, W. A. Benjamin, Inc., New York, 1966.
[L2] Loève, M., *Probability Theory I, II*, 4th edition, Springer-Verlag, New York, 1978.
[M1] Marsden, J. E., *Elementary Classical Analysis*, W. H. Freeman and Co., San Francisco, 1974.

[M2] Meyer, P.-A., *Probability and Potentials*, Blaisdell Publ. Co., Waltham, Mass., 1966.

[N1] Neveu, J., *Mathematical Foundations of the Calculus of Probability*, Holden-Day, Inc., San Francisco, 1965.

[N2] Neveu, J., *Discrete-Parameter Martingales*, North Holland Publ. Co., Amsterdam, 1975.

[P1] Parthasarathy, K. R., *Probability Measures on Metric Spaces*, Academic Press, New York, 1967.

[P2] Protter, P., *Stochastic Integration and Differential Equations*, Springer-Verlag, Berlin, 1990.

[R1] Revuz, D., *Markov Chains*, North Holland Publ. Co., Amsterdam, 1975.

[R2] Revuz, D. and Yor, M., *Continuous Martingales and Brownian Motion*, Springer-Verlag, New York, 1991.

[R3] Royden, H. L., *Real Analysis*, 3rd edition, MacMillan, New York, 1988.

[R4] Rudin, W., *Principles of Mathematical Analysis*, McGraw-Hill, New York, 1953.

[S1] Stein, E. M. and Weiss, G., *Introduction to Fourier Analysis on Euclidean Space*, Princeton University Press, Princeton, N. J., 1971.

[S2] Stroock, D. W., *Probability Theory, an Analytic View*, Cambridge University Press, Cambridge, 1993.

[T] Titchmarsh, E. C., *Theory of Functions*, 2nd edition, Oxford University Press, Oxford, 1939.

[W1] Wheeden, R. L. and Zygmund, A., *Measure and Integral*, Marcel Dekker, New York, 1977.

[W2] Williams, D., *Probability with Martingales*, Cambridge University Press, Cambridge, 1991.

[Z] Zacks, S., *The Theory of Statistical Inference*, John Wiley & Sons, Inc., New York, 1971.

INDEX

Absolutely continuous function on $[a,b]$, 76
absolutely continuous probability, 57
absolutely continuous with respect to μ, 66
absolutely convergent, 54
absolutely summable, 146
adapted, 229
almost isomorphic measure spaces, 206
analyst's distribution function, 200
approximate identity, 191
approximation to the identity, 191
Archimedean property, 2
arithmetic mean, 142
at most countable, 15
atoms, 224, 230
axiom of the least upper bound, 2

Backward martingale, 248
Banach algebra, 134
Banach space, 146
Bessel's inequality, 155
betting system, 232
Bolzano–Weierstrass property, 28
Boolean algebra, 6
Boolean ring, 45
Borel function, 36
Borel's strong law of large numbers, 205
Borel–Cantelli lemma, 148
bounded interval, 4
bounded variation, 71

Cantor discontinuum, 8
Cantor set, 8
Cantor's diagonal argument, 15

Cantor–Lebesgue function, 84
Cartesian product, 15
Cauchy sequence, 146
Cauchy–Schwarz inequality, 59, 216
central limit theorem: i.i.d. case, 275
central limit theorem: Lindebergh condition, 280
central limit theorem: triangular arrays, 281
Cesaro mean, 162
characteristic function of a set, 30
characteristic function, 79, 263
Chebychev's inequality, 169
closed, 14
closed interval, 4
closed monotone class, 134
closure, 15
compact, 20
compact support, 150
compact support in an open set, 150
complete measure space, 105
complete metric space, 146
complete orthonormal system, see orthonormal system
completing the measure, see completion
completion, 47, 105
completion of $\mathfrak{B}(\mathbb{R})$, 81
complex conjugate of a complex number, 289
conditional distribution function of X given $Y = y$, 221
conditional expectation of X, 212
conditional probability of E given A, 47
conditionally convergent, 54

293

conditionally independent, 226
conjugate indices, 141
continuity theorem, characteristic function, 269
continuous at $x_0 \in E$, 36
continuous at $x_0 \in \mathbb{R}$, 36
continuous measure, 83
continuous on E, 36
continuous on $\mathbb{R}$, 36
converge in distribution, 255
converge in law, 255
converge weakly, 251, 255
convergence in measure, 168
convergence in probability, 168
convergence **P**-a.e., 167
convergence μ-a.e, 167
converges to $+\infty$, 2
converges to B, 2
converges to x, 146
converges to 0 in L^p, 145
converges to 0 in L^p-norm, 145
converges uniformly, 134
converges weakly, 251
convex, 144
convex set, 214
convolution kernels, 120
convolution of integrable functions, 157
convolution of two probabilities, 108
countable, 15
countable additivity, 9
countably additive, 9
countably generated, 248
countably subadditive, 21

Dedekind cut, 2
dense, 16
Dini derivatives, 194
Dirac measure at a, 26
Dirac measure at the origin, 26
Dirichlet's kernel, 162
discrete distribution function, 57
discrete measure, 83
distribution function, 11

distribution of a random vector X, 88
distribution of X, 37
Doob decomposition, 243
Doob's L^p-inequality: countable case, 239
Doob's maximal inequality: finite case, 237

Egororov's theorem, 168, 202
empirical distribution function, 250
equicontinuous, 283
ergodic probability space, 94
essential supremum, 144
essentially bounded, 144
excessive function, 231
expectation, 30
expectation of X, 38
exponentially distributed (with parameter λ), 56

$\mathfrak{F}$-measurable, 32
$\mathfrak{F}$-simple function, 30
Fatou's lemma, 39
Féjer kernel, 160
Féjer's theorem, 163
filtration, 225, 229
finite expectation, 42
finite measure, 44
finite measure space, 45
finite signed measure, 61
finitely additive probability space, 6
Fourier inversion theorem, 271
Fourier series, 154
Fourier transform, 79
Fubini's theorem for integrable random variables, 100
Fubini's theorem for Lebesgue integrable functions on $\mathbb{R}^n$, 131
Fubini's theorem for L^1-functions, 106

INDEX

Fubini's theorem for positive functions, 105
Fubini's theorem for positive random variables, 99

G.l.b., *see* greatest lower bound
gamma distribution, 56
geometric mean, 142
Glivenko–Cantelli theorem, 261
greatest lower bound, 2

Hahn decomposition, 64
half-open interval, 4
Hardy–Littlewood maximal function, 187
harmonic function, 231
Heaviside function, 26
Heine–Borel theorem, 19
Helly's first selection principle, 269
Hermite polynomial, 80
Hilbert space, 154, 216
Hölder's inequality, 142
homogeneous Markov chain, 119

I.i.d. sequence, 170
image measure, 138
image of **P**, 37
improper Riemann integral of f over $[0, +\infty)$, 54
increasing process, 243
independent collections of events, 89
independent events, 89
independent family of classes of events, 132
independent family of events or sets, 111
independent family of random variables, 111
independent, finite collection of random variables, 94
indicator function of a set, 30
inequality of Cauchy–Schwarz, 139

inf, *see* infimum
infimum, 20
infinite product of a sequence of probability spaces, 118
infinite product set, 113
infinite series, 3
infinitely differentiable function, 79, *see also* smooth
initial distribution, 122
inner product, 59
integers, 1
integrable random variable, 41, *see also* non-negative integrable random variable.
integral, 30
integral of X, 38
integration by parts, 135
integration by parts: functions, 281
integration by parts: series, 241
interval, 4
inverse of a distribution function, 259
inversion theorem, characteristic function, 272
irrational, 1
isomorphic measure spaces, 206

Jensen's inequality, 144
Jensen's inequality, conditional expectation, 219
joint distribution, finite-dimensional, 110
Jordan decomposition, 65

Khintchine's weak law of large numbers, 172
Kolmogorov's 0-1 law, 111
Kolmogorov's criterion, 179
Kolmogorov's inequality, 178, 237
Kolmogorov's strong law of large numbers, 177, 240
Kolmogorov's strong law: backward martingale, 249

Kronecker's lemma, 241

L^2-bounded, 238
L^2-martingale convergence theorem, 238
Laplace transform, 79
law of a random vector X, *see* distribution of a random vector X
law of X, *see* distribution of X
least squares, 218
least upper bound, 1
Lebesgue decomposition theorem for measures, 68
Lebesgue integral, 48
Lebesgue measurable sets, 51
Lebesgue measure, 46
Lebesgue measure on $\mathfrak{B}(\mathbb{R}^n)$, 104
Lebesgue measure on E, 48
Lebesgue measure on $\mathbb{R}^n$, 104
Lebesgue point, 188
Lebesgue set, 188
Lebesgue's differentiation theorem, 188
left continuous, 11
left limit, 11
liminf, *see* limit infimum
limit infimum, 33
limit supremum, 33
limsup, *see* limit supremum
Lindebergh condition, 280
Lipschitz truncation of a random variable at height $c > 0$, 181
locally integrable function, 189
lower bound, 1
lower semi-continuous, 191
lower semi-continuous at x_0, 191
lower sum, 49
lower variation, 65
l.u.b., *see* least upper bound
Lusin's theorem, 202

Marginal distributions, 89
Markov chain, 119
Markov chain with transition probability **N**, initial distribution **P**$_0$, 127
Markov process, 225
Markov process with transition kernel **N**, 225
Markov property, 127, 226
Markovian kernel, *see* transition kernel
martingale, 229
martingale convergence theorem: countable case, 246
martingale transform, 232
maximal function, *see* Hardy–Littlewood maximal function
maximal function: martingale, 239
maximal function: submartingale, 237
measurable, 86
measurable function, 32
measurable space, 32
measure, 44
measure space, 44
measure zero, 105
metric, 146
metric space, 146
metric space E, 48
Minkowski's inequality, 142
modulus of a complex number, 289
monotone class, 25
monotone class of functions, 134
monotone class theorem, Dynkin's version, 92
monotone class theorem for functions, 133
monotone class theorem for sets, 25
monotone rearrangement, 260
multivariate normal distribution, 107
mutually singular, 65

INDEX

ν-negative set, 62
ν-positive set, 62
n-dimensional random vector, 88
natural numbers, 1
nearest neighbours, 120
neighbourhood of x_0, 36
non-increasing rearrangement, 260
non-Lebesgue measurable set, 49
non-negative integrable random variable, 38
non-negative measure, 44
norm, 58
normal number, 171
normally summable, 146
normed vector space, 146
null function, 41
null set, 104, *see also* null function
null variable, 41

Occurrence of a set prior to a stopping time, 234
open, 14
open ball in a metric space about a of radius ϵ, 146
open interval, 4
open relative to E, *see* open subset of E
open set in a metric space, 146
open subset of E, 131
open subset of $\mathbb{R}^2$, 87
open subset of $\mathbb{R}^n$, 87
optional stopping theorem: finite case, 235
orthogonal system, 154
orthonormal system, 154
outer measure, 21

2π-periodic integrable function, 158
pth absolute moment, 140
pth moment, 140
pairwise independent random variables, 172
parallelogram law, 216
Parseval's equality, 155

Parseval's relation, 266
Parseval's theorem, 156
partition π, 49
path, 127
Poisson distribution, 13
Poisson distribution with mean $\lambda > 0$, 38
Polish space, 224
predictable process, 232
principle of mathematical induction, 7
principle of monotone convergence, 39
probability, 6, 9
probability density function, 56
probability space, 9
product of a sequence of σ-algebras, 113
product of complex numbers, 289
product of n probabilities, 102
product of n probability spaces, 102
product of two probabilities, 99
product of two probability spaces, 99
product topology, 286
projection operator, 218

Quantile transformation, 259

Radon–Nikodym derivative of **P** with respect to λ, 57
Radon–Nikodym theorem: measures, 67
Radon–Nikodym theorem: signed measures, 68
random variable, 32
random vector, 86
rational numbers, 1
real numbers, 1
regular Borel measure, 151
regular conditional probability, 223
Riemann integrable, 50

Riemann integral of φ over $[a,b]$, 50
Riesz–Fischer theorem, 147
Riesz representation theorem, 217
right continuous, 11
right limit, 11

Scheffé's lemma, 186
second axiom of countability, 87
separable, 224
sequence of elements from, 2
set of Lebesgue measure zero, 104
σ-additive, 9
σ-additive probability on a Boolean algebra, 17
σ-algebra, 9
σ-algebra generated by $\mathfrak{C}$, 14
σ-algebra of Borel subsets of E, 131
σ- algebra of Borel subsets of $\mathbb{R}$, 14
σ-algebra of Borel subsets of $\mathbb{R}^2$, 87
σ-algebra of Borel subsets of $\mathbb{R}^n$, 87
σ-algebra of Lebesgue measurable subsets of $\mathbb{R}^n$, 104
σ-algebra of $\mathbf{P}^*$-measurable sets, 24
σ-field, 10
σ-finite measure, 44
σ-finite measure space, 45
σ-finite signed measure, 61
signed measure, 61
simple function, 30
singular distribution function, see Cantor–Lebesgue function
singular measure, 57
smooth, 79, see also infinitely differentiable function
square summable sequences, 59
standard deviation, 140
standard measurable spaces, 224
stochastic integral, 232
stochastic process, 110

stopping time, 234
strong law of large numbers, 170
submartingale, 229
subsequence, 2
sufficient σ-field, 226
sufficient statistic, 227
sup, see supremum
supermartingale, 229
supremum, 20
symmetric difference, 78

Theorem of dominated convergence, 43
theorem of dominated convergence, equivalent form, 44
theorem of dominated convergence, first version, 43
three-series theorem, 242
topology, 14
total variation, 65, 71
trajectory, 127
transition kernel, 120
translation invariance, 49
translation invariant, 49
triangle inequality, 5, 146, 216
triangular array, 280
trigonometric polynomial, 153
truncation of a random variable at height c, 180

Unbounded interval, 4
uniform closure, 134
uniform convergence, 134
uniform distribution on [0,1], 12
uniform norm, 153
uniformly closed, 134
uniformly equicontinuous, 283
uniformly integrable, 181
uniqueness theorem, characteristic function, 268
unit normal distribution, 13
unit point mass at a, 26
unit point mass at the origin, 26

universally measurable sets, 82
upcrossing, 245
upper bound, 1
upper sum, 49
upper variation of ν, 62

Vanishing at infinity, Borel function, 110
variance, 140
Vitali cover, 195

Vitali covering lemma, 195

Weak law of large numbers, 170
weak $L^1(\mathbb{R})$, 187
weak type, 187
Weierstrass approximation theorem, 175

Young's inequality, 142, 201